# Model Systems in
# Behavioral Ecology

# MONOGRAPHS IN BEHAVIOR AND ECOLOGY

Edited by John R. Krebs and Tim Clutton-Brock

# Model Systems in Behavioral Ecology

Integrating Conceptual, Theoretical, and Empirical Approaches

Edited by Lee Alan Dugatkin

Princeton University Press
Princeton and Oxford

Library of Congress Cataloging-in-Publication Data
Model sysyems in behavioral ecology: integrating
conceptual, theoretical, and empirical approaches /
edited by Lee Alan Dugatkin.
p.  cm. — (Monographs in behavior and ecology)
Includes bibliographical references (p. ).
ISBN 0-691-00652-0 (alk. paper) — ISBN 0-691-00653-9
(pbk. : alk. paper)
1. Animal behavior.  2. Animal ecology.  I. Dugatkin,
Lee Alan, 1962– .  II. Series.
QL751 .M577  2001
591.5—dc21    2001021990

British Library Cataloging-in-Publication Data is available

This book has been composed in Times Roman

Printed on acid-free paper. ∞

www.pup.princeton.edu

Printed in the United States of America

(Pbk.)
10  9  8  7  6  5  4  3  2  1

10  9  8  7  6  5  4  3  2  1 (Pbk.)

This book is dedicated to the intellectual father of modern behavioral ecology, the late Dr. W. D. Hamilton. He is sorely missed in more ways than one.

# Contents

PART III  Bird Model Systems

# Preface

Lee Alan Dugatkin

This is a book about the history and the future of behavioral ecology. In particular, it shows the beauty of using "model systems" in integrating conceptual, theoretical, and empirical perspectives in the discipline of behavioral ecology. More generally, it advocates this integrative approach as a way to tackle questions in any area of science.

Over the past thirty-five years, behavioral ecology has evolved from an initial hodgepodge of ideas drawn from other areas in biology into a legitimate and respected discipline within the sciences. Initially centered primarily (but not exclusively) around one idea—kin selection—behavioral ecology has developed into an area replete with conceptual, theoretical, and empirical advances on numerous fronts. From a field that primarily imported ideas and gave them a "new twist," behavioral ecology is now consistently exporting new ideas to psychology, anthropology, mathematics, and economics, not to mention other areas of biology. For example, behavioral ecology unabashedly swiped game theory from mathematicians and economists. Adopting the integrative approach taken here, we then spit back out "evolutionary game theory," which has now made its way into mainstream journals in economics, political science, and mathematics.

What started as a one-way street now flows in both directions. In short, behavioral ecology has grown up and its maturation is a function of integrating empirical, theoretical, and conceptual questions to a host of widely divergent issues (e.g., foraging, mating, safety, sex-ratio, cooperation, aggression, etc.). During the maturation process, behavioral ecologists learned a great deal. Much of what we learned, as well as much of what established behavioral ecology as a serious endeavor, has come from work in a series of "model systems." By "system" I mean a species (or group of closely related species) in which behavioral ecologists have undertaken detailed studies in an attempt to address fundamental (and often numerous) questions of interest. What ties these model systems together is that the scientists studying them inevitably take a *three-pronged approach*—conceptual, theoretical, and empirical—to any question they address in such systems.

We will examine twenty-five such model systems in a somewhat unusual manner—unusual in the sense of combining a more relaxed tone to the biography of the researchers who will navigate us through our model systems, with the more analytical tone that chapters take when we get down to the hard integrative science, and back again to a more relaxed tone when the future is discussed. In so doing, we will see the utility of integrating conceptual, theoretical, and empirical perspectives not only in thoroughly understanding a given system, but more important as the foundation for a view of behavioral ecology that constantly produces new, interesting areas for further exploration.

In choosing contributors to this volume, an effort was made not only to pick well-known scientists who have made fundamental contributions to behavioral ecology by using conceptual, theoretical, and empirical tools in studying their systems, but also to cover a taxonomically broad group of species. As such, it contains chapters on leaf-cutter ants, dungflies, various butterflies, honeybees, wasps, stalk-eyed flies, social spiders, Mexican jays, wood ducks, barn swallows, red-winged blackbirds, bluejays, swordtails, sticklebacks, bluehead wrasse, tree frogs, two species of lizards, coyotes, wild dogs, cheetahs, dolphins, bonnet macaques, chimpanzees, and gorillas. In addition, a broad suite of behavioral issues such as foraging, aggression, cooperation, mate choice, kinship, parasite resistance, territoriality, parent/offspring conflict, and reproductive suppression are all represented in various chapters peppered throughout the volume.

## A Discussion of Terms

"Begin at the beginning," the king said very gravely, "and go on till then you come to the end: then stop."

**Lewis Carroll, *Through the Looking Glass***

Before proceeding further, we need to define three critical terms used throughout this volume—*conceptual, theoretical,* and *empirical.* These terms can mean different things to different people and so to avoid any confusion, let's tackle them up front. Readers may have a slightly different definition of what each term means, indeed the contributors to this volume do as well, but at least we will all be working from the same general framework.

By conceptual perspective, I mean a perspective that is driven primarily by a broad-based concept per se. Broad-based conceptual approaches tend to draw bits and pieces generated in many different subdisciplines and unite them in a new, cohesive, and thought-provoking manner. Of course, observation and experimentation play an indirect role in concept generation, but usually a broad-based concept itself is not directly tied to any *specific* observation or experiment. Broad-based conceptual questions underlie all the chapters in this book, be they concepts in sex change, species interactions, antipredator behavior, coalition formation, or many other behavioral venues.

Major conceptual advances tend not only to generate new experimental work, but reshape the way a discipline looks at itself. In this sense they are akin to the scientific revolutions that Kuhn (1962) so eloquently describes in his classic *The Structure of Scientific Revolutions.* The clearest case for a concept that has made behavioral ecologists rethink the basic way they approach their science is Hamilton's inclusive fitness theory. The notion that an individual's total fitness is a combination of its direct (personal) and indirect (mediated by its interactions with relatives) fitness is both simple and extraordinarily powerful. It has made behavioral ecologists rethink the way they do virtually every study imaginable. If you attend a presentation at a behavioral ecology (or animal behavior) conference in which someone has not mentioned inclusive fitness in one way or another, you can almost bet the bank that an audience member will ask how "relatedness" might affect his results.

Needless to say, the mathematical theory underlying inclusive fitness is critical, as mathematics gives the ideas behind inclusive fitness a universal platform, not to mention a sort of respectability that is associated with complex equations. That being said, I would wager that while Hamilton's publications on inclusive fitness (particularly Hamilton 1964) are the most cited papers in the field of behavioral ecology, most behavioral ecologists have not read Hamilton's papers in detail (if at all), and even fewer could vouch for the mathematics in them. But, "inclusive fitness thinking" is so pervasive in behavioral ecology that every first-year graduate student in the field can tell you the basic ideas underlying Hamilton's papers, and that is a fine sign that a conceptual idea has made its mark.

Not unexpectedly, conceptual questions are the ones that often spur researchers toward their model systems, and this is often the case with the contributors to this volume. For example, Jerram Brown's chapter on Mexican jays makes it clear that while Brown wanted to get out in the field and empirically examine kin selection, it was the conceptual work that drove him to his system to begin with. Conceptual issues on coevolution drew Pierce into a lifelong association with butterflies and ants. The conceptual advances in work on alternative mating systems lured Ryan toward swordtails. And it need not be the case that conceptual work be in the finished stages for it to lure in researchers. Reeve's chapter demonstrates that the drive to *create* a conceptual framework for uniting heretofore unrelated areas of social behavior can draw you into studying wasps, while a desire to *create* a field of virtual ecology can draw the likes of Kamil and Bond to continue studying crypsis in bluejays.

A theoretical approach to an issue involves the generation of some sort of model of the world. Often theoreticians see a large empirical database that needs sound theory to support it, but it need not work that way. In a letter to A. R. Wallace, Charles Darwin noted that "I am a firm believer that without speculation there is no good and original observation." Good theory can be an antecedent or a precedent to lots of data collecting and hypothesis testing. Normally, theoretical work in behavioral ecology centers on mathematical models, in the sense of equations, computer simulations, and graphical representation. Verbal models, set up with the same structure as mathematical models (just without the mathematical notation), are in principle possible, but relatively rare. Such models, when they do exist, blur the line between theoretical and conceptual and often are criticized for being susceptible to "hand waving" (a term theoreticians use for models that are vague and tied to the persuasiveness of the presenter, rather than something intrinsically valuable in the mathematics of a model).

One unfortunate rift between those who primarily do theoretical work and those who focus on empirical studies revolves around whether any given theory matches the "real world" sufficiently to spur on tests. For example, I often hear behavioral ecologists dismiss a new model by claiming that is "just theory." This is unfortunate, to say the least. Just as there are many empirical studies that do little to truly advance our understanding of social behavior, no doubt there are many mathematical models that lead us to dead ends. That is just the nature of the science. There is nothing inherently less interesting about a new model than a new empirical finding.

Theoreticians are not interested in mimicking the natural world in their models, but rather in boiling a complex issue down to its barest ingredients in an attempt to make specific predictions. In that sense, the criticism that a particular theory doesn't match the details of system X will often be true, but irrelevant. A good theory will whittle away the details of specific systems, but just enough to allow for general predictions that will apply to many systems. We will see that many times in this volume, but more than that, we will get a picture of *how* and *why* this comes about, not just that it happens.

As a case in point about the power of theory in shaping behavioral ecology, consider the optimal foraging "boom" of the 1980s (Stephens & Krebs 1986). The first meeting I attended as a graduate student was the International Behavioral Ecology meeting at Simon Fraser University in 1988. Optimal foraging theory was clearly the "hot topic," in both presentations and in the halls between presentations. Yet the truth is that psychologists had been studying foraging behavior in laboratory animals for decades, and ethologists too had done much work on foraging in natural systems. In behavioral ecology, however, it was optimal foraging *theory* (e.g., the marginal value theorem; Charnov 1976) that spurred on the meteoric rise of animal foraging studies. Once a sound theoretical evolutionary framework existed for making predictions about what animals should eat and where they should eat it, behavioral ecologists went to work testing the many predictions generated from optimal foraging theory.

A similar case can be made for the increase in controlled experiments on cooperation and altruism in animals (Dugatkin 1997). Although cooperation had been the subject of much discussion since the early works of Kropotkin (1908) and Allee (1931, 1951), it was not until the pathbreaking models of Trivers (1971) and Axelrod and Hamilton (1981) that controlled studies of cooperation among unrelated animals began to skyrocket. The catalyst was, of course, the Prisoner's Dilemma game. Again, as with foraging, once theory was in place, empirical work was bound to follow, and so it was with cooperation and the Prisoner's Dilemma. Not only were studies of reciprocity spurred on by the introduction of a game theory model, but the Prisoner's Dilemma work created an atmosphere in which other theories of cooperation were easily generated.

Naturally, good theoretical work is also often related to some new conceptual approach. Zahavi's (1975) original "handicap principle" might be thought of in conceptual terms. Subsequent to this conceptual advance, however, much in the way of mathematical models has helped better elucidate the ideas Zahavi expounded, not to mention generating new testable predictions for the handicap principle.

In this volume we will see how sound theory has lured many researchers into the model systems they have chosen. The simple, but powerful, evolutionary version of the rock-papers-scissors children's game drew Sinervo into a long-term relationship with lizards, while tantalizing mathematical models of mate choice convinced Wilkinson that fruit flies were worth investigating, even if it meant taking some time away from his blood-swapping vampire bats. Early models of sex change and life history led Warner to live in the coral reefs, while models of parentage and effort allocation convinced Westneat that red-wing blackbirds were the perfect system to muck about with. We will see that the value of a novel new theory is not just in its testability, but in how it reshapes the very nature of the empirical experiments we perform.

The vast majority of studies in behavioral ecology are neither strictly conceptual nor theoretical, but empirical. The strength of such studies lies in direct proportion to the extent that they either put new concepts and models to the test, or generate new models and ideas. For behavioral ecology, as for all science, empirical studies are valued not so much for the facts they uncover, as for what they tell us about our hypotheses. Not surprisingly, then, this is the dominant way in which contributors to this volume handle their empirical side.

Naturally, there is some variation in how much effort each contributor devotes to the theoretical, conceptual, and empirical aspects of his or her model system, but regardless of where each contributor lies in this continuum, empiricism is a tool for hypothesis testing, not plain old data gathering. This perspective does not in any way diminish an important factor driving the empirical side of many contributors to *Model Systems*, namely, a passionate love, even admiration, for the creatures they study. A passionate love for some nonhuman species underlies many of the systems we will examine, but it need not always be present.

Having been raised in the heart of New York City, where the major wildlife one encounters (if you don't count cockroaches) is limited to squirrels and pigeons, I tend to see the world through conceptual and theoretical, rather than empirical glasses.

I often tell my students a true story. Until I was about twelve, I thought that sparrows were baby pigeons, and it was, I argue, a reasonable deduction. I encountered two classes of birds—smaller (sparrows) and larger (pigeons). Not having seen pigeon nests, I assumed the little birds were the babies of the bigger ones. So, when putting together this volume I must admit that I was surprised that some of the behavioral ecologists I most admire have been enamored with the creatures they study since childhood. When I learned that Geoff Parker's fantastic empirical and theoretical work on dung-flies was an outcrop of his childhood study of these creatures in the meadows around his house, I was amazed. After reading Tom Seeley's passionate, lifelong love of social insects, I wanted to go out and work with bees myself. Richard Connor's fortitude to study dolphins at all costs made me view both Connor and the dolphins in a different light. Not only do different behavioral ecologists approach empirical studies differently, the very nature of what constitutes "empirical" is itself evolving.

With the advent of powerful computers on the desks of nearly all modern behavioral ecologists, we probably need to amend the standard way of thinking of empirical studies (laboratory and field experiments) to include experiments in cyberspace (i.e., computer simulations). This will no doubt cause the line between empirical and theoretical work to blur somewhat. For example, there are now many long-term experiments on cooperation using digital organisms in which population structure, genetics, and behavior evolve simultaneously inside a computer (Nowak & Sigmund 1992, 1998; Crowley et al. 1995). Investigators can track different starting conditions, dynamic changes through time, and end conditions in digital organisms in such cyber-populations.

Should the above studies be labeled empirical or theoretical? The answer may be the stuff philosophers debate, but my point here is simply to raise the issue as an interesting interaction between theoretical and empirical work— an intersection that behavioral ecologists can ill afford to ignore. Another interesting outgrowth of cyberspace for empiricists is that in addition to examining multigeneration evolution on a desktop, one can create specific cyberorganisms meant to interact with biological organisms. Kamil and Bond's chapter on "virtual ecology" demonstrates the potential for this sort of interaction wonderfully.

## The Questions

In the real world it is more important that a proposition be interesting than that it be true. The importance of truth is, that it adds to interest.

Alfred North Whitehead, *Process and Reality*

I readily admit that the division among conceptual, theoretical, and empirical approaches that I have outlined can be blurry. That being said, I think that separating them can provide some heuristic benefits when examining how good science is undertaken. In any case, for a discipline to be productive there needs to be new empirical, theoretical, and conceptual work constantly underway. I believe that this is the case for behavioral ecology, and this volume attempts to show how the interaction among these approaches can help propel behavioral ecology into the future. Before looking at some of the details about how this volume is structured, let me first address two broad issues that touch on the integration of theoretical, empirical, and conceptual approaches in behavioral ecology.

First, let's tackle what might be thought of as a "chicken and egg" issue. Is there a hierarchy to theoretical, empirical, and conceptual approaches? In other words, does one approach drive a second, which in turn motivates a third? Does the process start with a broad conceptual advance, followed by new theory that is then put to the test empirically? Or is it a new startling empirical finding that generates new theory, which in turn promotes a new conceptual synthesis? Or perhaps some other permutation of theoretical, empirical, and conceptual approaches is most common? Both below and throughout this volume, we will see that there is no clear answer to this question—there are multiple paths to significant discovery. Some contributors to this volume have taken one path, others a different one. And, as will soon become clear, serendipity plays a large role in shaping the path of leading investigators in behavioral ecology.

Second, regardless of any hierarchy of approaches, there are many ways that one can examine the integration of empirical, theoretical, and conceptual approaches in behavioral ecology. One approach at integration might start by delineating areas under serious investigation. In behavioral ecology, we might have cooperation, aggression, mating, foraging, and the like on such a list. Then, using many study systems that span the gamut in both terms of the species studied and the experimental protocol employed, we can examine how empirical, theoretical, and conceptual work has shaped our understanding of the items on our list. Krebs and Davie's volumes over the past twenty-plus years do a masterful job at just such integration (see Krebs & Davies 1997 for the latest of these). But the Krebs and Davies approach, powerful as it is, isn't the only way to integrate.

Here, I am advocating a different approach to integration. We will examine how empirical, theoretical, and conceptual approaches have interacted in shaping behavioral ecology by studying twenty-five model systems, developed and fine-tuned by some of the most respected researchers in behavioral ecology.

There is something special about working in a "model system"—something that can't come out of a single study, no matter how significant it may be—and each of the contributors describes what that something special is in their own words. To bring to the reader the true power of integrating conceptual, theoretical, and empirical approaches within a model system contributors were instructed to touch on the following critical issues: how and why they ended up in the system they did; how they integrate conceptual, theoretical, and empirical work; how the general approach they adopt can be useful to others, particularly those at the start of their sortie into behavioral ecology; and lastly, where they see their system going in the future. This is a tall order for a single chapter, and the results speak much about the power of behavioral ecology.

## Finding a System

I happened to discover behavioral ecology as a senior in college, while finishing a degree in history. I had no idea that behavioral ecology and animal behavior even existed as scientific disciplines, let alone that one could make a career in these areas. I was lucky that a friend of mine gave me a book he used in an undergraduate course at Cornell—the book was Alcock's *Animal Behavior*—and from that moment on I was hooked. But it could have been different. Any one of a million small things might have caused my friend to not give me Alcock's book, and I honestly have no idea what I'd be doing now if that had happened.

I still recall that when I first fell in love with behavioral ecology and decided that this was what I wanted to focus my life's work on, I was completely in the dark about how scientists get started. I was reading all about well-established researchers who had undertaken novel, pathbreaking work in evolution and behavior, but I kept asking myself, "How on earth did they get started?" It all looked so easy reading about the past twenty years of some great scientists' work and the difference it made, but where did it all start? Were they always interested in science? How did they end up working in the system they did, asking the questions they did, and what separated the people I was reading about from the scores of people that slipped through the cracks and never made it? These are not only philosophical questions, they are the types of questions that beginning graduate students often ponder and can have a real impact on whether they continue in the field or look elsewhere, and if they stay, what they do. Wouldn't it be great if a guide existed to lead young behavioral ecologists in the right direction? One of the numerous goals of this volume is to act as such a guide.

Readers, particularly graduate students, may be surprised at how the contributors to this volume ended up where they did. Naturally, raw intellect and a good work ethic played a large role in bringing the contributors to this volume into behavioral ecology, and more specifically into the model system in which they each work. It could hardly be otherwise. Yet other factors played a huge role as well. One such factor was mentors; a good mentor, at almost any stage of their scientific careers, seemed to play a critical role in shaping the work of many in this volume. More generally, an air of intellectual excitement over behavioral ecology permeated the air in the mid-1970s—the time that many (but certainly far from all) behavioral ecologists we will learn about were in their formative years as scientists. Readers might wish to make up a score card and see how many contributors mention *Sociobiology: The New Synthesis* as influencing their decision to enter the field. For the contributors to this volume, four intellectual mentors were foremost in shaping their ideas—E. O. Wilson, W. D. Hamilton, John Maynard Smith, and Robert Trivers. Some in this volume (Pierce, Seeley, Reeve, Holldobbler, Milinski, Wilkinson, Parker) were lucky enough to work under or with one of these individuals; most others were simply affected by their contributions.

Luck, serendipity, and just being at the right place at the right time also played a major role in leading many of today's behavioral ecologists to science. By definition, no one can control these factors, but it may be refreshing and enlightening, particularly to those just starting out, to know that even some of the leaders in their field are subject to these forces, just like everyone else.

## Integration

How and why researchers ended up in their particular model system is only the tip of the iceberg in terms of the questions addressed here. Equally, if not more important, is for contributors to describe how they integrate conceptual, theoretical, and empirical perspectives into their work. When answering this question, we will see that there are numerous ways to integrate these approaches, and the contributors to this volume span the gamut on how they have accomplished this feat. For example, on one end of the continuum, we have those who use empirical work primarily to test well-established theories, while other contributors are more concerned with mining new hypotheses from their system. Dividing up the pie differently, contributors show great variation in the degree to which they test concepts and theories in the field versus the more controlled, but less natural, environment of the lab.

We will also see that contributors differ dramatically in the extent to which they are focused on using their system to address a specific issue, a suite of related questions, a suite of unrelated (but all interesting) questions, or a metatopic in behavioral ecology. The empirical/conceptual/theoretical axis can be divided up differently, but the variation among contributors would not disappear. Combining the variation on all these dimensions, a picture emerges of multiple pathways to effective use of models. That being said, there are consistencies that run through the work of all the contributors.

First and foremost, contributors give serious consideration to all three components of behavioral ecology—theoretical, conceptual, and empirical. It is one thing to say that you recognize the importance of all these components, but it is another thing entirely to build them into the system in which you work in just the right manner. Second, most contributors to this volume display an open mind to having their system lead them in unexpected directions. Be they empirical, theoretical, or conceptual new directions, model systems sometimes take on a life of their own and the researchers we are looking at here recognize that and are adept at regrouping and heading in new directions, when they appear to be fruitful. Lastly, many contributors are truly interdisciplinary in their approach to evolution and social behavior. They derive their theoretical, conceptual, and empirical ideas not only from evolutionary biology and ecology, but from psychology, physiology, endocrinology, anthropology, mathematics, sociology, political science, computer science, information theory, systematics, genetics, and economics as well.

Contributors to this volume have demonstrated an ability to integrate conceptual, theoretical, and empirical work in their model systems, but they have typically done more than that. In addition, they have demonstrated that their model systems are often the springboards for new directions in behavioral ecology. At some level this is not surprising since the model systems we examine are so well-studied that the researchers have a deep knowledge of that with which they work. This is a good starting point for new directions. That being said, it is not particularly easy to leave the comfort zone of the well-established to move into uncharted waters. In any case, the systems we examine here are an ideal breeding ground for the "hot" topics of behavioral ecology to mature and rise to the surface and so one mandate to contributors is to tell the reader where they hope the future work in their model system is headed and why they believe the path they foresee is critical in furthering our knowledge of fundamental questions in behavioral ecology.

## Jerry's Class

Just as my passion for behavioral ecology was taking root in 1985, I had the pleasure of being a student in Jerram Brown's course on this subject at SUNY Albany. Jerry was giving his introductory lecture on the history of behavioral ecology when he put up a figure on the board (figure P.1) that I still remember to this day (I double-checked my notes though, just to be sure).

I don't wish to debate whether this is the definitive way to depict the history of behavioral ecology, as it really doesn't matter for the point I want to make. And that is, as an impressionable young scientist-to-be, I was struck by this simple figure. For basic as it was, it suggested to me in a clear, concise form that behavioral ecology was a rich discipline, and one that was primed to make a major contribution to many different fields. Any discipline that drew its core principles from as widely diverse, but important areas as

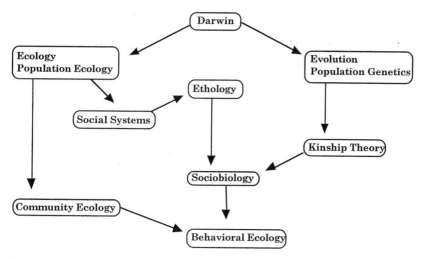

Figure P.1. Jerram Brown's (1986) classroom schematic of behavioral ecology's position in relation to other disciplines.

ecology, evolution, ethology, community and population ecology, and population genetics (plus all of the others areas mentioned earlier in this chapter), I was convinced, would be pivotal in developing a truly fundamental understanding of biology. I am to this day certain, that young and naive as I was when I took Jerry's class, I was correct.

# References

Allee WC, 1931. Animal aggregations. Chicago: University of Chicago Press.

Allee WC, 1951. Cooperation among animals. New York: Henry Schuman.

Axelrod R, Hamilton WD, 1981. The evolution of cooperation. Science 211:1390–1396.

Charnov EL, 1976. Optimal foraging, the marginal value theorem. Theo Pop Biol 9:129–136.

Crowley PH, Provencher L, Sloane S, Dugatkin LA, Spohn B, Rogers L, Alfieri M, 1995. Evolving cooperation: the role of individual recognition. Biosystems 37:49–66.

Dugatkin LA, 1997. Cooperation among animals: an evolutionary perspective. New York: Oxford University Press.

Hamilton WD, 1964. The genetical evolution of social behaviour. I and II. J Theo Biol 7:1–52.

Krebs J, Davies N (eds), 1997. Behavioral ecology: an evolutionary approach (4th ed.). Sunderland, MA: Sinauer Associates.

Kropotkin P, 1908. Mutual aid (3rd ed.). London: William Heinemann.

Kuhn T, 1962. The structure of scientific revolutions. Chicago: University of Chicago Press.

Nowak MA, Sigmund K, 1998. Evolution of indirect reciprocity by image scoring. Nature 393:573–577.

Nowak M, Sigmund K, 1992. Tit for tat in heterogeneous populations. Nature 355:250–252.

Stephens D, Krebs J, 1986. Foraging theory. Princeton: Princeton University Press.

Trivers RL, 1971. The evolution of reciprocal altruism. Q Rev Biol 46:189–226.

Zahavi A, 1975. Mate selection—a selection for a handicap. J Theo Biol 53:205–214.

# Acknowledgments

First and foremost, I would like to extend my heartfelt thanks to all the contributors to this volume. It has been an honor for me to work with some of the best behavioral ecologists of our day—I have learned much from each and every chapter in this book. I also wish to thank Sam Elworthy, my editor at Princeton University Press. Sam did much to make this volume what it is, and it has been a pleasure to work with him. Maria den Boer also did a terrific job copyediting the manuscript—quite a challenge when you realize this volume was about 900 typed pages when copyedited.

Lastly, I want to thank the loves of my life, my wife Dana, and my six-year-old son, Aaron. As usual, Dana did more proofreading than anyone should be expected to do, and never complained once. As for Aaron, the glimmer in his eyes and his wonderful smile helped me put everything in perspective.

L.A.D.
20 November 2000

PART I    Insect & Arachnid Model

Systems

# 1 Golden Flies, Sunlit Meadows: A Tribute to the Yellow Dungfly

Geoff A. Parker

## Introduction: A Personal History of an Obsession

One of my earliest recollections, long before schooldays, is a hazy mix of English wildflowers, long grass, and a myriad insects. Engrossed, I meandered in the outfield of a village cricket field, nestling amid the Cheshire pastures. The summer insects held my total concentration until I gradually became aware that my father (batsman and sometime captain of his team) was submitting tense and urgent demands from the far wicket. My entomological rapture had unwittingly lured me into the line of the bat, a sin of considerable magnitude for a small child. My father's requests were executed swiftly, and I was transported in my mother's arms to my proper place among the drowsy little throng awaiting tea by the cricket pavilion.

That could well have been my first contact with the common yellow dungfly, *Scatophaga stercoraria*. It certainly wasn't my last. Within weeks of beginning school, I began a (still greatly valued) friendship with the son of a dairy farmer in the same village as the cricket field, and the golden flies became a part of my childhood.

Essentially and exclusively a naturalist, I progressed through school with some academic success, but no real commitment to anything other than biology, and with an emerging antipathy to mathematics, which for some bizarre reason I saw as its antithesis. Natural history/biology became the focus, nurtured by two inspired teachers who helped generate its magic. At home, I sought woodlice under the *Aubrietia*, reared caterpillars and tadpoles, cherished many pets, boiled up dead frogs, rabbits, and hares to reconstruct their skeletons. My father, a research chemist/microbiologist, taught me quite a lot of natural history without my ever realizing it, and my mother (who knew British wildflowers) showed great patience. Home was warmth, love, and infinite security, and I didn't know then how very lucky I was.

It was unclear what road I should follow. Although my father was an industrial research scientist, it was not immediately obvious that I could study natural history for a living. A career as a vet seemed a possibility, especially in view of the farming associations. But for some reason I still don't fully understand, I found myself being interviewed for a place in the medical school at Bristol University. I began as a medical student in October 1962 and within two weeks had changed my registration to zoology. The preclinical tutor counseled earnestly, and the dean of medicine delivered a sharp rebuke—they thought I was mad. They were quite right financially, but that decision is one I never regretted.

The zoology course at Bristol then had animal diversity, physiology, phylogeny, and ecology, but passed almost without mention of behavior or evolution. The final honors year (1964–65) was based on research projects—no less than four in all—and here I found my utopia. The first was completed in the summer vacation before the start of the course. I devised an activity meter to monitor the diurnal rhythms of the carabid beetle, *Nebria brevicollis*. It consisted of a sort of seesaw in a plastic sandwich box, which tripped a make-and-break in an electrical circuit. The beetle's activity (probably much increased by the stimulus of the rocking seesaw) was measured with a relay counter, thanks to assistance from my physicist brother Antony, who was then constructing an early ESR machine as part of his Ph.D. There were peaks of activity at dawn and dusk, but I became preoccupied with a theoretical problem: How would the movement rate of a predator become adapted to the movement rate of its prey, and vice versa? It was my first dealing with an evolutionary game. I consulted widely—biology, physics, and chemistry professors—but made only a little progress toward a solution. It was a problem I never forgot, and one I revisited two decades later in the rather different guise of meetings between males and females ("the mobility game"; Hammerstein & Parker 1987).

The second project concerned the sexual behavior of the blowfly, *Protophormia terrae-novae*. Pheromones were all the rage and I worked very hard to show that males recognize females by a "contact chemical." That Christmas vacation, jars and tanks containing flies were arranged around the dining room at home, experiments continued relentlessly on the dining table, and once again my mother showed patience. The work was later published (Parker 1968a) but the main thing I learned was that experiments in the lab (or dining room) often pose more questions than answers: I desperately needed to know what happened in nature, in order that I could speculate on the adaptive function of what I had observed. Then followed a project on the strength of the pigeon's wing bones and tendons in relation to their function, closely supervised by the biophysicist, Colin Pennycuick. The question was essentially an evolutionary one: What mechanical safety factor evolves in a species constrained to economize on anatomical robustness in the interests of flight performance? This work was also published (Pennycuick & Parker 1966), and the project had taught me that not only that mathematics could relate to biology, but that I was mathematically illiterate. I began to suspect that I needed to become otherwise. Colin could explore and develop speculations using algebra; by this technique the power of reasoning about adaptation could be greatly extended.

The final project was suggested by the distinguished entomologist, the late

Howard E. Hinton, a truly remarkable man by any standards. It concerned mating behavior in dungflies and was to be exclusively a field study; I didn't want to experience the *Protophormia* frustrations once again. The past rose up to become the future. I knew this fly was for me. And it was; it offered the magic and familiarity of the cattle pastures. I was instantly captivated. With amazement and considerable relief, that summer I won the biology degree prize, and so more than qualified for the Ph.D. grant that Howard Hinton had promised.

In those glorious days, in zoology, supervisors asked their potential Ph.D. students what they would like to study. We barely discussed it—I would naturally continue with dungflies.

## The Dungfly Mating System

In spite of its incredible advantages as a study system, few studies had been directed toward dungfly sexual behavior before 1965. Two or three deserve special mention. One of the first was the naturalist R.A.F. de Réamur (1740), who gave detailed descriptions of its behavior and biology, and a second was Cotterell (1920), who focused on its remarkable feeding habits. A wonderful treatise on the ecology and biology of Diptera associated with cattle and their excrement was published by the Danish zoologist, Ole Hammer (1941), based on many years of fieldwork. Hammer's monograph (still a much-cherished possession) has held a special enchantment for me ever since Howard Hinton kindly gave me a copy in 1966.

One great merit of the dungfly mating system is that it is ubiquitous— everywhere in the Western world there are herds of pasturing cattle or sheep; the common yellow dungfly abounds for much of the year. A second major benefit is that the mating system is replicated as a series of discrete units, the cow pats, around the pasture. It is the archetypal patchy habitat. The patches (droppings) are produced frequently in a fair-sized herd; they can be labeled easily and inspected regularly at will. The third bonus is that everything is readily observable around each pat; measurements and observations can be taken as if in a natural laboratory. Fourth, what a world awaits the observer. It is hard to envisage a better model system for studying the operation of male-male competition: the operational sex ratio (Emlen & Oring 1977) is highly male-biased, and the behavior is indeed spectacular. The competition between males is fierce and intense: if dungflies were the size of red deer they would be the subject of a thousand books and nature films.

The males arrive quickly and often in considerable numbers, the gravid fewer females more cautiously, and pairing takes place on or around the dropping (Hammer 1941; Foster 1967; Parker 1970a). The flies are distributed with respect to wind direction: they fly in from downwind and often overshoot the cow pat; most of the searching occurs in the upwind surrounding grass within a meter of the dropping, or on the dung surface (Parker 1970b). Copulation occurs in the surrounding grass if pairs meet there. Pairs meeting on the dung surface copulate there if the density of searching males is low (Fig. 1.1), but tend to emigrate to the downwind surrounding grass if the density on the dropping is high (Parker 1971). Toward the end of copulation, if mating has been occurring in the grass, the male flies the female to the dung surface. On termination of genital contact the male raises his abdomen, but remains attached to the female with his front tarsi grasping the thorax-abdomen junction of the female (Fig. 1.2). He guards the female from the attentions of searching males while she lays her mature eggs in the cow pat, using a series of specialized defensive reactions (Parker 1970c), and often leaning over toward the attacker, which he deflects away with his middle legs while keeping his front legs attached to the female and his hind legs on the dropping (Fig. 1.3). Most of these interactions last only a second, but should the attacker grasp the female, a protracted struggle ensues for possession of the female (Fig. 1.4). The attacker attempts to pry himself in between the original male and the female, and if he should succeed in "takeover," he mates with the female. After the female has completed oviposition, she performs a series of side-to-side movements, the guarding male steps off, and she flies upwards and away from the dropping until she has matured her next egg batch.

Dungflies are remarkable in that not only do the adults eat from the dung surface and obtain nectar from flower-heads; they also capture other insects (mostly other Diptera), and obtain protein reserves by sucking the hemolymph after paralyzing the prey by puncturing the cervical region (Cotterell 1920; Réamur 1740).

To observe what is going on, the best procedure is to lie down on an old coat placed "side-wind" to the dropping. The only equipment I used was a stopwatch, ruler, thermometer, tape recorder, glass vials, entomological pins, a notebook, and a pencil. Thus the final bonus of dungfly behavioral ecology is that it has to be one of the most inexpensive systems to study.

I did not realize it, but to someone just starting out on a career in behavioral ecology (which I was, although the term was not in common usage then) the benefits of this easily located natural laboratory are truly immense. To go searching in dangerous far-off lands for rarely seen large mammals may be exciting, and will probably attract the media—it is a well-worked route to become a celebrity. But there is perhaps a trade-off. If your aim is an intellectual exploration into a model system that stimulates thinking about concepts, that allows collection of large data sets capable of developing and testing theories, that is readily available and accessible, and that does not require the hassle of government authorizations or licenses, think about dungflies. The only real peril is the occasional undetected bull, and the only irritations are the rain and the curiosity of unhabituated cows, dogs, and small children.

It is probably much more due to the intrinsic benefits of this uniquely tractable mating system than to any special merits of my own that the dungfly model has contributed to the development of concepts in behavioral ecology. It has played its part in generating concept and theory, and providing empirical support for the optimality approach in evolutionary biology. The work has not only become a feature of many behavioral ecology textbooks, but has also been interesting enough to attract the interest of philosophers of science (Kitcher 1985; Sober 1993; Myers 1990; Ruse 1997, 1999). I hope that the following sections will serve to outline some of the main areas of its contribution in terms of integration of concept, theory, and empiricism.

At the start of my research career in 1965, few ethologists and ecologists had been inclined to think in depth about the mechanism of natural selection. The prevailing sentiment then was that adaptation in some way or other served to be "of survival value to the species." A fellow student, R. R. Baker, and I reacted to this notion, and I strove in my Ph.D. study of the dungfly mating system to make detailed quantitative analyses of behavior in terms of the selective benefits arising from the strategies played by individuals. Despite some adverse criticism, I felt convinced that this approach was more satisfactory intellectually, and also appeared better to explain the facts observed. At that time, the "group selection" debate had hardly surfaced, although a handful of others were independently engaged in research activities with a similar focus, namely, that individual benefits were much more likely to shape adaptation than group or species benefits. A decade later, this movement had gained momentum in an explosion that coalesced into the discipline now known as behavioral ecology/sociobiology. The discipline is based on theoretical models, derived from the premises of natural selection, which serve to make predictions about behavioral strategies, that can later be tested empirically by field or laboratory observation. The aim is to understand adaptation in terms of the selective forces that have shaped it (Parker & Maynard Smith 1990).

Many of my contributions to behavioral ecology relate to cases where the fitness value of a given strategy depends on the strategies currently being "played" by other members of the population. J. Maynard Smith and G. R. Price established a conceptual framework for dealing with this frequency dependence in the mid-1970s; one seeks an evolutionarily stable strategy or ESS (Maynard Smith & Price 1973; Maynard Smith 1982), which is simply a strategy that, once fixed in the population, cannot be invaded evolutionarily by any rare alternative (mutant) strategy. But this came too late for my Ph.D. studies; I had little to guide my philosophical contemplations of the cow pats.

Figure 1.1. Copulation. From Parker (1970c).

Figure 1.2. Passive phase guarding. From Parker (1970c).

Figure 1.3. Guarding against an attacker. From Parker (1970c).

Figure 1.4. A struggle: attacker *left*, guarding male *right*. From Parker (1970c).

## The "Equilibrium Position" and the Ideal Free Distribution

Having found the best of all possible opportunities, I very nearly threw it away. The first year of my Ph.D. studies got me nowhere. After completing four detailed projects in my honors year, three years of graduate study seemed like an eternity. In those days, especially in entomology, many field-based studies were exhaustive, systematic surveys of "the ecology and biology of . . ." I entered a year of frenzied counting, measuring, and weighing of everything I could think of relating to dungflies. I would not be out-done—I would be even more exhaustive and systematic. But what was I trying to do? I had endless data, but no questions. Without a question, data gathering is meaningless. That summer (1966) was one of soul-searching and decision. I would go back to my honors project, to mating behavior. I would ask questions about dungflies in terms of the (then almost abandoned) Darwinian theory of sexual selection, and I would gather the data necessary to test whether male-male competition could account for the adaptations I saw. Apart from a classic paper by Bateman (1948), sexual selection had long been out of favor as an explanation of behavior, probably due to Huxley's writings in the 1930s (e.g., Huxley 1938).

I had learned a very important lesson; one I should have noted from the honors projects. All this seems so obvious now; it was less so then.

I became preoccupied with the time constraints on searching males. Time could be costed in terms of eggs fertilized—for a male dungfly each minute is worth around 0.23 eggs. There were several immediate questions. Why is copulation so long—commonly thirty to forty minutes—when many other flies can copulate within seconds? Why do males guard their female after copulating until she has laid her eggs? What behavior will be the outcome of competitive mate searching?

I thought about the incidence of the two sexes around the dropping. If males were in competition for females, then the distribution in space and time of females should determine the distribution in space and time of males. I fitted out a commercial van with a bed and cupboards, and so on, in which I would live in the field, at least for some months. I labeled over a hundred cow pats as each one hit the ground, during two days in August 1966, and counted the number of flies of each sex in each behavioral state as often as I could during the "life" of each dropping. When I averaged the number of flies present with time after the dropping was deposited (Fig. 1.5a), I could see that males arrived very quickly, with maximum numbers occurring some twenty minutes after the start. Numbers decline gradually, so that very few are left after five hours. There are always notably less females, and the peak of female numbers occurs much later (around fifty minutes). Why?

A male's prospects at any given time depend (partly) on how many females are arriving. I could work out the arrival rate of females from the number present, since I knew how long (on average) each one stayed at the dropping. This showed that the rate of arrival is fastest at the start, declining roughly exponentially with time. One could imagine that this arose because of the local depletion of females "waiting" for the cue of fresh dung, and the decline in usefulness of the dung as an oviposition site. Could the same logic explain the incidence of males, or was their incidence shaped by the arrival rate of females, that is, by sexual selection?

If a male has a continual "choice" of fresh droppings, surely the best strategy would be to get to each one as fast as possible and then to stay just a short time—this way the highest input rate of females could always be exploited. But staying a short time at each means that travels between drop-

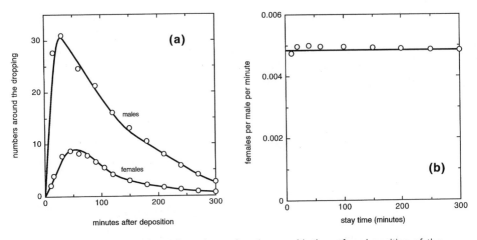

Figure 1.5. a. Changes in total numbers of each sex with time after deposition of the dropping. Upper curve = males; lower curve = females. After Parker (1970b). b. Mating rates achieved in relation to increasing male stay times at the dropping, assuming that males arrive on average after four-minute travel times between droppings. After Parker (1970b).

pings must be made more often, reducing the gain rate. Also a male's prospects depend on how many other males are present. I decided to calculate, in terms of mean eggs fertilized per minute of reproductive activity (the "egg gain rate") the payoffs achieved by males staying different lengths of time at a dropping ("stay times"). It seemed to me that if one stay time had higher success than others, more males should play this strategy until its value declines, until male-male competition produces a distribution of stay times such that all the stay times had equal success. I gave this distribution the rather awkward term *equilibrium position* (Parker 1968b, 1970b). After calculating that the mean time to find a new dropping was around four minutes, I was able to calculate that all stay times did indeed appear to have closely similar overall success rates if the travel times were included in the calculations (Fig. 1.5b). Sexual selection through male-male competition could indeed explain the typical decline of male numbers through time after deposition of the dropping! As I looked at the distribution of males around the pat—on the surface and in the surrounding grass—it looked as if a similar equilibrium may apply to the spatial distribution of males as well as the temporal distribution.

I could see that the "equilibrium position" in time and space had many implications in biology that extended beyond dungfly mate searching, but little did I realize that an entirely general theorem was about to appear. The "ideal free distribution" or IFD (Fretwell & Lucas 1970; Fretwell 1972) described just such a situation in very general terms. And independently, rather similar ideas had occurred to Orians (1969) and Brown (1969). I collaborated on a general interpretation of competitive foraging with R. A. Stuart, an electrical engineer at Liverpool with whom I traveled to work in the 1970s. If competitors are equal, and resources (e.g., female dungflies) are items that are continuously input into patches of awaiting competitors (e.g., male dungflies), then numbers of competitors in a patch should show "input matching," that is, they should match or track the input rate of items to the patch (Parker & Stuart 1976; see also Parker 1978a, 1978b). Input matching studies advanced with M. Milinski's ingenious laboratory experiments on competitive food foraging in sticklebacks (1979), in which prey items were input at different rates at each end of the tank. The fish showed close conformity with input matching, and Milinski-type experiments were to become a pervasive part of the IFD literature from that point on.

I could see that cases where competitors searched in patches for resources that declined as they were consumed (where there was no continuous input) must be different: the best strategy would be for all competitors to leave simultaneously at a certain time (Parker 1978a, b, 1979). The best strategy would be always to exploit the best patch, unless there was some form of interference (reduction in gain rate due to time spent on interactions with competitors). W. J. Sutherland (1983) was to make an important advance by applying the model of Hassell and Varley (1969) to show how IFD could operate under interference. We later collaborated to formalize the distinction between "continuous input" and "interference" systems and to consider how unequal competitive abilities would affect IFD predictions (Sutherland & Parker 1985; Parker & Sutherland 1986). Unequal competition is the rule rather than the exception, and applies to both the dungfly and stickleback cases as well as most others (Milinski & Parker 1991). Our approach suggested that where the ratios of payoffs of the different phenotypes change across patches, the phenotype distribution should be truncated among the patches: a rough interpretation is that the best phenotypes will occupy the best patches. But where the ratios of payoffs stay constant across patches, many equilibrium distributions are possible, and for some types of interaction (e.g., kleptoparasitism) there may be no equilibrium (Pulliam & Caraco 1984; Parker & Sutherland 1986). During the past decade or so, there has been considerable interest in modifying or developing new approaches to IFD with unequal competitors.

Although the original work (Parker 1968a, 1970b) attracted attention, it was criticized many years later (Curtsinger 1986) for its absence of statistical testing of a fit between the data and the predictions. This was true—but it was usual for statistics to be rudimentary or even absent in the 1960s. Anyone who can remember the "adding machines" of that time will understand why. A major problem was that obtaining an estimate of the error on the predicted values was not a trivial matter. The calculations involved many parameters, each of which had been determined in nature. These could each be given a mean and a variance, but how one calculated the combined error on predictions derived from these many parameters was obscure. Indeed, I had pondered on this problem as a research student. It applies still to many optimality studies in behavioral ecology. However, we (Parker & Maynard Smith 1987) were later able to test the observations against the predictions in a different way, and to show that they did not differ significantly from the predictions.

I continued work on the distribution of males around the dropping in my first years on the staff of the Zoology Department at the University of Liverpool (where I have remained ever since). The calculation of the equilibrium spatial distribution at each time after deposition involved more fieldwork and some quite complex computations. Males can gain eggs from newly arriving gravid females (often in the upwind surrounding grass), or from takeovers during oviposition (on the dung surface). The relative availability of these two female categories varies with time: incoming females predominate early on, then follows a peak of oviposition. The probability of catching a newly arriving female in the surrounding grass (versus the dung surface) increases with the number of males searching in the grass, which itself varies with time after deposition and with the total number of searching males (Fig. 1.6a). I was able to compute the equilibrium proportion of males searching in the upwind surrounding grass in relation to (a) time after dropping deposition, and (b) the total density of searching males (Parker 1974a). The predictions (Fig. 1.6b) matched the observations quite well, suggesting a fine-tuning of male search behavior to the two cues, time after deposition and competitor density. My first external research talks (in 1973 at Edinburgh and 1974 at Oxford) were on dungflies and IFD, and I visited the eminent evolutionary biologist John Maynard Smith in Sussex (1974)—a high point in my life—to discuss the new concept of evolutionarily stable strategies.

A detailed history of IFD, although fascinating, is not appropriate here (see, e.g., Milinski & Parker 1991; Tregenza 1995; Sutherland 1996), but ideal free approaches have been used to interpret a vast array of strategic competition problems in time and space, including seasonal incidence, the timing of life-history switches such as sex-change, migration patterns, metamorphosis, and a host of other problems. The common yellow dungfly played a role in the development and understanding of this most general of concepts, and I admit that when I think of modeling animal distributions, I find it hard not to think in terms of dungflies.

I am currently looking again at the August 1966 data set with my research student, Steve Hutchinson, to see if the temporal IFD can still be detected if droppings are split into categories relating to the total number of dungflies present. Results so far look encouraging.

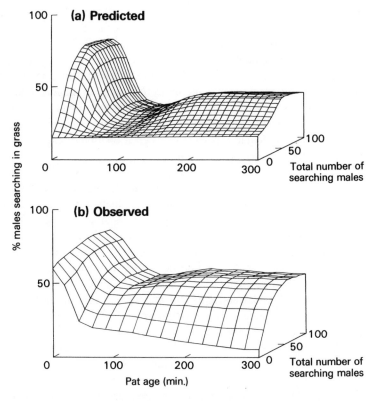

Figure 1.6. Proportion of the total searching males at the dropping that occur in the surrounding grass. After Parker (1974a). a. Predicted profile. b. Observed profile.

## Copula Duration and the Marginal Value Theorem

By 1968 I had been able to convince myself, at least, that sexual selection was indeed a power to be reckoned with—indeed, that it was hugely undervalued as a force in evolution. In 1969 I began new work on paternity. At Bristol, I had "labeled" sperm using the well-known "sterile-male" technique. I exposed males to 10 krad. gamma irradiation from a $^{60}$Co source; this induced dominant lethal mutations in the sperm so that eggs fertilized by them failed to hatch. I had mated irradiated males to gravid females that had been captured and mated in the field, to establish the proportion of eggs fathered by the last male to mate. This had been a pilot study; much more work was needed. I began an extensive project, without realizing that invertebrates have the huge advantage that no license is needed for experiments.

The results (Parker 1970f) were that, for females mated at least once previously, the last male to mate fertilized most (80 to 85 percent) of the eggs about to be laid, and the paternity of the remainder of the batch was allocated in the frequencies that would have applied before the last mating. This pattern continued throughout ten successive batches laid by the female, and was compatible with the notion that the last male's sperm displace most of the sperm stored in the female's sperm stores, then sperm mix randomly before fertilization. By separating pairs during copula, I found that a male's paternity gains showed diminishing returns with time copulating; the last male to mate had achieved a paternity expectation of around 50 percent within ten to fifteen minutes of copulation.

Half of the resulting paper (Parker 1970f) presented results on paternity, egg production, and fertility; the other half became a detailed analysis of the evolution of copula duration. Why do males copulate for so long a time? Alternatively, why do they stop copulating before they achieve full paternity? There is an obvious trade-off between the number of different females that can be mated, and the number of eggs gained from each one. It also appeared that the best copula duration depended (but only weakly) on the copula duration of other males. I constructed a detailed model which included the effect of takeovers, gains in subsequent egg batches, and the proportion of virgin females arriving at the dropping. Yet again, I sought to maximize the male's egg gain rate over a long time period of reproductive activity. The analyses were complex and carried out by computer. The natural copula duration was close to that predicted through sexual selection in respect of sperm competition and mate searching.

Still fascinated by the notion of gain rate maximization, I began work on a purely theoretical project on time investment strategies. If a male's probability of mating showed diminishing returns with time spent courting or guarding a given female, there should be an optimal time to quit—a time when it paid to leave in favor of finding other females. I talked over the problem with my father, and he saw a parallel in industrial economics. The situation resembled a process where there was a fixed "setting up" time before the process could begin, and where the cumulative yield showed diminishing returns with the time that the process ran. To maximize the yield rate, the optimal run time was given graphically by a tangent drawn to the curve of cumulative yield with time, the tangent starting from a point on the time axis to the left of the origin a distance equal to the setting up time. I wrote the paper using this principle, but because in those early days calculus eluded me, I only came close to the correct solution for how to optimize gains with females each of different value—this was formulated correctly by Charnov (1976) in his famous marginal value theorem or MVT (see also Parker & Stuart 1976). The paper went off to *American Naturalist*, to be rejected rather cursorily some months later. I then sent it to *Behaviour*, and after a long delay it was accepted, although the reviewer clearly felt rather bemused with the concept. It languished in press for a further year or so, and eventually appeared (Parker 1974b) in the year that the same tangent graph appeared in *American Naturalist* in the form of Smith and Fretwell's (1974) classic paper on the trade-off between offspring size and number.

I soon realized that the tangent method could be used to deduce the optimal copula duration in dungflies, since the frequency dependence involved was very weak. This time, one plotted the paternity gains with time $t$ spent copulating, and drew the tangent to this curve from a distance to the left of the origin equal to the mean time taken to find a female plus the guarding time. With dungflies as the inevitable background, I collaborated with my engineer friend, Bob Stuart, on a general approach to the problem of optimal exploitation of patchily distributed resources—I could get to (or close to) solutions by intuition, but Bob could prove them formally. He encouraged me to learn calculus. We (Parker & Stuart 1976) independently derived Charnov's marginal value theorem (Charnov 1976). The match between the predicted and observed copula durations was close (Parker & Stuart 1976; Parker 1978a, b) and became a popular example of MVT (Fig. 1.7a shows how it was depicted in the first edition of Krebs and Davies" *Introduction to Behavioural Ecology*, 1981).

Dungfly copula duration research did not fossilize in the 1970s. There have been many recent developments, which are still continuing. Paul Ward, Leigh Simmons, and I collaborated on a series of projects to extend the original work. We added elements of frequency-dependence, and considered what the optimal copula durations would be if males could differentiate between virgins and nonvirgins, which they appear not to be able to do (Parker et al. 1993). Testis volume decreases in size due to the demand for sperm during copulation (some sperm are stored in the testis), and copula duration was found to remain constant over several successive matings on the same day, and to decrease with dietary restriction. Further, the number of successive copulations that a male would perform decreased with dietary restriction (Ward & Simmons 1991). I devised models in which a male alternates between expenditure of reserves on reproductive activity around droppings, and replenishment of reserves by feeding on insects and nectar away from droppings. The models were able to generate qualitative explanations for these effects (Parker 1992).

We were becoming more interested in the mechanism of sperm competition (Parker et al. 1990). Dungfly displacement appeared to fit a model in which sperm flow into the female at a constant rate during copulation, with direct volumetric displacement of her stored sperm, and instant sperm random mixing (Parker & Simmons 1991). This model predicted that the proportion of eggs gained by the last male to mate ($P_2$) should follow exponentially diminishing returns with copulation time $t$, a form which had long been known to give a good fit to the data (see Fig. 1.7a). We found that large males show higher rates of displacement than small males (Simmons & Parker 1991), as estimated by the shape constant, $c$, in the best fit to the curve

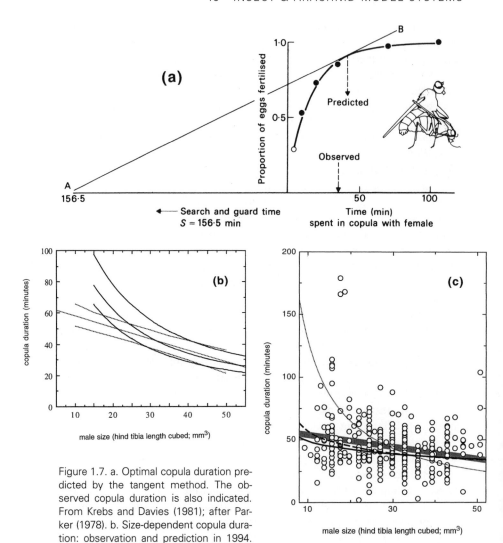

Figure 1.7. a. Optimal copula duration predicted by the tangent method. The observed copula duration is also indicated. From Krebs and Davies (1981); after Parker (1978). b. Size-dependent copula duration: observation and prediction in 1994. The observed regression is shown by the broken line and the prediction from the direct sperm displacement model by the continuous curve (both with ± 0.95 percent confidence limits). After Parker and Simmons (1994). c. Size-dependent copula duration: observation and prediction in 2000. The data set is increased (many of the points are not shown because they coincide), and the thick shaded line shows the best linear regression. The predictions from the indirect sperm displacement model are shown as the continuous and the broken curves that lie close to the regression line (these represent plausible prediction limits based on assumptions made about the biophysics of sperm flow from the male). Both of these fit well for all sizes of male. The 1994 prediction, based on direct displacement, is shown as the thin dotted curve. Based on the data presented in Parker and Simmons (2000).

of $P_2(t) = 1 - e^-c^t$, where $c$ is the proportion of the total stored sperm that is displaced in one time unit. Larger males also had an advantage in takeovers (Sigurjónsdóttir & Parker 1981), and so they had a lower travel time between females. We (Parker & Simmons 1994) modeled the two effects (displacement, takeovers) drawing on past field and lab data: both predicted negative relations between copula duration and male size, the displacement effect exerting the stronger influence. When both effects were included, the predicted relation between copula duration and size was quite close to that observed, although it is less good for very small males (Fig. 1.7b). Eric Charnov had reviewed our paper and noticed that the optimal solution would be for $ct$ to remain roughly constant whatever the male size. This arises if $c$ is directly proportional to size, and the setting up time $\mu$ (time to find and guard a female) is inversely proportional to size. For the general foraging problem, we can replace "male size" with "forager quality"—good foragers will have faster rates of uptake ($c$) from the patch, and shorter travel times ($\mu$) between patches, so that by analogy, $ct$ should remain equal across all forager qualities (Charnov & Parker 1995, 1999). Foragers of all qualities should quit a patch after the same cumulative gain. In support, despite the differences in $c$, $\mu$, and $t$, we found in dungflies no difference in $P_2$ between males of different sizes.

Our most recent work on sperm displacement in dungflies has focused further on mechanisms. Although the pattern of paternity gain with time, $P_2(t)$, fit the direct displacement model, it was hard to see how direct displacement could occur—the male has no obvious processes on the aedagus that could be inserted into the spermathecal ducts. Leigh Simmons and Paula Stockley mated females with two males, each with a different radio-label, to identify the amounts of sperm from each male in the female's spermathecae and bursa at different times during the copulation of the second male. The results (Simmons et al. 1999) indicate that sperm displacement is indirect, not direct. The male inputs sperm to the bursa, which acts as a receiver vessel. They then transfer across the spermathecal ducts to the storage vessels, the spermathecae. Hosken and Ward (2000) have substantiated this process by histological examination and suggest that the exchange across the spermathecal duct is achieved by the actions of a pulsatory device located in the apex of each spermatheca. Using a computer simulation of the indirect process, we have predicted the $P_2(t)$ relation under indirect displacement, which is very similar to that for direct displacement (Simmons et al. 1999). The indirect model can be used to predict the optimal copula duration in relation to male size—the fit is now rather better than that for the direct model, especially for small males (Parker & Simmons 2000); our ability to predict the adaptation has gradually improved as we have learned more about the details of the mechanism (Fig. 1.7c).

The recent data has also allowed us to predict how males should respond to the egg content of females (Parker et al. 1999). It appears that males cannot assess female size (and hence gravid egg content), but can assess the proportion of eggs that remain at the time of takeover during oviposition, possibly by the resistance of the female abdomen. Copula duration correlates positively with proportion of eggs remaining, in a manner that fits the model predictions.

All the copula duration work outlined above concerns optimality predictions made from the male perspective. Since establishing his own group in Zürich, Paul Ward has examined the possibility that female dungflies control certain aspects of paternity by "cryptic choice" (Eberhard 1985, 1996, 1998), and there is some evidence for this (Ward 1998). A fascinating possibility is that the three separate spermathecae may have evolved to allow female manipulation of ejaculates (Hellriegel & Ward 1998; Ward 1998; see also Otronen et al. 1997).

## Animal Contest Theory and Dungfly Struggles

I cannot resist a brief mention of the link between dungflies and the early days of animal contest theory which marked the start of the ESS concept. I had stared hard and long at dungfly struggles for many years; it was inevitable that I should puzzle over the evolution of animal contests, and had indeed discussed this during my Ph.D. with fellow research student, R. R. Baker. I had noticed how interactions between searching males and paired males could result in a struggle (Parker 1970c), which I had observed and recorded, but I had no theory on which to base an analysis. It was obvious that the best thing to do when competing for a female would depend on what one's opponent would do, and on the asymmetries between the costs and benefits of the fight for the two opponents, but I lacked a theory to make predictions about how males should behave in a fight. I began work on contest theory, guided mainly by intuition, and focusing mainly on the asymmetries inherent in a contest between two opponents for a fixed resource such as a mating prospect with a female. I had a manuscript in preparation around the time that John Maynard Smith's seminal papers (Maynard Smith 1972; Maynard Smith & Price 1973) appeared on ESS; it was hard to map my analysis onto the new ESS logic. It appeared (Parker 1974c) soon after Maynard Smith's (1974) paper on the "war of attrition" contest between two symmetric opponents.

This paper, if now it perhaps would not stand up to rigorous scrutiny in terms of modern ESS theory, had some merit in considering simultaneously—as did Maynard Smith's analyses—the best strategy for each opponent and how biological asymmetries may operate between contestants. I proposed that the contestants should assess each other's "fitness budgets" or values of winning ($V_A$, $V_B$) and rates of cost expenditure during the contest ($c_A$, $c_B$): after assessment, the opponent with the lower value for $V/c$

(i.e., the one that would use up its fitness budget first, should a long contest occur) should retreat. This rule (see also Popp & DeVore 1979) proved to be close to that derived several years later in a much more correct ESS analysis of the "asymmetric war of attrition" (Hammerstein & Parker 1982). I am still not certain whether this should be ascribed to luck or to intuition.

John Maynard Smith kindly invited me down to Sussex in 1974. We discussed ESS and the war of attrition, and dungfly stay times as a mixed ESS. Sussex was the major center for evolutionary biology in Britain at that time. It was, for me, a very special occasion. We began collaboration on a project to apply the new ESS logic to asymmetric contests. The paper appeared two years later (Maynard Smith & Parker 1976) and became an ISI "citation classic" in 1989. I have to say that my contribution was not greatly significant; it was indeed an asymmetric contest. But a continuing interaction with John Maynard Smith has been for me, both personally and scientifically, one of the most rewarding aspects of the past quarter of a century.

I eventually looked at my data on dungfly struggles in collaboration with the mathematician, E. A. Thompson, while spending a year (1978–79) as a visiting fellow of King's College, Cambridge. There was a superficial fit to Maynard Smith's symmetric war of attrition, but closer examination showed that struggles resulting in takeover were longer than those that did not (Parker & Thompson 1980). At that time, I supervised a talented Icelandic research student, Hrefna Sigurjónsdóttir—her project was to look at dungfly behavior in relation to body size (something that was being done independently in the United States; Borgia 1979). We showed that larger males had a distinct size advantage in takeovers (Sigurjónsdóttir & Parker 1981).

In addition to their role in contest theory, dungflies undoubtedly had a role in shaping models of sexual conflict (Parker 1979).

## Sperm Competition and Allied Phenomena

Three dungfly studies undoubtedly provided the stimulus for the development of the concept of sperm competition (competition between the ejaculates of two or more males for the fertilization of a given set of eggs; Parker 1970d) and its evolutionary consequences. Not only did the copula duration study suggest that males optimally displace previously stored sperm (1970a), but in addition, males protect their paternity by postcopulatory guarding (1970c) and by emigrating from the dropping to avoid takeovers (1971). The opposing elements of protection of self's paternity and ousting of sperm from previous males suggested general adaptive principles. My thoughts coalesced in the article in *Biological Reviews* in 1970 which is now often seen (e.g., Birkhead & Møller 1998) as the founding work for this rapidly developing field. Sperm competition forms a sort of postcopulatory version of male-male competition, whereas sperm selection by females forms the postcopulatory parallel of precopulatory choice of males by females (see, e.g., Birkhead 1998; Smith 1998; Eberhard 1998). The debt that sperm competition biology owes to the dungfly is considerable (see, e.g., Smith 1998; Birkhead 2000).

While looking at guarding in male dungflies, I had realized (from both observation and the literature) that there were many other cases of female guarding—both pre- and postcopulatory. They could be contact guarding (where the male maintains contact with the female) or noncontact guarding (where he simply stays close by). The best strategies for a male seemed to be (1) to displace or otherwise outcompete previously stored sperm , but simultaneously, (2) to prevent this occurring to his own sperm. These two conflicting selective forces suddenly seemed to explain a series of adaptations. When able to achieve a mating with an already-mated female, it was most common for males to achieve high levels of paternity, as in dungflies. The prevalence of postcopulatory guarding, mating plugs, and prolonged copulations seemed candidates for biological chastity belts of one form or other (Parker 1970d). Precopulatory guarding seemed a likely candidate for reducing the risk of sperm competition, as did substances in the ejaculate that induced unreceptivity to further matings ("monogamy") in females. I felt I could relate a series of behavioral and other adaptations to the fact that competing ejaculates could overlap in the female tract, often for long periods of time.

In addition to continuing the dungfly work, I investigated sperm competition and behavior in sepsids, locusts, and *Drosphila* during the 1970s, and supervised Ph.D. students working on *Gammarus* and *Asellus*. A history of the development of sperm competition as a discipline is given by R. L. Smith (1998). The 1970 review attracted little attention for almost a decade, until two American entomologists, Bob Smith (1979) and Jon Waage (1979), independently produced their classic works on sperm competition in (respectively) the water bug, *Abedus*, and the damselfly, *Calopteryx*. It appears also to have stimulated work on relative testis size. Increased relative testis size was attributed explicitly to the action of sperm competition independently by fish biologists (Robertson & Choat 1974; Warner & Robertson 1978; Robertson & Warner 1978), and by primatologists (Clutton-Brock & Harvey 1977; Short 1979; Harcourt et al. 1981). A symposium on sperm competition was held in Tucson, Arizona, in 1980, organized by Bob Smith. I stayed on after the conference to be hosted very generously by Bob and his wife, Jill, a gifted science writer. They led me into an entrancing new world of desert habitats—so different from my familiar green meadows that they seemed like a distant planet. The Tucson meeting eventually generated a collection of papers (Smith 1984) that acted as a catalyst for the discipline. It was, however, mainly the 1990s that marked my own return to the subject. During the past decade I have attempted to advance the general theory of sperm allocation under sperm competition, in a series of "sperm competition games" (reviewed in Parker 1998), and—some thirty years after my initial foray—to analyze detailed models of the sperm competition mechanism and its evolutionary consequences in the dungflies.

## Postscript

A central aim of this essay has been to show how detailed studies of a tractable system in behavioral ecology can generate not only fascinating conclusions about the how and why of adaptation in that species—it can also launch a series of questions about much more general phenomena. I hope I have shown that the dungflies helped to stimulate thinking about a range of entirely general problems in behavioral ecology, and often provided some good evidence for theories. Although I have since worked on many other problems that probably had little to do with dungflies, those early eight to nine years of field research were probably the most rewarding and idea-generating time of my research life.

It is paradoxical that my early interest in natural history, which promoted a disinterest—even dislike—of mathematics at school, and which resolved in a passion for golden flies in sunlit meadows, was to culminate in a career as a theoretical biologist. The seeds were sown in the undergraduate honors projects, when I sought solutions that were not easily deduced, other than by formal mathematical models. All my intensive fieldwork was done between 1965 and 1973, and most of my time since then has been spent modeling, computing, and manipulating algebra. While I still model dungfly adaptation, virtually all the empirical work is done now by collaborators, and all that remains of my love affair with dungflies is a stab of nostalgia on sunny days, when some relic emotion beckons me back to the pastures to gather data.

It has been fun to ponder how and why the golden fly has become so intimately entwined with my life. I can only feel disbelief at the good fortune I have had—the luck that as a child I became interested in natural history and studied dungflies, the luck that I began my research career just at the time when its scientific embodiment, behavioral ecology, was about to explode, generating such new vistas, and perhaps most of all, the luck that I could play a part in that explosion. Fate has delivered me an odd mix of passions—jazz, exhibition poultry, and behavioral ecology. My children will remember me when they hear a Dixieland band, the crowing of a bantam cock, or as they walk past a cow pat. But the greatest of these three passions is the last—and the golden flies will always be my icon.

Writing this essay has felt rather like writing a part of one's own obituary, but I would greatly prefer that it should act to stimulate a student interested in natural history to follow a similar journey. Behavioral ecology and the yellow dungfly have, for me, been an Odyssey of intrigue and excitement, and even now, endless inviting paths stretch into a future that I know I will never witness. It has all been terrific fun. For any student seeking a similar journey into the magic of insect mating systems, no better start could be made than by reading the remarkable treatise by Thornhill and Alcock (1983), with the Krebs and Davies editions (1978, 1984, 1991, 1997) to help broaden the vision.

## Acknowledgments

I should like to express my immense gratitude to all those who played—directly or indirectly—their parts in the dungfly story. The Richardson family and their dairy herd; Mrs. Jones at primary and Mr. Newman at secondary school; the distinguished entomologist H. E. Hinton; my dear late wife Sue (contemporary Bristol zoology student and dungfly field assistant); the important stimulus of Robin Baker; the entrancing legacy of Ole Hammer; my much more than marginally valued friend Eric Charnov; the incomparable John Maynard Smith; Bob Smith and Jill for the magic of Tucson 1980; and of course, my dungfly research colleagues over the decades: Hrefna Sigurjónsdóttir, Paul Ward, Leigh Simmons, Paula Stockley, Nic Seal, and Steve Hutchinson. And finally, my parents, who live on within me for those sunny, half-remembered outdoor days of my childhood, and for so very much more.

## References

Bateman AJ, 1948. Intra-sexual selection in *Drosophila*. Heredity 2:349–368.

Birkhead TR, 1998. Cryptic female choice: criteria for establishing female sperm choice. Evolution 52:1212–1218.

Birkhead TR, 2000. Promiscuity: an evolutionary history of sperm competition. Cambridge, MA: Harvard University Press.

Birkhead TR, Møller AP (eds), 1998. Sperm competition and sexual selection. San Diego: Academic Press.

Borgia G, 1979. Sexual selection and the evolution of mating systems. In: Reproductive competition and sexual selection in the insects (Blum MS, Blum NA, eds). New York: Academic Press; 19–80.

Brown JL, 1969. The buffer effect and productivity in tit populations. Am Nat 103:347–354.

Charnov EL, 1973. Optimal foraging: some theoretical expectations. Ph.D. thesis, University of Washington.

Charnov EL, 1976. Optimal foraging: the marginal value theorem. Theor Popul Biol 9:129–136.

Charnov EL, Parker GA, 1995. Dimensionless invariants from foraging theory's marginal value theorem. Proc Natl Acad Sci USA 92:1446–1450.

Charnov EL, Parker GA, 1999. Knowledge-independent invariance rules for copula duration in dung flies. J Bioeconom 1:191–203.

Clutton-Brock TH, Harvey P, 1977. Primate ecology and social organisation. J Zool 183:1–39.

Cotterell GS, 1920. The life history and habits of the yellow dung-fly (*Scatophaga stercoraria*); a possible blow-fly check. Proc Zool Soc Lond (1920):629–647.

Curtsinger JW, 1986. Stay times in *Scatophaga* and the theory of evolutionarily stable strategies. Am Nat 128:130–136.

Eberhard WG, 1985. Sexual selection and animal genitalia. Cambridge, MA: Harvard University Press.

Eberhard WG, 1996. Female control: sexual selection by cryptic female choice. Princeton: Princeton University Press.

Eberhard WG, 1998. Female roles in sperm competition. In: Sperm competition and sexual selection (Birkhead TR, Møller AP, eds). London: Academic Press; 91–116.

Emlen ST, Oring LW, 1977. Ecology, sexual selection, and the evolution of mating systems. Science 197:215–223.

Foster WA, 1967. Co-operation by male protection of ovipositing female in the Diptera. Nature 214:1035–1036.

Fretwell SD, 1972. Populations in a seasonal environment. Princeton: Princeton University Press.

Fretwell SD, Lucas HL, 1970. On territorial behaviour and other factors influencing habitat distribution in birds. Acta Biotheor 19:16–36.

Hammer O, 1941. Biological and ecological investigations on flies associated with pasturing cattle and their excrement. Vidensk Meddr dansk naturh Foren 105:1–257.

Hammerstein P, Parker GA, 1982. The asymmetric war of attrition. J Theor Biol 96:647–682.

Hammerstein P, Parker GA, 1987. Sexual selection: games between the sexes. In: Sexual selection: testing the alternatives (Bradbury JW, Andersson M, eds). Chichester: John Wiley & Sons Ltd; 119–142.

Harcourt AH, Harvey PH, Larson SG, Short RV, 1981. Testis weight, body weight and breeding system in primates. Nature 293:55–57.

Hassell MP, Varley GC, 1969. New inductive population model for insect parasites and its bearing on biological control. Nature 223:1133–1136.

Hellriegel B, Ward PI, 1998. Complex female reproductive tract morphology: its possible use in post-copulatory female choice. J Theor Biol 190:179–186.

Hosken DJ, Ward PI, 2000. Copula in yellow dung flies (*Scathophaga stercoraria*): investigating sperm competition models by direct observation. J Theor Biol (in press).

Huxley JS, 1938. The present standing of the theory of sexual selection. In: Evolution (DeBeer G, ed). Oxford: Clarendon Press; 11–42.

Kitcher P, 1985. Vaulting ambition: sociobiology and the quest for human nature. Cambridge, MA: MIT Press.

Krebs JR, Davies NB (eds), 1978, 1984, 1991, 1997. Behavioural ecology: an evolutionary approach. Editions 1, 2, 3, and 4. Oxford: Blackwell.

Krebs JR, Davies NB, 1981. An introduction to behavioural ecology. Oxford: Blackwell.

Maynard Smith J, 1972. On evolution. Edinburgh: Edinburgh University Press.

Maynard Smith J, 1974. The theory of games and the evolution of animal conflicts. J Theor Biol 47:209–221.

Maynard Smith J, 1982. Evolution and the theory of games. Cambridge: Cambridge University Press.

Maynard Smith J, Parker GA, 1976. The logic of asymmetric contests. Anim Behav 24:159–175.

Maynard Smith J, Price GR, 1973. The logic of animal conflict. Nature 246:15–18.

Milinski M, 1979. An evolutionarily stable feeding strategy in sticklebacks. Z Tierpsychol 51:36–40.

Milinski M, Parker GA, 1991. Competition for resources. In: Behavioural ecology: an evolutionary approach, 3rd ed. (Krebs JR, Davies NB, eds). Oxford: Blackwell; 137–168.

Myers G, 1990. Writing biology: texts in the social construction of scientific knowledge. Madison: University of Wisconsin Press.

Orians G, 1969. On the evolution of mating systems in birds and mammals. Am Nat 103:589–603.

Otronen M, Reguera P, Ward PI, 1987. Sperm storage in the yellow dungfly, *Scathophaga stercoraria:* identifying the sperm from competing males in separate female spermathecae. Ethology 103:844–854.

Parker GA, 1968a. The sexual behaviour of the blowfly, *Protophormia terrae-novae* R.-D. Behaviour 32:291–308.

Parker GA, 1968b. The reproductive behaviour and the nature of sexual selection in *Scatophaga stercoraria* L. (Diptera: Scatophagidae). Ph.D. thesis, University of Bristol.

Parker GA, 1970a. The reproductive behaviour and the nature of sexual selection in *Scatophaga stercoraria* L. (Diptera: Scatophagidae). I. Diurnal and seasonal changes in population density around the site of mating and oviposition. J Anim Ecol 39:185–204.

Parker GA, 1970b. The reproductive behaviour and the nature of sexual selection in *Scatophaga stercoraria* L. (Diptera: Scatophagidae). II. The fertilization rate and the spatial and temporal relationships of each sex around the site of mating and oviposition. J Anim Ecol 39:205–228.

Parker GA, 1970c. The reproductive behaviour and the nature of sexual selection in *Scatophaga stercoraria* L. (Diptera: Scatophagidae). IV. Epigamic recognition and competition between males for the possession of females. Behaviour 37:113–139.

Parker GA, 1970d. The reproductive behaviour and the nature of sexual selection in *Scatophaga stercoraria* L. (Diptera: Scatophagidae). V. The female's behaviour at the oviposition site. Behaviour 37:140–168.

Parker GA, 1970e. The reproductive behaviour and the nature of sexual selection in *Scatophaga stercoraria* L. (Diptera: Scatophagidae). VII. The origin and evolution of the passive phase. Evolution 24:774–788.

Parker GA, 1970f. Sperm competition and its evolutionary effect on copula duration in the fly *Scatophaga stercoraria* L. J Insect Physiol 16:1301–1328.

Parker GA, 1970g. Sperm competition and its evolutionary consequences in the insects. Biol Rev 45:525–567.

Parker GA, 1971. The reproductive behaviour and the nature of sexual selection in *Scatophaga stercoraria* L. (Diptera: Scatophagidae). VI. The adaptive significance of emigration from the oviposition site during the phase of genital contact. J Anim Ecol 40:215–233.

Parker GA, 1974a. The reproductive behaviour and the nature of sexual selection in *Scatophaga stercoraria* L. (Diptera: Scatophagidae). IX. Spatial distribution of fertilization rates and evolution of male search strategy within the reproductive area. Evolution 28:93–108.

Parker GA, 1974b. Courtship persistence and female-guarding as male time investment strategies. Behaviour 48:157–184.

Parker GA, 1974c. Assessment strategy and the evolution of fighting behaviour. J Theor Biol 47:223–243.

Parker GA, 1978a. Searching for mates. In: Behavioural ecology: an evolutionary approach, 1st ed. (Krebs JR, Davies NB, eds). Oxford: Blackwell; 214–244.

Parker GA, 1978b. Evolution of competitive mate searching. Ann Rev Ent 23:173–196.

Parker GA, 1979. Sexual selection and sexual conflict. In: Sexual selection and reproductive competition in insects (Blum MS, Blum NA, eds). New York: Academic Press; 123–166.

Parker GA, 1992. Marginal value theorem with exploitation time costs: diet, sperm reserves, and optimal copula duration in dung flies. Am Nat 139:1237–1256.

Parker GA, 1998. Sperm competition and the evolution of ejaculates: towards a theory base. In: Sperm competition and sexual selection (Birkhead TR, Møller AP, eds). London: Academic Press; 3–54.

Parker GA, Maynard Smith J, 1987. The distribution of stay times in Scatophaga: reply to Curtsinger. Am Nat 129:621–628.

Parker GA, Maynard Smith J, 1990. Optimality theory in evolutionary biology. Nature 348:27–33.

Parker GA, Simmons LW, 1991. A model of constant random sperm displacement during mating: evidence from Scatophaga. Proc R Soc Lond B 246:107–115.

Parker GA, Simmons LW, 1994. The evolution of phenotypic optima and copula duration in dungflies. Nature 370:53–56.

Parker GA, Simmons LW, 2000. Optimal copula duration in yellow dungflies: ejaculatory duct dimensions and size-dependent sperm displacement. Evolution (in press).

Parker GA, Simmons LW, Kirk H, 1990. Analysing sperm competition data: simple models for predicting mechanisms. Behav Ecol Sociobiol 27:55–65.

Parker GA, Simmons LW, Stockley P, McChristie DM, Charnov EL, 1999. Optimal copula duration in yellow dungflies: effects of female size and egg content. Anim Behav 57:795–805.

Parker GA, Simmons LW, Ward PI, 1993. Optimal copula duration in dungflies: effects of frequency dependence and female mating status. Behav Ecol Sociobiol 32:157–166.

Parker GA, Stuart RA, 1976. Animal behaviour as a strategy optimizer: evolution of resource assessment strategies and optimal emigration thresholds. Am Nat 110:1055–1076.

Parker GA, Sutherland WJ, 1986. Ideal free distributions when individuals differ in competitive ability: phenotype-limited ideal free models. Anim Behav 34:1222–1242.

Parker GA, Thompson EA, 1980. Dung fly struggles: a test of the war of attrition. Behav Ecol Sociobiol 7:37–44.

Pennycuick CJ, Parker GA. 1966. Structural limitations on the power output of the pigeon's flight muscles. J Exp Biol 45:489–498.

Popp JL, DeVore I, 1979. Aggressive competition and social dominance theory: synopsis. In: The great apes (Hamburg DA, McCown ER, eds). Menlo Park, CA: Benjamin-Cummins; 316–338.

Pulliam HR, Caraco T, 1984. Living in groups: is there an optimal group size? In:

Behavioural ecology: an evolutionary approach, 2nd ed. (Krebs JR, Davies NB, eds). Oxford: Blackwell; 122–147.

Réamur R. A. F. de, 1740. Mémoires pour servir á l'histoire des insectes. V:377–379.

Reimann JG, Moen DJ, Thorson BJ, 1967. Female monogamy and its control in the housefly, *Musca domestica*. J Insect Physiol 13:407–418.

Robertson DR, Choat JH, 1974. Protogynous hermaphroditism and social systems in labrid fishes. Proc 2nd Int Symp Coral Reefs 1:217–225.

Robertson DR, Warner RR, 1978. Sexual patterns in the labroid fishes of the Western Carribean, II. The parrotfishes (Scaridae). Smithsonian Contrib Zool 255:1–26.

Ruse M, 1997. Monad to man: the concept of progress in evolutionary biology. Cambridge, MA: Harvard University Press.

Ruse M, 1999. Mystery of mysteries: is evolution a social construction. Cambridge, MA: Harvard University Press.

Short RV, 1979. Sexual selection and its component parts, somatic and genital selection, as illustrated by the great apes. Adv Study Behav 9:131–158.

Sigurjónsdóttir H, Parker GA, 1981. Dung fly struggles: evidence for assessment strategy. Behav Ecol Sociobiol 8:219–230.

Simmons LW, Parker GA, 1991. Individual variation in sperm competition success of yellow dung flies, *Scatophaga stercoraria*. Evolution 46:366–375.

Simmons LW, Parker GA, Stockley P, 1999. Sperm displacement in the yellow dung fly, *Scatophaga stercoraria*: an investigation of male and female processes. Am Nat 153:302–314.

Smith RL, 1979. Repeated copulation and sperm precedence: paternity assurance for a male brooding water bug. Science 205:1029–1031.

Smith RL, 1998. Foreword. In: Sperm competition and sexual selection (Birkhead, TR, Møller, AP, eds). London: Academic Press; xv–xxi.

Smith, RL (ed), 1984. Sperm competition and the evolution of animal mating systems. London: Academic Press.

Smith CC, Fretwell SD, 1974. The optimal balance between size and number of offspring. Am Nat 108:499–506.

Sober E, 1993. Philosophy of biology. Oxford: Oxford University Press.

Sutherland WJ, 1983. Aggregation and the "ideal free" distribution. J Anim Ecol 52:821–828.

Sutherland WJ, 1996. From individual behaviour to population ecology. Oxford: Oxford University Press.

Sutherland WJ, Parker GA, 1985. Distribution of unequal competitors. In: Behavioural ecology: the ecological consequences of adaptive behaviour (Sibly RM, Smith RH, eds). Oxford: Blackwell; 255–273.

Thornhill R, Alcock J, 1983. The evolution of insect mating systems. Cambridge, MA: Harvard University Press.

Tregenza T, 1995. Building on the ideal free distribution. Adv Ecol Res 26:253–307.

Waage J, 1979. Dual function of the damselfly penis. Science 203:916–918.

Ward PI, 1998. A possible explanation for cryptic female choice in the yellow dung fly, *Scathophaga stercoraria* (L.). Ethology 104:97–110.

Ward PI, Simmons LW, 1991. Copula duration and testes size in the yellow dung fly, *Scathophaga stercoraria* (L.): effects of diet, body size, and mating history. Behav Ecol Sociobiol 29:77–85.

Warner RR, Robertson DR, 1978. Sexual patterns in the labroid fishes of the Western Carribean, I. The wrasses (Labridae). Smithsonian Contrib Zool 254:1–27.

# 2  A Feeling and a Fondness for the Bees

Thomas D. Seeley

Every creature on earth—whether great or small, furry or scaly, of land or of sea—presents us with beautifully adaptive behavior that can pique our curiosity. Who among us is not intrigued by the intelligence of ravens, the playfulness of otters, and the savagery of slave-making ants? Given the countless wonders to explore in the natural world, why is it that I aim my research efforts mainly toward one small part of nature, at only one species: the honey bee? How can it be professionally adaptive for a behavioral ecologist to give years of his or her life to the patient study of a single species? What can I tell someone who is starting a career in behavioral ecology and delights in studying nature deeply through a specific model system? What makes a species a model system for behavioral research? These are the questions that I will attempt to answer in this essay.

## Why and How I Became Interested in Honey Bees

Some biologists orient their work around general questions, while others are guided by an aesthetic appreciation for a particular taxonomic group. I definitely belong in the latter set. Ever since childhood, I have been fascinated by the societies of insects, especially colonies of honey bees. I grew up in a rural community called Ellis Hollow, located a few miles east of Ithaca, New York, where I spent much time alone exploring the woods, abandoned fields, swamps, and creeks. My favorite find was about a mile from home, atop the big hill in front of our house, down a dirt road that led to an old farmhouse. Here I discovered a couple of bee hives that someone had set up in a sunny spot beside a field of goldenrod. I loved visiting these hives. Sitting beside one, I could see bees landing heavily at the entrance with loads of brightly colored pollen, I could hear the hum of bees fanning their wings to ventilate their nest, and I could smell the aroma of ripening honey. That thousands of insects could live together so densely and so harmoniously, and could build delicate wax combs filled with delicious honey, was an incomparable wonder that left a deep impression. No less impressive was what I saw when I lay in the tall grass beside these hives: thousands of humming bees crisscrossing the blue summer sky like shooting stars.

Almost ten years later, before heading off to Dartmouth College to start my undergraduate studies, I began to pursue in earnest my fascination with the bees. I spent the summer working as a helper for the late Roger A. Morse, professor of apiculture at Cornell University, who directed an off-campus laboratory devoted to honey bee studies. (This job was arranged by my kind mother, who understood my interest in the bees and was a good friend of Roger Morse's wife.) Mostly I painted hives, mowed grass, and ran errands, but sometimes I helped conduct the experiments and do the bee-keeping. I loved driving one of the lab's green pickup trucks to a beeyard, lighting a straw-filled bee smoker, donning a bee veil (and *only* a veil; I was taught that real beekeepers do not need a bee suit or gloves), and going through the hives, checking each one for disease and a vigorous queen, and maybe giving it another "super" of combs. Through such experiences I began to accumulate the broad, descriptive knowledge of honey bees that forms the foundation for my scientific work on this system. To this day, I feel an incalculable debt to Roger Morse for introducing me to the scientific study of honey bees.

During my college years I organized my studies based on the dim understanding that I would become, like my first mentor, Professor Morse, a student of the bees. Because chemical communication was one of the hottest topics in behavioral biology in the early 1970s, I decided to major in chemistry and master the skills needed to decipher the chemical signals of the bees. Early on, however, I had nagging doubts about the wisdom of being a biologist focused on honey bees. I knew I liked the bees, especially their complex sociology, but I wondered if I could address "Important Questions in Biology" by studying honey bees as a behavioral biologist. Should I instead pursue some other branch of biology with greater luster—I considered marine ecology, molecular genetics, and protein biochemistry—and handle my love for the bees by being a hobby beekeeper? The answer came suddenly one Friday night during my sophomore year. While scanning the new arrivals shelf in the biology library at Dartmouth, I found Edward O. Wilson's (1971) book *The Insect Societies*, and upon reading its first chapter, "The Importance of Social Insects," my misgivings about my career plans evaporated. The book's opening words, so important to me then, still inspire me today:

> Why do we study these insects? Because, together with man, hummingbirds, and the bristlecone pine, they are among the great achievements of organic evolution. Their social organization—far less than man's because of the feeble intellect and absence of culture, of course, but far greater in respect to cohesion, caste specialization, and individual altruism—is nonpareil. The biologist is invited to consider insect societies because they best exemplify the full sweep of ascending levels of organization, from molecule to society.

Ever since reading Wilson's book I have felt that deepening our understanding of honey bee colonies is a scientifically important endeavor. By shedding light on why these bees form such thoroughly cooperative colonies, and of how they achieve their high cooperation, I help us understand what

John Maynard Smith and Eörs Szathmáry (1997) have called a "major transition in evolution." This is the transition from solitary individuals to social groups as loci of selection, and thus of adaptation. As one of only a handful of such transitions in the history of life on Earth, understanding it is an important challenge in biology.

Only through detailed studies of particular species that exemplify this evolutionary transition can we understand how it was actually achieved. Of course, some species will be more tractable for experimental analysis than others. So when I started my graduate studies at Harvard University, under the supervision of Bert Hölldobler, whose elegant investigations of the inner workings of ant colonies I found inspiring, I wondered once again whether honey bees were the best choice for a study system. To help settle this issue, I consciously compared honey bees to several other social insect groups with highly integrated colonies (superorganisms)—including yellowjacket wasps and stingless bees—and concluded that honey bees really are the best for my purposes. That I could easily keep colonies of honey bees close at hand, peer inside them using observation hives, manipulate the contents of a colony's nest, label a colony's workers for individual identification, and even control a colony's genetic structure, convinced me that the honey bee is the species of choice for my scientific work. Perhaps, though, all this deliberation was just a rationalization of a decision that I had actually made many years before.

## Integration of Theoretical and Empirical Issues

Many behavioral ecologists seem to place theory ahead of their empirical work. They master the theory regarding some general behavioral topic—such as optimal foraging theory, sex ratio theory, or kin selection theory—and then they look for an organism with which to test the hypotheses that are framed by this theory. I confess that I proceed the other way around. Robert Frost once said that for him the writing of a poem starts out as a feeling, as a lump in his throat. Likewise for me the doing of a research project starts out as a feeling, as a sense of emptiness in my understanding of some aspect of the social life of the bees. Only later, after I have started watching the bees and have begun to understand what they are doing, do I turn to the relevant theory to help me organize my thoughts and place my work in a larger context.

I respect greatly the usual style of behavioral ecological research, wherein theory guides the topics that one probes. This approach guarantees that one works on topics of broad interest and that one sees behavioral phenomena as embodying general concepts. Certainly finding deep concepts, such as Hamilton's Rule, that provide compact, general explanations for behavior is the ultimate goal of our scientific work. Nevertheless, I come at research problems from the opposite direction, starting with specific behaviors rather than general theories. I do so not out of an urge to be a contrarian, but simply because this is how my curiosity works. Some puzzling behavior or product of behavior, especially something that I have seen firsthand (*not through another person*) and directly (*not through a videorecording*), is what excites my curiosity. Personal examples include seeing how bees fan their wings in response to a puff of carbon dioxide, how a scout bee walks all about the interior of a potential nest cavity, how workers crawl about their nest excitedly after contacting their queen, how the foraging sites advertised by the waggle dances in a colony change markedly from day to day, how worker-laid eggs do not appear in hives (even in regions not visited by the queen), how a nectar forager can suddenly switch from waggle dancing to tremble dancing, and how a colony's comb building suddenly starts up after days or weeks without any such building.

I think that the research approach of starting with specific observations about favored animals—what Niko Tinbergen (1985) has so nicely called "Watching and Wondering"—even holds some advantages over starting with general theories about abstract entities. For one thing, it helps you get a quick start in research, for you skip the process of hunting for a "suitable" study species. More important, the observation-first approach helps you tap into entirely new behavioral phenomena, ones not mentioned in the textbooks and for which there may be little or no relevant theory. Such phenomena offer, I feel, the best opportunity to do truly creative and pioneering work. They also let you work in uncrowded territory where, at least for a while, you needn't worry about somebody "scooping" you. Perhaps the most important bit of advice that I received from Edward O. Wilson while I was a graduate student was this: "When choosing a thesis topic, carefully assess where the biggest scientific battles are being waged, where the intellectual action is the hottest, then move as fast as you can in the opposite direction." If you listen to your aesthetic sense and allow your curiosity about animals to be your guiding star, the odds are high that soon you will find yourself gazing over a region of scientific terrain which remains as yet uncharted. This is, for example, how Rüdiger Wehner (1982) began exploring navigation in desert ants, how Emmett Duffy (1996) discovered eusociality in a coral-reef shrimp, and how Bernd Heinrich (1999) started probing the minds of ravens.

I will now share in some detail one example from my own work that illustrates how studies that are launched from specific empirical observations can give rise to general conceptual advances. My example centers on the curious fact that the workers in a honey bee colony with a queen refrain from laying eggs, hence are genetic dead ends, even though they retain functional ovaries. This fact is, of course, central to understanding the evolution of a colony as a unit of function, that is, to the making of the evolutionary transition from solitary individual to coherent colony. Just as the somatic cells (also genetic dead ends) of a metazoan body toil selflessly to enable their body to produce gametes, so the worker bees of a queenright honey bee colony toil almost as selflessly to enable their colony to produce new queens and males. In both cases, entities (single cells and single bees) that were capable of independent replication of their genes before the transition can replicate their genes only as part of a larger whole (a multicellular organism or an insect colony) after it.

As a graduate student, I was keenly interested in the system of chemical communication between the queen honey bee and her workers whereby the queen produces a pheromonal signal that spreads among the workers and, it was thought, prevents the workers from laying eggs and rearing queens. This communication system was studied in the 1950s and 1960s by Colin Butler (1954) of the Rothamsted Experimental Station, whose papers on the subject I had pored over in connection with my undergraduate studies in chemistry. I was especially intrigued by the facts that this signal has just one sender but several thousand receivers, that although it is a chemical signal it is not dispersed throughout a hive as an airborne odor, and that if you remove the queen from a colony the workers will begin rearing a replacement queen within a few hours. Evidently, each worker receives a signal from the queen every hour or so as long as Her Majesty is alive and well. But how on earth is this accomplished? Having read all the papers by Butler and others on this topic, I knew that this puzzle remained unsolved. I decided to have a go at it. Like anyone else who has ever peered inside an observation hive, I had seen the striking behavior of worker bees around their queen when she is stationary: they press toward her forming a tight retinue, and then they lick her, feed her, and brush her with their antennae. Surprisingly, no one had examined the behavior of these bees after leaving the retinue, when almost certainly they would be carrying some of their queen's pheromones. Might these workers do something that helps disperse the queen's chemical signal? By labeling and watching individual workers in contact with their queen, I discovered that those workers that make extensive contact with the queen appear to behave as messengers dispersing the queen's signal. They walk more rapidly, have more contacts with nestmates, and perform fewer other tasks in the thirty minutes following queen contact than do randomly chosen workers of the same age (Seeley 1979). A subsequent study by Ken Naumann and his colleagues (1991), elegantly performed with radionuclide-labeled queen pheromone, has provided direct support for the messenger bee hypothesis.

The discovery of messenger bee behavior was interesting in itself, but it also led to advances on broader issues. This came about because the existence of messenger bee behavior raised questions about the role of the queen's chemical signal. The standard functional interpretation of this signal at the time was one of "queen control," that is, this signal is the means by which the gently despotic queen keeps the workers from laying eggs that would produce males. But I had just found that workers show specific behavioral adaptations to help disperse the queen's putative "control" pheromones. How could such worker adaptations evolve and be maintained if they do not serve the genetic interests of the workers? This led me to scrutinize the literature on the effects of the queen's pheromones on the workers. I was intrigued to learn that several older studies (reviewed in Seeley 1985; see also Willis et al. 1990) indicated that the queen's pheromones are neither necessary nor sufficient for inhibiting the workers' ovaries. The queen's pheromones do, however, strongly inhibit the workers from rearing additional queens. Thus it was clear that the function of this signal is to inform the workers that they have a queen. It is now recognized that in many species of social insect the pheromones of queens serve the genetic interests of both the queen and the workers, by signaling the queen's presence to the workers, and that these pheromones do not really control worker reproduction (Woyciechowski & Lomnicki 1989; Keller & Nonacs 1992).

So if not the queen's pheromones, then what is it that prevents worker reproduction in honey bee colonies? In the mid-1980s, I recognized that the answer to this puzzle was likely to be mutual inhibition by workers, or what is now known as worker policing. "If [a worker] attempted to lay eggs in her mother's nest, her nestmates would probably maul her just as they maul laying workers in queenless colonies. Most of any worker's nestmates are her half sisters; they are expected to prefer rearing their mother's sons ($r = 0.25$) rather than those of a half-sister ($r = 0.125$)" (Seeley 1985, 28). This work set the stage for Francis Ratnieks (1988) to develop the formal theory of worker policing. This theory, in turn, stimulated experiments that showed that workers do actually police one another, both by mauling workers with active ovaries (Visscher & Dukas 1995) and by destroying worker-laid eggs (Ratnieks & Visscher 1989).

Thus we can see that even though many, perhaps most, empirical studies in behavioral ecology are tests of hypotheses deduced from general theory, it is also possible for empirical studies to lead inductively to the creation of new theory. It seems to me that both research approaches help us understand the lives of animals, and that which approach one adopts (if not both!) is mainly a matter of personal style.

## Why Focus on a Single Species? Why Seek a Model System?

I feel that in the field of animal behavior, the best contributions have come from individuals who have devoted years of their lives to careful, patient investigation of single species. To note just a few examples in which an individual has persistently probed one "model system" with great success, consider Jane Goodall (1986) and chimpanzees, Vincent Dethier (1976) and blow flies, Niko Tinbergen (1953) and herring gulls, Karl von Frisch (1967) and honey bees, Rüdiger Wehner (1982) and desert ants, John Hoogland (1995) and prairie dogs, Hans Kruuk (1995) and river otters, Glen Woolfenden (Woolfenden & Fitzpatrick 1984) and Florida scrub jays, Tim Clutton-Brock (Clutton-Brock et al. 1982) and red deer, and Bernd Heinrich (1999) and ravens. Of course, extremely important contributions have also come from comparative studies involving closely related species, but I believe that this approach works best once a profound knowledge of one species has been acquired and so a solid baseline for the comparative work has been set.

What constitutes a model system (species) for behavioral research? Given that a prerequisite to studying an animal's behavior is being able to observe the fine details of its behavior as it lives in its natural world, for a species to be a model system you must be able to get and stay close to its members living in the wild. It is also essential that, while making your observations, you can recognize particular individuals (through natural or artificial markings) and do not disturb them by your presence. Also, a model system species must be amenable to experimental work, that is, to delicate alterations of the animal itself, the animal's environment, or both. Unless you can perform manipulations of specific properties of the animal or its environment, you will be unable to conduct controlled experiments. And without experiments, your analysis of the causes and the effects of the animal's behavior will be crippled. Finally, for a species to be a model system for behavioral-ecological studies, you must be able to measure the lifetime reproductive success of individuals in nature, or at least reasonable proxies of reproductive success, so that you can see what behavioral factors contribute to variation in reproductive success.

Let me illustrate how all these criteria can be fulfilled by a single species. I will, of course, refer to my favorite animal for scientific work, the honey bee. First, with respect to observing individuals living under natural conditions, honey bees make things extremely easy. They thrive inside hives with glass walls ("observation hives") which allow one to peer inside a colony and see the normally hidden in-hive activities of the bees. And when outside the hive, bees are—like most insects—not at all shy in the presence of humans. Also, although tiny compared to birds, mammals, and reptiles, honey bees are large enough to be given either paint marks or plastic tags, rather like miniature license plates, for individual recognition (Frisch 1967). Furthermore, many aspects of the lives of bees are accessible to experimentation. The bee's entire foraging process, for instance, is amenable to experimental manipulation. We can precisely alter the nutritional needs and the foraging opportunities of the bees, and can closely monitor the behavior of individuals to see how they adaptively respond to changing conditions both inside and outside the hive (Seeley 1995). I like too the fact that honey bee colonies, like plants, are fixed in location, for this makes it relatively easy to measure their patterns of survivorship and reproduction (production of queens and drones). This enables us to examine the differences in reproductive success between colonies that natural selection uses to produce adaptations in honey bees (Seeley & Visscher 1985).

Consider now why there is such a large payoff to concentrating on one species (your personal "model system") and giving it meticulous attention. First, there is the reality that it takes much time to get to know another species of animal, that is, to touch the world and the travails of beings that are not humans. Perhaps I am just very slow in the uptake, but it took me, for example, nearly ten years of experience with the bees before I realized how strongly the availability of the bee's main food, nectar, varies from day to day. It then took me several more years to appreciate fully how much the bee's diverse communication signals—including the waggle dance, the tremble dance, and the shaking signal—are all adaptations for coping with the variability in a colony's food supply (Seeley 1992; Seeley et al. 1998). And even now, after more than thirty years of close involvement with the bees, I still know little about the importance of diseases as agents of natural selection on bees living in the wild (Sherman et al. 1998). One of the early behavioral ecologists, F. Fraser Darling (1938), eloquently expressed how getting to know the behavior and the world of another species is both difficult and time consuming: "How surely it has been borne upon me that the glimpses of minutes, hours, days or even weeks, which a life of bird watching as a hobby have given, are inadequate for an interpretation or solution of the deeper problems of evolution, natural selection, and survival in the bird world! We need time, time, time and a sense of timelessness. Our pictures of behaviour must be detailed in time equally with those of space" (p. 78).

A second benefit to focusing on one model system is that it helps you assemble a large store of good research ideas, from which you can then select the best ones for your full-blown research projects. For me, research ideas rarely come by wandering vaguely about wondering what would be good to study. Most new ideas come instead unexpectedly when I see some surprising detail while my mind is engaged in performing an experiment, doing background reading for a specific project, or taking care of my bees. Consider one example: how I came up with my hypothesis regarding the function of the bees' tremble dance. I learned about this puzzling behavior in the fall of 1974 when, as part of my general reading as a graduate student, I slowly read Karl von Frisch's 1967 masterwork, *The Dance Language and Orientation of Bees*, from cover to cover. At the time it seemed likely to me that the tremble dance is a communication behavior, but I had no good ideas about its information content or about the role it plays in a colony's functioning. It was only many years later, while my mind was focused on performing an experiment and the bees unexpectedly began performing this dance right before me, that the hunch came to me that this communication signal helps remove a bottleneck in a colony's nectar acquisition process by signaling the need for additional nectar receivers (Seeley 1989, 191). Over my thirty years of working and playing with the bees, I have experienced many such thought-provoking surprises. And because ideas arise much faster than they can be pursued, my notebooks and brain contain many thoughts that potentially could be expanded into major research projects. I feel certain that my store of research ideas would be less well stocked if I had not stayed focused on the bees.

The third major benefit that I see arising from concentrating on one subject is that this approach allows you to assemble a large set of research methods that work with your particular study species. Having a large assortment of tools is, in itself, not important, but having a full "toolbox" that can help you perform your research more efficiently and with greater analytic power is important. Let me illustrate this with a personal example: my efforts, spanning some twenty-five years, to understand how a swarm of bees chooses a nesting site. This process is a kind of plebiscite in which the bees appear to make a "democratic" choice among multiple sites that are advertised through waggle dances performed by scout bees on the surface of the swarm. It is a spectacular biological phenomenon that has fascinated me ever since my freshman year in college, when I read Martin Lindauer's (1961) lovely summary of his pioneering work on this topic. Indeed, my fascination was so strong that in 1974, when I needed to choose a topic for my Ph.D. thesis research, I chose the bees' househunting process. However, at this early stage in my career I lacked the tools, both technical and conceptual, needed to unravel this decision-making process. In hindsight, I now realize that I lacked three tools in particular: (1) I did not know how to label for individual identification each of the thousands of bees in a swarm, (2) I did not possess high-resolution videorecording equipment and a slow-motion video player, and (3) I did not see clearly how to think about the general issue of distributed decision making. With the limited research skills that I did possess I was able to investigate the process by which a scout bee evaluates a potential dwelling place, and so completed my Ph.D. thesis (Seeley 1977; Seeley & Morse 1978). I then turned my attention to a more tractable topic, the organization of the food-collection process of a honey bee colony. And during the fifteen years that I devoted to this rich subject I gradually (and unintentionally) acquired the technical and conceptual tools needed to crack the ever-beckoning mystery of how a swarm functions as a collective decision-making agent. Recently I have taken up this puzzle again and have been delighted to find that now, unlike twenty-five years before, I can make solid progress toward solving it (e.g., Seeley & Buhrman 1999). I am sure that my experience of gradually acquiring better and better research tools relevant to a particular study species, and thereby developing a greater ability to analyze the behavior found in this species, is common among behavioral ecologists that have patiently studied a single species for many years.

## Future Endeavors

Even though no one can predict accurately his or her future, I believe that for the "foreseeable" future my scientific work will continue to focus on unraveling the functional organization of honey bee colonies. This is a line of study that suits who I am and provides an important contribution to biology.

All biologists acknowledge that natural selection has built adaptive units at the levels of single cells and multicellular organisms, but fewer accept that natural selection has also built adaptive units at the level of groups of organisms. The resistance to the view that groups can function as adaptively organized entities stems from the widespread rejection of group selection as an important force in evolution, because to recognize the existence of group-level adaptive units requires that one accept that natural selection can act at the level of groups. The wholesale rejection of group selection is, however, an error. This is demonstrated both theoretically and empirically. Multilevel selection theory, which partitions selection into within-group fitness variation (= individual selection) and between-group fitness variation (= group selection), makes clear that group selection can be a potent evolutionary force in group-structured species, especially those whose groups are assemblages of kin (Dugatkin & Reeve 1994; Bourke & Franks 1995; Sober & Wilson 1997). The empirical evidence that natural selection has actually produced adaptive units at the group level comes mainly from the biology of social insects. Here we find striking examples of kin groups—that is, colonies—functioning as coherent wholes in which the members contribute (largely) harmoniously to the ultimate goal of propagating the group's genes. Moreover, the members of these colonies possess morphological, physiological, and behavioral specializations that serve the efficient functioning of the colony to which they belong. The best examples are the elaborate colonies of such social insects as leaf-cutter ants (Hölldobler & Wilson 1990), fungus-growing termites (Lüscher 1955), and honey bees (Seeley 1995). Even Richard Dawkins (1989, 256) has acknowledged that colonies of honey bees possess sufficient functional integration to qualify as "vehicles of replicator survival."

I want, however, to go beyond fostering the basic recognition that a group, like a cell or an organism, can be an adaptively organized entity. I want to give us a deeper perception of this reality by revealing the richness of the functional organization of social insect colonies. This will require carefully analyzing their inner workings—the communication processes, feedback controls, labor specializations, and behavioral tunings—that improve the efficiency of how a colony of social insects operates. In working toward this goal, I will probably continue to study the honey bee, the social insect that I know best, because I find that nature is subtle and complex. A single bee, for example, is a marvelous bundle of subtle adaptations, and the complexity of a whole hive equals the sophistication of one bee multiplied many times over.

For the next several years I will continue studying how a swarm of bees acts as a decision-making agent capable of skillfully selecting a good nesting site. I will do so because a honey bee swarm choosing a dwelling place is the most spectacular example I know of an animal group functioning as a unit. If I can succeed in laying bare the mechanisms underlying a swarm's decision-making process, then I think we will have an extremely compelling example of adaptive organization at the group level. I should confess, though, that this subject also attracts me because it promises a hard but deeply satisfying adventure with the bees.

Eventually I will feel that the swarm project has been "completed" and I will switch to another topic: whatever is fueling my curiosity most strongly at the time. If the new topic involves some kind of animal other than honey bees (bumble bees? yellow jacket wasps? siphonophores?), I know it will have to be a species that I love. I will be living with it for several years and if I don't love it my interest will fade for I cannot sustain a purely intellectual interest. I know too that I will do my best to know this animal's world. I agree wholeheartedly with David Lack (1965, 24) who wrote: "If you are studying a particular species, try to enter as much as possible into its way of life, and if your friends start saying that you are becoming very like a Rook, or a Grebe, or whatever species it is, you are progressing."

Whatever species it is that I will be studying years and years hence, I know that I will start investigating it by broadly observing the animal's behavior. Only after I have made a thorough, descriptive reconnaissance of the phenomenon under study will I proceed to experiments. Rarely is it useful to alter the circumstances of an animal's life before you are thoroughly familiar with its normal circumstances. If you do so, it is likely that your experimental results will be extremely hard to interpret properly. I do not do experiments with the bees for the sake of experimenting, but to answer questions suggested by prior observations on the natural life of the bees.

Ever since my days as a graduate student, I have always been directly involved in the most laborious parts of my investigations—such as labeling thousands of bees and transcribing hours upon hours of videotape—and I suspect that this will continue well into the future. Most scientists, as they mature, delegate more and more of the seemingly routine work to others, and I have slowly begun to do this with respect to video transcription, which sometimes requires more time than I can arrange. But I do not consider any part of an investigation routine. In part this is because I pride myself on doing my own work, but mainly it is because I feel that the talent that I have as a student of animal behavior resides mainly in my capacity to observe animals, and to process and interpret what I have seen. If ever I tried to do a study in which I did not observe the bees myself, and instead delegated this to a helper, I think I would fail. Without the intimate knowledge of the bees that I get by observing the fine details of their behavior firsthand, I would not have the insights that lie at the heart of scientific discovery.

## Closing Thoughts

For me, honey bees are a model system for research on questions of animal social behavior. There is no other species that rivals them with respect to offering both supraorganismal unity and scientific tractability. In studying them, I have felt the passion of curiosity, the pain of failure, and the pride of discovery. They have enabled me to leave, in a sense, the conflictual society of our own species and join the cooperative society of a far different species. I suspect that the bees' mysteries will engage me for a lifetime.

## References

Bourke AFG, Franks NR, 1995. Social evolution in ants. Princeton: Princeton University Press.

Butler CG, 1954. The method and importance of the recognition by a colony of honeybees (*A. mellifera*) of the presence of its queen. Trans Roy Ent Soc London 105:11–29.

Clutton-Brock TH, Guinness FE, Albon SD, 1882. Red deer: behavior and ecology of two sexes. Chicago: University of Chicago Press.

Darling FF, 1938. Bird flocks and the breeding cycle: a contribution to the study of avian sociality. Cambridge: Cambridge University Press.

Dawkins R, 1989. The selfish gene. Oxford: Oxford University Press.

Dethier VG, 1976. The hungry fly: a physiological study of the behavior associated with feeding. Cambridge, MA: Harvard University Press.

Duffy E, 1996. Eusociality in a coral-reef shrimp. Nature 381:512–514.

Dugatkin LA, Reeve HK, 1994. Behavioral ecology and levels of selection: dissolving the group selection controversy. Adv Study Behav 23:101–133.

Frisch K von, 1967. The dance language and orientation of bees. Cambridge, MA: Harvard University Press.

Goodall J, 1986. The chimpanzees of Gombe: patterns of behavior. Cambridge, MA: Harvard University Press.

Heinrich B, 1999. Mind of the raven: investigations and adventures with wolf-birds. New York: HarperCollins.

Hölldobler B, Wilson EO, 1990. The ants. Cambridge, MA: Harvard University Press.

Hoogland JL, 1995. The black-tailed prairie dog: social life of a burrowing mammal. Chicago: University of Chicago Press.

Keller L, Nonacs P, 1992. The role of queen pheromones in social insects: queen control or queen signal? Anim Behav 45:787–794.

Kruuk H, 1995. Wild otters: predation and populations. Oxford: Oxford University Press.

Lack D, 1965. Enjoying ornithology. London: Methuen.

Lindauer M, 1961. Communication among social bees. Cambridge, MA: Harvard University Press.

Lüscher M, 1955. Der Sauerstoffverbrauch bei Termiten und die Ventilation des Nestes bei *Macrotermes natalensis* (Haviland). Acta Tropica 12:289–307.

Maynard Smith J, Szathmáry E, 1997. The major transitions in evolution. New York: Oxford University Press.

Naumann K, Winston ML, Slessor NK, Prestwich GD, Webster FX, 1991. Production and transmission of honey bee queen (*Apis mellifera*) mandibular gland pheromone. Behav Ecol Sociobiol 29:321–332.

Ratnieks FLW, 1988. Reproductive harmony via mutual policing by workers in eusocial Hymenoptera. Am Nat 132:217–236.

Ratnieks FLW, Visscher PK, 1989. Worker policing in the honeybee. Nature 342:796–798.

Seeley TD, 1977. Measurement of nest cavity volume by the honey bee (*Apis mellifera*). Behav Ecol Sociobiol 2:201–227.

Seeley TD, 1979. Queen substance dispersal by messenger workers in honeybee colonies. Behav Ecol Sociobiol 5:391–415.

Seeley TD, 1985. Honeybee ecology: a study of adaptation in social life. Princeton: Princeton University Press.

Seeley TD, 1989. Social foraging in honey bees: how nectar foragers assess their colony's nutritional status. Behav Ecol Sociobiol 24:181–199.

Seeley TD, 1992. The tremble dance of the honey bee: message and meanings. Behav Ecol Sociobiol 31:375–383.

Seeley TD, 1995. The wisdom of the hive: the social physiology of honey bee colonies. Cambridge, MA: Harvard University Press.

Seeley TD, Buhrman SC, 1999. Group decision making in swarms of honey bees. Behav Ecol Sociobiol 45:19–31.

Seeley TD, Morse RA, 1978. Nest-site selection by the honey bee. Insectes Soc 25:323–337.

Seeley TD, Visscher PK, 1985. Survival of honeybees in cold climates: the critical timing of colony growth and reproduction. Ecol Ent 10:81–88.

Seeley TD, Weidenmüller A, Kühnholz S, 1998. The shaking signal of the honey bee informs workers to prepare for greater activity. Ethology 104:10–26.

Sherman PW, Seeley TD, Reeve HK, 1998. Parasites, pathogens, and polyandry in honey bees. Am Nat 151:392–396.

Sober E, Wilson DS, 1997. Unto others: the evolution of altruism. Cambridge, MA: Harvard University Press.

Tinbergen N, 1953. The herring gull's world: a study of the social behaviour of birds. London: Collins.

Tinbergen N, 1985. Watching and wondering. In: Studying animal behavior: autobiographies of the founders (Dewsbury DA, ed). Chicago: Chicago University Press; 430–463.

Visscher PK, Dukas R, 1995. Honey bees recognize development of nestmates' ovaries. Anim Behav 49:542–544.

Wehner R, 1982. Himmelsnavigation bei Insekten: Neurophysiologie und Verhalten. Zürich: Orell Füssli.

Willis LG, Winston ML, Slessor KN, 1990. Queen honey bee mandibular pheromone does not affect worker ovary development. Canadian Ent 122:1093–1099.

Wilson EO, 1971. The insect societies. Cambridge, MA: Harvard University Press.

Woolfenden GE, Fitzpatrick JW, 1984. The Florida scrub jay: demography of a cooperative-breeding bird. Princeton: Princeton University Press.

Woyciechowski M, Lomnicki A, 1989. Worker reproduction in social insects. Tr Ecol Evol 4:146.

# 3 Peeling the Onion: Symbioses between Ants and Blue Butterflies

Naomi E. Pierce

My first encounter with entomology was not a success. As a sophomore in college, I was attracted by a listing in the course catalog for "Terrestrial Arthropods," taught by Charles L. Remington. I knew little about insects, but I could spend hours watching a line of ants running along a sidewalk. The introductory lecture had me hooked, but then came the laboratory practical: cockroach vivisection. The cockroaches in question were not the familiar denizen of kitchen cabinets, but rather *Gromphadorhina portentosa*, the Madagascar hissing roach. True to their name, these blackish-brown insects, the size of a baby's fist, can hiss by driving air out of their spiracles when disturbed. The sound of that hissing made my hair stand on end. They had the tangy odor of warm armpits, and I found them hideous.

At that first lab, we were each expected to pick up our roach at the front of the room, take it back to our lab bench, and dissect it. I loitered at the end of the line, and when the time came, stood for a long time looking into the bin of cockroaches. The teaching assistant wasn't impressed: "Well, go ahead—just pick it up!" I made a quick lunge, but my target scuttled deftly out of my grasp, its tibial spurs rasping against my palm. That was that. The teaching assistant was by now completely exasperated. "Come on, everyone is waiting!"

I burst into tears, seized my books, and ran the entire six blocks down "Science Hill" to my dormitory. There, I discovered a note on my door telling me that I'd been accepted into a special seminar on "Personal Journalism" taught by Loudon Wainwright, a writer I greatly admired, who for many years ran a thoughtful column in the back pages of *Life* magazine. I felt I could never show my face in Terrestrial Arthropods again.

Years later, I remember proudly presenting Mr. Wainwright with my first published paper over lunch in New York: "Parasitoids as selective agents in the symbiosis between lycaenid butterfly caterpillars and ants." He was about sixty then, a big man with a bushy white beard, a little like a Santa Claus. He glanced down at the paper and gave me a wry smile: "Catchy title." I realized at that moment how far I had come from my escape into personal journalism. While cramming the title with every term I could imagine that might be useful to a computer abstracting service, I hadn't given style a moment's thought.

So I'm definitely not someone who was committed to studying insects since childhood. I still feel slightly unsettled when confronted with large insects with hairy legs, and I am interested to observe that my three-year-old daughters switch from delight to horror if I place a live swallowtail butterfly on my hand and show them its black legs.

However, I developed a passion for butterflies when I returned to Terrestrial Arthropods a year after my disgrace, and it's a passion that has held me ever since. I can no longer step outdoors without finding myself straining to catch a glimpse of that elusive flutter of wings. Part of their appeal is aesthetic, but butterflies offer more than that. As another of my undergraduate professors, G. Evelyn Hutchinson, once explained to me in an offhand remark, "The wings of a butterfly are the only place where the laws of evolution are printed in color on a single page." I have never overcome my dislike of killing them, but there are few activities I prefer to chasing and observing butterflies. I have surprisingly frequent dreams of discovering new species with intricate and impossible wing patterns, or migrating with flocks of metallic blue *Morphos*.

I have reviewed more technical aspects of my research elsewhere (Pierce 1987, 1989; Pierce et al. 1991; Pierce & Nash 1999). Here I focus on how I first came to study ants and butterflies in the family Lycaenidae, and on some of my early experiences as an experimental field ecologist. I try to explain how each new finding contributed to or extended our conceptual understanding of ecology and evolution. I describe reasons why lycaenids, their attendant ants, host plants, and natural enemies have proved to be a model for the study of insect/plant interactions, chemical communication, mutualism, biodiversity, conservation, and the evolution of complex life history traits. Where possible, I delineate the features that I continue to find intriguing about the natural history of the system, and how they might be valuable in addressing further questions about adaptation and evolution.

## Choosing a System

I wanted to do my Ph.D. on insect/plant interactions. I was impressed by Paul Ehrlich and Peter Raven's proposal that butterflies and their host plants coevolved, and that this process of coevolution had, in fact, shaped most of organic life as we know it (Ehrlich & Raven 1964). However, when I started my degree at Harvard, two of the three entomologists in the department, E. O. Wilson and Bert Holldobler, specialized in ants. At that time, Wilson had just published *Sociobiology*, and the evolution of intraspecific cooperation, particularly the origin of altruistic behavior, was a topic of intense discussion among my fellow students. Bob Trivers was still at Harvard, and W. D. Hamilton was spending the year there as a distinguished visitor. I remember meeting Hamilton at the Estabrook Woods near Concord, Massachusetts, where he had been digging insects out of rotting wood, and with a glint in his eyes, held out both hands dripping with brown slime toward me. "Forgive me if I do not shake your hand," he said softly.

My choice of a thesis topic was motivated then by a combination of passion and pragmatism. Although everything about ants and the evolution of complex social behavior was fascinating, I couldn't shake my initial inclination toward butterflies. This was only semirational: I can't explain why butterflies seem so much more captivating than anything else, but the fact remains that they do. I recognized early on that it was important to select a topic that would sustain my interest long enough to be able to complete a dissertation.

After graduating from college, I had received a year-long traveling fellowship to Australia to visit Ian Common's laboratory at the Commonwealth Scientific Industrial Research Organization (CSIRO) in Canberra, where Charles Remington was then on sabbatical. A friend that I made at CSIRO, Roger Kitching, had told me about a family of butterflies whose caterpillars associated with ants. This seemed to present a unique opportunity to combine all of my developing interests in a single system. I was fortunate to be supported on both fronts: Robert Silberglied and Bert Holldobler were my thesis advisors at Harvard. Silberglied was immensely knowledgeable about all things involving Lepidoptera, and Holldobler had pioneered the study of "myrmecophilous" interactions, analyzing the relationship between staphylinid beetles and the ants whose nests they inhabit (Holldobler 1971).

As I have at almost every critical juncture in my professional life, I wrote to Charles Remington to seek his advice, this time about the feasibility of studying ant-associated butterflies in the Lycaenidae for my dissertation. He replied straight away to say that there were several suitable species to be found near the Rocky Mountain Biological Laboratory in Colorado, not far from where I grew up. This convinced me, and I wrote a proposal to study the evolution of interspecific cooperation between caterpillars of the family Lycaenidae and ants. My initial approach to this problem was not so much why as how: I was interested in the mechanisms involved in initiating and maintaining interspecific interactions, including the signals involved in interspecies communication. This provided an accessible entry point into a complex system.

I had the problem in mind, but I had yet to see the actual insects involved. I headed out to Colorado. I still didn't have a driver's license, partly because I rarely had access to a car. I had lived away from home since I was thirteen, and the general expense was more than my family or I could afford while I was in college. However, my father, a geologist who enjoys any kind of field expedition, drove me over to Red Rocks Park, where we had had some success collecting butterflies with a graduate student from Yale, Bob Pyle, and a high school student named Mark Epstein on a Fourth of July butterfly count. I found my first lycaenid caterpillar feeding on alfalfa beside the parking lot of Red Rock's gigantic outdoor amphitheater. I knew what it was the moment I saw it, and I can feel that thrill of recognition even today. It was a late instar of *Plebejus melissa,* the Orange-Margined Blue, feeding on flowers, and it was being assiduously tended by several small black ants.

Shortly afterwards, I moved up to the Rocky Mountain Biological Laboratory (RMBL) just outside Crested Butte to find a system that I could study for my thesis. My inability to drive became a serious liability, but fortunately, the youngest son of family friends, Paul Mead, had just graduated from high school and was willing to join me as a field assistant for the summer, along with his truck.

At RMBL, it was my great fortune to be taken under the wing, figuratively and almost literally, of an expert on lycaenid butterflies, Paul Ehrlich. During my first few days at the Lab, he took Paul Mead and me flying in his small plane over a number of field sites that hosted populations of *Glaucopsyche lygdamus,* the Silvery Blue. The caterpillars of this species were tended by numerous species of ants, and Ehrlich and his students had analyzed the relationship between the butterflies and their host plants, several different species of lupine (Breedlove & Ehrlich 1968; Dolinger et al. 1973). Ehrlich then drove us to some of the best localities, and essentially laid the groundwork for our summer's research. Another ecologist interested in coevolution, John Downey, had made the first intensive study of the ant associates of another lupine-feeding lycaenid several years earlier, and he also provided invaluable information about the behavior of the ants and caterpillars. He put me in touch with R. E. Gregg, who painstakingly helped me identify the attendant ants.

## Cost/Benefit Analysis of a Mutualism

From the outset, I had decided to do a cost/benefit analysis of the association between *G. lygdamus* and its attendant ants. However, my initial approach focused entirely on the butterflies. While at first it seemed obvious that the caterpillars were providing food for ants through secretions from specialized glands, it was less clear what the ants were doing for the caterpillars. Earlier work suggested that the caterpillars benefited primarily because ants did not attack them; normally, ants are serious predators for lepidopteran larvae. However, it also seemed possible that attendant ants were more than just appeased, that they actually protected the lycaenids from their enemies.

After several messy attempts, Paul and I settled on using sticky barricades made from a viscous, gooey substance called Tanglefoot © to exclude ants from tending caterpillars in the field. We compared untended caterpillar survivorship with that of tended counterparts. Such "ant exclusion" experiments are standard fare in ecology field courses today, but we were among the first to apply these techniques to a large sample of plants under field conditions. I owe the suggestion of this approach to Bob Robbins, now at the Smithsonian, who gave me a number of helpful ideas early in my graduate work. In experiments at three field sites over two years, we found that ant-tended caterpillars of *G. lygdamus* were four to twelve times more likely to survive to pupation than their untended counterparts (Pierce & Mead 1981; Pierce & Easteal 1984). The ants were primarily effective against a suite of parasitoids that attacked the larvae and pupae. Paul and I were thrilled and more than a little bit amazed when our paper describing this result, with its dismally uncatchy title, was accepted for publication in *Science* in 1981. In his book, *Curious Naturalists*, Niko Tinbergen argued that the most important thing a young scientist can do is to gather data, because the data helps establish self-confidence, and this was certainly my experience.

In the meantime, however, I had also learned some of the disadvantages of working with this system. *G. lygdamus* had only one brood a year, and population numbers could vary dramatically from one year to the next. We tried to augment our experimental possibilities by studying two populations that occurred at different altitudes: one near Gunnison, Colorado, became active at least a month earlier than the one near Crested Butte, and this meant that we could conduct two sets of field experiments in one season. However, the fluctuation in numbers, while interesting, presented difficulties for someone working on the time frame of a Ph.D. For example, in the summer of 1982, my collaborator from Australia, Roger Kitching, came out to Crested Butte for the field season, and we recall seeing exactly three individuals the entire summer. This extreme fluctuation was not unusual: Ehrlich and his colleagues had previously documented the extinction of the population near Crested Butte (Ehrlich et al. 1972).

Moreover, we were never able to induce males and females of *G. lygdamus* to mate with each other, and this clearly restricted the kinds of experiments we could do. Larger butterfly species can often be hand-paired, but small and delicate Lycaenidae are usually not so obliging. The host plants, perennial species of lupines, also presented difficulties. We could not easily pot them and move them around the habitat.

## A Tractable Experimental System

By working as a teaching assistant in one or two courses every semester, and acting as a tutor in one of the residential houses on campus, I finally saved enough money for a return visit to Australia in December 1979. Roger Kitching had moved from CSIRO to Griffith University in Brisbane, where he had started to study another species of lycaenid whose caterpillars associate with ants (Kitching 1976, 1983). In contrast to the North American lycaenids I had become familiar with, whose caterpillars were tended only intermittently by many species of ants, the caterpillars of *Jalmenus evagoras* are constantly tended, and specifically by only a few species of ants in the genus *Iridomyrmex*. One of Roger's students, Martin Taylor, had a small culture of these butterflies in a field house on the roof of Griffith's School of Australian Environmental Studies. I could hardly believe my eyes. Instead of the single, slug-like caterpillar gingerly caressed by half a dozen eager ants to which I was accustomed, here were clusters of spiny black caterpillars and pupae, seething with so many ants that it was difficult to distinguish them beneath the moving layer of legs, bodies, and antennae.

Martin and a friend, Mark Elgar, then took me to Mount Nebo, a location just outside Brisbane, to see the butterflies in the field. The road, now so familiar, is nauseatingly winding, and my car-sick stomach spent much of the journey telling me that this was not the place for a long-term study. By the time we reached the field sites, I made a headlong scramble out of the car to keep from throwing up all over my colleagues. Despite this inauspicious introduction, my relationship with *J. evagoras* was love at first sight.

As at RMBL, I was again fortunate to be befriended at Mount Nebo by an invaluable ally. Charmaine Lickliter had built her own house and lived "on the mountain" for many years. She knew everything about self-reliance. She showed me how to design a gravity-feed toilet (useful if the electricity is cut), raised all her own food and flowers, and recounted fascinating outback lore such as the uses of emu oil and how it could slip out of glass containers. Her back yard was a haven for kookaburras, cockatoos, and shy little wallabies called paddymelons. She acted as though it was perfectly normal for a single woman who couldn't even drive (I finally learned the first summer there) to spend the year living in a large, upside-down watering tank while studying the behavior of butterflies and ants. She helped me in countless ways over the years, even providing her own paddock as a breeding ground for the butterflies. She never seemed to tire of daily accounts of insect happenings, and her support and friendship made an intangible but enormous difference to my ability to do fieldwork there every summer for the next ten years.

In contrast to *G. lygdamus*, *J. evagoras* afforded a marvelously tractable system. Host plants and ants are easily cultured in the laboratory, designated butterfly matings are not difficult to achieve, and the species overwinters in the egg stage. This means that eggs can be transported in small vials, stored in a cold room, and brought out of diapause when host plants and ants are readily available. All stages aggregate, which makes them easier to find and introduces some interesting complications to their ecology and evolution. Moreover, rather than crawling off into the leaf litter to pupate in hidden locations, the full-grown caterpillars of *J. evagoras* pupate in the open, like clusters of grapes lined up along the stems of their *Acacia* host plants. Pupae are tended by ants, and this has been a helpful feature in studying the relationship between *J. evagoras* and its associated ants. For example, to learn more about the chemicals involved in attracting ants, we can take surface washings from pupae that cannot regurgitate or defecate in the process.

Moreover, because individuals pupate in locations that are highly visible to us as well as to potential mates, it has been possible to analyze their mating behavior with the kind of detail seldom possible in field studies of an insect.

## Lycaenids, Ants, and Their Host Plants

Research on lycaenid/ant associations led me back in a somewhat unexpected way to my original interest in insect/plant interactions, and more specifically to butterfly/plant coevolution. One of the most striking findings from our work on *J. evagoras* and its associated ants started with the observation that caterpillars secret amino acids, the building blocks of proteins, as well as simple sugars as rewards for attendant ants (Pierce et al. 1987; Pierce 1985, 1989). It seemed likely that these amino acids and the nitrogen they contained played a central role in the currency of exchange between these mutualistic partners (Pierce 1989; Baylis & Pierce 1993; Pierce & Nash 1999). The growth rate of colonies allowed to forage on caterpillar secretions far exceeded that of their counterparts raised without access to these secretions (Nash 1989; Pierce & Nash 1999). Moreover, caterpillars raised on plants that had been enriched through the application of fertilizer attracted more ants and survived longer than their counterparts on plants that had not received treatment. Laboratory analysis confirmed that the foliage of these fertilized plants had a higher content of nitrogen, phosphorous, and other minerals, and female butterflies preferred to lay eggs on these high-quality plants (Baylis & Pierce 1991).

In their paper on coevolution, Ehrlich and Raven mentioned more than once that "in the Lycaenidae, ants . . . may further modify patterns of food plant choice" (p. 588), and their prediction proved to be true. Moreover, it helped to account for some of the diversity of host plant use exhibited by the Lycaenidae, much of which appeared to be unexplained by their central paradigm, which rested upon the importance of plant secondary chemistry in governing the relationships between butterflies and their host plants. They emphasized that toxic secondary metabolites could exclude competition from other herbivores and potentially provide both attractants and a means of defense for butterflies that could detoxify and/or sequester them; the butterflies could then radiate in this new adaptive zone.

Our findings suggested that in the Lycaenidae, host plant chemistry could also play a role in mediating the "enemy free space" surrounding a caterpillar (e.g., Atsatt 1981). In this case, however, it is the nutritional quality of the host plant that is critical in determining a caterpillar's ability to maintain its defense force. By feeding on plants that are sufficiently rich in protein, the caterpillar can satisfy not only its own needs, but also those of its attendant ants. If this premise is true, one might expect Lycaenidae that reward ants with amino acid secretions to feed only on plants or plant parts that are also rich in proteins. Indeed, the Lycaenidae are well-known as a group for their predilection for nitrogen-rich parts of plants, including terminal foliage, flower, and seed pods (Mattson 1980). Furthermore, a survey of a great number of species, each scored for its degree of ant association and its choice of host plant, revealed a striking association between being tended by ants and feeding on legumes (Pierce 1985; Fiedler 1991, 1995). Not only are legumes rich in nitrogen (one reason why beans are considered nutritious), but unlike most other kinds of plants, legumes have symbiotic bacteria in their roots that can fix atmospheric nitrogen. Thus one might expect them to vary less over evolutionary time in their composition of nitrogen compared with other plants.

However, the general significance of the strong correlation between ant association and legume feeding is still unresolved. It is possible that the association of these two traits in the Lycaenidae might simply be the result of historical accident. For example, if a proto-lycaenid fed on legumes and also happened to be ant-associated, the descendants of this lycaenid might be legume-feeding and ant-associated without this correlation having a particular functional significance. However, if one could show that legume feeding and ant association had evolved together in a number of independent instances, it would provide support for the hypothesis that these two characteristics are in fact functionally linked. Our limited knowledge of the evolutionary history of the Lycaenidae has made it hard to assess how many times ant association and legume feeding have evolved together.

Partly as a consequence, members of my laboratory and I have been working toward estimating the phylogeny of the Lycaenidae using molecular characters (e.g., Braverman 1989; Blair 1995; Mignault 1996; Taylor et al. 1993; Campbell 1998; Campbell et al. 2000; Rand et al., in press). The foundation for this task was laid by Colonel John Eliot. In 1973, he published a classification of the family based on morphological characters, delineating the major groups and suggesting their historical relationships. In sampling our taxa, this work has been invaluable, and I have been impressed and humbled by his careful and insightful analysis.

## Life History Evolution: Parasitism Arising from Mutualism

Another key finding from research on *J. evagoras* and its ant associates was the demonstration that the caterpillars pay a dramatic metabolic cost for maintaining an ant guard (Pierce et al. 1987; Baylis & Pierce 1992). Larvae raised in the greenhouse without attendant ants pupate at a much larger size than their tended counterparts, and these pupae become larger adults. Since size is correlated with fecundity in females (Hill & Pierce 1989), and lifetime mating success in males (Elgar & Pierce 1988; Hughes et al. 2000), ant association therefore represents a significant cost for these butterflies. The degree of fine-tuning involved in meting out this cost is impressive: caterpillars even modify the amount of secretion they produce per capita depending upon their social context. When they are in groups, they are able to maintain a threshold level of ant guard at a lower cost, and they therefore secrete correspondingly fewer droplets per capita as rewards for ants (Axén & Pierce 1998).

The significant cost to the caterpillars of cooperating with ants is interesting because selection should favor any participant in the interaction that can still reap the benefits at a minimum cost. For example, if the chemicals necessary to fool ants into tending caterpillars are cheaper to produce than nutritious rewards, selection should favor parasitism of ants by lycaenids rather than the mutualism observed in this system.

This switch from mutualism to parasitism has evolved repeatedly in the Lycaenidae (Cottrell 1984; Pierce 1995). For example, species in at least two genera, *Maculinea* and *Lepidochrysops*, are "phytopredaceous." Caterpillars spend their early instars feeding on plants, and then drop to the ground, where they are picked up by workers of their host ant species and carried into the brood chamber of the nest. Here, chemically camouflaged and undetected by the adult ants, they consume the helpless brood. The caterpillars of other Lycaenidae, such as the species of the Australian genus *Acrodipsas*, never go through a plant-eating stage, but spend their entire lives consuming ants. In an even more sophisticated twist on this theme, some lycaenids, such as the Japanese species *Niphanda fusca*, enter the ant nest, but have mastered the signals made by brood to elicit regurgitations from adult ants. These "cuckoo" species are fed entirely on ant regurgitations through a process called trophallaxis.

Convergent origins of parasitism, including both carnivorous and cuckoolike behaviors, are exhibited by individual species in a number of genera whose other members are all plant-eating and apparently mutualistic. These include representatives from *Spindasis, Ogyris, Arhopala,* and *Chrysoritis* (Pierce 1995). The reverse relationship, that of mutualism arising from parasitism in these myrmecophilous relationships, has yet to be documented.

## Species-Specificity and Chemical Communication

The species-specific relationship between certain species of Lycaenidae and ants parallels the host plant specialization exhibited by many species of phytophagous insects. Research to date has only scratched the surface of possible mechanisms involved. How do ants recognize the lycaenids with which they associate, and vice versa? And just as the evolution of host plant specialization in phytophagous insects remains a conundrum, a satisfying evolutionary explanation for why some lycaenids are allied with only a single species of ant whereas others are generalists is likewise unknown.

Given the importance of ants for their survival, it seemed reasonable that females of *J. evagoras* might use ants as cues in finding suitable host plants upon which to lay eggs. Nevertheless, I was surprised when our experiments showed that females not only use attendant ants as cues in laying eggs, but can tell the difference between different species (Pierce & Elgar 1995; Pierce & Nash 1999). I was even more astonished when my student, Ann Fraser, and a postdoctoral associate, Tom Tregenza, recently showed that females can tell the difference between various populations of attendant ants, and are more likely to lay eggs on plants containing workers from their natal populations (Fraser 1997 and unpublished results). How females distinguish between different populations, and the possible selective advantages of such fine-tuned behavior, remain to be determined.

Another open area for research on species-specificity is the biochemistry of the signals involved in ant/caterpillar recognition. The lycaenids that are tended by ants secrete substances that appease ants and gain favorable recognition. In the case of species-specific interactions, these signals would appear to be highly specialized. A pairwise analysis of the surface secretions of a suite of lycaenid caterpillars and their respective ant associates could provide considerable insight into the signals used by ants in species-specific recognition, and possibly in brood recognition (Pierce 1989). Many lycaenids, including *J. evagoras*, also stridulate to attract attendant ants, and these interspecific acoustical signals may contribute significantly to the fine-tuning of their interactions (Travassos & Pierce, in press).

Since all known attendant ants species have alternative food sources, it seems unlikely that there has been any kind of "coevolution" between ants and lycaenids in the Ehrlich/Raven sense of the term. However, some of our work on the phylogeny of different genera with species-specific ant associations has indicated strong conservatism on the part of the lycaenids for ants within a particular subfamily (Pierce & Nash 1999 and unpublished results). It would appear that once a relationship has been established, speciation is more likely to occur within the association than outside. In other words, ant-associated taxa tend to have ant-associated sister groups, and sister groups tend to be associated with related ant species. Perhaps this is because the numerous complex behaviors and biochemical mechanisms necessary to achieve specificity in the first place influence the evolutionary trajectory of a particular lineage. In this respect, the ant fauna might be considered a template against which the lycaenids have diversified.

## Biogeography

I was struck by the qualitative difference between the *Jalmenus*/ant relationship and the *Glaucopsyche*/ant relationship. These species may reasonably be regarded as representative of their respective continents, although they are both on the high end of their respective ant-associated spectra. Lycaenids such as *J. evagoras*, have been described as "obligately" ant-associated, in the sense that caterpillars and pupae are never found without ants, and their survivorship is negligible if attendant ants are experimentally excluded. Their relationship also involves a high degree of specificity: while juveniles of *J. evagoras* are known to associate with several species of ants, all are in the genus *Iridomyrmex*. Other species of Lycaenidae in Australia exhibit even greater specificity. For example, *Hypochrysops ignita* has been observed feeding on seventeen different host plant families, yet the larvae are tended by only one species of ant, *Papyrius* sp. (*nitidus* group). At one of our field sites in Australia, *J. evagoras* and two of its close relatives, *J. ictinus* and *J. daemeli*, all co-occur on the same host plant species, brigalow (*A. harpophylla*). However, each associates exclusively with its own species of attendant ants.

This kind of obligate ant association, frequently combined with species-specificity and ant-dependent oviposition, is essentially unknown among North American taxa, and uncommon in the Palearctic. However, more than a third of the myrmecophilous Lycaenidae in Australia have obligate associations with ants, and such associations are likewise well-developed in South Africa. The records available from India, while sketchy, hint at a number of strongly ant-associated taxa. This distribution of ant-associated Lycaenidae led me to speculate about whether it might have been generated by a Gondwanaland-Laurasia split in ant-associated and non-ant-associated lineages (Pierce 1987). For this pattern to be explained by such a faunal split, the evolutionary history of the Lycaenidae would also have to reflect this division, and the pattern of ant association should track it.

Ideally, to investigate this pattern further, we would want to conduct a detailed analysis of phylogeny, ant association, and biogeographic distribution for each clade within the Lycaenidae. Our understanding of the phylogeny of the Lycaenidae is still too limited to permit such a detailed analysis. However, only one of the thirty-three tribes of Lycaenidae recognized by Eliot (1973) is unique to the Holarctic, and all others have representatives in biogeographic regions derived from both Gondwanaland and Laurasia. Thus a simple phylogenetic explanation—in this case, two main lineages, one obligately ant-associated and the other not, one Gondwanan and the other Laurasian—is not tenable. In addition, current ideas about the origin of butterflies suggest that the Gondwanan/Laurasian split occurred before the diversification of the lycaenids. Nevertheless, the pattern may well reflect biogeographic history if the lycaenids have responded to an ancient dichotomy in critical aspects of their biology such as the distribution of host plants (*Acacia*, for example) or attendant ants.

The biogeographic distribution of highly ant-associated Lycaenidae remains a truly fascinating pattern that begs explanation. After all, the preponderance of obligately ant-associated lycaenids in Australia and South Africa is not the result of a distinctive biogeographic distribution of a single taxon. It reflects the distribution of a suite of species interactions, including obligate ant associations (both parasitic and mutualistic ones), host plant affiliations, and selection pressures exerted by parasitoids and predators.

## Ant Association and Its Evolutionary Consequences: Speciation and Extinction

Ant association has an important impact on the demography of lycaenids such as *J. evagoras*. Both larvae and pupae require appropriate species of attendant ants as well as suitable host plants in order to survive. Their populations are often small, localized, and patchily distributed (Smiley et al. 1988; Taylor et al. 1993; Costa et al. 1996). As a result, lycaenids are likely candidates for a peripheral isolate model of speciation, whereby evolutionary change is concentrated in small, marginal, isolated, or semi-isolated populations.

Such a propensity for speciation may in turn lead to amplified rates of diversification. However, small population sizes may also contribute to the negative component of the evolutionary demographic, extinction. Many species of Lycaenidae are recognized as endangered, and these taxa have frequently featured as emblems for conservation biology. Examples include the Large Blue in the United Kingdom, the Arionides Blue in Japan, the Karner Blue and Xerces Blue in the United States, the Brenton Blue in South Africa, and Illidge's Blue in Australia. As our research has indicated, highly specialized lycaenids such as these are likely to be more sensitive to environmental perturbations because their life histories are so complex. Moreover, both theory and observation have shown that small, inbred populations are prone to extinction (Saccheri et al. 1998). Lycaenid butterflies and ants may therefore provide a model not only for understanding mechanisms generating diversity, but processes leading to the loss of diversity.

## Final Thoughts on Finding a System and an Approach

I doubt I would have continued to study lycaenids if their interactions hadn't proved to be so multidimensional. I never set out to study or establish a model system; the work just progressed. My interests were simple to start with—an assessment of the costs and benefits of the interaction for each partner of a putative mutualism. Exploration itself was part of the goal, and each new peel of the onion revealed a fresh and equally fascinating layer. In particular, I gained an appreciation for the different levels at which this and all other natural systems can be analyzed. And to a greater extent than I would have ever anticipated, I learned that a detailed understanding of natural history and behavior is essential in considering broader issues. One of my postdoctoral advisors, Dick Southwood, exhorted students to "Know thy barnacles!" with reference to Darwin's classic study. By this he meant that a deep knowledge of a particular group of organisms assists enormously in providing a general understanding of the principles of ecology and evolution.

Model systems usually have a suite of desirable traits, depending on the questions they are used to explore. A geneticist might favor organisms that have a small genome size and fast generation time, and afford the possibility of manipulating the genetic background. A behavioral ecologist might favor organisms with complex interactions and close relatives exhibiting diverse life histories. In both cases, experimental tractability can be critical. For a behavioral ecologist, it usually helps to study organisms where individuals and/or units of selection can easily be identified, especially when trying to measure something as elusive as lifetime reproductive success.

My students and I have discussed whether it might be better to start with a systematic and phylogenetic framework for a particular group before working on specific aspects of behavioral ecology. My own development took the reverse approach: I was drawn to studying life history evolution through my interest in lycaenid behavioral ecology. But because history counts so much in understanding biological systems, I have wondered at times whether it wouldn't have been better to have started by building the structure upon which to hang the questions. This would argue for systematics first.

However, such logic ignores one thing: the importance of a passionate interest. For myself, I would always advocate setting sails with the subject closest to the heart before spending time collecting navigational charts. It was my field experience with living animals that informed the kinds of questions I wanted to ask about their evolutionary history. Without that initial inspiration, I might not have persisted, especially in difficult times. And it was the many forms of support from friends, family, and colleagues that helped me develop ideas, deepen my interest, and in many cases, simply do the work. I still rely on good friends, and reveries of caterpillars seething with ants and butterflies winking on and off in the Australian bush to get me through another faculty meeting and another Massachusetts winter.

## Acknowledgments

I thank Andrew Berry, Brian Farrell, Ann Hochschild, David Lohman, John Mathew, Tomi Pierce, and Jim Schwartz for their helpful comments on the manuscript, and Lee Dugatkin for his advice, patience, and unflaggingly friendly support.

# References

Atsatt PR, 1981. Lycaenid butterflies and ants: selection for enemy-free space. Amer Nat 118:538–654.

Axén A, Pierce NE, 1998. Aggregation as a cost reducing strategy for lycaenid larvae. Behav Ecol 9:109–115.

Baylis M, Pierce NE, 1991. The effect of host plant quality on the survival of larvae and oviposition behaviour of adults of an ant-tended lycaenid butterfly, *Jalmenus evagoras*. Ecol Entomol 16:1–9.

Baylis M, Pierce NE, 1992. Lack of compensation by final instar larvae of the myrmecophilous lycaenid butterfly, *Jalmenus evagoras*, for the loss of nutrients to ants. Physiol Entomol 17:107–114.

Baylis M, Pierce NE, 1993. The effects of ant mutualism on the foraging and diet of lycaenid caterpillars. In: Caterpillars: ecological and evolutionary constraints on foraging (Stamp NE, Casey TM, eds). New York: Chapman and Hall; 404–421.

Blair MP, 1995. Ecology, evolution and molecular phylogenetics of myrmecophily within the Theclini (Lepidoptera: Lycaenidae) based on nucleotide sequences of the mitochondrial gene, Cytochrome Oxidase sub-unit I. Senior thesis, Harvard University.

Braverman JMN, 1989. DNA sequence variation and evolutionary radiation in the Australian genus *Jalmenus* (Lepidoptera: Lycaenidae). Senior thesis, Princeton University.

Breedlove DE, Ehrlich PR, 1968. Plant-herbivore coevolution: lupines and lycaenids. Science 162:671–672.

Campbell DL, 1998. Higher-level phylogeny and molecular evolution of the Riodinidae (Lepidoptera). Ph.D. thesis, Harvard University.

Campbell DL, Brower AVZ, Pierce NE, 2000. Molecular evolution of the *wingless* gene and its implications for the phylogenetic placement of the butterfly family Riodinidae (Lepidoptera: Papilionoidea). Mol Biol Evol 17:684–696.

Costa JT, McDonald JH, Pierce NE, 1996. The effect of ant association on the population genetics of the Australian butterfly *Jalmenus evagoras* (Lepidoptera: Lycaenidae). Biol J Linn Soc 58:287–306.

Cottrell CB, 1984. Aphytophagy in butterflies: its relationship to myrmecophily. Zool J Linn Soc 79:1–57.

Dolinger PM, Ehrlich PR, Fitch WL, Breedlove DE, 1973. Alkaloid and predation patterns in Colorado lupine populations. Oecologia (Berl) 13:191–204.

Ehrlich PR, Breedlove DE, Brussard PR, Sharp MA, 1972. Weather and the 'regulation' of subalpine populations. Ecology 53:243–247.

Ehrlich PR, Raven PH, 1964. Butterflies and plants: a study in coevolution. Evolution 18:586–608.

Elgar MA, Pierce NE, 1988. Mating success and fecundity in an ant-tended lycaenid butterfly. In: Reproductive success: studies of selection and adaptation in contrasting breeding systems (Clutton-Brock TH, ed). Chicago: University of Chicago Press; 59–75.

Eliot JN, 1973. The higher classification of the Lycaenidae (Lepidoptera): a tentative arrangement. Bull Br Mus (Nat Hist) Entomol 28:375–505.

Fiedler K, 1991. Systematic, evolutionary and ecological implications of myrmecophily within the Lycaenidae (Insecta: Lepidoptera: Papilionoidea) Bonner Zoologisches Monographien 31.

Fiedler K, 1995. Lycaenid butterflies and plants: is myrmecophily associated with particular hostplant preferences? Ethol Ecol Evol 7:107–132.

Fraser AM, 1997. Evolution of specialization in lycaenid butterfly-ant mutualisms. Ph.D. thesis, Harvard University.

Hill CJ, Pierce NE, 1989. The effect of adult diet on the biology of butterflies, 1: The common imperial blue, *Jalmenus evagoras*. Oecologia 81:249–257.

Holldobler B, 1971. Communication between ants and their guests. Scient Amer 224:86–93.

Hughes L, Chang BS-W, Wagner D, Pierce NE, 2000. Effects of mating history on ejaculate size, fecundity, longevity, and copulation duration in the ant-tended lycaenid butterfly, *Jalmenus evagoras*. Behav Ecol Sociobiol 47:119–128.

Kitching RL, 1976. The ultrastructure of the eggs of *Jalmenus evagoras* (Donovan) (Lepidoptera: Lycaenidae). Aust Entomol Mag 3:42–44.

Kitching RL, 1983. Myrmecophilous organs of the larvae of the lycaenid butterfly *Jalmenus evagoras* (Donovan). J Nat Hist 17:471–481.

Mattson WJ, 1980. Herbivory in relation to plant nitrogen content. Ann Rev Ecol Syst 11:119–161.

Mignault AA, 1996. Proposed genealogical relatedness among butterflies of the Australian genus *Jalmenus* (Lepidoptera: Lycaenidae)—a case study in phylogentic inference, ecology and biogeography of a rapidly evolving system. Senior thesis, Harvard University.

Nash DR, 1989. Cost-benefit analysis of a mutualism between lycaenid butterflies and ants. D.Phil. thesis, Oxford University.

Pierce NE, 1985. Lycaenid butterflies and ants: selection for nitrogen-fixing and other protein rich food plants. Am Nat 125:888–895.

Pierce NE, 1987. The evolution and biogeography of associations between lycaenid butterflies and ants. In: Oxford surveys in evolutionary biology (Harvey PH, Partridge L, eds). Oxford: Oxford University Press; 4:89–116.

Pierce NE, 1989. Butterfly-ant mutualisms. In: Toward a more exact ecology (Grubb PJ, Whittaker J, eds). Oxford: Blackwell Scientific; 299–324.

Pierce NE, 1995. Predatory and parasitic Lepidoptera: carnivores living on plants. J Lepid Soc 49:412–453.

Pierce NE, Easteal S, 1986. The selective advantage of attendant ants for the larvae of a lycaenid butterfly, *Glaucopsyche lygdamus*. J Anim Ecol 55:451–462.

Pierce NE, Elgar MA, 1985. The influence of ants on host plant selection by *Jalmenus evagoras*, a myrmecophilous lycaenid butterfly. Behav Ecol Sociobiol 16:209–222.

Pierce NE, Kitching RL, Buckley RC, Taylor MF, Benbow K, 1987. Costs and benefits of cooperation between the Australian lycaenid butterfly, *Jalmenus evagoras* and its attendant ants. Behav Ecol Sociobiol 21:237–248.

Pierce NE, Nash DR, 1999. The Imperial Blue: *Jalmenus evagoras* (Lycaenidae). In: The biology of Australian butterflies (Monographs on Australian Lepidoptera, vol. 6) (Kitching RL, Sheermeyer E, Jones R, Pierce NE, eds). Sydney: CSIRO Press; 279–315.

Pierce NE, Nash DR, Baylis M, Carper ER, 1991. Variation in the attractiveness of lycaenid butterfly larvae to ants. In: Ant-plant interactions (Cutler D, Huxley C, eds). Oxford: Oxford University Press; 131–143.

Rand DB, Heath A, Suderman T, Pierce NE. In press. Phylogeny and life history evolution of the genus *Chrysoritis* within the Aphnaeini (Lepidoptera: Lycaenidae), inferred from mitochondrial cytochrome oxidase I sequences. Mol Phyl Evol.

Saccheri I, Kuussaari M, Kankare M, Vikman P, Fortelius W, Hanski I, 1998. Inbreeding and extinction in a butterfly metapopulation. Nature 392:491–494.

Smiley JT, Atsatt PR, Pierce NE, 1988. Local distribution of the lycaenid butterfly, *Jalmenus evagoras*, in response to host ants and plants. Oecologia (Berl) 76:416–422.

Taylor MFJ, McKechnie SW, Pierce NE, Kreitman ME, 1993. The lepidopteran mitochondrial control region: structure and evolution. Mol Biol Evol 10:1259–1272.

Travassos MA, Pierce NE. 2000. Acoustics, context and function of vibrational signalling in a lycaenid butterfly-ant mutualism. Anim Behav 60:13–26.

# 4

# In Search of Unified Theories in Sociobiology: Help from Social Wasps

Hudson Kern Reeve

What exactly are behavioral ecologists and sociobiologists trying to accomplish? What, if anything, would have to occur for a behavioral ecologist to conclude that the field had nearly completed its major tasks? These questions are rarely explicitly entertained, but I suspect that the ensemble of answers one would obtain from behavioral ecologists would be immensely variable, in contrast, say, to answers collected for the parallel question posed to physicists.

My answer to the above questions is: the ultimate aim for behavioral ecologists is to explain and predict behavior, given its ecological, social, and genetic context, from general, quantitative, evolutionary principles. Moreover, we seek to know exactly how these evolutionary principles are interconnected as parts of a truly unified mathematical evolutionary theory of behavior. This answer is structurally no different from the answer that I suspect most physicists would give to their corresponding question, their answer probably being framed in terms of unification of the fundamental forces or successful integration of quantum physics and general relativity.

However, my interactions with many of my colleagues, particularly at scientific meetings, have indicated to me that my answer is by no means the modal one in behavioral ecology! As I describe below, my "unificationist" perspective sits as one extreme along a continuum of views about the maximum level of theoretical unification possible in behavioral ecology. At the other extreme is what I will call "causal pluralism," the belief that there will be nearly as many (largely isolated, often unique) explanations as there are behavioral phenomena to be explained. Intermediate is the (perhaps majority) view that, while much behavior can be seen as approximating that predicted by some general (although often disconnected) evolutionary principles, there is sufficient "noise" in behavior that these principles cannot be taken too seriously in their quantitative details.

In this essay, I will attempt to describe why I have come to reject the intermediate, and certainly the causal pluralist, view in favor of a fully unapologetic unificationist view. This movement in my thought has been fueled partly by theoretical reasons (some of which I will allude to in this essay), but mainly through my empirical study of cooperation in social wasps. I will summarize the key developments in my empirical study of wasps that led me down the unificationist path and to my current plea for a renewal of serious attempts to unify and test evolutionary theories of behavior in quantitative detail.

## Youthful Dreams

When I was a teenager in Abilene, Texas in the early 1970s, very little about my world (especially the world of human social behavior) made sense to me, and whatever didn't make sense to me caused me great discomfort. I would find temporary relief in the wonderful scientific expositions of mysterious phenomena such as plate tectonics, quasars, and animal navigation in the pages of *Scientific American*, which I eagerly awaited to arrive each month at the local bookstores. However, such pleasant distractions didn't compare in intensity to the pulse-pounding rapture that I experienced one night when viewing a U.S. Public Broadcasting System *Nova* episode on the possibility of grand unified theories in physics. The idea that all physical phenomena from electricity to black holes to radioactivity could eventually be understood within a single mathematical framework was absolutely entrancing. Empirically grounded theoretical unification had long been a goal in the physical sciences, and it seemed that training for a career in theoretical physics would be the natural course of action for me.

However, I was greatly distracted by what seemed to me to be a deep fissure in our scientific knowledge. Our knowledge of the physical world was growing exponentially, but our understanding of human behavior seemed embarrassingly rudimentary, especially given that we, as the very objects of study, should have special insight into why we behave as we do (as opposed to other conceivable ways). There clearly were regularities in my own behavior, as well as in that of my family members and friends, that seemed just as ripe for scientific analysis as did the regular radio emissions of pulsars! *Why* did we, or any animal for that manner, behave in such odd but predictable ways, instead of alternative ways?

So began my trek into psychology and the philosophy of mind, continually spurred by the search for at least the basic tools from which to construct a general scientific theory of animal behavior. For a short time, I thought that the behaviorist approach of B. F. Skinner was what I had been searching for. In his books *The Behavior of Organisms* and *Science and Human Behavior*, Skinner at least had seriously posed the possibility of a rigorous science of behavior. However, I found the behaviorist paradigm ultimately unsatisfying: behaviorist explanations typically involved analysis of behavior into different variants of classical and operant conditioning, but the question of *why* these variants occurred, as opposed to alternative variants, or to various kinds of unlearned behavior, was never adequately addressed.

Slowly, I was inching closer to evolutionary biology, which of course is capable of answering the kind of "why" questions that interested me. The two key events for me were (1) being introduced as an undergraduate to the behavioral ecologist George Gamboa, then at the University of Texas at San Antonio (now at Oakland University in Rochester, Michigan) and (2) reading Dawkins's *The Selfish Gene* (1976). George quickly introduced me to the evolutionary approach to behavior and kindly involved me in some of the early studies of nestmate recognition in social wasps, studies that were ultimately motivated by the prediction from Hamilton's (1964) kin selection theory that agonism should be more infrequent among relatives than among nonrelatives, everything else being equal. Dawkins's seminal book convinced me that evolutionary biology was exactly what I had been looking for. It could potentially predict which phenotypes, including behavioral variants, should be seen in the long run as a consequence of the inevitable selective competition among the genes underlying those variants (or underlying the general rules that specify those variants in context-specific ways). That is, evolutionary biology could provide answers to the "phenotype existence" questions (*sensu* Reeve & Sherman 1993) about behavior that had long been nagging me. Wilson's magnificent *Sociobiology* (1975) convincingly clinched the case that virtually any aspect of social behavior could be dissected with an evolutionary probe. Moreover, evolutionary biology was undergirded by a sophisticated mathematical theory, thanks to the classic work of Fisher, Haldane, and Wright, and, more recently, of Hamilton and Maynard Smith.

Hamilton's kin selection theory (1964), together with the game-theoretic methods introduced by Maynard Smith (1982), jointly provided a firm mathematical foundation for a rigorous evolutionary theory of social behavior, to which my interests quickly gravitated. This joint theory held tremendous promise in quantitatively explaining not only the paradoxical occurrence of reproductive altruism, as beautifully explicated by West-Eberhard (1975), but also sex-ratio evolution, dispersal, foraging strategies, and any aspect of cooperation and conflict among organisms (as exhibited in the subsequent theoretical articulations of Grafen, Hammerstein, Parker, and many others). This prior theoretical work was for me was a map to the Holy Grail, a high-level mathematical theory from which to derive *testable* explanations of all aspects of animal sociality. All that was needed was a method for deriving predictions from the theory (via "lower-level" theories) for specific organisms in specific social contexts, as well as a strategy for rigorously quantitatively testing these predictions with these organisms. That both were possible was dramatically exemplified to me by Brockmann, Grafen, and Dawkins's (1979) elegant study of nest digging and entering as alternative tactics in mixed ESS in digger wasps and by Riechert's (1978) fascinating quantitative tests of game theory in territorial spider interactions. The way seemed clear for pursuit of my youthful dreams.

## The Rise of Causal Pluralism

My early, almost unbounded, optimism for extending and testing general models of social evolution gradually faded as I attended talks at conferences and meetings in my field. In a typical empirical talk, the central theoretical constructs of kin selection and game theory were mentioned only in passing or at best "tested" in rather vague, qualitative ways. Worse yet, the theories were sometimes rejected because the empirical data were inconsistent with a highly simplistic prediction derived from just one part of the theory (or with a prediction not validly derived at all). Here is an example: the theoretical "atom" of kin selection theory, Hamilton's Rule, clearly involves at least three terms (cost $c$, benefit $b$, and genetic relatedness $r$), combined as $rb > c$ in the condition for favored altruism. (Actually, the inequality involves four parameters since the relatedness to self, i.e., 1.0, is absorbed into the value $c$.) However, it was not uncommon to hear even in the 1990s that kin selection could not explain a particular case of cooperation because the "relatedness was not high." Of course, even a tiny relatedness can be decisive in the evolution of altruistic cooperation if it is greater than the ratio of cost to benefit $c/b$. As a second example, I recently heard a talk at a conference in which the speaker nicely showed that cooperation occurred among unrelated queens of an ant species, probably because of some mutualistic advantages accrued by being in such associations. However, the punch line of the talk was that "This study shows that relatedness isn't everything" (which drew approving applause), as if the main take-home message was that Hamilton's kin selection theory was incomplete in some way. To me, this was akin to a physicist asserting that Galileo's supposed "Leaning Tower of Pisa" experiment demonstrates that "mass isn't everything," as a swipe against Newtonian mechanics. Of course, Galileo's finding that mass did not affect an object's time to impact is *predicted* jointly by Newton's first law of motion and his gravitational law under the special conditions of the experiment (force $= ma = GMm/r^2$, where $m$ is the object's mass, $a$ is its acceleration, $G$ is the gravitational constant, $M$ is the earth's mass, and $r$ is the distance from the earth's center; the latter equation implies that $a = GM/r^2$, i.e., independence of the object's mass and its downward acceleration). Likewise, the evolution of cooperation among nonrelatives is fully predicted by Hamilton's Rule for the case of mutualism, that is, $c < 0$ (because $rb > c$ as seen from $rb = 0*b = 0 > c$) and depends crucially on the organism's relatedness to itself. Now, remembering an inequality with three (really four) terms hardly requires great intellectual prowess, so it may seem mysterious to a beginning graduate student that professional behavioral biologists could repeatedly commit such fundamental errors. How could this happen?

I have come to believe that such fuzzy thinking (or at least fuzzy writing and speaking!) results from a peculiar reward system in my field that asymmetrically rewards evidence for versus evidence against a general quantitative theory of social behavior. Evidence against a prediction of a general theory, no matter how inappropriate or simplistic the prediction, is usually enthusiastically accepted and even specially honored with prizes, jobs, and grants. To be honest, negative results probably reinforce the somewhat self-flattering belief that the subject matter of one's field is too complicated to be understood through any general framework—the comforting implication is that every one of us is desperately needed to understand our particular complex corner of the complex biological universe. In contrast, evidence in support of a general theory is usually weighed with far more critical scrutiny, the apparent belief being that erroneous support for a general theory is more costly than erroneous falsification of the theory. I am of the opinion that *rejection errors* can be just as costly as *acceptance errors* and that extreme care should be taken to derive the appropriate theoretical prediction from the theory to be tested. The special virtue of extracting and testing *quantitative* predictions of a general theory is that one is forced to consider all of the factors that are predicted to affect the behavioral phenomenon of interest. All of the variables in the theory that affect the predicted outcome must be measured or otherwise accounted for, even if only crudely.

The unfortunate effect of the asymmetrical reward system together with nonquantitative, often vague, testing of the general theories has been that (among some empiricists, at least) one theoretical construct after another has been increasingly viewed with suspicion, if not totally abandoned. In place of appeals to such theories has risen a kind of causal pluralism, that is, the idea that each particular behavioral system has its own particular kind of explanation or unique combination of low-level explanations, rather than being a particular instantiation of a very general theory. Often these explanations are couched in terms of the unique hardware (morphology or physiology) of the study organism, guaranteeing that the explanations will be limited in scope. (In extreme cases, explanations based on natural selection are dispensed with completely, and appeals to supposedly alternative processes or entities such as "emergent phenomena" are intended to be explanatory [see critique in Reeve & Sherman 1993].) What we are left with is a patchwork quilt of behavior stories, that is, nothing resembling the quantitative, general explanatory, and predictive frameworks in the physical sciences. Causal pluralism may appeal to those with artistic inclinations, that is, those who think that a patchwork quilt is just fine and indeed quite beautiful to behold in its totality. Causal pluralism may even now be "politically correct" because it eliminates the numerous territorial disputes that arise when a general construct is imperialistically applied across taxonomic boundaries. However, I fear that causal pluralism would be the death of my scientific discipline if left unquestioned. Fortunately, as I argue below, I think we are on the verge of a quantum leap in our ability to evaluate truly general quantitative, sociobiological theories.

The rise of causal pluralism has coincided with the development of low-level theories that stress the roles of historical contingency and constraints on selection in the evolution of behavior. These two views are naturally connected: the less important is natural selection in shaping behavior, the less likely it is that we can generate a unified evolutionary theory of behavior, because the nonselective forces of evolution like genetic drift typically lead to indeterminate, highly variable outcomes. For example, an increasingly popular view of the organism appears to be that it is a somewhat awkward system of interconnected toaster-style units, the connections being relatively fixed and constraining the operation of the mutually connected units (these are the "genetic correlations" of some evolutionary quantitative geneticists). The operation of one unit is viewed as often preventing the optimal operation of another. In "sensory exploitation" models of female mating preferences (Ryan 1998), for example, the toaster-units that control the female's *nonmating* preferences, say food preferences, also dispose a female to mate with males that resemble the preferred food. (I point out briefly here that high interconnectedness could actually *facilitate* the optimal operation of one of the interconnected units, for example, if the connection weightings are facultatively modulated according to the stimulus context, but that is a subject for another paper [see also Reeve & Dugatkin 1998].)

The "toaster-network" view described above is comfortably mated to phylogenetic approaches to understanding behavior, particularly the "historical ecology" research program (Brooks & McLennan 1991; see critique by Reeve & Sherman, 2001). Because the toaster units are frequently operating suboptimally due to their constraining network connections, the behaviors they generate will clearly bear the stamp of historical accident. There are infinitely many ways in which a toaster can be suboptimal, whereas there is only one or few ways it which it can be optimal (for a given context), so sharing of the same suboptimal characteristics will be a reliable sign that two taxa share a common evolutionary history, including the particular historical accident that constrained the shared phenotype.

I am strongly skeptical of the toaster-network view and frankly have little regard for the associated causal pluralism. Of course, these sentiments are not evidence against either, and it must be admitted that there is a nonzero probability that both views are the correct ones. However, now it is time to reveal why social wasps have convinced me that the latter probability is quite close to zero.

## Social Wasps and Reproductive Skew

My early experiments with George Gamboa (and Janet Shellman-Reeve and David Pfennig) on nestmate recognition in social paper wasps of the genus *Polistes* sealed my commitment to the evolutionary study of social behavior.

That social wasps indeed could recognize their nestmate relatives and that they learned recognition cues from their nests quickly after their emergence as adults (i.e., ensuring that such cues are reliable indicators of the current group of nestmates), all made beautiful sense in light of Hamilton's brilliant kin selection theory (Gamboa et al. 1986). Here, for me, was the hard evidence that the theory could bring the biological world into better focus. I went on to study the role of the queen as a pacemaker of colony activity, and queen-worker conflict over worker industriousness, both in social wasps and, with Paul Sherman at Cornell University, in naked mole-rat societies.

Working both with social wasps and with naked mole-rats lit up a question in me that has shaped my empirical research since 1990. In naked mole-rats, a single female out of up to hundreds produces *all* of the offspring in the colony. However, in social paper wasps, it was known by the 1980s that the dominant female in a group of co-nesting spring-foundresses *shares* the reproduction to some degree with her co-foundresses (Reeve 1991). The vexing question was: How could there be such tremendous variation in the reproductive skew, that is, the degree of reproductive domination by the dominant, across animal societies, vertebrate or invertebrate? What factors might explain this variation in skew?

The first interesting theoretical answer to this question was proposed by Sandra Vehrencamp (1979, 1983) and Steve Emlen (1982). They suggested that, under some circumstances, a dominant might be favored to "pay" a subordinate to stay and cooperate in the group, leading to reproductive sharing within the group. Francis Ratnieks and I used a flexible version of Hamilton's Rule together with game theory to explore the exact conditions under which a dominant would yield to the subordinate either a staying incentive (the minimal reproductive payment required to prevent the subordinate from leaving and nesting solitarily) or a peace incentive (the minimal reproductive payment required to prevent the subordinate from fighting to the death for complete control of the group's resources) (Reeve & Ratnieks 1993). The result of this theoretical analysis was not only a theory of reproductive partitioning within a social group, but also a theory of when groups should or should not form at all.

The exciting feature of these "optimal skew" or "transactional" theories is that they integrate information on (1) the expected reproductive success of a solitarily breeding subordinate, (2) the expected success of a lone dominant, (3) the expected total success of the group if the subordinate joins, (4) the fighting ability of the subordinate relative to the dominant, and (5) the genetic relatedness between the dominant and the subordinate, to produce a *quantitative* prediction of the reproductive skew. Transactional theory is thus a terrific candidate for a unifying evolutionary theory of animal societies, in that it shows exactly how ecological, social, and genetic factors should combine to determine the reproductive partitioning within any animal society (Keller & Reeve 1994). Each society might have its own unique skew given a unique combination of values for the above variables, but all would be both explainable and even predictable in advance from a single comprehensive theory. Moreover, to the extent that reproductive partitioning determines the scope for within-group conflict, transactional theory also is a quantitative theory for the evolution of intrasocietal conflict and aggression (Reeve & Keller 1997; Reeve & Ratnieks 1993; Reeve 2000). In the years since the original transactional models were advanced, there has been a theoretical explosion of extensions and alternatives to transactional theory (e.g., Cant 1997; Johnstone et al. 1999; Kokko & Johnstone 1999; Reeve & Keller 1995; Reeve et al. 1998; Reeve 1998; Reeve 2000), all sharing the goal of providing general explanations for the properties of animal societies with a rigorous, quantitative evolutionary framework.

The development of transactional theory has not been universally greeted with enthusiasm, as one might expect from the momentum toward causal pluralism. In particular, I have received several kinds of admonitions from my colleagues, some silly but some to be taken seriously. A silly objection is that, because transactional theory takes multiple ecological, genetic, and social variables as inputs, the theory can always be rescued from seemingly inconsistent data by appealing to effects of one of the unmeasured variables, that is, the theory is unfalsifiable. This is like saying that Newton's first law of motion is less falsifiable because an object's acceleration is determined by two factors (the external force and the object's mass) instead of just one (a similarly silly objection could be raised about Hamilton's Rule itself). In fact, the theory becomes *more* easily falsified because more relationships are predicted among the theoretically connected variables (thus, we could keep mass constant and see if varying force affects acceleration in the predicted ways *or* we could keep the force constant and vary the mass and examine those effects on acceleration). In other words, the more variables that are quantitatively integrated in a theory being tested, the more constraints that are placed on the values of unmeasured variables, each such constraint in effect generating a new prediction. The *least* falsifiable theory would be one of the form: A causes B with random noise, that is, A → B + "Noise," because one can always appeal to the "Noise" to save the theory from a failed test—the more unified the theory, the more the noise disappears and the less room there is to rescue a bad theory. So the way out of the imagined problem is to measure (at least crudely) all of the relevant variables *or* at least compare contexts which are known to differ primarily in only one of the variables (although what is predicted may depend on the values of one or more of the fixed variables). Almost by definition, a unifying theory must be one that integrates information from multiple sources.

A second, more worrisome, objection is that animal societies are far too complex to be reducible to the few factors entering into transactional theory. This may well be correct, of course, but the only way to find this out is to test the theory with quantitative data. It is precisely how well transactional theory explains social life in paper wasp societies that convinces me that this worry so far is unfounded, as will be detailed below. One should remember that an understanding of the behavior of complex systems does not necessarily require complex explanations (and complexity does *not* equal indeterminacy). This is why natural selection theory is so potent—it shows how complex biological systems will be attracted to relatively few stable equilibria (adaptive peaks), infinitely fewer than the total number of conceivable states of these systems.

Social paper wasps of the genus *Polistes* are the perfect organisms for testing transactional models of reproductive skew. They live in small societies in which all adults are easily marked with enamel paint for study, and all adult-adult interactions and even all adult-brood interactions can be videotaped with little disruption because the nest combs are not enveloped. At the end of study, all adults (foundresses and workers) and the brood from previously mapped cell locations can be collected for genetic analyses. Moreover, colonies are found throughout the world and can be locally numerous, occurring commonly in shrubs or on the wooden eaves of sheds and barns. Thus, social paper wasps are "model systems" in the sense that there are few practical problems associated with their study. Of course, for a unificationist such as me, *all* social organisms are "model systems" in the sense that their study should shed light on general evolutionary principles. However, aside from the practical advantages of their study, the especially tantalizing feature of paper wasp societies for tests of transactional models was the genetic evidence that dominant foundresses share reproduction with subordinates. Are dominants conceding staying or peace incentives to subordinates?

In the early 1990s, Peter Nonacs (now at UCLA) and I, while conducting postdoctoral research in E. O. Wilson's lab at Harvard University, realized that for the transactional model of reproductive skew to work, the subordinate must have some way of detecting reproductive cheating by the dominant foundress. That is, what prevents a dominant foundress from only *pretending* to yield a staying incentive, for example, by allowing the subordinate to lay eggs, but later eating those eggs and replacing them with her own? This was not far-fetched, because reciprocal egg eating had been well documented among *Polistes* foundresses (review in Reeve 1991). We hypothesized that if the transactional (or "social contract") model were correct, the subordinate should respond to egg eating by leaving the colony or otherwise retaliating against the dominant.

Thus, Nonacs and I carried out a series of controlled experiments involving removal of eggs from *P. fuscatus* combs after the foundresses had been collected, then returning the foundresses to the nest and observing their subsequent behavior (Reeve & Nonacs 1992). To our amazement, subordinates responded to removal of eggs with marked increases in aggression toward the dominant foundress. Moreover, the increase in the subordinate's aggression was significantly greater than the increase in the dominant's aggression (this was somewhat puzzling to us, because presumably both the subordinate's and dominant's eggs had been removed in our experiment). Finally, we found that these effects depended acutely on the kind of eggs removed. When worker-destined eggs (which we presumed would give rise to largely nonreproductive individuals) were removed, the subordinate's aggression hardly changed, with the above retaliatory aggression being observed only when late female or male eggs, that is, eggs destined to develop future reproductive brood, were removed. This seemed definitive proof of the kind of cheating-detection and retaliation predicted by the transactional skew models, but a number of alternative hypotheses (such as that the subordinate's increased aggression merely reflected heightened alertness in the perceived presence of an external brood parasite) had to be systematically eliminated.

However, this conclusion didn't go unchallenged. Joan Strassmann of Rice University, another excellent wasp sociobiologist, raised a new possibility: perhaps egg removal had the effect of telling the subordinate that the dominant was weak and vulnerable to being overthrown (Strassmann 1993). Maybe *that* was why the subordinate had elevated her aggression toward the dominant. Fortunately, we had a way of testing this hypothesis with our data, because dominant queens replaced the removed eggs at bimodally distributed rates: some dominants didn't replace any of the removed eggs during the post-treatment videotaping, but others quickly laid new replacement eggs. If the "weak queen" hypothesis were correct, the subordinates should have been most aggressive toward dominants that had replaced eggs at below the median rate; on the contrary, if the "retaliatory aggression" hypothesis were correct, subordinates should have been most aggressive toward the dominants that replaced eggs at above the median rate. Indeed, the latter result was decisively supported (Reeve & Nonacs 1993). (It should be mentioned to the beginning students reading this essay that is never entirely fun to have alternative hypotheses hurled at one's seemingly well-polished conclusions, but this is the hallmark of healthy scientific inquiry, at least when the alternative hypotheses, like Strassmann's weak queen hypothesis, are well-grounded theoretically.) Thus, it appeared that the social contract hypothesis and the attached transactional model of skew had won the day.

## Recent Surprises

In recent years, Nonacs and my graduate students Phil Starks and John Peters have continued to study foundress associations of *P. fuscatus* in the field and have performed genetic (microsatellite) analyses to infer the actual reproductive skew among co-foundresses. We were in for a series of surprises, each of which indicated that we had achieved at best a glancing blow on the truth in the egg removal experiments.

The first surprise was that we found evidence that many if not most "workers" were actually females that quickly dispersed from their natal nest to enter early diapause, in at least a few cases appearing as foundresses the following year (Reeve et al. 1998). Thus, early female eggs *were* reproductively valuable to subordinate foundresses, although perhaps not quite as valuable as late female or male eggs. The second major surprise was that the dominant and subordinate shared reproduction only in the production of early females; in contrast, the dominant produced about 93 percent of the late females and males (Reeve et al. 2000). That is, the dominant's reproductive skew clearly increased over the course of the colony cycle.

The latter result was very puzzling in light of the egg removal experiments. The subordinate had produced only a tiny fraction of the late female and male eggs, yet her aggression increased much more when these eggs were removed than when the early female eggs (nearly 50 percent of which were hers) were removed. Why did the subordinate not retaliate more aggressively when the early female eggs (now known to be reproductively valuable!) were removed? The key to resolving this puzzle comes from an analysis of the dominant's changes in aggression when early versus late eggs were removed. When late eggs were removed, the dominant's aggression increased, although not significantly and not as much as the subordinate's aggression. Importantly, however, when the early eggs were removed, the dominant's rate of aggression nose-dived by 61 percent (Reeve & Nonacs 1997). In other words, the dominant markedly and significantly decreased its aggression toward the subordinate precisely when the subordinate was in danger of losing its presumed staying incentive, that is, the reproductive payment conceded by the dominant to the subordinate to prevent the subordinate from leaving. This could explain why the subordinate failed to retaliate when its early eggs were removed—the dominant in effect "short-circuited" the subordinate's retaliation by drastically lowering its own aggression. In contrast, when late eggs are removed, that is, when the subordinate has only a small staying incentive, the dominant does not lower its aggression, presumably because now it is favored to compete for access to almost all of the newly empty cells, and the subordinate challenges the dominant for access to those cells. Egg removal at this stage stimulates especially large increases in beta's aggression not only because of the enhanced egg-laying competition with a now aggressively competing alpha, but also because beta may be retaliating for losing her (albeit small) staying incentive.

We still needed to know *why* the reproductive skew increased over the colony cycle. The crucial clue was provided by the nature of the grouping advantage among co-foundresses. Both demographic evidence from our study population (Reeve & Nonacs 1997) and comparative analyses (Reeve 1991) indicated that the major advantage of co-founding with another adult female resides in the reduced risk that all colony adults will by chance die before the first workers emerge. Intriguingly, this "survivorship insurance" model predicts that the reproductive skew should increase over the colony cycle: as a multiple-foundress colony approaches the time when the first workers will emerge, the probability of survival of this colony increases relative to that of a single-foundress colony started from scratch by a departing subordinate. Thus, a subordinate's theoretical minimal staying incentive decreases, and the reproductive skew increases, over the course of the colony cycle (Reeve et al., 2000). This elegantly explains why a subordinate produces a substantial fraction of the early eggs but very few of the late eggs—the subordinate is paid "up front" by the dominant, that is, at the beginning of the colony cycle, because this is when the solitary nesting option is most attractive to the subordinate and thus it requires the largest fraction of the reproduction to stay.

The above transactional model for reproductive skew neatly explains some other aspects of polistine social biology. For example, Gamboa and Stump (1996) showed that aggression by both the dominant and the subordinate foundresses increases as the colony cycle progresses. Early in the cycle, the dominant should concede a staying incentive to the subordinate, and little aggression is expected during this reproductive transaction. Late in the cycle, the dominant needs to concede very little to keep the subordinate around, so each engages in open warfare over egg-laying rights. Even more puzzling had been the little-cited Gamboa and Dropkin (1979) observation that dominants actually behave submissively to subordinates very early in the foundress association. They concluded that "in early foundress associations, queens frequently exhibit behavior characteristic of subordinates in mature hierarchies." This is exactly what would be expected when a reproductive payment is being transferred from the dominant to the subordinate early in the colony cycle. (It is interesting to speculate that the Gamboa and Dropkin result had never been cited much because it upset the then prevalent view that foundresses are *always* engaged in a tug-of-war for egg-laying rights.)

We regard the above results as impressive support for the transactional model of skew in *Polistes* wasps, but perhaps even more telling support has come from a quantitative analysis of the relationships between reproductive skew and colony productivity and co-foundress genetic relatedness. The transactional skew models predict that the subordinate's staying incentive should decrease (the skew should increase) as the colony productivity increases (because the more productive the colony, the less attractive the leaving option), and this is exactly what we found (Reeve et al. 2000). In addition, the theory predicts that the staying incentive should decrease (the skew should increase) as the genetic relatedness between the dominant and the subordinate increases because a subordinate that is more closely related to the dominant automatically has greater genetic incentive to stay and help the dominant. Again, the data strongly supported this prediction, and, moreover, the observed magnitudes of the skews closely matched those theoretically predicted from the survivorship insurance model and our demographic data combined with transactional skew theory (Reeve et al. 2000).

Thus, the picture that emerges from our new data (and integration of old, sometimes previously ignored, data) is that social wasps are finely tuned by natural selection to adaptively monitor their continually changing ecological and social circumstances, as well as the behavior of their social partners, and to engage in sophisticated reproductive transactions when it is beneficial to pay partners to become more cooperative. Since social wasps can do this, it seems nearly impossible that other social organisms could not also have this capability (obviously, humans have developed a remarkably advanced ability to monitor and execute sophisticated reproductive transactions). The idea that even *insects* can engage in high-order, highly context-dependent, inclusive fitness-maximizing reproductive transactions takes us a very long way from the "toaster-network" view of animal behavior.

## A Renewed Search for Unified Theories of Social Evolution

The data generated from *Polistes* as a model system for testing transactional skew theory has convinced me that there are huge rewards awaiting those who take general models of social evolution seriously, down to their quantitative details. Fine-grained tests of these theories (whether or not the existing theories ultimately are judged correct) would at last propel behavioral ecology and sociobiology beyond causal pluralism and into the upper reaches of scientific rigor previously thought reserved exclusively for the physical sciences. There may be some painful casualties (perhaps the toaster-network view among them), but I have little doubt that behavioral ecologists and sociobiologists pursuing such a path someday will have developed a highly successful, unified account of social evolution. I predict that the scientists who accomplish this feat (hopefully some who read this essay) will be accomplished in mathematics and experimental design as well as being keen observers of natural history. The days of causal pluralism will have ended, just as has long been the case in physics. Instead, fuzzy tests of disconnected, qualitative theories will have been replaced by sharply discriminating, quantitative tests of empirically grounded, general mathematical theories. This will not be good news for some, but for many it may be stirring.

## References

Brockmann HJ, Grafen A, Dawkins R, 1979. Joint nesting in a digger wasp as an evolutionarily stable preadaptation to social life. Behaviour 71:203–245.

Brooks DR, McLennan DA, 1991. Phylogeny, ecology, and behavior. Chicago: University of Chicago Press.

Cant MA, 1997. A model for the evolution of reproductive skew without reproductive suppression. Anim Behav 55:163–169.

Dawkins R, 1976. The selfish gene. Oxford: Oxford University Press.

Emlen, ST, 1982. The evolution of helping. II. The role of behavioral conflict. Am Nat 119:40–53.

Gamboa GJ, Dropkin JA, 1979. Comparison of behaviors in early versus late foundress associations of the paper wasp *Polistes metricus* (Hymenoptera: Vespidae). Can J Entomol 111:919–926.

Gamboa GJ, Reeve HK, Pfennig D, 1986. The evolution and ontogeny of nestmate recognition in social wasps. Ann Rev Entomol 31:432–454.

Gamboa GJ, Stump KA, 1996. The timing of conflict and cooperation among cofoundresses of the social wasp *Polistes fuscatus* (Hymenoptera: Vespidae). Can J Zool 74:70–74.

Hamilton WD, 1964 . The genetical evolution of social behavior, I & II. J Theor Biol 7:1–52.

Johnstone RA, Woodroffe R, Cant MA, Wright J, 1999. Reproductive skew in multimember groups. Am Nat 153:315–331.

Keller L, Reeve, HK, 1994. Partitioning of reproduction in animal societies. Trends Ecol Evol 9:98–102.

Kokko H, Johnstone RA, 1999. Social queuing in animal societies: a dynamic model of reproductive skew. Proc Roy Soc (Lond), Series B 266:571–578.

Maynard Smith J, 1982. Evolution and the theory of games. Cambridge: Cambridge University Press.

Reeve HK, 1991. *Polistes*. In: The social biology of wasps (Ross K, Matthews R, eds). Ithaca: Cornell University Press; 99–148.

Reeve HK, 1998. Game theory, reproductive skew, and nepotism In: Game theory and animal behavior (Dugatkin L, Reeve HK, eds). Oxford: Oxford University Press; 118–145.

Reeve HK, 2000. A transactional model of within-group conflict. Am Nat 155:365–382.

Reeve HK, Dugatkin LA, 1998. Why we need game theory. In: Game theory and animal behavior (Dugatkin L, Reeve HK, eds). Oxford: Oxford University Press; 304–311.

Reeve HK, Emlen ST, Keller L, 1998. Reproductive sharing in animal societies: reproductive incentives or incomplete control by dominant breeders? Behav Ecol 9:267–278.

Reeve HK, Keller L, 1995. Partitioning of reproduction in animal societies: mother-daughter versus sibling associations. Am Nat 145:119–132.

Reeve HK, Keller L, 1997. Reproductive bribing and policing as mechanisms for the suppression of within-group selfishness. Am Nat (special issue on "Multi-level Selection") 150:S42–S58.

Reeve HK, Nonacs P, 1992. Social contracts in wasp societies. Nature 359:823–825.

Reeve HK, Nonacs P, 1993. Weak queen or social contract? Reply. Nature 363:503.

Reeve HK, Nonacs P, 1997. Within-group aggression and the value of group members: theory and a field test with social wasps. Behav Ecol 8:75–82.

Reeve HK, Peters JP, Nonacs P, Starks P, 1998. Dispersal of first "workers" in social wasps: causes and implications of an alternative reproductive strategy. Proc Nat Acad Sci USA 95:13737–13742.

Reeve HK, Ratnieks FLW, 1993. Queen-queen conflict in polygynous societies: mutual tolerance and repoductive skew. In: Queen number and sociality in insects (Keller L, ed). Oxford: Oxford University Press; 45–85.

Reeve HK, Sherman PW, 1993. Adaptation and the goals of evolutionary research. Q Rev of Biology 68:1–32.

Reeve HK, Sherman PW, 2001. Optimality and phylogeny: a critique of current thought. In: Optimality and adaptation (Orzack S, Sober E, eds). Cambridge: Cambridge University Press.

Reeve HK, Starks PT, Peters JM, Nonacs P, 2000. Genetic support for the evolutionary theory of reproductive transactions in social wasps. Proc Roy Soc (Lond), Series B 267:75–79.

Riechert SE, 1978. Games spiders play: behavioral variability in territorial disputes. Behav Ecol Sociobiol 3:135–162.

Ryan MJ, 1998. Sexual selection, receiver biases, and the evolution of sex differences. Science 281:1999–2003.

Strassmann JE, 1993. Weak queen or social contract? Nature 363:502.

Vehrencamp SL, 1979. The roles of individual, kin and group selection in the evolution of sociality. In: Social behavior and communication (Marler P, Vandenbergh J, eds). New York: Plenum Press; 351–394.

Vehrencamp SL, 1983. A model for the evolution of despotic versus egalitarian societies. Anim Behav 31:667–682.

West-Eberhard MJ, 1975. The evolution of social behaviour by kin selection. Quart Rev Biol 50:1–33.

Wilson EO, 1975. Sociobiology, the modern synthesis. Cambridge, MA: Harvard University Press.

# 5

# Genetic Consequences of Sexual Selection in Stalk-Eyed Flies

Gerald S. Wilkinson

My interest in using stalk-eyed flies, family Diopsidae, to study the genetic consequences of sexual selection developed while I was a postdoctoral fellow at the University of Edinburgh in 1984. I began postdoctoral work on *Drosophila melanogaster* at the University of Sussex in the laboratory of Brian Charlesworth, an eminent evolutionary geneticist. Partway through my twelve-month fellowship, he moved to the University of Chicago so I asked Linda Partridge, who had recently been conducting studies of life history evolution with *Drosophila*, if I could spend the remainder of my fellowship time in her lab in Edinburgh. She graciously agreed.

I went to England without any prior lab experience studying flies. In fact, I began graduate studies as a committed field biologist. After taking a two-month course in tropical ecology sponsored by the Organization for Tropical Studies in 1978, I initiated a five-year project on the social behavior of the common vampire bat, *Desmodus rotundus*. During this time I lived in Costa Rica for three years and gathered evidence that female vampire bats reciprocally exchange blood when a group member has failed to feed successfully during the night (Wilkinson 1984). Although females are sometimes related to the recipient of these blood meals (Wilkinson 1985), the primary benefit of reciprocal food sharing lies in increasing individual survivorship of group mates (Wilkinson 1988). While I enjoyed studying vampire bats, I realized that inferring causation can be difficult when the results consist of correlations from observational field studies. I also realized that much of the relevant theory underlying the evolution of social behavior is based on genetic models, the results of which are often taken for granted without ever questioning the underlying assumptions. I decided, therefore, to pursue postdoctoral work that would give me experience designing and conducting experiments aimed at testing assumptions or predictions of genetic models. I reasoned that together with my thesis work on vampire bats, these experiences would allow me to pursue and advise a wide range of field and laboratory studies in behavioral ecology when I eventually obtained a faculty position.

The fly experiments I conducted in Scotland were motivated by two theoretical papers that had recently been published on sexual selection and the evolution of female choice. In 1981 Russell Lande had published a quantitative genetic model (Lande 1981) which made the startling prediction that a line of equilibrium is expected because sexual selection by female choice for a male ornament will eventually be opposed by natural selection against the trait. In the absence of direct selection on females, an infinite combination of selection pressures can result in equilibrium, hence the line. Under some genetic conditions, for example, when the genetic regression between the trait and preference is particularly high, the male trait and female preference will increase at geometric rates and produce what Fisher (1958) called "runaway" sexual selection. In 1982 Mark Kirkpatrick published qualitatively similar results using a two-locus, haploid genetic model that had first been explored by O'Donald (O'Donald 1962, 1980). These models substantiated Fisher's insights and not only motivated my work, but also inspired many other theoretical and empirical studies (see Andersson 1994 for review).

My experiments with *D. melanogaster* were designed to test the equilibrium assumption, that is, that some male traits are constantly under equal but opposing forces of natural and sexual selection. I reasoned that if you prevent flies from competing for or choosing their mates by pairing randomly chosen males with females, then sexual selection would be eliminated, but natural selection would still occur. Consequently, by comparing populations of flies that had either been allowed to mate in large cage populations or were pair-mated I could measure the intensity of natural selection in the absence of sexual selection. Because previous work had indicated that body size influenced mating success (Ewing 1961; Partridge & Farquar 1983), I measured the length of the thorax to determine if selection was occurring. In replicated experiments I found consistent evidence for sexual selection on thorax length in males (Wilkinson 1987). Large body size increases development time and exposes larvae longer to inhospitable media (Botella et al. 1985). Thus, natural selection apparently favors smaller bodied males while sexual selection favors larger bodied males. After observing matings and conducting female choice experiments I concluded that much of what I had observed was consistent with larger males outrunning smaller males for access to receptive females (Partridge et al. 1987), rather than females choosing to mate with larger males. I began, therefore, looking for an alternative system more appropriate for testing female choice models of sexual selection.

## Model System Discovery

Even though I had only been working with flies for less than a year, I was already convinced that an insect would be preferable for further study. Flies seemed particularly attractive because generation times are short and captive breeding is often simple and inexpensive. When I first saw a picture of a diopsid stalk-eyed fly (Burkhardt & de la Motte 1983), I immediately knew I had found the perfect system. In contrast to *Drosophila*, where few mor-

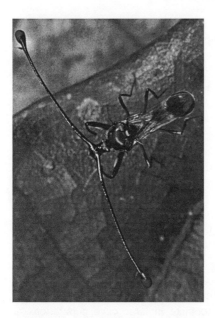

Figure 5.1. Male *Teleopsis belzebuth* from Brunei. Photograph by Mark Moffett.

phological characters exhibit sexual dimorphism, in many species of diopsids (Burkhardt & de la Motte 1985) males have dramatically longer eye stalks than females (Fig. 5.1). Furthermore, Dietrich Burkhardt at the University of Regensburg had been successful at breeding several species. One day at lunch in Edinburgh I mentioned my growing interest in diopsids to Arthur Ewing, a faculty member who studied courtship song in *Drosophila*. Arthur proceeded to describe stalk-eyed flies he had seen at a fishing camp on the side of Mount Kenya during a recent holiday. Within a month I had booked a cheap flight from London to Nairobi. The day after I arrived, just as he had described, I found stalk-eyed flies in abundance along several streams in the forests on Mount Kenya (Fig. 5.2). During two weeks at the site, I failed to observe any matings probably because it was the dry season and because I failed to go out and observe the flies at dawn. I did, however, manage to bring some back to Scotland and discover that they could be bred in the lab.

My fellowship ended in Scotland soon thereafter and I moved to the University of Colorado to begin a second postdoctoral fellowship. Even though I had received funding to study behavior genetics in mice, I hoped I could also begin developing methods for culturing stalk-eyed flies. Unfortunately, I was unable to obtain permission to import diopsids into Colorado. The species I had collected from Kenya fed on decaying plant material, but at least one diopsid species is a minor pest of maize (Tan 1967). The number one export crop in Colorado is corn, so the state entomologist decided it was not worth risking the importation of any diopsid. Therefore, I had to wait until I joined

Figure 5.2. Nocturnal aggregation of *Diasemopsis collaris* from Thego National Forest, Kenya.

the faculty at the University of Maryland in 1987 before I could return to Africa and collect more flies to establish my own breeding colony.

After several more years of experimentation, a postdoctoral associate, Paul Reillo, and I developed methods for breeding flies consistently and in quantity. We also traveled to Southeast Asia and collected several species that had been studied by Burghardt and his associates. The Southeast Asian diopsids proved to be much easier to breed than any of the African species and in several cases exhibited dramatic differences in sexual dimorphism for eye span. I chose the most reliable breeder, *Cyrtodiopsis dalmanni*, to use in an artificial selection study designed to test several key assumptions of the sexual selection models. This decision turned out to be fortuitous, because these experiments led to unforeseen discoveries, as I describe below, that have fostered new theoretical work.

## Genetic Correlation between Trait and Preference?

A central prediction shared by all coevolutionary models of sexual selection by female choice, that is, Fisherian runaway (Kirkpatrick 1982; Lande 1981) and good genes (Iwasa et al. 1991; Pomiankowski 1988), is that a genetic correlation, if the model is based on quantitative traits, or linkage disequilibrium, if the model is based on single-locus traits, should develop between the male trait and the female preference. Most models of sexual selection by female choice assume that autosomal genes affect female mating behavior and are physically separate from autosomal genes influencing the male trait. A genetic correlation develops because of nonrandom mating and is necessary for the evolution of the female preference when it is not under direct selection. Thus, finding evidence for a genetic correlation between a male trait and a female mating preference is a testable prediction of these models.

Genetic correlations can be estimated in two ways: family comparisons or artificial selection. Family comparisons are used to estimate causal components of variance either from an analysis of variance among full or half-siblings, in which sire and dam are random effects, or a regression analysis of offspring on parents (Falconer 1981). For example, if a male trait correlates with his daughter's preference, a genetic correlation can be inferred. Family comparisons are attractive because they provide estimates in one generation and, therefore, may be less influenced by adaptation to a laboratory environment. However, considerable statistical error is associated with these estimates (Klein 1974). Consequently, unless the correlation is high, very large sample sizes (often two hundred or more families) are required to demonstrate that a correlation differs from zero. Because coevolutionary models of sexual selection could operate even when the genetic correlation between trait and preference is very small, I decided to use artificial selection rather than a family comparison. I also thought that selected lines might be useful for future studies because they would provide a replicable source of individuals that differ genetically for a sexually selected trait.

Before embarking on any artificial selection study several questions must be answered. What trait should be selected? How many animals should be bred in each line? How many replicate lines should be maintained? I used my burgeoning experience with stalk-eyed flies to answer each of these questions. Direct evidence for a genetic correlation would be obtained if female preferences changed as a consequence of selecting on a male trait or vice versa. Thus, either male eye span or female preference could be chosen for selection. At the time I began this study we had not found a reliable method for scoring female preferences. We attempted to replicate experiments (Burkhardt & de la Motte 1988) where model males were glued to strings and female preferences were scored as female approaches to models with surgically lengthened or shortened stalks. However, we were unable to obtain repeatable preferences using this method, so I decided to select for male eye span rather than female preference. Because eye span correlates phenotypically with body size (Burkhardt & de la Motte 1985), I was concerned that selecting only on eye span would simply create large or small flies. Therefore, I decided to select on the ratio of eye span to body length in males. By temporarily immobilizing flies with $CO_2$ under a microscope, we could quickly measure adult flies from digitized images with a computer program.

The time needed for measuring and the space available for breeding effectively dictated the total number of flies we could breed and select each generation. When we began this experiment I estimated we could process about 500 flies every 8 weeks. To find evidence for a correlation I needed to select for both increased and decreased eye span. Replicate lines are essential to distinguish effects of selection from genetic drift, and control lines are needed to assess unexplained environmental variation shared each generation by all lines. Given these constraints, I opted to measure 50 males and choose 10 for breeding each generation in each of 6 lines. These males were mated to 25 randomly chosen females in 2 control lines, 2 lines selected for long relative eye span, and 2 lines selected for short relative eye span. Because I only exerted selection on males, net selection on autosomal genes was one-half that experienced by males, that is, 40 percent. Assuming a normal distribution of trait values, truncation selection of 40 percent corresponds to an overall selection intensity of 0.7 standard deviation units each generation (Falconer 1981). This intensity of net selection is about double what we measured in the field, as the covariance between the number of females roosting at night with a male and his relative eye span divided by the standard deviation in male relative eye span (Wilkinson & Reillo 1994).

I quickly discovered that selection on relative eye span in males was having the desired effect of changing eye span but not body length. Because eye span in females was also exhibiting change, I knew that a genetic correlation between the sexes existed for eye span. Under an assumption of equal heritabilities for eye span in each sex, I estimated the between-sex correlation for relative eye span to be $0.39 \pm 0.07$ (Wilkinson 1993). Selecting on the ratio of eye span to body length also caused the slope of the phenotypic regression between eye span and body length to change (Wilkinson 1993). Thus, by selecting on the ratio of eye span to body length we had changed allometric relationships between eye span and body length in a way that mirrored the differences between species (Burkhardt & de la Motte 1985; Baker & Wilkinson, in press). These results demonstrated that allometric relationships can be molded by selection and do not act as constraints on evolutionary change, *contra* Gould (1989).

Although I was eager to know if morphological change in males was accompanied by behavioral change in females, I waited for 2 years (13 generations) before testing female mating preferences. By waiting I figured I would be more likely to detect change if the genetic correlation between trait and preference was small. For example, after 10 generations of selection with intensity of 0.66 (Wilkinson 1993), heritabilities of 0.35, and a genetic correlation of 0.1, a female preference should have changed by 0.5 standard deviation units.

To score female preference I put two males on opposite sides of a transparent perforated partition in a small cage. I used selected-line males that were matched for body length but differed in eye span by 15 percent. Several females were then introduced and allowed to explore the cage. Because the holes in the partition were greater in diameter than female eye span but smaller than male eye span, females could pass unimpeded to either side of the cage while males were restricted to one side. Initially, we scored female preference as the proportion of five nights a female spent on the same side of a partition as a male (Wilkinson & Reillo 1994), because most matings in the wild occur in aggregations (Lorch et al. 1993). In subsequent experiments (Wilkinson et al. 1998a) we measured preference as the proportion of copulations by a female with each male over a ten-day period. The repeatability, which sets an upper limit to the heritability for a trait, of the preference was 0.4 and 0.3 for *C. whitei* and *C. dalmanni*, respectively, the two closely related sexually dimorphic species.

As predicted by coevolutionary models of sexual selection, we found evidence for a genetic correlation between a female's mating preference and male eye span (Wilkinson & Reillo 1994). Females from lines with decreased eye span preferred to mate with males that had shorter eye stalks while females from lines with increased eye span preferred to mate with males that had longer eye stalks. Contrary to prediction, however, females from lines with increased eye span did not differ in their preferences from outbred, unselected females. A possible explanation for this apparent contradiction is that we only provided a single choice to females, that is, a 15 percent difference between two males. Subsequent experiments, where we systematically varied the difference between two males, indicate that *C. dalmanni* and *C. whitei* females can detect a difference in eye span of 7 percent, and exhibit increasingly strong preferences as the eye span difference between males increases (Wilkinson et al. 1998a). Thus, perhaps we would have detected a difference between unselected and selected line females if we had offered them a series of males that differed progressively in eye span. Nevertheless, this result remains one of a handful of examples in which a genetic correlation between a female mating preference and a male trait has been documented (Bakker 1993; Houde 1994).

## Good Genes for Viability?

Distinguishing between the Fisher process and good genes models of sexual selection has proven difficult. Most studies that have claimed to have demonstrated one process over the other have focused on finding evidence for a correlation between a male ornament and offspring survival. Several recent studies on birds (Hasselquist et al. 1996; Møller 1994; Norris 1993; Petrie 1994; Sheldon et al. 1997) have found such a relationship. However, offspring survival is only one component of fitness, so even when a positive correlation is detected, the possibility remains that there are trade-offs which would cause other fitness components to covary negatively with male ornaments (Boake 1986). Nevertheless, my initial approach to this problem was to determine if the genes that influence eye span are linked to genes that influence offspring survival from egg to eclosion as an adult *C. dalmanni*.

I used the lines selected for long and short eye span after twenty generations of selection to test for differences in egg hatchability, development time, and pupal eclosion success. To minimize any confounding effects of inbreeding that might differ between lines, I crossed the replicate lines before quantifying these fitness components. I decided to cross flies from each line to flies from one of the control lines to test genes with known effects on eye span in similar genetic backgrounds. In retrospect, this crossing scheme had the undesirable effect of decreasing any potential difference between the lines. A better method for these experiments would have been to cross the replicates within each selection treatment. Nevertheless, this experiment produced significant and informative results. In addition, I scored survival of twenty adult flies from each selected line after seventeen generations of selection. I did not use line crosses for scoring adult survival because the time associated with such work was prohibitive. Flies routinely survive more than six months.

The fitness component analysis revealed that changing eye span influenced both development and survival (Fig. 5.3). No difference between the lines was found in the proportion of eggs that successfully developed into pupae ($\chi^2 = 2.5$, 2 df, P = 0.29) after being transferred from agar in groups of 50 to 50 ml of food. On average, about 30 percent of eggs hatched in each treatment (Fig. 5.3a). In contrast, development time at 25°C differed significantly between lines ($F_{2,625} = 6.0$, P < 0.003) as well as between males and females ($F_{1,625} = 71.4$, P < 0.0001). Post hoc tests revealed that flies carrying genes for long eye span took longer to develop than either control-line flies or flies carrying genes for short eye span (Fig. 5.3b). In most flies, rapid development is assumed to be under strong directional selection because larvae are usually more vulnerable than pupae or adults. In addition, pupal eclosion success differed between lines ($\chi^2 = 32.7$, 2 df, P < 0.0001). Flies carrying genes for long eye span did not eclose as successfully as control-line flies or as flies carrying genes for short eye span (Fig. 5.3c). Finally, mean age of death differed between lines (repeated measures ANOVA, $F_{1,58} = 4.02$, P = 0.023). Short eye span–line flies did not survive as long as control-line flies or flies carrying genes for long eye span (Fig. 35.d). Thus, longer eye span increased larval mortality while shorter eye span increased adult mortality.

While these results obviously depend on captive rearing conditions, it seems likely that development rate and pupal eclosion success would exhibit similar, if not more extreme, differences in the wild since these measurements reflect developmental processes. We know that there are more than six thousand neurons that run the length of the eyestalks in these flies (Burkhardt & de la Motte 1983; Seitz & Burkhardt 1974) and that during the development of the eye-antenna imaginal disc the stalks are present before any neurons have begun to project from either the eye bulb into the brain or vice versa (Buschbeck & Hoy 1998). Consequently, it seems likely that increasing eye stalk length could create developmental problems that would

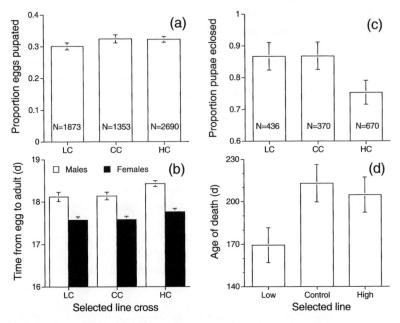

Figure 5.3. Effect of genes for increased or decreased eye span on fitness components in *Cyrtodiopsis dalmanni*: (a) egg eclosion and larval survival, (b) development time, (c) pupal eclosion success, and (d) adult survival. In a–c, flies from lines selected for de-creased (L) or increased (H) eye span were mated to control line flies after twenty generations of selection. In d, selected line flies were used after seventeen generations of selection. See text for methods.

increase mortality during metamorphosis. Thus, in contrast to a conventional good genes process, increasing eye span does not appear to improve off-spring viability, at least under laboratory conditions. Nevertheless, we dis-covered that a good genes process can operate by a mechanism quite differ-ent from anything previously imagined.

## Selfish Genes and Sexual Selection

As part of a series of experiments designed to determine the magnitude of change in genetic variation and covariation for morphological traits that have or have not been under sexual selection (Pomiankowski & Møller 1995; Rowe & Houle 1996), we conducted large-scale half-sib breeding studies of 6 species of stalk-eyed flies (Wilkinson & Taper 1999) and of the replicate lines selected for long and short eye span after 22 generations of selection. In these experiments one male was mated to 6 or more females. Our goal was to measure 2 male and 2 female offspring from each of 4 females mated to one male and then replicate this mating scheme to create 100 half-sib familes for each line and species. Such large samples are necessary to determine if genetic variation or covariation differs between populations (Shaw 1991). As we began this work I instructed my research assistant, Lili Crymes, to count the number of males and females produced by each female because I thought progeny sex ratios might provide information about differential mortality between the lines. To my surprise, Lili told me that she was collecting more males than females from the lines selected for long eye span and more females than males from the lines selected for short eye span. This observation, in combination with results that we were obtaining at the same time regarding sex chromosome meiotic drive (Presgraves et al. 1997), made me realize that we had inadvertently stumbled upon a novel mechanism for sexual selection: perhaps females choose males that alter their progeny sex ratio and thereby increase their production of grandchildren. Understanding how this mechanism might operate requires an understanding of meiotic drive.

Sex chromosome meiotic drive is used to describe cases where one of the sex chromosomes is transmitted to more than half of the gametes. The most common mechanism for such non-Mendelian inheritance involves differential development and survival of X-bearing sperm relative to Y-bearing sperm. Sex chromosome meiotic drive can lead to the production of all female progeny by male carriers in species with male heterogamety. X chromosome meiotic drive has now been documented in twenty-one different species of *Drosophila* (Jaenike 1996, 1999; Lyttle 1991) and has also been reported to occur in a number of other fly species, as well as in some mammals and fishes (Hurst et al. 1996). Previously reported cases in butterflies appear to be due to cytoplasmic agents that kill males (Jiggins et al. 1999). Y chromosome drive is much less common than X chromosome drive, perhaps because it can rapidly cause population extinction (Hamilton 1967), and has only been reported in some mosquitoes (Wood & Newton 1991). A driving X chromosome, $X^d$, is a selfish genetic element that can spread itself at the expense of genes on the Y chromosome and the autosomes. Autosomal genes do not benefit from the presence of $X^d$ because half of the time they occur in Y-bearing sperm, which are destroyed. For this reason, genetic modifiers on autosomes, as well as on Y chromosomes, are favored by selection if they ameliorate or suppress the effects of drive. Because sex chromosome drive can cause population extinction if one sex is lost, either selection against $X^d$, genetic suppression, or both must occur to maintain sex chromosome meiotic drive in a population (Hamilton 1967).

Very soon after we first brought the two sexually dimorphic Southeast Asian species, *C. dalmanni* and *C. whitei*, into the lab we discovered that some males produced strongly female-biased sex ratios and that this sex ratio bias was caused by an X-linked factor. However, it took several more years of experiments to show that the mechanism of sex ratio change is caused by improper elongation of the heads of some sperm, and that other genetic factors on the Y chromosome and on the autosomes act to suppress this X-linked drive (Presgraves et al. 1997). Evidence for Y-linked or autosomal suppression of X chromosome drive has been found in *D. mediopunctata* (Carvalho & Klaczko 1993; Carvalho & Klaczko 1994), *D. subobscura* (Hauschteck-Jungen 1990), *D. affinis* (Voelker 1972), *D. simulans* (Cazemajor et al. 1997), *D. paramelanica* (Stalker 1961), and *D. quinaria* (Jaenike 1999). In some of our crosses we even found sex ratios of males that were carrying $X^d$ to exhibit male bias, that is, up to 75 percent of the progeny of such males were sons. This observation is unique among sex chromosome drive systems, but it potentially explains the sex ratio changes that we observed among the selected lines. Apparently, by increasing eye span we also increased the frequency of genetic modifiers that counter the effect of drive and cause the production of male-biased offspring (Wilkinson et al. 1998b).

These observations led us to propose that sex chromosome meiotic drive could provide a novel mechanism for maintaining genetic variation in fitness that can be exploited by choosy females (Wilkinson et al. 1998b). The basic idea is that sex chromosome meiotic drive will cause the population sex ratio to deviate from 1:1. If the drive polymorphism is maintained by selection (Curtsinger 1980), then the sex ratio will equilibrate and remain constant at some proportion of males below 50 percent. In the two sexually dimorphic *Cyrtodiopsis* species the proportion of males carrying the driving X chromosome in wild populations is 17 to 30 percent (Wilkinson et al. 1998b). Consequently, females that preferentially mate with males that either produce unbiased or male-biased sex ratios will leave more grandchildren (Fisher 1958). The feasibility of this scenario is bolstered by experiments with *D. mediopunctata*, which have shown that sex chromosome meiotic drive can maintain heritable variation in the sex ratio (Varandas et al. 1997). While this proposal has generated interest (Hurst & Pomiankowski 1998), it has also been met with skepticism (Reinhold et al. 1999). However, a recent population genetic model (Lande & Wilkinson 1999) confirms that meiotic drive can enhance the evolution of female preference for a male ornament.

This conclusion comes from comparing the evolution of female choice for ornamental males in the presence and absence of meiotic drive. To keep the model as simple as possible, we assume that a sex-limited male ornament is influenced by an X-linked locus while female choice for ornamented males is determined by an autosomal locus. In the absence of meiotic drive and without natural selection on either the ornament or preference, the introduction of choice and ornament alleles at low frequencies into a population leads to evolution of the male trait, but no correlated change in the female preference. If selection maintains a stable polymorphism for $X^d$ and the male ornament is enhanced by an allele on the nondriving X chromosome, then evolution of the male trait occurs more rapidly and is accompanied by evolution of the female preference due to sex ratio selection. In a population with a female-biased sex ratio, females that mate with nondriving males produce more grandchildren than average because sons contribute a higher proportion of automsomal genes to the next generation than daughters. However, in the absence of recombination on the X chromosome, female choice sexual selection will eventually eliminate $X^d$ from the population. On the other hand, if recombination is rare, that is, occurs at rates comparable to what has been observed among paracentric inversions (Powell 1992), such as are often associated with sex chromosome meiotic drive, then trait and preference can coevolve until a recombination event moves the ornament allele from the nondriving X to $X^d$. At that point, further evolution of the preference ceases and the drive frequency returns to its equilibrium value. If a new mutation influencing eye span arises on the nondriving X, then further evolution of the trait and preference can occur. Thus, meiotic drive can catalyze the evolution of female preferences.

This model makes the prediction that the X chromosome should contribute disproportionately to the expression of the male ornament. Recent evidence indicates that the X chromosome does exert considerable influence on male eye span (Wolfenbarger & Wilkinson 2000). Reciprocal crosses between the lines selected for long and short eye span after thirty-two generations of selection show that the X chromosome explains $25 \pm 6$ percent of the difference between the parental lines in relative male eye span. Measurement of the X chromosome from metaphase chromosome preparations indicates that the X comprises only 7 percent of the genome in males (Wolfenbarger & Wilkinson 2000). Additional work is in progress to determine if autosomal or Y-linked genes also contribute to eye span variation.

A second prediction made by the meiotic drive sexual selection model is that selection should act against the drive chromosome in order for it to persist as a stable polymorphism in the population. We are currently investigating if males or females carrying the drive chromosome differ in fertility or survival in comparison to individuals that carry the nondriving chromosome. We have preliminary evidence that $X^dY$ males that carry the drive chromosome do not fertilize as many eggs as XY males (Fry & Wilkinson, in preparation). Similar results have been reported for *Drosophila pseudoobscura* (Wu 1983) and *D. quinaria* (Jaenike 1996). By using *C. whitei* males that differ in body color, we have been able to score sperm precedence of females mated on successive days and have found that $X^dY$ males fertilize only about 5 percent of female eggs, independent of the order of mating. In previous work we found that the precedence of sperm from the second male to mate with a female, that is, P2, averages 0.5 as long as males mate with females after an hour or more (Lorch et al. 1993). Thus, these results suggest that either $X^d$-carrying sperm are competitively inferior to X- and Y-carrying sperm, or that females somehow preferentially use X- and Y-bearing sperm, rather than $X^d$-bearing sperm, for fertilization. We are in the process of designing studies to investigate these fascinating alternatives.

# Evolution of Exaggerated Eye Stalks

A resurgence of interest in the pattern (Harvey & Pagel 1991), rather than the process, of evolution has, I believe, led behavioral ecologists to more closely examine the role of evolutionary history in explaining species-level variation than has been characteristic of the field over the past twenty years. With regard to sexual selection, a number of investigators (Burley 1985; Endler & McClellan 1988; Ryan et al. 1990) have suggested that female preferences may arise through preexisting biases that influence behaviors in other contexts. If males evolve traits to exploit sensory biases of females, then the evolution of female preferences should be decoupled from, rather than coevolve with, the evolution of male ornaments. Thus, if one reconstructs the evolutionary history of a preference and an ornament in a group of organisms, the sensory exploitation expectation is that the female preference should have evolved earlier in the lineage than the male ornament (Basolo 1990, 1995).

A somewhat different idea, termed chase-away sexual selection (Holland & Rice 1998), has recently been proposed and yields a similar prediction. In a number of species (Partridge & Hurst 1998) evidence now indicates that conflict occurs between the sexes over many aspects of the mating process. For example, in some cases the process of mating can be detrimental to female survival due to the transfer of male proteins (Chapman et al. 1995; Holland & Rice 1999), which presumably are produced to augment competition with other males' sperm. Thus, females might be expected to evolve resistance to male courtship, which should then favor the evolution of more elaborate male display in order to overcome female resistance. Female preferences must, therefore, precede the evolution of male ornaments for this process to occur.

We have considered these non-coevolutionary alternatives for eye span exaggeration by developing a phylogenetic hypothesis for thirty-three species of diopsid flies using three mitochondrial and three nuclear genes (Baker et al. 2000) and then examining the behavior of a few closely related species. By giving females two males with different eye spans and observing which males mate, we measured female preference as the proportion of matings that occur with the male exhibiting the longer eye span. This work shows that a female preference for long eye stalks does not exist in the sexually monomorphic Southeast Asian species, *Cyrtodiopsis quinqueguttata* or *Teleopsis quadriguttata*, which lie basal to the two sexually dimorphic *Cyrtodiopsis* species (Wilkinson et al. 1998a). Thus, the phylogenetic evidence is consistent with the artificial selection studies in supporting coevolution between the male trait and female preference.

Our work still begs the question of which came first—the ornament or the preference. For now, this question remains unanswered. However, the absence of a preference as well as no evidence for sex chromosome meiotic drive in the sexually monomorphic congener (Wilkinson et al. 1998b), is consistent with the meiotic drive system arising prior to the evolution of female preferences. We have also found evidence for sex chromosome meiotic drive in an African species with sexually dimorphic eye span in another genus (Lande & Wilkinson 1999). Thus, the tantalizing possibility that eye span exaggeration indicates sex chromosome behavior in other lineages of flies cannot be ruled out. For meiotic drive to be more than a curiosity, though, one would expect to find cases of sex-linked inheritance of sexually dimorphic traits in other groups of organisms. Interestingly, such a pattern has recently been reported (Reinhold 1998). Furthermore, a number of poeciliid fish that exhibit female choice for male ornaments or body size also display sex-linked inheritance of these traits, that is, guppies, *Poecilia reticulata* (Houde 1992); platyfish, *Xiphophorus maculatus* (Kallman & Borkoski 1977); sailfin mollies, *X. latipinna* (Ptacek & Travis 1997; Travis 1994); and pygmy swordtails, *X. nigrensis* (Zimmerer & Kallman 1989). In at least one case, the guppy, there is also evidence that sex chromosome meiotic drive occurs (Farr 1981).

Unfortunately, obtaining evidence for the presence or absence of sex chromosome meiotic drive can be complicated by the existence of genetic suppressors. *Drosophila simulans* provides an instructive example. Sex chromosome meiotic drive was recently discovered (Mercot et al. 1995) in this cosmopolitan, well-studied species when flies from Tunisia were crossed with flies from the Seychelles. The F1 males produced highly distorted sex ratios because they lacked the appropriate genetic modifiers that suppress drive. Subsequent study has revealed that $X^d$ chromosomes occur at frequencies up to 60 percent in populations throughout the world (Atlan et al. 1997). Thus, considerable work is required to test for the presence of sex chromosome meiotic drive in other species.

Before concluding, I must acknowledge that my focus on female choice has ignored the very important role eye stalks likely play in resolving contests among diopsids, as well as among other flies with head projections (Grimaldi & Fenster 1989; Wilkinson & Dodson 1997). Examination of male fighting among *Cyrtodiopsis* species has shown that sexually dimorphic flies use eye span to resolve contests while sexually monormophic species do not (Panhuis & Wilkinson 1999). Because eye span correlates highly with body size and, therefore, provides an honest indicator of larval condition (David et al. 1998; Wilkinson & Taper 1999), males, as well as females, should be expected to use eye span for assessment. We have found strong evidence indicating that eye stalks play important roles for both male contest resolution and female mate choice, as expected if eye stalks function as both ornaments and assessment signals (Berglund et al. 1996). However, additional work is needed to determine if both mechanisms of sexual selection are involved in the evolution of sexual dimorphism in other lineages in this family as well as in other families of flies in which males exhibit eye stalks (Wilkinson & Dodson 1997).

## Conclusions

I believe my research on stalk-eyed flies illustrates some of the benefits associated with trying to test key assumptions or predictions from currently controversial theories using an appropriate animal model. After some perseverance, we discovered that stalk-eyed flies are easier to rear and study in the lab than almost any other appropriate vertebrate or invertebrate species, with the notable exception of some *Drosophila* species. While sophisticated genetic experiments are not easily conducted with stalk-eyed flies, they are arguably better for studying sexual selection than most *Drosophila* because the male ornament and female behavior are easy to quantify, can be changed readily by selection in the lab and field, and exhibit extraordinary variation among related species.

On the other hand, our discovery of sex chromosome meiotic drive and its relationship to sexual selection provides a valuable lesson for how scientific advances are made. While we now have theoretical confirmation that biased sex chromosome transmission can influence sexual selection, no one conceived of this possibility before we initiated our work. While serendipity may be unpredictable, it is possible to remain open to alternative interpretations as well as refrain from becoming too invested in any single hypothesis under study. One of the best pieces of advice my own advisor gave me was never to ignore "cheap" data. Such philosophy encourages data collection that may seem tangential to the purpose of a current study, but could inform future work.

Testing model assumptions and predictions is a necessary part of science. However, most major advances in behavioral ecology, as well as in other fields of science, have not been made this way. Rather, discovery of unexpected behavioral patterns or associations, such as that between haplodiploidy and reproductive altruism, has led to the development of an explanation for those patterns, that is, kin selection (Hamilton 1964). Thus, future advances in the field will depend on discovering behavioral patterns that fail to conform to conventional theories. My best advice, therefore, is to be alert for the unexpected. Such discoveries may well turn out to be more important than any preconceived ideas under examination.

## Acknowledgments

My work on stalk-eyed flies would not have been possible without the continued financial support of the National Science Foundation. I am partic-

ularly grateful for the efforts of my past and current research assistants—
Sharvari Bhatt, Lili Crymes, Alicia Kawecki, and Mara Sanchez, graduate
students—Pat Lorch, Daven Presgraves, Ahmad Hariri, Catherine Fry, and
Sarah Toll, postdoctoral associates—Paul Reillo, Marion Kotrba, LaReesa
Wolfenbarger, and John Swallow, and over twenty undergraduate students
who have contributed to this work.

# References

Andersson M, 1994. Sexual selection. Princeton: Princeton University Press.

Atlan A, Mercot H, Landre C, Montchamp-Moreau C, 1997. The sex-ratio trait in *Drosophila simulans*: geographical distribution of distortion and resistance. Evol 51:1886–1895.

Baker RH, Wilkinson GS, in press. Phylogenetic analysis of eye stalk allometry and sexual dimorphism in stalk-eyed flies (Diosidae). Evol, in press.

Baker RH, Wilkinson GS, DeSalle R, 2001. The phylogenetic utility of different types of molecular data used to infer evolutionary relationships among stalk-eyed flies (Diopsidae). Syst Biol 50:1–20.

Bakker TCM, 1993. Positive genetic correlation between female preference and preferred male ornament in sticklebacks. Nature 363:255–257.

Basolo AL, 1990. Female preference predates the evolution of the sword in swordtail fish. Science 250:808–810.

Basolo AL, 1995. Phylogenetic evidence for the role of a pre-existing bias in sexual selection. Proc R Soc Lond B Biol Sci 259:307–311.

Berglund A, Bisazza A, Pilastro A, 1996. Armaments and ornaments: an evolutionary explanation of traits of dual utility. Biol J Linn Soc 58:385–399.

Boake CRB, 1986. A method for testing adaptive hypotheses of mate choice. Am Nat 127:654–666.

Botella LM, Moya A, Gonzalez MC, Mensua JL, 1985. Larval stop, delayed development and survival in overcrowded cultures of *Drosophila melanogaser*: effect of urea and uric acid. J Ins Physiol 31:179–185.

Burkhardt D, de la Motte I, 1983. How stalk-eyed flies eye stalk-eyed flies: observations and measurements of the eyes of *Cyrtodiopsis whitei* (Diopsidae, Diptera). J Comp Physiol 151:407–421.

Burkhardt D, de la Motte I, 1985. Selective pressures, variability, and sexual dimorphism in stalk-eyed flies (Diopsidae). Naturwissenschaften 72:204–206.

Burkhardt D, de la Motte I, 1988. Big 'antlers' are favoured: female choice in stalk-eyed flies (Diptera, Insecta), field collected harems and laboratory experiments. J Comp Physiol A 162:649–652.

Burley N, 1985. The organization of behavior and the evolution of sexually selected traits. Ornith Monogr 37:22–44.

Buschbeck EK, Hoy RR, 1998. Visual system of the stalk-eyed fly, *Cyrtodiopsis quinqueguttata* (Diopsidae, Diptera): an anatomical investigation of unusual eyes. J Neurobiol 37:449–468.

Carvalho AB, Klaczko LB, 1993. Autosomal suppressors of *sex-ratio* in *Drosophila mediopunctata*. Heredity 71:546–551.

Carvalho AB, Klaczko LB, 1994. Y-linked suppressors of the *sex-ratio* trait in *Drosophila mediopunctatat*. Heredity 73:573–579.

Cazemajor M, Landre C, Montchamp-Moreau C, 1997. The sex-ratio trait in *Drosophila simulans*: genetic analysis of distortion and suppression. Genetics 147:635–642.

Chapman T, Liddle LF, Kalb JM, Wolfner MF, Partridge L, 1995. Cost of mating in *Drosophila melanogaster* females is mediated by male accessory gland products. Nature 373:241–244.

Curtsinger JWaF, MW, 1980. Experimental and theoretical analysis of the "sex-ratio" polymorphism in *Drosophila pseudoobscura*. Genetics 94:445–466.

David P, Hingle A, Greig D, Rutherford A, Pomiankowski A, Fowler K, 1998. Male sexual ornament size but not asymmetry reflects condition in stalk-eyed flies. Proc R Soc Lond B 265:2211–2216.

Endler JA, McClellan T, 1988. The processes of evolution: toward a newer synthesis. Ann Rev Ecol Syst 19:395–421.

Ewing AW, 1961. Body size and courtship behaviour in *Drosophila melanogaster*. Anim Behav 9:93–99.

Falconer DS, 1981. Introduction to quantitative genetics. New York: Longman.

Farr JA, 1981. Biased sex ratios in laboratory strains of guppies, *Poecilia reticulata*. Heredity 47:237–248.

Fisher RA, 1958. The genetical theory of natural selection. New York: Dover.

Gould SJ, 1989. A developmental constraint in *Cerion*, with comments on the definition and interpretation of constraint in evolution. Evol 43:516–539.

Grimaldi D, Fenster G, 1989. Evolution of extreme sexual dimorphisms: structural and behavioral convergence among broad-headed Drosophilidae (Diptera). Am Mus Nov 2939:1–25.

Hamilton WD, 1964. The genetical evolution of social behavior. J Theor Biol 7:1–51.

Hamilton WD, 1967. Extraordinary sex ratios. Science 156:477–488.

Harvey PH, Pagel MD, 1991. The comparative method in evolutionary biology. Oxford: Oxford University Press.

Hasselquist D, Bensch S, von Schantz T, 1996. Correlation between male song repertoire, extra-pair paternity and offspring survival in the great reed warbler. Nature 381:229–232.

Hauschteck-Jungen E, 1990. Postmating reproductive isolation and modification of the sex-ratio trait in *Drosophila subobscura* induced by the sex chromosome gene arrangement $A_{2+3+5+7}$. Genetica 83:31–44.

Holland B, Rice WR, 1998. Chase-away sexual selection: antagonistic seduction versus resistance. Evol 52:1–7.

Holland B, Rice WR, 1999. Experimental removal of sexual selection reverses intersexual antagonistic coevolution and removes a reproductive load. Proc Natl Acad Sci USA 96:5083–5088.

Houde AE, 1992. Sex-linked heritability of a sexually-selected character in a natural population of guppies, *Poecilia reticulata* (Pisces: Poeciliidae). Heredity 69:229–235.

Houde AE, 1994. Effect of artificial selection on male colour patterns on mating preference of female guppies. Proc R Soc Lond B 256:125–130.

Hurst LD, Atlan A, Bengtsson BO, 1996. Genetic conflicts. Q Rev Biol 71:317–364.

Hurst LD, Pomiankowski A, 1998. The eyes have it. Nature 391:223–224.

Iwasa Y, Pomiankowski A, Nee S, 1991. The evolution of costly mate preferences. II. The "handicap" principle. Evol 45:1431–1442.

Jaenike J, 1996. Sex-ratio meiotic drive in the *Drosophila quinaria* group. Am Nat 148:237–254.

Jaenike J, 1999. "Sex-ratio" meiotic drive and the maintenance of Y chromosome polymorphism in *Drosophila*. Evol 53:164–174.

Jiggins FM, Hurst GDD, Majerus MEN, 1999. How common are meiotically driven sex chromosomes? Am Nat 154:481–483.

Kallman KD, Borkoski V, 1977. A sex-linked gene controlling the onset of sexual maturity in female and male platyfish (*Xiphophorus maculatus*), fecundity in females and adult size in males. Genetics 898:79–119.

Kirkpatrick M, 1982. Sexual selection and the evolution of female choice. Evol 36:1–12.

Klein TW, 1974. Heritability and genetic correlation: statistical power, population comparisons, and sample size. Behav Genet 4:171–189.

Lande R, 1981. Models of speciation by sexual selection on polygenic traits. Proc Natl Acad Sci USA 78:3721–3725.

Lande R, Wilkinson GS, 1999. Models of sex-ratio meiotic drive and sexual selection in stalk-eyed flies. Genet Res 74:245–253.

Lorch P, Wilkinson GS, Reillo PR, 1993. Copulation duration and sperm precedence in the Malaysian stalk-eyed fly, *Cyrtodiopsis whitei* (Diptera: Diopsidae). Behav Ecol Sociobiol 32:303–311.

Lyttle TW, 1991. Segregation distorters. Ann Rev Genetics 25:511–557.

Mercot H, Atlan A, Jacques M, Montchamp-Moreau C, 1995. Sex-ratio distortion in *Drosophila simulans*: co-occurrence of a meiotic drive and a suppressor of drive. J Evol Biol 8:283–300.

Møller AP, 1994. Male ornament size as a reliable cue to enhanced offspring viability in the barn swallow. Proc Natl Acad Sci USA 91:6929–6932.

Norris K, 1993. Heritable variation in a plumage indicator of viability in male great tits, *Parus major*. Nature 362:537–539.

O'Donald P, 1962. The theory of sexual selection. Heredity 17:541–552.

O'Donald P, 1980. Genetic models of sexual selection. New York: Cambridge University Press.

Panhuis TM, Wilkinson GS, 1999. Exaggerated male eye span influences contest outcome in stalk-eyed flies. Behav Ecol Sociobiol 46:221–227.

Partridge L, Ewing A, Chandler A, 1987. Male size and mating success of *Drosophila melanogaster*: the roles of male and female behaviour. Anim Behav 35:555–562.

Partridge L, Farquar M, 1983. Lifetime mating success of male fruitflies (*Drosophila melanogaster*) is related to their size. Anim Behav 31:871–877.

Partridge L, Hurst LD, 1998. Sex and conflict. Science 281:2003–2008.

Petrie M, 1994. Improved growth and survival of offspring of peacocks with more elaborate trains. Nature 371:598–599.

Pomiankowski A, 1988. The evolution of female mate preferences for male genetic quality. Oxf Surv Evol Biol 5:136–184.

Pomiankowski A, Møller AP, 1995. A resolution of the lek paradox. Proc R Soc Lond B 260:21–29.

Powell JR, 1992. Inversion polymorphisms in *Drosophila pseudoobscura* and *Drosophila persimilis*. In: Drosophila inversion polymorphisms (Krimbas CB, Powell JR, eds). Boca Raton, FL: CRC Press; 73–126.

Presgraves DC, Severence E, Wilkinson GS, 1997. Sex chromosome meiotic drive in stalk-eyed flies. Genetics 147:1169–1180.

Ptacek MB, Travis J, 1997. Mate choice in the sailfin molly, *Poecilia latipinna*. Evol 51:1217–1231.

Reinhold K, 1998. Sex linkage among genes controlling sexually selected traits. Behav Ecol Sociobiol 44:1–7.

Reinhold K, Engqvist L, Misof B, Kurtz J, 1999. Meiotic drive and the evolution of female choice. Proc R Soc Lond B 226:1341–1346.

Rowe L, Houle D, 1996. The lek paradox and the capture of genetic variance by condition-dependent traits. Proc R Soc Lond B 263:1415–1421.

Ryan MJ, Fox JH, Wilczynski W, Rand AS, 1990. Sexual selection by sensory exploitation in the frog *Physalaemus pustulosus*. Nature 343:66–67.

Seitz G, Burkhardt D, 1974. Bau und optische Leistungen des Komplexauges der Stielaugenfliege *Cyrtodiopsis dalmanni* Wiedemann. J Comp Physiol 95:49–59.

Shaw RG, 1991. The comparison of quantitative genetic parameters between populations. Evol 45:143–151.

Sheldon BC, Merilä J, Qvarnström A, Gustafsson L, Ellegren H, 1997. Paternal genetic contribution to offspring condition predicted by size of male secondary sexual character. Proc R Soc Lond B 264:297–302.

Stalker HD, 1961. The genetic systems modifying meiotic drive in *Drosophila paramelanica*. Genetics 46:177–202.

Tan KB, 1967. The life-histories and behaviour of some Malayan stalk-eyed flies (Diptera, Diopsidae). Malay Nat J 20:31–38.

Travis J, 1994. Size dependent behavioral variation and its genetic control within and among populations. In: Quantitative genetic studies of the evolution of behavior (Boake CRB, ed). Chicago: Chicago University Press; 165–187.

Varandas FR, Sampaio MC, Carvalho AB, 1997. Heritability of sexual proportion in experimental sex-ratio populations of *Drosophila mediopunctata*. Heredity 79:104–112.

Voelker R, 1972. Preliminary characterization of "sex ratio" and rediscovery and interpretation of "male sex ratio" in *Drosophila affinis*. Genetics 71:597–606.

Wilkinson GS, 1984. Reciprocal food sharing in vampire bats. Nature 309:181–184.

Wilkinson GS, 1985. The social organization of the common vampire bat. II. Mating system, genetic structure, and relatedness. Behav Ecol Sociobiol 17:123–134.

Wilkinson GS, 1987. Equilibrium analysis of sexual selection in *Drosophila melanogaster*. Evol 41:11–21.

Wilkinson GS, 1988. Reciprocal altruism in bats and other mammals. Ethol Sociobiol 9:85–100.

Wilkinson GS, 1993. Artificial sexual selection alters allometry in the stalk–eyed fly *Cyrtodiopsis dalmanni* (Diptera: Diopsidae). Genet Res 62:213–222.

Wilkinson GS, Dodson G, 1997. Function and evolution of antlers and eye stalks in flies. In: The evolution of mating systems in insects and arachnids (Choe J, Crespi B, eds). Cambridge: Cambridge University Press; 310–328.

Wilkinson GS, Kahler H, Baker RH, 1998a. Evolution of female mating preferences in stalk-eyed flies. Beh Ecol 9:525–533.

Wilkinson GS, Presgraves DC, Crymes L, 1998b. Male eye span in stalk-eyed flies indicates genetic quality by meiotic drive suppresion. Nature 391:276–278.

Wilkinson GS, Reillo PR, 1994. Female preference response to artificial selection on an exaggerated male trait in a stalk-eyed fly. Proc R Soc Lond B 255:1–6.

Wilkinson GS, Taper M, 1999. Evolution of genetic variation for condition dependent traits in stalk-eyed flies. Proc R Soc Lond B 266:1685–1690.

Wolfenbarger LL, Wilkinson GS, 2001. Sex-linked expression of a sexually selected trait in the stalk-eyed fly, *Cyrtodiopsis dalmanni*. Evol 55:103–110.

Wood RJ, Newton ME, 1991. Sex-ratio distortion caused by meiotic drive in mosquitoes. Am Nat 137:379–391.

Wu C-I, 1983. Virility deficiency and the sex-ratio trait in *Drosophila pseudoobscura*. I. Sperm displacement and sexual selection. Genetics 105:651–662.

Zimmerer EJ, Kallman KD, 1989. Genetic basis for alternative reproductive tactics in the pygmy swordtail, *Xiphophorus nigrensis*. Evol 43:1298–1307.

# 6 The Behavioral Ecology of Stridulatory Communication in Leaf-Cutting Ants

Bert Hölldobler and Flavio Roces

We became interested in the life and natural history of ants very early in our careers. In fact, with Bert it started in his boyhood, when he was about seven years old. During one of the nature walks he used to do with his father, he turned over a lime stone rock that sheltered a colony of the carpenter ants *Camponotus ligniperdus*. What a magnificent spectacle! The large queen with her fat, shiny blackish-brown abdomen, the frantically rushing workers, grabbing the white grublike larvae and carrying them down the subterranean channels of the nest. A whole society had revealed itself for an instant, then trickled away into the lower regions of its elaborate nest in the soil. From that moment on BH was interested in ants. He cultured colonies in his parents' home, observed their behavior, took notes, drew the ants, and tried to learn as much about the local species as was possible. His first teacher in myrmecology was his father, a physician and ardent zoologist, who had a particular interest in ants and especially in the many other arthropods that live in ant societies. At that time Karl Hölldobler was an internationally known authority on these social parasites, called myrmecophiles.

BH's fascination with the diversity of insect societies continued to flourish when he entered university and studied biology and chemistry. As a graduate and postdoctoral student he studied several physiological and ecological aspects of social insect behavior. In fact, ants turned out to be ideal model systems for the study of communication mechanisms and the workings of insect societies. Thus, BH had the good fortune to never get out of his "bug period" (Hölldobler & Wilson 1994).

When BH was a professor at Harvard University, in the mid-1980s, he received letters from an Argentinean student, Flavio Roces, who was interested, among other topics, in the analysis of the social thermoregulation and circadian rhythms in carpenter ants. He asked for advice and explored the possibilities of joining BH's research group at Harvard. FR was a student of Josué Núñez at the University of Buenos Aires in Argentina, and as is typical for Núñez and some of his bright students, their approach to solving scientific questions often did not follow the conventional routes. FR, for example, also worked at the University of Buenos Aires on the foraging behavior of the leaf-cutting ant *Acromyrmex*. His findings in part challenged the general view of optimization of foraging patterns in social insects. In 1993 he finally joined BH's newly established research group at the University of Würzburg. At that time we were in the process of developing a major interdisciplinary research program jointly with botanists and ecologists, to study the impact of the leaf-cutting ants' herbivory on the competitive interactions of plants in the canopy region of neotropical forests. FR's interests in the analysis of the energetic costs and social organization of foraging matched excellently with Bert's interest in the analysis of communication signals and the function of their multimodal composition. A most fruitful collaboration on the study of several aspects of leaf-cutting ant foraging commenced and continues to this day. A facet of this joint work will be presented in this chapter.

The leaf-cutting ants of the genera *Atta* and *Acromyrmex* are among the dominant herbivores of the neotropics, consuming far more vegetation than any other group of animals. They are unique in the animal kingdom in their ability to utilize freshly cut leaves and other vegetation for the rearing of fungus. Colonies of these leaf-cutting ants subsist entirely on the symbiotic fungus and on the sap of plants (Littledyke & Cherett 1976).

In spite of the problems the leaf-cutters can cause for agriculture, it would be a mistake to think of them solely as pest insects. During millions of years of evolution, they have become an integral part of the ecosystems of the New World tropics. They prune the vegetation, stimulate new plant growth, break down vegetable material, and turn and enrich the soil.

Leaf-cutting ants are among the most advanced of all social insects. A mature colony consists of a single, large queen, which is the only reproductive, and hundreds of thousands of workers. The collective worker caste is composed of an array of physical subcastes by combining extreme size variation with moderate allometry. In *Atta sexdens*, for example, the head width varies eightfold and the dry weight two hundred-fold from the smallest minors to the huge majors (Wilson 1980a, b). This complex caste system, which gives rise to an elaborate organization of division of labor, represents an adaptation for collecting and processing of fresh vegetation. Fresh leaves require a whole series of special operations before they can be converted into fungal substrate. They must first be harvested from the tree and carried to the nest, then chopped into fine pieces, next chewed and treated with enzymes, and finally incorporated into the garden comb. In addition, the fungus must be provided with constant care after it sprouts on a substratum. The *Atta* workers organize these many activities of foraging and gardening in the form of an assembly line, in which different physical subcastes perform certain tasks especially well (Wilson 1980a, b). All these features make the leaf-cutting ants a model system for the study of the evolution of caste and of the adaptive nature of colony demography (Wilson 1985).

Such elaborate modes of division of labor and the organization of the highly effective foraging columns that move along established routes from the nest to harvesting sites in the tree canopy 200 to 300 meters distant depend on a sophisticated communication system that involves a combination of chemical and mechanical signals. Only recently have researchers paid renewed attention to such multimodal communication in animals (Hauser 1996; Hölldobler 1995, 1999; Partan 1998), and the leaf-cutting ant genus *Atta* became for us an ideal model system for the experimental analysis of the design and ecological significance of such cross-modal signals. This will be the topic of our chapter.

## Long-Distance Foraging Trails

Foraging by a social insect colony is a complex process that results from individual activities and collective decisions emerging from the interactions among colony members. The success of social foraging largely rests on the ability of a colony to allocate foragers quickly to the exploitation of newly discovered food sources. Through recruitment communication, successful foragers inform their nestmates about the existence of a food source and orient them to its location. As a consequence, the temporal development of the recruitment process might be of fundamental importance, since in a competitive environment, colonies should try to monopolize as quickly as possible the discovered resources (Hölldobler & Wilson 1990).

Social communication in ants is mediated primarily by chemical signals (Hölldobler & Wilson 1990; Billen & Morgan 1997). The principal recruitment and trail pheromones in leaf-cutter ants originate from the poison gland, and, depending on the species, their most effective compound is either methyl 4-methylpyrrole-2-carboxylate or 3-ethyl-2,5-dimethylpyrazine (Morgan 1984). In general, only minute amounts of the pheromone are needed at a given time to elicit a highly effective trail-following behavior in the ants. For example, Tumlinson et al. (1971), the discoverers of methyl-4-methylpyrrole-2-carboxylate as the trail substance of *Atta texana*, estimated that one milligram of this substance (roughly the quantity in a single colony), if laid out with maximum efficiency, would be enough to lead a column of ants three times around the earth. The chief disadvantage of such a chemical system is the slowness of fade-out. When using pheromones alone, ants cannot transmit a rapid sequence of signals in the manner of vocalizations or quickly changing visual signals. In order to replace signals, they must wait until the active space of the pheromone expands to maximum diameter, then shrinks back to the point of emission (Wilson & Bossert 1969). However, in many cases this property has been turned to the advantage of the insects. A long-standing active space is needed, for example, in the employment of colony odor, alarm pheromones, or trail substances. In leaf-cutter ants the slowly fading trail pheromones are deposited along the trunk routes that lead to the trees where leaf fragments are being harvested by the ants, and also on the branches and twigs on the tree that are frequented by the foraging ants. Thus, the foraging leaf-cutters continuously perceive the chemical trail signal, and any additional signal that mediates, for example, short-range recruitment at the foraging site, would be most effective if transmitted through a different sensory channel. In the following section we will describe such a superposition of a mechanical signal upon slowly fading chemical signals.

## Short-Range Recruitment

Most harvesting by leaf-cutting ants occurs in the canopy of the trees. Here one can often observe collectives of ants cut fragments out of particular leaves until nothing is left except a few leaf veins, while other leaves nearby remain almost untouched (Fig. 6.1). We hypothesized that those leaves intensely frequented by the foragers are more desirable, perhaps because they are more tender or less loaded with secondary plant compounds than other leaves. We also assumed that the ants are able to summon fellow foragers to these quality leaves by employing special short-range recruitment signals. The experimental tests of these hypotheses were only possible in the laboratory, where we cultured colonies of *Atta cephalotes* in artificial nests and offered them a variety of leaves in the foraging arena. Here we noticed that a number of workers cutting a leaf fragment were raising and lowering their gasters, a motion pattern identical with that performed by *Atta* workers when producing stridulatory vibrations (Markl 1968).

These ants possess a stridulatory organ that consists of a cuticular file on the first gastric tergite and a scraper situated on the postpetiole. By rubbing the file against the scraper the ants produce stridulatory vibrations. In order to verify that workers engaged in leaf-cutting actually produced stridulatory signals, noninvasive laser-Doppler-vibrometry was employed (Michelsen & Larsen 1978). This enabled us to record vibrational signals on the leaf surface, whenever a leaf-cutting worker stridulated. Such vibrations were transmitted onto the leaf surface mainly through the ant's mandibles, and propagated as substrate-borne vibrations. They consisted of long series of repetitive pulse trains ("chirps"), each pulse resulting from the impact of the scraper on a ridge of the file. The signal repetition rate varied between two and twenty chirps per second. Typical waveforms recorded from the leaf being cut are presented in Figure 6.2.

However, not all ants stridulated while cutting a fragment out of a leaf. In fact, when ants were offered leaves of different qualities, the proportion of workers that stridulated during cutting differed remarkably. In a typical experimental series we presented privet leaves of two different grades of toughness ("thin" and "thick" leaves), because it is known that leaf toughness affects leaf choice in the field. In addition, in another set both kinds of leaves were first soaked in sugar solution and then dried, because sugarcoat-

Figure 6.1. Harvesting site of leaf-cutting ants of *Atta* sp., showing that certain leaves have been almost entirely removed, except the leaf stalk, while other leaves remained untouched.

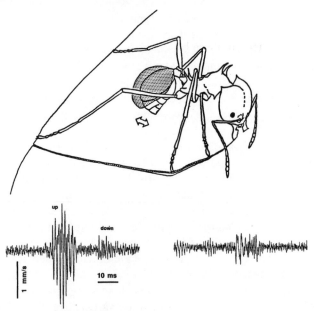

Figure 6.2. Schematic illustration of an *Atta* worker cutting a leaf and simultaneously stridulating by moving the gaster up and down. Stridulation signals produced by an *Atta cephalotes* worker, recorded by laser-Doppler-vibrometry (as velocity of the leaf's vibrations). Measurements performed 20 millimeters away from the cutting site. *Left*: Substrate-borne vibrations transmitted mostly through the mandibles during cutting; *right*: vibrations transmitted onto the substrate through the legs when the mandibles do not touch the leaf (from Roces et al. 1993).

ing enhances the attractiveness of leaves to the ants. While for the thick leaves about 40 percent of the workers stridulated, this value increased significantly up to 70 percent for workers cutting thin leaves. When the quality of the two kinds of leaves was enhanced because of sugarcoating, almost all workers were observed to stridulate while cutting, irrespective of the differences in the mechanical properties of the material being cut. From these observations we concluded that the production of stridulatory vibrations correlates with the qualities of the leaves (Roces et al. 1993).

Next we asked, do ants that stridulate communicate leaf quality to their nearby nestmates? Since vibrations produced at the cutting site propagate along the plant, we postulated that they might play a role in attracting workers to this site, acting as short-range recruitment signals. To test this hypothesis, foraging workers were confronted with a choice between two twigs of privet: a "stridulating" (test) twig and a "silent" (control) one, and the ants' preferences were recorded. From previous work by Markl (1965) we knew that workers of *Atta* species do not respond to airborne components of the stridulation sound, but are highly sensitive to vibrations propagated through the substrate.

Each twig in our experimental setup was connected to a membrane of a loudspeaker that served as a vibrator, and stridulations were played back to the one or the other side (Fig. 6.3). Any potential chemical cues that could affect the choice were excluded. The results were unequivocal: significantly more ants chose the vibrating stem. In control experiments, where both stems were "silent," the distribution of the workers choosing the one or the other did not differ from the ratio 1:1. Interestingly, when we offered the ants the choice between a "silent" stem marked with trail pheromone and an unmarked "stridulating" stem, significantly more ants chose the "silent" (marked) branch. However, when both stems were marked with trail pheromone but only one was vibrating, again the "stridulating" stem was preferred (Fig. 6.4, Roces & Hölldobler, unpublished). From these results we can draw two conclusions. First, the substrate-borne stridulatory vibrations alone (in the absence of recruitment pheromones) can act as short-range recruitment signals. Second, when given a choice more *Atta* foragers respond to recruitment pheromones than to substrate-borne stridulatory vibrations, but the effectiveness to the recruitment pheromone is significantly enhanced when the chemical signal is combined with the vibrational signal. Under natural conditions, nearby workers would respond to the stridulatory vibrations transmitted through the plant material by orienting toward the source of the vibrations and subsequently joining in leaf-cutting.

There is another phenomenon connected with stridulation in leaf-cutting ants, which we have to consider. As indicated, most of the energy of the stridulatory vibrations is transmitted into the substrate through the mandibles. Since workers preferably stridulate when cutting thin and tender leaves, we hypothesized that these stridulatory vibrations might mechanically facilitate the cutting process in the manner of vibratomes (Tautz et al. 1995).

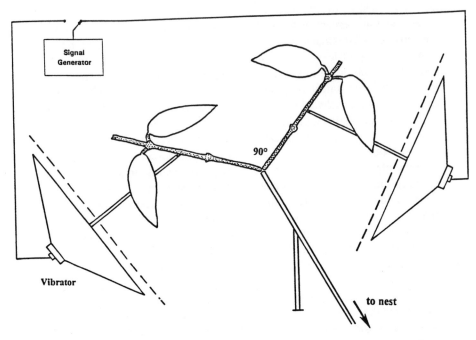

Figure 6.3. Experimental arrangement used during the choice experiments (from Roces et al. 1993).

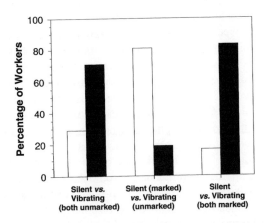

Figure 6.4. Percentage of recruited leaf-cutting ant workers that selected a "vibrating" or a "silent" stem of privet (with or without pheromonal marking) in the choice experiments, using the setup presented in Figure 6.3. Results of three different experimental series are presented, in which a total of 200, 100, and 150 workers were tested, respectively. The stems were marked with a solution of 0.05 ng/ml of the synthetic trail pheromone 4-methylpyrrole-2-carbolylic acid (Sigma) in hexane, producing a pheromone trail with 1 milliliter of the solution over 5 centimeters. In each series, black-and-white bars were statistically different (p < 0.05 after $G_T$ values from log-likelihood G-test, for goodness of fit to the ratio 1:1).

Vibratomes are instruments used in histology to produce very thin sections through soft material. The vibrations of the microtome-knife stiffen the material being cut and thereby facilitate smooth sections. Have the leaf-cutter ants in the course of evolution invented the same technique? To answer this question we conducted a detailed study of the cutting behavior in leaf-cutting ants. In *Atta cephalotes*, the two mandibles play different roles during the cutting process. While one mandible actively moves, the other remains almost fixed (cutting mandible). The steps in one bite are as follows (Fig. 6.5): the motile mandible is opened (abducted) and anchored with its tip to the leaf tissue. The cutting mandible is not abducted, but held steady. During the opening of the motile mandible, the cutting mandible is pushed against the leaf by lateral head movements. Next the motile mandible is closed (abducted), pulling the cutting mandible farther against the leaf, which increases the incision. In this phase also the abducting mandible moves deeper into the leaf surface, thus preparing the way for the cutting mandible. As soon as both mandibles meet, the cycle starts again by abduction of the motile mandible.

We analyzed the temporal relation between mandible movements and stridulation by videotaping the cutting behavior and simultaneously recording the vibrational signals from the leaf surface with the aid of laser vibrometry. As it turned out, the ant did not stridulate continuously. Stridulation occurred most often when the mandibles were closing, that is, when the cutting mandible was moved through the plant tissue (Fig. 6.5). Further measurements revealed that this stridulation generated complex vibrations of the mandibles. The high vibrational acceleration of the mandible (up to three times the gravitational force at peak acceleration at about 1,000 hertz), appears to stiffen the material to be cut.

Next we detached mandibles from dead ants, mounted them on a force meter, and pushed them slowly against the edge of a leaf. We compared the force fluctuations caused when cutting with vibrating and nonvibrating mandibles, and found that the vibrating mandible clearly reduced the force fluctuations, thus facilitating a smoother cut through tender leaf tissue (Tautz et al. 1995).

## Evolution of Stridulatory Communication

These findings led us to an intriguing hypothesis with which we tried to explain the evolution of stridulation as a short-range recruitment signal. We postulated that the original function of stridulation was, and still is, to serve as a mechanical aid during cutting of soft plant tissue. Indeed, the vibratome effect caused by stridulation is only noticeable when tender leaves are being cut. But, as we have discussed above, tender leaves are also the most desirable vegetable material for the ants. We therefore reasoned that during the course of evolution, stridulation as a mechanical aid has acquired a second

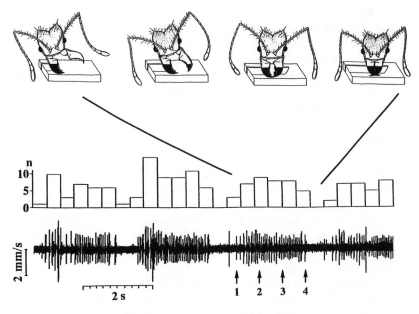

Figure 6.5. *Above*: Mandible and head movements during one cut into a tender leaf. *Below*: Stridulation during four bites. The histogram shows the number of chirps counted at a bin width of 400 ms. The trace underneath depicts an original laser vibrometry of stridulation on a leaf measured 2 centimeters away from the head of the ant. The arrows denote the temporal occurrence of the four cutting stages shown above (modified from Tautz et al. 1995).

ary function as a communication signal. Such an evolutionary process where a particular morphological, physiological, or behavioral trait attains a signal function in a communication context is called ritualization. In other words, ants stridulated when harvesting tender leaf fragments in order to facilitate the cutting. The stridulatory vibrations transmitted through the substrate were perceived by other workers in the close vicinity and attracted them to the desirable leaf. Thus stridulatory vibrations were ritualized to serve as short-range recruitment signals.

Yet, there is one observation that did not support this scenario. As we already reported, increasing the sugar contents of the leaves without changing their physical traits significantly increased the occurrence of stridulation in foraging workers. We therefore conducted a series of additional experiments designed specifically to test two competing hypotheses: (1) foragers stridulate to support their cutting behavior; (2) stridulation is employed as a short-range recruitment signal, and the mechanical facilitation is only an epiphenomenon caused by the use of stridulation as a communication signal.

One line of testing the alternative hypotheses would keep invariant the physical traits of the leaves, and manipulate foraging motivation, in order to change the thresholds at which recruitment communication is initiated. If stridulation is primarily used as a recruitment signal irrespective of the mechanical facilitation during cutting, it would be expected that the occurrence of stridulation by workers directly depends on foraging motivation, that is, it would be modulated by leaf nutritional contents, colony starvation, and so on, as previously observed for chemical recruitment signals (Roces & Núñez 1993).

Workers were presented with tender leaves of constant physical traits, and the production of stridulation was recorded in the following situations: first, when the leaves were smeared with chemicals in order to reduce their palatability; second, when workers were deprived of cutting leaves for different periods; finally, when workers were confronted with unfamiliar leaves after a period of feeding with familiar leaves. We know from previous work that unfamiliar, palatable leaves are readily preferred by leaf-cutting ants, and this "novelty effect" significantly increases the intensity of chemical recruitment (Roces & Hölldobler 1994).

Three lines of independent evidence supported the hypothesis that leaf-cutting ant workers stridulate during cutting in order to recruit nestmates, and that the observed mechanical facilitation of cutting represents an epiphenomenon of recruitment communication. First, it was observed that the percentage of *Atta cephalotes* workers that stridulated while cutting tender leaves coated with tannic acid solutions strongly depended on the concentration of tannic acid (Fig. 6.6). In other words, by keeping constant the physical features of the leaves, and by reducing their palatability, it was shown that the number of stridulating workers decreases with decreasing palatability of the harvested leaves, irrespective of the leaf's physical features. This negative effect of leaf palatability on the production of stridulation is inversely comparable to the positive effect on stridulation caused by sugarcoating (Fig. 6.6, inset).

Second, after intense feeding, no workers were observed to stridulate while cutting tender leaves, and the percentage of stridulating workers increased with the deprivation time. Under conditions of "harvesting-satiation," when *Atta cephalotes* colonies were intensively fed prior to the experiments, no workers at all were observed to stridulate when cutting tender privet leaves. The percentage of stridulating workers increased with harvesting deprivation, reaching a maximum of 100 percent five days after the last feeding episode.

Third, whatever kind of leaves used to feed the colony until satiation, a significantly higher percentage of workers stridulated when cutting unfamiliar leaves of similar physical features. This "novelty" effect of a food source makes evident the production of stridulation as a short-range recruitment signal, independent of the mechanical features of the leaves (Roces & Hölldobler 1996).

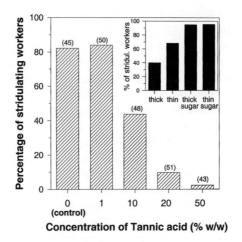

Figure 6.6. Percentage of *Atta cephalotes* workers that stridulated while cutting tender leaves of *Ligustum vulgaris* previously coated with different concentrations on an aqueous solution of tannic acid. *Inset*: percentage of *A. cephalotes* workers that stridulated while cutting leaves of different toughness, and the same kind of leaves coated with 20 percent sugar solution (from Roces et al. 1993; Roces & Hölldobler 1996).

In trying to outline the scenario in which the use of stridulation evolved, the initial hypothesis that stridulation in leaf-cutting ants first evolved as a mechanical aid during cutting and further as a communication signal was particularly tempting. Like many evolutionary hypotheses, this one was based on a plausible correlation rather than experimental proof, that is, the mechanical facilitation is only observed when stridulating ants cut tender leaves, and more nestmates are recruited when foraging workers cut tender leaves. If it is assumed that the use of stridulation as a mechanical aid has been favored during evolution, it should be expected that stridulation reduces time- and/or energy-costs of cutting, especially since leaf-cutting is a highly energetically expensive behavior. Although it was demonstrated that the mandibular vibrations generated by stridulation reduce force fluctuations during the cut of tender leaf tissue, there was no evidence that the total force employed is reduced, or that stridulating ants sever the leaf tissue faster than nonstridulating workers (Tautz et al. 1995). In addition, investigations on the metabolism of leaf-cutting ants demonstrated that there are no differences in metabolic expenditure between stridulating and nonstridulating ants while cutting tender leaves (Roces & Hölldobler 1996). It is unlikely that any energetic improvement (due to stridulation) on cutting mechanics would significantly reduce the impressive costs of cutting a leaf fragment. The metabolic rate of the mandibular muscles measured during leaf-cutting is extremely high (leaf-cutting costs are 31 times higher than resting costs), and comes close to that for the insect flight muscle, the metabolically most costly tissue known (Roces & Lighton 1995). Based on this experimental evidence we reject our original hypothesis that stridulatory signals employed in short-range recruitment communication by leaf-cutting ants derived from a vibratory mechanism that supports the leaf-cutting process (Roces & Hölldobler 1996).

## Context-Specific Responses to Stridulatory Signals

Stridulatory behavior has been observed in many ant species belonging to different subfamilies (Markl 1973). In some cases it has been shown to be effective in the context of defense and alarm behavior. In fact, Markl (1965) demonstrated that workers of *Atta* species stridulate whenever they are prevented from moving freely. For instance, when part of the colony is confined because of a cave-in of the nest, buried workers stridulate and attract nest-mates, which subsequently engage in rescue digging. Thus, it is more likely that such rather unspecific vibrational signals used in defense and alarm later acquired a function as a recruitment signal, and the observed mechanical facilitation during cutting represents an epiphenomenon of recruitment communication.

This is further supported by our finding that in *Atta cephalotes* stridulatory signals mediate communication in yet another ecological context. In leaf-cutting ants minim workers (the smallest worker subcaste) can be seen among foraging workers at the cutting sites, even though they are unable to cut leaf fragments. They usually walk around or stand with their mandibles opened and their antennae outstretched near workers engaged in cutting. At the cutting site and also along the foraging trail they often investigate leaf carriers by briefly climbing onto the carrier and its leaf fragment. In fact, many of them do not walk back to the nest, but ride ("hitchhike") on the leaf fragments being carried to the nest (Fig. 6.7). It has been demonstrated that they defend leaf carriers from attacks by parasitic phorid flies, which attempt to oviposit on the ants' bodies (Eibesfeldt & Eibl-Eibesfeldt 1967; Feener & Moss 1990).

Three different lines of evidence demonstrate that hitchhikers and leaf carriers communicate by using plant-borne stridulatory vibrations produced by the latter. First, there was a significant increase in the repetition rate of the stridulations produced by foraging workers as they maneuvered the leaf fragment into position to get ready for carrying the harvest to the nest. Even workers that did not stridulate during cutting were observed to stridulate as they loaded up the fragment. This is the moment when hitchhiking usually commences. Second, the leaf-borne stridulatory vibrations proved to be highly attractive for minim workers. They spend significantly longer times on an artificially "stridulating" than on a "silent" leaf, even in the absence of cutting workers. Finally, the occurrence of hitchhikers was significantly

Figure 6.7. Minim workers of *A. cephalotes* hitchhiking on a fragment carried by a forager along a foraging column (from Roces & Hölldobler 1995).

higher in leaf carriers foraging on stridulating than on silent leaves, indicating that stridulation motivated minims to search for loaded workers in order to mount them (Roces & Hölldobler 1995).

It is important to note, however, that in stridulating workers that load up or carry a leaf fragment, the vibrations are transmitted to the substrate through the workers' legs. Such vibrations show considerable attenuation in comparison with those transmitted mostly through the mandibles, when workers are actually engaged in cutting. Their amplitudes average $2 \times 10^{-6}$ centimeters, that is 4 to 5 times lower than those recorded during cutting activity (Roces et al. 1993). They lie near the sensitivity threshold measured electrophysiologically in leg nerves of *Atta* workers ($1.3 \times 10^{-7}$ centimeters in forelegs of minor workers, Markl 1970). Such vibrations can therefore only be detected from a maximal distance of 2 to 3 centimeters. However, Markl (1970) also demonstrated that minims are on average 3 to 4 times more sensitive to substrate-borne vibrations, making them considerably more responsive to leg-transmitted vibrations than larger workers.

## Conclusion and Outlook

The exploratory story we told in this chapter started with our noticing a simple behavioral pattern performed by *Atta* workers, namely, the rhythmic raising and lowering of the gaster while cutting a leaf fragment. One could have easily be inclined to ignore this seemingly unimportant behavior and look for more striking aspects in the ants' world to investigate. Would we have done this, we would have missed a major feature in the communication system of *Atta cephalotes*.

Chemical communication in *Atta* species has been studied from different points of view, that is, pheromone chemistry, information analysis, behavioral ecology (Morgan 1984; Jaffé & Howse 1979; see Hölldobler & Wilson 1990), and we have learned that chemical signals serve well in long-range communication and in orienting the masses of foragers to the harvesting sites. But at these sites leaf-cutting ants are highly selective in their harvest choice, even at the level of specific leaves of an individual plant. Both chemical and physical parameters of the vegetation are known to influence plant acceptability by the ants. Such plant traits undergo temporal and spatial changes; therefore, a given host plant presents a considerable variation of harvestable patches of leaves. The stridulatory signals produced by a worker during leaf-cutting are important for concentrating harvesting efforts at these palatable, high-quality patches. Our investigations on vibrational communication provide one of the possible explanations for previous observations that workers cutting leaves often attract other foragers from the immediate vicinity, and further indicate that the effects of stridulatory vibrations, despite their elementary and unitary character, depend on the social context where they are perceived.

Most of our analytical work was conducted with *Atta* colonies cultured in the laboratory. We plan to extend these studies to our field sites, where we will pay special attention to the interactions of stridulatory signals with short- and long-lasting pheromone trails. We will also incorporate grass-cutting ant species, where the harvesting process can be more easily studied in the field. Connected with these investigations is our analysis of energetic parameters that affect the decision making during harvesting at the individual as well as the colony level. Of particular interest for our future work is the study of the integration of individual behavior into colony functions, with communication being a major organizing factor.

As we have learned, in *Atta* species stridulation serves as an alarm- and rescue-signal; it also mediates short-range recruitment communication and communication between leaf carriers and their hitchhiking tiny guard ants that protect them from attacks by parasitic phorid flies. In each case the stridulatory vibrations alone elicit the context-specific response in the nestmates.

Stridulatory vibration signals employed during recruitment to food sources are known also from other ant species. However, in those cases they do not serve as behavioral releasers, but rather as modulators of chemical signals. Modulatory signals are devices for shifting the threshold for the releasing effectiveness of other stimuli (such as chemical signals), thus enhancing the behavioral response to them (Markl 1985). For example, in *Aphaenogaster cockerelli* or *A. albisetosus*, two myrmicine ant species common in the Southwestern United States, a forager, after discovering a prey object too large to be carried or dragged by a single ant, releases poison gland secretion into the air. Nestmates as far away as 2 meters are attracted and move toward the source. When a sufficient number of foragers have assembled around the prey, they gang-carry it swiftly to the nest. *Aphaenogaster* workers, in addition to releasing the poison gland pheromone, also regularly stridulate at the prey object. Ants perceiving the substrate-borne signals start to encircle the prey sooner, and they are likely to release the attractive poison gland pheromone earlier. Overall, both the recruitment of workers and the retrieval of the food object are advanced by one to two minutes as consequence of stridulation (Markl & Hölldobler 1978). This is certainly an adaptive trait, because *Aphaenogaster* must remove food from the scene before their formidable mass-recruiting competitors, including fire ants and the dolichoderine ant *Forelius pruinosus*, arrive in large numbers (Hölldobler at al. 1978). A similar link of stridulation to the recruitment process has been noted in species of the European harvesting ant genus *Messor*, which employ stridulatory vibrations in conjunction with odor trails laid with secretions from the Dufour's gland (Hahn & Maschwitz 1985; Buser et al. 1987; Baroni-Urbani et al. 1988). These studies report evidence demonstrating that chemical recruitment communication is enhanced by substrate-borne stridulatory vibrations, but neither in *Aphaenogaster* nor in *Messor* has it been shown that stridulatory signals alone elicit a recruitment response in the recipient ants. In these cases the substrate-borne vibrations are part of a multimodal signal lowering the response threshold of the receiver for the releasing component of the signal. Since in *Atta* species stridulatory vibrations also enhance the response to chemical recruitment signals, it is possible that it first evolved from an alarm- and defense-signal to become a modulatory signal in the recruitment process and subsequently has further evolved to function as an independent signal in short-range recruitment.

From all these more recent studies on multimodal signaling in ants and other social insects we have learned that communication systems are much more complex than previously thought. Chemical signals often consist of several compounds, sometimes even originating from different exocrine glands, whereby different components of complex pheromone mixtures may have different effects on the receiver. In addition, chemical signals can be combined with signals from different sensory modalities, thereby fine-tuning the receivers' responses and increasing the colony's efficiency (Hölldobler 1999). In future studies of communication mechanisms we have to pay closer attention to such accessory behavioral patterns like knocking, stroking, jerking, or stridulation, all of which have been observed to occur during chemical communication behavior in ants. Such cross-modal perception of composite signals has become of particular interest to researchers on human and nonhuman primates and on birds (Kuhl & Meltzoff 1982; Hauser 1996; Partan 1998), and, in fact, was already recognized by Darwin in his book, *The Expression of the Emotions in Man and Animals* (1872), where he noted that the power of communication by language is much enhanced by "the expressive movements of the face and body."

Many ant societies are stationary; like terrestrial plants they spend most of their adult lives fixed in one spot and produce winged reproductive forms to disperse away from the nest as the functional analogs of seeds. Ant workers comb the surrounding terrain, where they gather information, energy, and matter and bring these resources to the nest. All these activities are performed cooperatively by the usually sterile worker castes. Whereas a solitary animal can at any moment be in only one place and can be doing only one thing, a colony of social insects can be in many places by deploying its workers and can be doing many different things because of the size of the worker cohorts and their division of labor. Such a division of labor system can, however, only function by means of communication. An amazing array of signals and signal combinations are involved in social interactions within ant societies (Hölldobler & Wilson 1990), and examples of the behavioral analysis of such signals have been presented in this chapter. Such signals mediate mutualistic communication, that is, information transmitted by such signals is shared within the colony to the benefit of all members of the society. The cooperative functioning and collective fitness of the colony depend on such mutualistic communication. The social interactions mediated by such communication can be considered an important part of the "extended phenotype" of the colony (Dawkins 1982). In fact, communication and division of labor are key features of highly evolved eusocial organizations. They determine the efficiency of the society and they are shaped by natural selection, not only on the individual and kin levels, but also on the colony level. We can expect colonies in a population to show variations with respect to these features. Colonies compete with one another for resources; those colonies that employ the most effective recruitment system to defend and retrieve these resources, that exhibit the most powerful colony defense against competitors and predators, will be able to raise the largest number of reproductive females and males every year, and thus will have the greatest fitness within the population of colonies (Hölldobler 1999). Therefore, future investigations on adaptive significance of communicative interactions among members of an insect society should not exclusively focus on the benefits a signal sender gains by manipulating the behavior of the signal receiver (as proposed by Dawkins & Krebs 1978), but instead should consider the entire colony as the beneficiary of such communicative behavior.

# References

Baroni-Urbani C, Buser MW, Schilliger E, 1988. Substrate vibration during recruitment in ant social organization. Insectes Sociaux 35:241–250.

Billen J, Morgan ED, 1997. Pheromone communication in social insects: sources and secretions. In: Pheromone communication in social insects (Vander Meer RK, Breed MD, Espelie KE, Winston ML, eds). Boulder, CO: Westview Press; 3–33.

Buser MW, Baroni-Urbani C., Schilliger E, 1987. Quantitative aspects of recruitment to new food by a seed-harvesting ant (*Messor capitatus* Latreille). In: From individual to collective behavior in social insects (Pasteels JM, Deneubourg JL, eds). Basel: Birkhäuser Verlag; 139–154.

Darwin C, 1872. The expression of the emotions in man and animals. London: John Murray.

Dawkins R, 1982. The extended phenotype. Oxford: Freeman.

Dawkins R, Krebs JR, 1978. Animal signals: information or manipulation. In: Behavioral ecology (Krebs JR, Davies NB, eds). Oxford: Blackwell; 2282–2309.

Eibesfeldt I, Eible-Eibesfeldt E, 1967. Das Parasitenabwehren der Minima-Arbeiterinnen der Blattschneider-Ameise (*Atta cephalotes*). Z Tierpsyhol 24:278–281.

Feener DH, Jr, Moss KAG, 1990. Defense against parasites by hitchhikers in leaf-cutting ants: a quantitative assessment. Behav Ecol Sociobiol 26:17–29.

Hahn M, Maschwitz U, 1985. Foraging strategies and recruitment behaviour in the European harvester *Messor rufitarsis* (F.). Oecologia 68:45–51.

Hauser MD, 1996. The evolution of communication. Cambridge, MA: The MIT Press.

Hölldobler B., 1995. The chemistry of social regulation: multicomponent signals in ant societies. Proc Natl Acad Sci 92:19–22.

Hölldobler B., 1999. Multimodal signals in ant communication. J Comp Physiol A 184:129–141.

Hölldobler B, Stanton RC, Markl H, 1978. Recruitment and food retrieving behavior in *Novomessor* (Formicidae, Hymenoptera). I: Chemical signals. Behav Ecol Sociobiol 4:163–181.

Hölldobler B, Wilson EO, 1990. The ants. Cambridge, MA: The Belknap Press of Harvard University Press.

Hölldobler B, Wilson EO, 1994. Journey to the ants. Cambridge, MA: The Belknap Press of Harvard University Press.

Jaffé K, Howse PE, 1979. The mass recruitment system of the leaf cutting ant, *Atta cephalotes* (L.). Anim Behav 27:930–939.

Kuhl PK, Meltzoff AN, 1982. The biomodal perception of speech in infancy. Science 218:1138–1141.

Littledyke M, Cherrett JM, 1976. Direct ingestion of plant sap from cut leaves by the leaf-cutting ants *Atta cepholotes* (L.) and *Acromyrmex octospinosus* (Reich) (Formicidae, Attini). Bull Entomol Res 66:205–217.

Markl H, 1965. Stridulation in leaf-cutting ants. Science 149:1392–1393.

Markl H, 1968. Die Verständigung durch Stridulationssignale bei Blattschneiderameisen. II. Erzeugung und Eigenschaften der Signale. Z Vergl Physiol 60:103–150.

Markl H, 1970. Die Verständigung durch Stridulationssignale bei Blattschneiderameisen. III. Die Empfindlichkeit für Substratvibrationen. Z Vergl Physiol 69:6–37.

Markl H, 1973. The evolution of stridulatory communication in ants. Proc VII Congr IUSSI, London: 258–265.

Markl H, 1985. Manipulation, modulation, information, cognition: some of the riddles of communication. In: Experimental behavioral ecology and sociobiology (Fortschritte der Zoologie, no. 31) (Hölldobler B, Lindauer M, eds). Stuttgart: Gustav Fischer Verlag; 163–194.

Markl H, Hölldobler B, 1978. Recruitment and food-retrieving behavior in *Novomessor* (Formicidae, Hymenoptera). II. Vibration signals. Behav Ecol Sociobiol 4:183–216.

Michelsen A, Larsen ON, 1978. Biophysics of the ensiferan ear. I. Tympanal vibrations in bushcrickets (Tettgonidae) studied with laser vibrometry. J Comp Physiol 123:193–203.

Morgan ED, 1984. Chemical words and phrases in the language of pheromones for foraging and recruitment. In: Communication (Lewis T, ed). London: Academic Press; 169–194.

Partan SR, 1998. Multimodal communication: The integration of visual and acoustic signals by Macaques. Ph.D. thesis, University of California, Davis.

Roces F, Hölldobler B, 1994. Leaf density and a trade-off between load-size selection and recruitment behavior in the ant *Atta cephalotes*. Oecologia 97:1–8.

Roces F, Hölldobler B, 1995. Vibrational communication between hitchhikers and foragers in leaf-cutting ants (*Atta cephalotes*). Behav Ecol Sociobiol 37:297–302.

Roces F, Hölldobler B, 1996. Use of stridulation in foraging leaf-cutting ants: mechanical support during cutting or short-range recruitment signal? Behav Ecol Sociobiol 39:293–299.

Roces F, Lighton JRB, 1995. Larger bites of leaf-cutting ants. Nature 373:392–393.

Roces F, Núñez JA, 1993. Information about food quality influences load-size selection in recruited leaf-cutting ants. Anim Behav 45:135–143.

Roces F, Tautz J, Hölldobler B, 1993. Stridulation in leaf-cutting ants: short-range recruitment through plant-borne vibrations. Naturwissenschaften 80:521–524.

Tautz J, Roces F, Hölldobler B, 1995. Use of a sound-based vibratome by leaf-cutting ants. Science 267:84–87.

Tumlinson JH, Silverstein RM, Moser JC, Brownlee RC, Ruth JM, 1971. Identification of the trail pheromone of a leaf-cutting ant, *Atta texana*. Nature 234:348–349.

Wilson EO, 1980a. Caste and division of labor in leaf-cutter ants (Hymenoptera: Formicidae: *Atta*). I. The overall pattern in *A. sexdens*. Behav Ecol Sociobiol 7:143–156.

Wilson EO, 1980b. Caste and division of labor in leaf-cutter ants (Hymneoptera: Formicidae: *Atta*). II. The ergonomic optimization of leaf cutting. Behav Ecol Sociobiol 7:157–165.

Wilson EO, 1985. The principles of cast evolution. In: Experimental behavioral ecology and sociobiology (Fortschritte der Zoologie, no. 31). Sunderland, MA: Sinauer Associates, Inc.; 307–324.

Wilson EO, Bossert WH, 1969. A general method for estimating threshold concentrations of odorant molecules. J Insect Physiol 15:597–610.

# 7

# Understanding the Evolution of Social Behavior in Colonial Web-Building Spiders

George W. Uetz

## A Blueprint for Research:
## Social Behavior in an Asocial Animal Model

For over twenty years, my students, colleagues, and I have been interested in the social behavior of colonial web-building spiders in the orb weaver genus *Metepeira* (Fig. 7.1). Our research with this unique model system has allowed testing of numerous hypotheses about costs and benefits of group-living and the evolution of sociality. This project had its origins in scientific skepticism—that is, I had read reports of what I considered highly unlikely behavior in some tropical spiders, and I needed to be convinced that the behavior described was indeed sociality and not some artifact or aggregation phenomenon. Spiders are generally solitary, asocial predators by nature, and are often highly aggressive toward conspecifics, so upon reading about "colonies" of spiders resembling those of social insects, I was curious. These reports were from early literature recountings of tropical expeditions, including Darwin's, and the descriptions of huge webs containing hundreds to thousands of coexisting and possibly cooperating spiders seemed unrealistic (Darwin 1845, cited in Shear 1970; Simon 1891; Hingston 1932; Shear 1970; Wilson 1971).

In 1975, when I was in graduate school, E. O. Wilson's book, *Sociobiology*, had an enormous impact on an entire generation of students. Wilson synthesized ideas from animal behavior, population biology, and evolution, and applied them to the social behavior of animals. Wilson's ideas created a renaissance of interest in social insects and other arthropods, centered on William Hamilton's (1964) earlier ideas about haplodiploidy (the genetic mechanism of sex determination in Hymenoptera) and its role in kin selection. In his book, Wilson (1975) mentioned some of the spider species described in earlier works as well as some European literature I was unaware of, and I was intrigued. At a meeting of the American Arachnological Society that same year, I heard papers on spider sociality presented by Ruth Buskirk on *Metabus gravidus* (see Buskirk 1975), Yael Lubin on *Cyrtophora* spp. (see Lubin 1974), and J. Wesley Burgess on *Metepeira spinipes* and *Mallos gregalis* (see Burgess 1978), each on various forms of sociality in

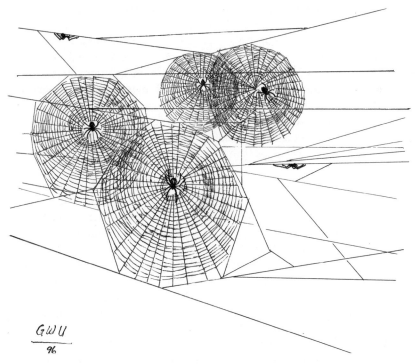

$\dfrac{G\,\omega\,U}{96}$

Figure 7.1. Portion of a colony of *Metepeira incrassata* from Fortin de las Flores, Veracruz, Mexico, showing several individuals with orb webs. Drawn from a photograph by the author.

spiders. Although my interests were in the community ecology of spiders, I jumped at the chance to go along with fellow grad student J. Wesley Burgess on an expedition to Mexico, when he asked me to provide him with the insights of a field ecologist (and, as I later learned, to serve as an additional driver).

Burgess was interested in applying mathematical models of animal spacing (adapted from nearest neighbor estimation techniques) to the three-dimensional structure of colonial webs of the orb weaver *Metepeira spinipes*, found in central Mexico. Upon our arrival in Mexico, however, I was struck by the fact that these spider "colonies" were mostly in agave plants, which appeared to be the only structures capable of supporting webs. During my work on orb weavers in forest habitats, I frequently encountered multiple spider webs attached to a common structure, and I wondered whether spider "sociality" simply represented aggregative dispersion patterns in response to environmental constraints. Thus, our first order of business was to test the null hypothesis of aggregation due to patchy habitat, before we could consider the alternative hypothesis that aggregations of web-building spiders were exhibiting a form of social behavior. This task was a challenge, because the spiders only occurred grouped in habitat patches, that is, a typical mapping of dispersion against a standard Poisson distribution could not be used, as there were no zero values. So, we adapted a technique from primate research (Cohen 1971—the zero-truncated Poisson distribution) and were able to resolve this issue and reject the null hypothesis. Moreover, we learned from a number of field and laboratory studies that while these spiders are not randomly distributed, they also reaggregate when separated and vary spacing within colonies depending on ecological conditions, suggesting some form of social behavior (Burgess 1978; Uetz & Burgess 1979; Burgess & Uetz 1982).

After this trip, I accepted a position at the University of Cincinnati, and Burgess moved to a post-doc at Stanford studying fish schools, and later to a degree in psychiatry studying aggregative behavior in autistic children. Thus, I "inherited" the project, and the responsibility for finishing up the manuscript(s). Needing more data, I obtained a grant from the National Geographic Society to return to Mexico with UC colleague Tom Kane and (new) grad student Gail Stratton. From this beginning, and over two dozen trips to Mexico in the past two decades, my colleagues and I have found a wide range of social behaviors in a novel arthropod taxon without haplodiploidy. Our work on colonial web-building spiders in the genus *Metepeira*, and tropical *M. incrassata* in particular, has allowed the testing of numerous hypotheses about the ecological factors involved in the evolution of social behavior in a unique model system (see Uetz & Hieber 1997 for a review).

## Building a Model System:
## Integrating Natural History, Theory, and Empirical Work

Our approach has integrated conceptual, theoretical, and empirical work with natural history observation through several cycles of the development of the project. In particular, there are three areas in which studies with these colonial web-building spiders have provided valuable insights to important questions in behavioral ecology: (1) phenotypic plasticity in group size and spacing related to environment; (2) risk-sensitivity and group foraging; (3) balancing fitness trade-offs inherent in group-living, especially those related to predation.

While studies on birds and mammals and even insects may draw upon a rich literature as prelude to developing a model system as a means of testing theory, this is not true for spiders. At first, we needed to collect a great deal of descriptive data on basic biology, behavior, and distribution. I found this phase of the work uniquely frustrating, because many reviewers of manuscripts and grant proposals shared my initial skepticism about spider social behavior, and with every attempt to publish or obtain funding, I found it necessary to explain the colonial spider model system "from scratch," so to speak. Moreover, descriptive studies, even of unknown taxa with promising attributes as model systems, are clearly less competitive for grant support. It seemed that my enthusiasm for what to me (and to Darwin, I might add) was an *amazing* phenomenon—social behavior in otherwise rapacious spiders— was not shared by the scientific community. Were it not for the continued support of the National Geographic Society, whose primary interests are in exploration of the world's biodiversity, little more would be known about spider social behavior. Happily, with their initial support, and later with support from the National Science Foundation, we have been able to gain enough data to use this fascinating model system to test numerous theories about the evolution of social behavior originally developed by others working with birds and mammals. Science works in mysterious ways.

## Laying a Foundation: Environment and Plasticity in Group Size and Spacing

One of the first things we noticed during our travels through central Mexico was that colonial spiders exhibited great phenotypic plasticity—that is, the size and spacing of colonies varied tremendously with habitat and availability of prey resources (Fig. 7.2). Colonies ranged in size from small groups of two to three individuals in desert habitats to groups of tens to hundreds in more mesic sites, to thousands and tens of thousands in tropical rainforest habitats (Uetz et al. 1982). Experimental tests manipulating prey availability and "transplanting" colonies in the field suggested prey availability played a large role in determining group size, interindividual spacing within colonies, and life history parameters over the gradient of habitats sampled (Uetz et al. 1982; Benton & Uetz 1986; Uetz et al. 1987). The unresolved question, however, was whether this range of phenotypic plasticity reflected steps along pathways in the evolution of spider sociality proposed earlier by Shear (1970), that is, aggregations of (unrelated) individuals in areas of high prey availability or ecotypes with different levels of social tolerance.

Given the extremes of group size and spacing observed over a geographic range encompassing divergent habitat types and numerous potential isolating barriers, we wondered whether there might be distinct geographic ecotypes, with varying levels of social aggregation. We set about to explore this in two ways: (1) raising spiders from different populations in the laboratory in a "common garden" experiment under identical controlled conditions; and (2) using gel electrophoresis to examine genetic differences between populations. Results of these studies were highly interesting, and somewhat contradictory. First, the "common garden" experiments revealed that beneath the variation in nearest neighbor spacing in the field, there were distinct reaction norms typical of each population tested (Uetz & Cangialosi 1986; Cangialosi

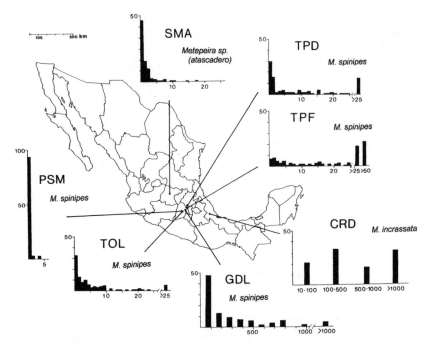

Figure 7.2. Frequency distributions of group size for populations of three *Metepeira* species in various locations in Mexico. *M. atascadero*: SMA—San Miguel de Allende, Gto.; *M. spinipes*: TPD, TPF—Tepotzotlan, Mex.; PSM—Parque Sierra Morelos, Mex.; TOL—Toluca, Mex.; GDL—Guadalupe Lake, Mex.; *M. incrassata*: CRD—Cordoba/Fortin de las Flores, Ver. (from Uetz et al. 1982; © Science, AAAS; used with permission).

& Uetz 1987). Under identical, controlled conditions, with ad-lib food availability, spiders from three source populations—desert grassland (San Miguel de Allende), agricultural valley of central Mexico (Tepotzotlan), and tropical rainforest/agriculture (Fortin de los Flores)—exhibited significant differences in interindividual spacing behavior and aggressive behavior (Uetz & Cangialosi 1986; Cangialosi & Uetz 1987). Under conditions of ad-lib prey, interindividual spacing was reduced, suggesting reduced competition and higher degrees of tolerance among colony members, a prerequisite for the evolution of social behavior in spiders (Rypstra 1986). Analyses of geographic differentiation among these same populations using six enzymatic loci with polyacridamide gel electrophoresis revealed that despite some differences in the degree of heterozygosity among populations, very little geographic differentiation was seen among the three sites. Measures of genetic similarity were similar to those for continuously distributed populations of single species (Uetz et al. 1987).

Our preliminary conclusion was that subpopulations with different levels of social tolerance, reflected in spacing and aggressive behavior patterns, represented ecotypic variation in social behavior. This conclusion was supported by field studies of aggressive interactions among individuals, which differ between the desert and tropical rainforest habitats (Uetz & Hodge 1990; Hodge & Uetz 1992). However, one aspect of raising spiders under optimal laboratory conditions is that variation in adult size, morphology, and color pattern is greatly reduced, and in our case this revealed differences that called into question whether these populations represented a single species. Despite the electrophoretic evidence above, a detailed examination of specimens by several taxonomists confirmed that these populations are in fact three species: *Metepeira incrassata* (Fortin de las Flores), *M. spinipes* (Tepotzotlan), and *M. atascadero* (a new species from San Miguel de Allende described by Piel 1997). While what we found did not resolve issues in the manner we expected, results of these studies pushed the project in new directions. With three species from different habitats exhibiting divergent levels of social behavior, we could examine the role of ecological factors shaping the evolution of social behavior in this atypically social animal. Then, we could derive predictions based on theoretical models proposed for other social animals and test them with field observation and experimentation.

## Adding a Framework: Group Foraging and Risk-Sensitivity

Given the variation in social grouping with environment and prey availability, it seemed logical to address the most probable causative factor, and so we looked first into prey capture behavior to gain insight into the mechanisms involved in reduction of competition and tolerance. In order to collect data on both prey availability and individual prey capture rates, we made observations of all insects flying into and captured by spiders within a volume of colony of 1 m$^3$ (or less, in the case of small colonies) in thirty-minute intervals throughout the day. Observers were assigned to colonies and time periods at random, and we amassed an enormous data set for each of the three species populations. Analyses of the data revealed an interesting pattern: insects escaping from one web were often caught by the web of a neighboring spider. This "ricochet effect" (Fig. 7.3) appears to be an important advantage of grouping webs, as spiders in colonies captured more and larger prey than their solitary counterparts (Uetz 1989).

However, as often happens in science, one idea leads to another, and attempts to explain counterintuitive results provide challenges that lead to conceptual breakthroughs. A logic flaw in my thinking was pointed out to me during a question and answer session after a presentation, and while this disturbed me greatly it also led to an important finding. The increase in prey associated with the ricochet effect was, in a sense, a paradoxical explanation for the evolution of tolerance and sociality in different habitats: that is, if grouping webs provides spiders with more prey, why do spiders *only* group

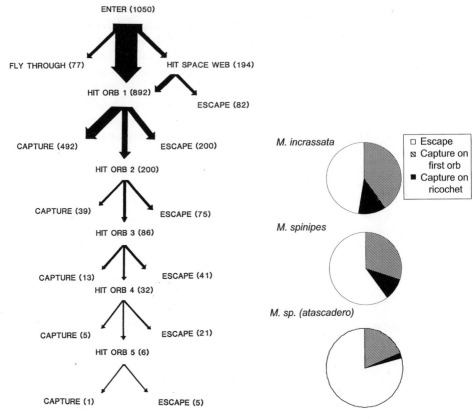

Figure 7.3. A. Demonstration of the ricochet effect, showing the fate of insects flying into a colony of *Metepeira incrassata* in Fortin de las Flores, Mexico (from Uetz 1989; © Springer-Verlag, used with permission). B. Proportions of prey captured on the first orb hit, upon ricochet, and escaping from colonial webs of three *Metepeira* species.

webs together in high-prey environments, and not in poor-prey habitats? It would seem that if increased prey was the driving force behind colonial web-building, sociality should be a lot more common than it is. This seeming paradox troubled me until I happened to hear two papers by Ron Pulliam and Tom Caraco, introducing models of "risk-sensitivity" in foraging flocks of birds. Their theories were based on the idea that foraging is influenced not only by the mean amount of food obtained, but also by its variance, which affects the risk of starvation (Thompson et al. 1974; Caraco 1981; Pulliam & Milikan 1982; Caraco & Pulliam 1984; Pulliam & Caraco 1984; Real & Caraco 1986). Risk-sensitivity models are based on the assumption that foraging in flocks reduces variance in individuals' food intake (via a number of mechanisms), and predict that: (1) foraging solitarily (the risk-prone strategy) should occur where prey are in habitats where the rate of return, on

Table 7.1

Comparison of Prey Capture Rates and Variance Parameters in Three Species of
Colonial *Metepeira* from Mexico

| Species | x̄ Prey/Spider/Hr. | St. Dev. | Coeff. Var. (%) |
|---|---|---|---|
| *Metepeira* sp. *(atascadero)* | | | |
| Solitary | 0.71 | 1.11 | 155.83 |
| Grouped | 1.28 | 1.38 | 107.82 |
| *Metepeira spinipes* | | | |
| Solitary | 1.30 | 1.26 | 97.0 |
| Grouped | 1.80 | 1.18 | 66.0 |
| *Metepeira incrassata* | | | |
| Solitary | 2.14 | 2.76 | 129.0 |
| Grouped | 6.58 | 2.43 | 37.0 |

average, is equal to or below their energetic needs; (2) foraging in groups (the risk-averse strategy) should arise only where prey availability is greater than needed.

The contrasting habitats and group-living arrangements of colonial *Metepeira* provided an ideal opportunity to test the assumptions and predictions of risk-sensitivity models (Uetz 1988a, b; Uetz & Hodge 1990; Uetz 1996; Caraco et al. 1995). Field studies comparing prey capture rates of solitary and grouped spiders in the same habitat, as well as laboratory studies with caged spiders (in groups of varying size), confirmed that prey capture variance was lower for spiders in groups than their solitary counterparts (Table 7.1) (Uetz 1988a, b, Uetz 1996). Field studies of the frequency of spiders foraging in groups also supported predictions of the Pulliam/Caraco models. In desert grassland habitats, where average available daily prey biomass is less than or equal to individual needs and of relatively high variability, *M. atascadero* forage mostly as solitaries. Under these circumstances, it appears better for the individual spider to forage alone and risk starvation on the chance it would encounter a site with above average prey availability, rather than to forage in a group and reduce variance in an already inadequate food level, and starve for certain. In contrast, the tropical montane rainforest/ agriculture habitat of *M. incrassata* is moist and moderate, and prey availability there is 2 to 3 times what individual spiders require for daily energetic needs, and variability in prey is more constant than the desert site. There, *M. incrassata* forage in a risk-averse manner in large groups, with a considerable margin of safety from starvation.

The relationship between habitat and prey availability, and the foraging constraints imposed by risk-sensitivity, provide an answer to the question of why sociality is rare in spiders (Gillespie 1987; Uetz 1988a, b, 1996; Rypstra 1989). Solitary foraging is the appropriate strategy for spiders in most habitats, because prey are highly variable, and food obtained by each spider is likely to be just enough or less than required. In contrast, some habitats have superabundant prey, for example, tropical forest sites, and spiders there link webs because it is advantageous to do so, and benefit from foraging socially. It is interesting to note that during El Niño years, when high levels of rain create a "bottom-up" trophic cascade of abundant prey, spiders in the Mexican desert (*M. atascadero*) and California coast (*M. spinipes*) forage more frequently in groups (Hieber & Uetz 1990; Uetz, unpublished). It is indeed possible that cyclic events like these or permanent climatic shifts in certain habitats may have set the stage for evolution of colonial web-building.

## Putting the Pieces Together: Fitness Trade-Offs Inherent in Group-Living

A considerable amount of theory in behavioral ecology concerns the fitness trade-offs faced by animals living in groups, especially those associated with predation risk (Hamilton 1971; Foster & Treherne 1981; Inman & Krebs 1987; Wrona & Dixon 1991). While the selective forces favoring group-living in many animal species are related to protection from predators, the sedentary nature, high density, and conspicuous silk of colonial web-building spiders would appear to make them especially vulnerable to predators and parasites (Lubin 1974; Rypstra 1979; Buskirk 1981; Smith 1982). However, these same attributes make colonial web-building spiders a unique model system for testing hypotheses about predation risk, antipredator mechanisms, and risk-balancing trade-offs, as attacks are common and it is possible to observe predator-prey encounters at both the group and individual levels.

With regard to attack by predators and parasites, it seems that both benefits and costs of group-living accrue to colonial web-building *Metepeira*. I convinced Craig Hieber, a specialist in the egg sac parasites of orb-weavers, to join in the research in the same way (and for some of the same reasons!) as Burgess interested me at the start of this project. We sampled a wide range of naturally occurring colony sizes in both *M. atascadero* and *M. incrassata*, and compared levels of egg sac predation/parasitism (Hieber & Uetz 1990). In the desert grassland habitat of San Miguel de Allende, typical years have low prey availability and egg sac production is low for *M. atascadero*, and the overall rates of egg sac parasitism (3.4 to 6.9 percent) are also low. In El Niño years, prey are abundant and (consequently) spider egg production is greater, and higher rates of egg sac parasitism (up to 40 percent) are seen. However, there appears to be no relationship between group size and the rate of parasitism of egg sacs in this species. In contrast, parasitism rates for *M. incrassata* are relatively constant from year to year, but there is a significant positive relationship between colony size and rates of parasitism (as high as 30 to 40 percent in the largest colonies), representing an increased cost in terms of egg loss to individual spiders (Hieber & Uetz 1990).

Colonial orb weavers also have to contend with other forms of social parasitism by spiders. Colonies of *M. incrassata* are frequently invaded by other spider species, including the large tropical orb weaver *Nephila clavipes* (Araneae: Tetragnathidae), and often resemble mixed-species "flocks" (Hodge & Uetz 1996). In her doctoral work, Maggie Hodge found that *Nephila* gain foraging advantages when associated with *Metepeira* colonies, and capture significantly more prey and a different prey spectrum than solitaries or single-species groups of *Nephila* (Hodge & Uetz 1996). In addition, when *Nephila* and other typically solitary species invade *M. incrassata* colonies, they frequently bring along kleptoparasitic spiders of the genus *Argyrodes* (Araneae: Theridiidae). Andrea McCrate studied these kleptoparasites for her M.S. thesis (McCrate 1989), and found that the burden of *Argyrodes* increases dramatically with *Metepeira* colony size. In experimental studies with caged colonies, McCrate also found that *Argyrodes* represent a twofold cost of group-living: not only do they steal prey, but they also frequently prey upon *Metepeira*, and represent a significant source of mortality when present.

While increased attacks from predators are a common cost of sociality for many animals, including spiders (Spiller & Schoener 1989), costs may be offset by gains from antipredator mechanisms inherent in group-living. Some of the best-known antipredator benefits to individuals in groups have been attributed to a set of hypothesized mechanisms referred to as the "encounter effect" and the "dilution effect" (Hamilton 1971) and collectively as "attack-abatement" (Turner & Pitcher 1986; Inman & Krebs 1987; Wrona & Dixon 1991). Both an "encounter effect" and a "dilution effect" operate against predatory wasps attacking *M. incrassata* colonies, but work in different ways and at different times in the predator's attack (Uetz & Hieber 1994). An "encounter effect" (Turner & Pitcher 1986) is evident in that the encounter rate between wasp predators and spider colonies increases with colony size, but that rate is far less than expected in the smallest and largest colonies. Solitary individuals and small colonies (less than ten) are never encountered by wasp predators, suggesting they are below some detection threshold. Additionally, rates of encounter do not increase beyond a level seen for mid-sized groups (approximately five hundred), suggesting that visual apparency does not increase linearly with colony size (i.e., the true size of a three-dimensional colony is masked from view). As individual spacing within a colony becomes more compact in larger colonies (Uetz et al. 1982; Uetz & Hodge 1990), spiders in large groups may benefit from increased protection because they are no more likely to be located than those individuals in smaller groups. These results are significant in themselves, as they demonstrate that the encounter effect alone is sufficient to confer fitness advantages on individuals living in large colonies (Inman & Krebs 1987).

When a predator encounters a colony, individual risk of attack and capture is predicted to be an inverse function of colony size by a numerical "dilution effect" (Hamilton 1971; Foster & Treherne 1981; Inman & Krebs 1987; Wrona & Dixon 1991). We collected data that supports this prediction, but our field observations revealed the relationship to be more complex than a simple mathematical rule. This complexity is caused by two different factors operating at the same time. First, wasps may attack more than one spider in a colony after it is located, which could offset any gain in fitness from a dilution effect. Second, despite multiple attacks, the capture success of wasp predators decreases with group size, creating an overall decreased risk for individual spiders living in groups.

Observations of wasp attacks suggested that spiders are forewarned of approach, and that spiders attacked later in a "trapline run" are more likely to escape by "bailing out" of their web into the vegetation below (Fig. 7.4a). On one of our field expeditions, Craig Hieber, Stim Wilcox, Jay Boyle, and I hypothesized that the network of interconnected silk webbing provides "early warning" of predator approach (Lubin 1974; Buskirk 1975; Uetz 1986; Hodge & Uetz 1992; Uetz et al., unpublished), analogous to the "many eyes" vigilance system seen in flocks of birds (Bertram 1978; Kenward 1978), or the "Trafalgar" effect demonstrated by ocean skaters (Treherne & Foster 1982). We tested the "early warning" hypothesis using both "high-tech" and "low-tech" means. Using a high-sensitivity microphone and a Marantz field-portable tape recorder, we recorded the wingbeat frequencies of caged wasp predators of *M. incrassata*. We digitized the sound on a laptop PC, analyzed the recordings with the Canary® Bioacoustic Software package, and found that the two most common wasp predators had wingbeat frequencies ranging from 92 to 115 Hz. We then used a widely available, inexpensive, field-portable, battery-operated vibration source (with a constant frequency of 95 Hz) as a simulated predator. Within colonies of different sizes, we approached spiders directly along a measured path, vibrating the silk lines of their web near the retreat. The distance at which spiders responded with defensive or evasive behavior was recorded, and varied significantly with colony size, position within the colony, and age/sex of the spider (three-way ANOVA: Colony size—$F_{5, 144} = 41.42$, $p < 0.001$; Position (core/periphery)—$F_{1, 144} = 51.88$, $p < 0.001$; Sex/age—$F_{2, 144} = 20.16$, $p < 0.001$). In a second experiment, we simulated attacks of "traplining" wasps by "buzzing" five adult female spiders in close proximity to each other with the vibration source in successive thirty-second intervals. The response behaviors as well as response latency dramatically changed across the attack sequence. While the first spiders attacked in this manner appeared caught off-guard, the third to fifth spiders in the sequence responded on average in one second or less by "bailing out" of their web into the vegetation below (Fig. 7.4b). Taken together, these results provide strong support for an "early warning" mechanism within colonial webs.

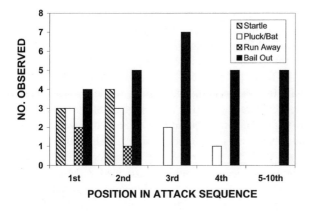

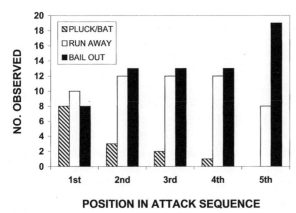

Figure 7.4. A. Responses of spiders to attacks by predatory wasps, based on their position in the sequence of attack. B. Responses of spiders to a simulated predator (95 Hz vibration source).

It is clear that as colonies grow larger, they attract the attention of predators, and while there are a number of mechanisms that may reduce individual risk, predation risk is an important and ever-present selection factor for colonial web-building spiders. Linda Rayor joined our research team as a postdoc, as she thought this system would be ideal for investigating how spiders deal with trade-offs between maximizing foraging success and minimizing predation risk. Previous studies had recognized a consistent size-based structure within colonies: larger individuals were more often found in the central core of colonies, while smaller spiders were found on the periphery. Was this structure simply a consequence of colony growth around founders, or something else? Hamilton (1971) first hypothesized a "selfish herd effect" whereby animals in the center of a group decrease their chances of predation by surrounding themselves with others. In her studies, Rayor first set about to test whether colony size structure was random or a simple consequence of growth by collecting whole colonies, sorting them by size, and introducing them to wooden frames in different order according to size. Her observations convinced us that the colony structure was a consequence of the dominance of larger individuals. In subsequent studies, we found that predator attack and capture rates varied with the spatial position of spiders within the three-dimensional volume of the colony. Risk of attack by several species of wasps was higher for spiders on the periphery, but lower in the core of the colony (Rayor & Uetz 1990, 1993). At the same time, spiders must face a difficult trade-off, as prey capture rates are clearly higher on the periphery than in the colony core (Fig. 7.5). As predation risk varies with size and age (risk is highest for young spiderlings and reproductive females), this foraging/predation risk trade-off leads to "cover-seeking" behavior and creates a size-biased spatial hierarchy predicted by selfish herd theory (Mooring & Hart 1992; Hart 1997). The characteristic three-dimensional structure of large *Metepeira* colonies arises from reproductive females seeking protection in the core and forcing smaller spiders to occupy the periphery, and results in higher survival rates and reproductive success for the largest females (Rayor & Uetz 1990, 1993).

## Completing and Testing: Dynamic Optimization Models and Field Experiments

What we have learned about sociality in these fascinating spiders has given us insights into the mechanisms involved in evolution of social behavior, and has enabled us to quantify fairly precisely the costs and benefits of group-living. After accumulating a great deal of data from field studies, I undertook a collaboration with Beth Jakob and Marc Mangel to construct reality-based dynamic optimization models about behavior of individual spiders, make predictions, and test them with field experiments.

The structure of *Metepeira* colonies, with larger spiders in the core and smaller spiders on the periphery, is a result of individual decisions by each spider about where to place its orb. Every morning, each spider must select a location for its orb web within the colony frame. The optimal location will depend on the trade-off between predation risk and prey availability between the core and periphery of the colony. Factors that may affect this decision are numerous, and include the amount of energy a spider has, its size (which determines its prey capture efficiency and risk of predation), the cost of moving through a colony, and how close a spider is to reproduction. Dynamic optimization models are ideal for predicting optimal decisions in complex systems such as this.

We constructed a dynamic model with the goal of predicting the optimal decisions of spiders from hatching until egg production. These decisions depend in part on state variables, or changeable characteristics of each spider. We included the state variables of energetic reserves, current location

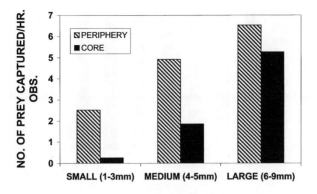

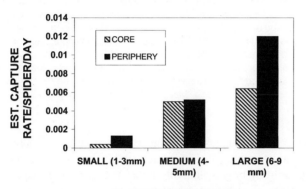

Figure 7.5. Trade-offs with spatial position and spider size within *M. incrassata* colonies: A. Prey capture (biomass/spider/30 min. obs.) in core and periphery locations for small (1–3 mm), medium (4–6 mm), and large (7–10 mm) spiders. B. Capture risk from predators in core and periphery locations for spider size classes (from Rayor & Uetz 1993 © Springer-Verlag, used with permission).

(because there is a cost of moving through the colony to change locations), days since molt (as spiders cannot molt several times in close succession), and instar (body size influences prey capture success and predation risk). Also included in the model were the probability of encountering prey or predators, the cost of moving through the colony, and the cost of molting, among other variables. We could estimate most of these variables with field or laboratory data. For example, field observations of spiders that were transplanted from one area of the colony to another during the morning web-building period showed that spiders that move lose about a half hour of foraging time as they find a new place to place their webs.

By varying the parameters of the model in different runs on the computer,

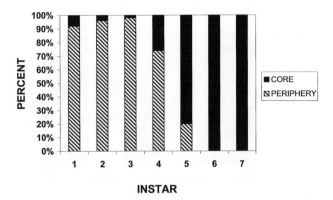

Figure 7.6. Illustration of dynamic model predictions, showing proportion of spiders expected to seek web sites in the core of colonies, based on their age and energetic state. Optimal choices for seven instars are shown, with model parameters based on field values of food availability, predation risk, and cost of moving (probability of aggressive encounters with larger or smaller spiders in a given position). Larger spiders, particularly those with higher energy levels, are predicted to seek the core (where predation risk is low) despite reduction in food availability. Smaller spiders, even those with high energy levels, are predicted to remain on the periphery, as the cost of moving (probability of aggression) is high (Jakob, Mangel, Porter, & Uetz, unpublished).

we are able to draw conclusions about which variables are more likely to be driving colony structure (Fig. 7.6). Interestingly, smaller spiders appear to be driven primarily by prey availability rather than predation risk. If we mimic an influx of predators by increasing predation risk in both core and periphery, we see that small spiders stay on the periphery even when predation risk increases up to sixfold. Larger spiders, in contrast, are predicted to move to the core even under low levels of predation, unless they are very low on energetic reserves. In an experiment in the field, we manipulated the energy level of the spiders and found that this is indeed the case: large spiders responded to starvation by moving to the periphery and to feeding by moving to the core, whereas small and medium spiders were most likely to be found on the periphery regardless of their energy level.

We also tested how spiders responded to a perceived predator attack. We monitored the orb web location of marked individuals for at least three consecutive days and calculated the average distance the orb web was moved from day to day. We then attacked these individuals with an artificial predator: a vibrator with an insect pin attached to it with which we approached spiders sitting on their orbs. The vibrator had a frequency of 95 Hz, well within the range of frequency of predatory wasp wingbeats (90 to 110 Hz; R. S. Wilcox, personal communication). Spiders appeared to treat the vibratory device as they would treat a wasp predator, by first waving their legs in a defensive manner, and finally by dropping from the orb on a silk line. Spiders, after the attack, returned to their web for the remainder of the foraging period. We located spiders on the day after they were attacked to see whether they had changed the locations of their orb webs. Spiders moved significantly farther after they were attacked by an artificial predator than before they were attacked (paired t-test on ln-transformed data; n = 29; p < 0.03). After buzzing, 20 of 29 spiders moved toward the core and 9 moved toward the periphery, which significantly differs from random movement (chi-squared goodness-of-fit test; p < 0.01).

As predictions of the model are well-supported by experimental field studies, we feel that the application of dynamic optimization modeling to this field system has been successful. Our research has confirmed the earlier conclusions of Rayor and Uetz (1990, 1993) that dynamic fitness trade-offs between resource acquisition and predation risk with spatial position influence the web location decisions of spiders and ultimately the structure of the group.

# New Construction: Future Research

So far, the picture that has emerged of the population dynamics and colony structure of colonial spiders suggests varying degrees in the permanency of colonial aggregations. Our research, as well as a wealth of natural history observations, has provided evidence for both a "parasocial" and a "subsocial" pathway in the evolution of group-living for *Metepeira* species. The weight of evidence for a parasocial pathway, that is, the origin of coloniality as a consequence of ecological factors, is more substantial. Originally proposed by Shear (1970) and supported by Buskirk (1981), Uetz (1988b), Rypstra (1989), and Piel (1997), this theory suggests that spiders aggregate in areas of high prey density (perhaps as a consequence of increased tolerance, or risk-averse behavior). Evolution of higher levels of sociality may be constrained by the negative impact of predators and parasitoids (Smith 1982; Hieber & Uetz 1990), or (in some cases) fostered by the high degree of relatedness within colonies and the potential for kin selection.

Evidence for the subsocial pathway, that is, the origin of coloniality owing to kin selection acting upon prolonged parent-offspring association or delayed dispersal of siblings, is based on the strong likelihood of inbreeding in aggregations, and is more circumstantial than substantive at this time. In other more highly social spider species, a number of studies using a variety of molecular techniques have documented high degrees of genetic relatedness within and between colonies of *Achaearanea wau* (Lubin & Crozier 1985), *Agelena consociata* (Roelofs & Riechert 1988), and *Anelosimus* sp. (Smith & Hagen 1996). Evidence for varying degrees of inbreeding in *Metepeira* is supported by electrophoretic data, but is as yet unconfirmed by more powerful molecular techniques. Estimates of heterozygosity for the relatively permanent aggregations of *M. incrassata* are lower than for seasonal and facultatively colonial *M. spinipes* and *M.atascadero* populations (Uetz et al. 1987).

However, observations of colonial *M. incrassata* through the rainy and dry seasons in tropical Fortin de las Flores also suggest that in most years, cyclic climate events (hurricanes, wind storms) have a destructive impact on colonies. The tendency of this species to re-form colonies after such events could foster fragmentation, mixing, and outbreeding. Alternating cycles of inbreeding and outbreeding could lead to considerable spatiotemporal isolation in the genetic structure of populations and colonies. Given the potential for high relatedness within colonies, and the possibility of kin selection (or even interdemic group selection), information on the genetic structure of populations of colonial spiders will be extremely valuable in understanding the evolution of social behavior in spiders. Thus, our future directions will involve using microsatellite DNA to determine the degree of genetic relatedness among colony members as a means of understanding the influence of genetic structure on the evolution of sociality.

Sadly, one impediment to continued study of colonial spiders has been the impact of urban and rural development on destruction of habitat and dependable field sites. For example, subsequent studies of the genetic structure of *M. spinipes* populations were rendered impossible, owing to the destruction of field sites by the expanding development of suburban Mexico City in the 1980s and 1990s. Continued monitoring of *M. incrassata* has also been difficult, as agricultural reform has led to large-scale clearing of coffee plantations for new varieties of coffee. Despite these setbacks, a recent discovery has allowed us to launch a new project, with potential to address unanswered questions.

Localized populations of *M. spinipes* in coastal California include groups of two to forty spiders in willow thickets along creeks and sloughs. Since the occurrence of these populations in riparian habitats along the coast is likely to be a consequence of bottom-up influences of El Niño (temporarily increased prey availability), there are three possible explanations: (1) risk-sensitivity— unrelated spiders attach webs together as a response to prey availability; (2) kin selection—increased prey has allowed improved survival or delayed dispersal of sib groups, such that colonies represent individuals from the same egg sac; (3) a combination of (1) and (2)—related individuals are preferentially grouping webs together to take advantage of increased prey. My UC colleague Ron DeBry and graduate student Gina Sagel are currently developing and using DNA techniques so that we may determine relatedness among colony members in these populations and test these alternative explanations.

## Evaluating Other Plans: Advice on Choosing a Model System

My advice to students of behavioral ecology in search of a model system is to choose a species and a project that captures your scientific curiosity, whether that curiosity is based in logic, skepticism, or simply fascination. This is important, because enthusiasm for the organism and question you work on is critical to sustain interest over the long term, as well as attract others as collaborators. In addition, drive, salesmanship, and cheerleading is often necessary to make a point, and it is much easier to "sell" a project you are passionate about. While some students may carve out a research niche by working in some unexplored aspect of a well-established model system, the opportunities for discovery of entirely new model systems—especially among invertebrate taxa—are enormous. If anything, my own fascination with colonial web-building spiders, and in trying to answer the question of why social behavior occurs where it shouldn't, should provide one example.

## Acknowledgments

This research was supported by National Science Foundation grants BSR-8615060 and BSR-9109970 and National Geographic Society Grants 3095-85 and 4428-90. I thank the Mexican Government, Direccion General de Conservacion y Ecologia de los Recursos Naturales, Direccion de Flora y Fauna Silvestres for permission to work in Mexico. I especially wish to thank Ana Valiente de Carmona and family, Becky Cotera, Blanca Alvarez, and Loli Alvarez-Garcia for the use of their properties as research sites. I am grateful for field assistance, manuscript review, and other forms of collaboration in research by Craig Hieber, Maggie Hodge, Linda Rayor, Beth Jakob, Karen Cangialosi, Stim Wilcox, Alison Mostrom, Sam Marshall, Adam Porter, Tom Kane, Mike Benton, Gail Stratton, Bob Hollis, Debbie (Fritz) Hayes, Andrea McCrate, David Kroeger, Jay Boyle, Steve Leonhardt, Veronica Casebolt, and Rebecca Forkner. Most important, I am grateful to my wife Kitty for tolerating my absence during periods of fieldwork.

## References

Benton MJ, Uetz GW, 1986. Variation in life-history characteristics over a clinal gradient in three populations of a communal orb-weaving spider. Oecologia 68:395–399.

Bertram BCR, 1978. Living in groups: predators and prey. In: Behavioral ecology: an evolutionary approach (Krebs JR, Davies NB, eds). Oxford: Blackwell; 64–96.

Burgess JW, 1978. Social behavior in group-living species. Symp Zool Soc Lond 42:69–78.

Burgess JW, Uetz GW, 1982. Social spacing strategies in spiders. In: Spider communication: mechanisms and ecological significance (Witt PN, Rovner JS, eds). Princeton: Princeton University Press; 317–351.

Buskirk RE, 1975. Coloniality, activity patterns and feeding in a tropical orb-weaving spider. Ecology 56:1314–1328.

Buskirk RE, 1981. Sociality in the Arachnida. In: Social insects, vol. II (Hermann HR, ed). New York: Academic Press; 281–367.

Cangialosi KR, Uetz GW, 1987. Spacing in colonial spiders: effects of environment and experience. Ethology 76:236–246.

Caraco T, 1981. Risk sensitivity and foraging groups. Ecology 62:527–531.

Caraco T, Pulliam HR, 1984. Sociality and survivorship in animals exposed to predation. In: A new ecology: novel approaches to interactive systems (Price PW, Slobodchikoff CN, Gaud WS, eds). New York: Wiley Interscience; 279–309.

Caraco T, Uetz GW, Gillespie RG, Giraldeau L-A, 1995. Resource-consumption variance within and among individuals: on coloniality in spiders. Ecology 76:196–205.

Cohen JE, 1971. Casual groups of monkeys and men. Cambridge, MA: Harvard University Press.

Darwin C. 1845. Voyage of the Beagle, noted in: Shear WA, 1970. The evolution of social phenomena in spiders. Bull British Arachnol Soc 1:65–76.

Foster WA, Treherne JE, 1981. Evidence for the dilution effect in the selfish herd from fish predation on a marine insect. Nature 293:466–467.

Gillespie RG, 1987. The role of prey availability in aggregative behavior of the orb weaving spider *Tetragnatha elongata*. Anim Behav 35:675–681.

Hamilton WD, 1964. The genetical evolution of social behavior. J Theor Biol 7:1–52.

Hamilton WD, 1971. Geometry for the selfish herd. J Theor Biol 31:295–311.

Hart BL, 1997. Behavioural defence. In: Host-parasite evolution: general principles and avian models (Clayton DH, Moore J, eds). London: Oxford University Press; 59–77.

Hieber CS, Uetz GW, 1990. Colony size and parasitoid load in two species of colonial *Metepeira* spiders from Mexico (Araneae: Araneidae). Oecologia 82:145–150.

Hingston RWG, 1932. A naturalist in the Guinea forest. New York: Longmans, Green.

Hodge MA, Uetz GW, 1992. Anti-predator benefits of single and mixed-species grouping by *Nephila clavipes* (L.) (Araneae:Tetragnathidae). J Arachnol 20:212–216.

Hodge MA, Uetz GW, 1996. Foraging advantages of mixed-species association between solitary and colonial orb-weaving spiders. Oecologia 107:578–587.

Inman AJ, Krebs J, 1987. Predation and group living. Trends Ecol Evol 2:31–32.

Jakob EM, Marshall SD, Uetz GW, 1996. Estimating fitness: a comparison of body condition indices. Oikos 77:61–67.

Kenward RE, 1978. Hawks and doves: factors affecting success and selection in goshawk attacks on wood pigeons. J Anim Ecol 47:449–460.

Lubin YD, 1974. Adaptive advantages and the evolution of colony formation in *Cyrtophora* (Araneae: Araneidae). Zool J Linn Soc 54:321–339.

Lubin YD, Crozier RH, 1985. Electrophoretic evidence for population differentiation in a social spider *Achaeranea wau* (Theridiidae). Insectes Sociaux 32:297–304.

McCrate A, 1989. Kleptoparasites: a cost of group-living for the colonial spider *Metepeira incrassata* (Araneae: Araneidae). Unpublished M.S. thesis, University of Cincinnati.

Mooring MS, Hart BL, 1992. Animal grouping for protection from parasites: Selfish herd and encounter–dilution effects. Behaviour 123:173–193.

Piel WH, 1997. The systematics of neotropical orb-weaving spiders in the genus *Metepeira* (Araneae, Araneidiae). Ph.D. thesis, Harvard University.

Pulliam HR, Caraco T, 1984. Living in groups: is there an optimal group size? In Behavioral ecology: an evolutionary approach. (Krebs JR, Davies NB, eds). Oxford: Blackwell; 122–147.

Pulliam HR, Milikan GC, 1982. Social organization in the non-reproductive season. In: Avian biology, vol. 6 (Farner DS, King JR, eds). New York: Academic Press; 45–87.

Rayor LS, Uetz GW, 1990. Trade-offs in foraging success and predation risk with spatial position in colonial spiders. Behav Ecol Sociobiol 27:77–85.

Rayor LS, Uetz GW, 1993. Ontogenetic shifts within the selfish herd: predation risk and foraging trade-offs with age in colonial web-building spiders. Oecologia 95:1–8.

Real L, Caraco T, 1986. Risk and foraging in stochastic environments. Ann Rev Ecol System 17:371–390.

Roeloffs R, Riechert SE, 1988. Dispersal and population-genetic structure of the cooperative spider, *Agelena consociata*, in west African rainforest. Evolution 42:173–183.

Rypstra AL, 1979. Foraging flocks of spiders: a study of aggregate behavior in *Cytophora citricola* Forskal (Araneae: Araneidae) in west Africa. Behav Ecol Sociobiol 5:291–300.

Rypstra AL, 1985. Aggregations of *Nephila clavipes* (L.) (Araneae, Araneidae) in relation to prey availability. J Arachnol 13:71–78.

Rypstra AL, 1986. High prey abundance and a reduction in cannibalism: the first steps to sociality in spiders (Arachnida). J Arachnol 14:193–200.

Rypstra AL, 1989. Foraging success of solitary and aggregated spiders: insights into flock formation. Anim Behav 37:274–281.

Shear WA, 1970. The evolution of social phenomena in spiders. Bull British Arachnol Soc 1:65–76.

Simon E, 1891. Observations biologiques sur les Arachnides. Ann Soc Entomol Fr 60:5–14.

Smith DR, 1982. Reproductive success of solitary and communal *Philoponella oweni* (Araneae: Uloboridae). Behav Ecol Sociobiol 11:149–154.

Smith DR, 1983. Ecological costs and benefits of communal behavior in a presocial spider. Behav Ecol Sociobiol 13:107–114.

Smith DR, 1985. Habitat use by colonies of *Philoponella republicana* (Araneae: Uloboridae). J Arachnol 13:363–373.

Smith DR, Hagen T, 1996. Population structure and interdemic selection in the cooperative spider *Anelosimus eximius*. J Evol Biol 9:589–608.

Spiller DA, Schoener TS, 1989. Effect of a major predator on grouping of an orb-weaving spider. J Anim Ecol 58:509–523.

Thompson WA, Vertinsky I, Krebs JR, 1974. The survival value of flocking in birds: a simulation model. J Anim Ecol 43:785–820.

Treherne JE, Foster WA, 1978. Group size and anti-predator strategies in a marine insect. Anim Behav 32:536–542.

Turner GF, Pitcher TJ, 1986. Attack abatement: a model for group protection by combined avoidance and dilution. Am Nat 128:228–240.

Uetz GW, 1986. Web building and prey capture in communal orb weavers. In: Spiders: webs, behavior, and evolution (Shear WA, ed). Stanford: Stanford University Press; 207–231.

Uetz GW, 1988a. Risk-sensitivity and foraging in colonial spiders. In: Ecology of social behavior (Slobodchikoff CA, ed). San Diego: Academic Press; 353–377.

Uetz GW, 1988b. Group foraging in colonial web-building spiders: evidence for risk sensitivity. Behav Ecol Sociobiol 22:265–270.

Uetz GW, 1989. The "ricochet effect" and prey capture in colonial spiders. Oecologia 81:154–159.

Uetz GW, 1996. Risk-sensitivity and the paradox of colonial web-building in spiders. Amer Zool 36:459–470.

Uetz GW, Burgess JW, 1979. Habitat structure and colonial behavior in *Metepeira spinipes* (Araneae: Araneidae), an orb-weaving spider from Mexico. Psyche 86:79–89.

Uetz GW, Cangialosi KR, 1986. Genetic differences in social behavior and spacing in populations of *Metepeira spinipes*, a communal-territorial orb weaver (Araneae: Araneidae). J Arachnol 14:159–173.

Uetz GW, Hieber CS, 1994. Group size and predation risk in colonial web-building spiders: analysis of attack-abatement mechanisms. Behav Ecol 5:326–333.

Uetz GW, Hieber CS, 1997. Colonial web-building spiders: balancing the costs and benefits of group-living. In: The evolution of social behaviour in insects and arachnids (Choe JC, Crespi BE, eds). Cambridge: Cambridge University Press; 458–475.

Uetz GW, Hodge MA, 1990. Influence of habitat and prey availibility on spatial organization and behavior of colonial web-building spiders. Nat Geogr Res 6:22–40.

Uetz GW, Kane TC, Stratton GE, 1982. Variation in the social grouping tendency of a communal web-building spider. Science 217:547–549.

Uetz GW, Kane TC, Stratton GE, Benton MJ, 1987. Environmental and genetic influences on the social grouping tendency of a communal spider. In: Evolutionary genetics of invertebrate behavior (Huettel MD, ed). New York: Plenum; 43–53.

Wilson EO, 1971. The insect societies. Cambridge, MA: Belknap Press.

Wilson EO, 1975. Sociobiology, the new synthesis. Cambridge, MA: Harvard University Press.

Wrona FJ, Dixon RW, 1991. Group size and predation risk: a field analysis of encounter and dilution effects. Am Nat 137:186–201.

PART II     Fish, Amphibian, & Reptile
            Model Systems

# 8

# Variation and Selection in Swordtails

Michael J. Ryan and Gil G. Rosenthal

## A View from the Field: 10 March 1999, Río Choy near Tamim, Mexico

We are at the birthplace of the Río Choy, in the foothills of the Sierra Madre of eastern Mexico. Water surges from deep underground, through the cave where we stand watching it cascade down a little waterfall. The water flows into a sunny blue pool flanked by ruins from the 1920s, when this *nacimiento* was a weekend resort. Myron Gordon, the eminent geneticist and ichthyologist, no doubt descended the long concrete staircase to the water, to peer at northern swordtails *Xiphophorus nigrensis*, courting in the shallows. He was followed by other geneticists, Klaus Kallman and Don Morizot, in the decades to follow.

Today the stairway's bottom half is a heap of rebar on the rocks, the diving board has collapsed, and they have melted down the great bronze plaque of mediocre poetry that was this place's centerpiece. The rock faces cast deep shade all day, and soft winds blow down the mountain and through dark stands of bamboo. On the surface, the *nacimiento* is a melancholy place.

Dive underwater and this changes. The section of staircase that has fallen underwater is now encrusted with a generous brown carpet of algae and tiny animals. A group of a dozen female *X. nigrensis* graze along this surface, taking little nibbles here and there. Among them is a large male, a brilliant blue on this sunny afternoon, who approaches a certain female and begins to court her. His dorsal fin fully raised, he executes a series of five rapid turns before her eyes. The lower rays of his caudal fin are extended into a long sword, black to us but probably bright to her. Unimpressed in any case, she returns to her feeding. The male, most likely seeking more promising ground, ventures into open water toward the next patch, trying to avoid the sharp teeth of the many Mexican tetras (*Astyanax mexicanus*) in his path. Sure enough, a large tetra nips off most of his long sword. He reaches his destination diminished but alive, ready to court again. He performs a brief display, similar to his courtship dance, to an intermediate-sized male, smaller than himself, followed by a brief chase. The intermediate male flees.

Our female, meanwhile, continues feeding. She is almost ready to mate, and would prefer to do so with a large male, a male who has waited five long months to mature into a robust, ornamented dancer. Instead, a bright yellow male, smaller than herself, surprises her as she grazes, forcibly inseminating her with no preliminaries whatsoever. Thirty days later, she gives birth to about two dozen young, who seek refuge and food among fine algal filaments. All of her sons will mature in only three months; like their father, they will be small and bright yellow throughout their adult lives. And like their father, once mature they will try to force themselves on females, abandoning any pretext of the seduction or cajoling characteristic of the large courting males.

We're at the birthplace of the Río Choy, the birthplace of *Xiphophorus nigrensis*, which lives nowhere else in the world but here. To see these things, we need only slip down the waterfall and out of the cave, into the light. Yet to truly understand these phenomena, we must admit that Plato had it right; we remain here, forced to discern pattern from the shadows of experiment and observation. They're pleasant little sirens, these bright fish, and the silhouettes they throw off yield moments of real revelation and insight into the wider workings of nature; but never, in their infinite variety, do they tolerate complacency. Each experiment, each new population we find in a forgotten ravine, yields its own share of surprises. Some of these surprises follow.

## Size Variation and the *P* Gene

The watchword for our studies of swordtails is *variation*—what does it mean, and how is it maintained? In nature, there is a striking disparity in size and style among these male swordtails. At one extreme are brilliantly blue, deep-bodied animals sporting large dorsal fins and swords, and engaging in the rapid figure eight that constitutes the courtship display. On the other extreme are small, slender, somewhat drably colored males lacking the elaborate fins and complex courtship of their larger counterparts. In fact, these smaller males have abandoned courtship altogether; they force themselves on females for sex rather than courting them.

Prior to the sociobiology revolution, reproductive behavior was viewed in the context of species specificity. The New Synthesis in evolutionary biology in the mid-twentieth century wrestled with the problem of the "species," and ethologists made critical contributions to this endeavor by documenting the species-specificity of both male courtship and female response to such behavior (Mayr 1982). Variation in reproductive behavior among males, as we see in the swordtails, was not given serious consideration. But sociobiology brought with it an emphasis on selection on individual variation within a population, and alternative mating systems presented a focal point for examining how such drastic behavioral variation can be maintained (Wilson 1975).

Alternative processes can account for qualitative variation in male mating behavior. Variation can result from phenotypic behavioral plasticity. For example, it can have an ontogenetic basis—younger males exhibit noncourting mating strategies until they achieve larger size (e.g., bullfrogs; Howard 1978). Others are merely "making the best of a bad situation" (Dawkins 1980). Environmentally induced variation has dealt them a poor hand; they lack the strength, fortitude, attractiveness, or charm to compete successfully for females. There are certainly a plethora of other environmental and gene-by-environment interactions that bias an animal's phenotypes, including but not restricted to alternative mating strategies (West-Eberhard 1989).

It is also possible that alternative mating behaviors can be strongly or even primarily influenced by genetic variation. If there is heritable genetic variation for courting versus noncourting strategies, both forms could be maintained in the population under certain conditions of spatial and temporal fluctuation in the intensity of selection. Perhaps a more likely process would be the maintenance of alternative mating behaviors in genetic equilibrium; that is, all strategies would have equal lifetime reproductive success. There are only a few mating systems in which alternative mating strategies appear to be in genetic equilibrium: swordtails, ruffs (Lank et al. 1995), and stomatopods (Shuster & Wade 1991). There were none when we were first enticed into the blue waters of the Río Choy.

A challenge to understanding the genetic basis of behavior of animals in the wild is having a system that is amenable to both quantitative or molecular genetic and behavioral studies in the lab, or having a system in which the genetic variation underlying the behavior is simple and readily identifiable in the wild. Few mating systems have these qualities; there are systems that are wonderful for laboratory genetic analysis of behavior, such as aggregative behavior in *C. elegans* (DeBono & Bargmann 1998) and mating behavior in fruit flies (Hall 1994). But these systems are not usually characterized by variation in mating strategy. The genetic variation underlying alternative mating strategies in the wild is rarely known. Swordtails, however, offer a clear exception.

The swordtails and platyfish, genus *Xiphophorus*, are members of the live-bearing family Poeciliidae. It is not clear if swordtails and platys are each monophyletic, but monophyly seems certain for both the genus as a whole and the northern swordtails (Rauchenberger et al. 1990; Borowsky et al. 1995; Morris et al. 2001; cf. Meyer et al. 1994). Species in this group are distributed throughout the Río Pánuco Basin in northeastern Mexico, in the foothills and areas just east of the Sierra Madre Oriental (Fig. 8.1). The consensus of a variety of phylogenetic analyses is that the group consists of nine species; these species are in three clades of three species each. The species we have studied most extensively are in the pygmy swordtail species group: *X. nigrensis, X. multilineatus, X. pygmaeus*.

Klaus Kallman and his colleagues have been studying the genetics of phenotypic variation in *Xiphophorus* for the past three decades (reviewed in Kallman 1989). These animals sport a variety of traits, such as color and

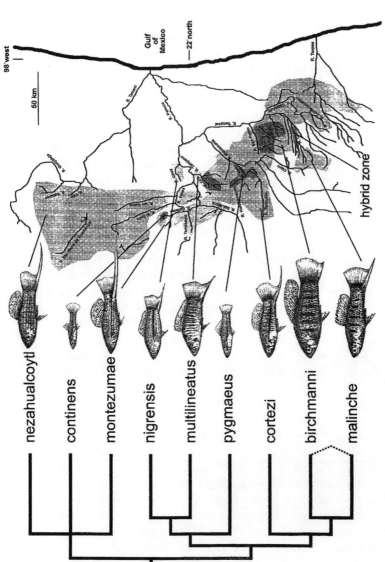

Figure 8.1. The Río Pánuco Basin swordtails. Phylogenetic hypothesis after Rauchenberger et al. (1990), Borowsky et al. (1995), and Morris et al. (2001). The tritomy in the *montezumae* group reflects the placement of *X. montezumae* and *X. nezahualcoytl* as sister species in Rauchenberger et al.'s (1990) hypothesis, and of *X. montezumae* and *X. continens* in the later phylogenies. Map and ranges after Rauchenberger et al. (1990), with range extensions based on recent collections by GGR. Darker areas indicate sympatry between two species; "hybrid zone" refers to *X. birchmanni* X *X. malinche*. Line drawings by Christopher Elmore.

melanophore patterns, whose genetics have been well-characterized. Many of these traits are exhibited as if they are under simple allelic control; the variation is often linked to the Y chromosome. One of the more astounding forms of variation is in body size. In poeciliid fish, male growth decreases drastically upon reaching sexual maturity. In *Xiphophorus nigrensis*, one of the species that has received most of our attention, the standard length (SL; snout to hypural plate, a basic measure of size in fish) of mature males can vary from 18 to 38 millimeters, as measured from the snout of the fish to the base of its tail. Size breeds true through the father. Large males produce large sons while small fathers produce small sons. The size of the mother has no detectable influence on how large their sons grow. Kallman's analysis reveals three size genotypes in *X. nigrensis*: small (> 26 mm SL), intermediate (26 to 31 mm), and large (< 31 mm). There is only slight overlap between the size classes, and the narrow-sense heritability for size in lab-bred animals is greater than 90 percent (Ryan & Wagner 1987; Kallman 1989; Dries et al. 2001).

The gene responsible for size variation in these fish is the pituitary (*P*) gene. It achieves its effect by activating the hypothalamic-pituitary-gonadal axis, resulting in the circulation of androgens in males which initiates sexual maturity and drastically halts growth. The earlier androgens are circulated, the sooner, and smaller, the males mature. In *X. nigrensis* there are three *P* alleles, *s*, *I*, and *L*. The *P* gene on the X chromosome appears fixed for *s*; females mature early but continue to grow, as do most poeciliid females.

## Advantages of the Swordtail System

As of the mid-1980s, the studies of Kallman and colleagues seemed not to have penetrated the literature in behavioral ecology, even though the maintenance of *P* gene variation was synonymous with the critical issue of the maintenance of behavioral variation. Kallman's work not only provided a motivation for our studies of behavioral ecology, but also elucidated an elegant genetic system whose logistics provide a number of advantages.

The *P* gene, as well as several pigment and melanophore patterns, is Y-linked and shows high heritability. Combined with the fact that females are live-bearers, this readily allows paternity analysis to estimate changes across generations in the frequencies of alleles underlying the traits. For these Y-linked traits, assessing the phenotype provides the male genotype. This can be estimated directly from field collections or surveys. Females can be collected and returned to the lab, and the genotype of their sons can then be quantified. Few systems allow one to document changes in allele frequencies of traits under sexual selection.

Swordtails tend to live in cool, clear, slowly moving streams in mountain foothills. They quickly habituate to a human observer, and underwater observation of their behavior offers no special challenges. Most of their behavior is restricted to within a meter or so of the surface, permitting easy observation by snorkelers. Males do not defend permanent territories but maintain individual distances while feeding on small invertebrates in plants and algae on rocks. Females move through these areas, where they are courted and mated by males.

Swordtail behavior can be studied in more detail in the lab. The males readily court and fight with one another in captivity, and females exhibit mating preferences in free-ranging tests or when males are confined behind glass partitions. A major advance in our studies occurred when it was shown that female swordtails, *X. helleri*, respond to video playbacks of courting males (Rosenthal et al. 1996). Subsequent studies have used frame-by-frame (Rosenthal & Evans 1998) and synthetic animations (see below), which allow almost unlimited degrees of freedom in varying stimuli. The study of visual signaling has lagged behind that of acoustic signaling, largely due to the relative ease with which acoustic parameters can be quantified and manipulated. The importance of these advances in stimulus presentation should not be underestimated.

Finally, knowledge of the phylogenetic relationships within northern swordtails provides the basis for comparative studies. The detailed phylogenetic analysis of the northern swordtails by Rauchenberger et al. (1990) was recently challenged by an analysis of Meyer et al. (1994) of the entire genus *Xiphophorus*. Recent studies by Borowsky et al. (1995) and Morris et al. (2001) largely support the original phylogenetic hypothesis advanced by Rauchenberger et al. (1990; Fig. 8.1).

We are not able to summarize the large number of studies we have conducted on sexual selection in swordtails. Instead, we will concentrate on two earlier series of studies that provide the foundation for much of our subsequent and future work in this system.

## Genetic Equilibrium and Alternative Mating Strategies

We first asked how the *P* gene polymorphism is maintained in wild populations of *X. nigrensis*. Populations are polymorphic for three *P* genotypes that result in distinct size classes. We conducted paternity analysis for females collected from a population in which we estimated the frequency of male size classes and, by extension, *P* genotypes (Ryan et al. 1990). Across generations, there was a significant decrease in the *P* alleles for small size relative to the *P* alleles for intermediate and large sizes, while the relative reproductive success of the intermediate and large males did not differ. The reproductive success of small males was only 44 percent that of intermediate and large males. This raised two obvious questions: What is the nature of selection acting against small males, and how are small males maintained in the population?

As noted above, males of different sizes vary in their mating behavior (Ryan & Causey 1989). Males below 26 millimeters rely almost exclusively on chasing after females in quick darting motions and rarely exhibit the courtship behavior shown by larger males (Fig. 8.2). Female mate choice studies using live males show a preference for intermediate and large males over small males but no preference between males of intermediate and large size (Ryan et al. 1990). These results mirror the differences in reproductive success seen in the wild. They do not, however, elucidate the relative contribution of body size and behavior to male attractiveness.

Male-male interactions also favor larger body size. In the field, larger males had greater access to females than did smaller males (Morris et al. 1992). Larger males also excluded smaller males from their territories more often than smaller males excluded larger males. In the laboratory, larger males blocked access to females more often than did smaller males. Large size usually provided an advantage in male-male competition regardless of the size classes of the males. In the female choice studies, however, the effect of male size was more categorical: females preferred males in the larger and intermediate size class to small males, but did not discriminate between males in the intermediate and large size classes.

Selection generated by female choice and male competition favors larger males over small males and, by extension, acts against the $s$ allele at the Y-linked $P$ locus. Size variation is under a strong genetic influence in this species; the alternative mating strategies are not conditional. There are several processes that could maintain genetic variation, including spatially or temporally varying selection. But one must remember that the differences in size at maturity among the size classes results from differences in time to maturity. Males of different sizes grow at similar rates but reach maturity at different times. Conveniently, males cease depositing otolith rings when they stop growing (Morris & Ryan 1990). Counts of otolith rings from males in the field showed that small, intermediate, and large males matured in the wild at average ages of 78, 90, and 124 days, respectively. Although small males have lower instantaneous reproductive success than larger males, that is, they are less likely to mate on any given day, they enter the reproductive lottery sooner. The costs to smaller males derived from sexual selection might be offset by advantages of time to sexual maturity. As the daily mortality rate in a population increases, the advantage to maturing earlier should increase as well.

Given our data on the relative reproductive success and time to maturity of small versus larger males, we calculated the daily mortality rate necessary to yield equal fitnesses among the genotypes (Ryan et al. 1992). That rate is 0.028/day. Despite considerable effort, we were not able to measure mortality rates in the field. But we used a more circuitous route to ascertain the

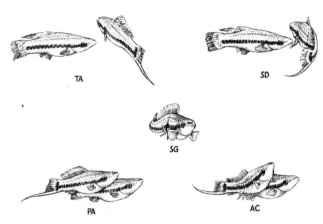

Figure 8.2. Mating behavior of *Xiphophorus nigrenis*: transverse approach (TA), sexual display (SD), sigmoid display (SG), parallel approach (PA) and attempted complation (AC).

validity of our mortality estimate. We expect population size to be constant over an ecologically relevant time span, thus the rate of population increase, $r$, should be zero. We combined data on the female fecundity schedule, which includes age at sexual maturity, brood size, and interbrood interval (Morris & Ryan 1992), with our estimate of daily mortality to calculate $r$. If our estimate were much lower than the true mortality rate, the population should drastically increase in population size: $r$ would be strongly positive. If our estimate of $r$ were too high, the population would go extinct: $r$ would be strongly negative. Using our empirical data on the female fecundity schedule and our estimate of daily mortality of 0.028/day we calculated an $r$ of 0.00. This suggests that the daily mortality rate needed to maintain the $P$ genotypes in equilibrium is the same mortality rate needed to maintain the population at a constant size.

Our model supports the hypothesis that $P$ genes are maintained in genetic equilibrium in *X. nigrensis*. But caution needs to be exercised. The 95 percent confidence intervals around the mortality rate generated from a bootstrap statistic are relatively large, 0.006 to 0.06/day. Furthermore, our analysis assumes that mortality does not vary among the genotypes or between the sexes. We can envision scenarios that would favor the survivorship of each genotype and sex. For example, large courting males might suffer greater mortality because of their conspicuousness, but smaller males might be more vulnerable because they are within the gape limit of more predators. Nevertheless, this simple model supports the contention that the variation in body size and the $P$ alleles that underlie this variation can exist in the wild because they result in very different types of males with very similar Darwinian fitness.

## When Size Doesn't Vary, Does It Matter?

South of the Río Choy, another *nacimiento* births another river, the Río Huichihuayán. It is a tributary of the Río Axtla, whose drainage is home to *X. pygmaeus*. The specific name refers to the unusually small size of males (Hubbs & Gordon 1943); males look and behave much like *s* males in *X. nigrensis*, with slender, swordless bodies and lacking courtship (Kallman 1989). Interspecific crosses suggest that *X. pygmaeus* are in fact fixed for the *s* allele. Parsimony analysis suggests that alleles for larger size classes were secondarily lost in *X. pygmaeus* (Morris & Ryan 1995). The major predators of swordtails, Mexican tetras *Astyanax mexicanus*, are far more abundant in *X. pygmaeus* habitat than elsewhere; predation pressure could have driven the *s* allele to fixation in an ancestor of *X. pygmaeus*.

Although the large, courting phenotype was lost in *X. pygmaeus*, females retained the ancestral preference for these sexually dimorphic males. Experiments by Ryan and Wagner (1987) showed that female *X. pygmaeus* preferred large, courting *X. nigrensis* males over conspecifics.

From the mid-1930s to 1982, ichthyologists and hobbyists made extensive collections throughout the range of *X. pygmaeus*. All of the males in museum collections were below 29 millimeters SL, and most were diminutive, below 24 millimeters SL. In 1988, we visited the *nacimiento* of the Río Huichihuayán, which no one had sampled before, and were surprised to find large males—some twice the size of the *X. pygmaeus* males previously known. A year later, we found these behemoths downstream, at the town of Huichihuayán. This site had been well-surveyed over the previous fifty years, and not a single large male had been found. Additional collections at several localities suggested a recent spread of large body size (Morris & Ryan 1995).

These large males were not like the larger males in other northern swordtails: like their smaller counterparts, they were slender-bodied, with short dorsal fins and lacking swords. We never observed them courting, either in the wild or in the laboratory. Behaviorally and morphologically, they were simply "blown-up" versions of small males. We were able to rule out a *P* gene system for the inheritance of this phenotype: large males produced only small sons, both in intraspecific crosses and in crosses to *X. nigrensis* (Morris & Ryan 1995; Dries et al. 2001). We have tried for over ten years to recover the large male phenotype in the laboratory, both in aquaria and in large outdoor ponds. So far, we have yet to produce a single captive-born large male *X. pygmaeus*. The genetic and environmental mechanisms producing this phenotype remain mysterious.

Females from la Y-Griega Vieja—a hamlet downstream from Huichihuayán, where large males are rare—have retained the ancestral preference for large males, preferring both large, courting *X. nigrensis* and the large *X.*

*pygmaeus* from upstream over their own small males. Yet at the upstream sites, where high frequencies of large males have been found, females fail to prefer the larger males. This loss of preference appears to be stable over time: females tested in 1988 and in 1993 showed identical patterns. These females also failed to prefer courting *X. nigrensis* over size-matched, noncourting conspecifics, suggesting that both the preference for large size and that for courtship had been secondarily lost (Morris et al. 1995).

Why have females from populations with large males lost the preference? The obvious suggestion is that there is a cost associated with mating with large males (Morris et al. 1995; Holland & Rice 1998). Yet large males produce normal, small male offspring in the laboratory. Females mated to large and to small males produce equal numbers of fry, and there is no detectable difference in the viability or fertility of offspring (Dries et al. 2001). While it is of course impossible to exclude the possibility that our methods are insensitive to small fitness differences in the offspring of large and small males, the lack of any *detectable* fitness cost of mating with large males makes it unlikely that loss of preference would have swept through the Huichihuayán population in at most sixteen years, or about forty-eight *X. pygmaeus* generations.

Recent data suggests that heterospecifics may play a role in the loss of size preference (Rosenthal & Ryan, in review). *X. pygmaeus* are sympatric throughout their range with *X. cortezi*, which share many characteristics with larger morphs of *X. nigrensis*. Male *X. cortezi* are on average much larger than male *X. pygmaeus*, ranging between 24 and 55 millimeters (Kallman 1989; personal observation), have deep bodies and swords, and perform courtship displays (Franck 1970). There is no evidence to suggest that the large *X. pygmaeus* males arose from a hybridization event with *X. cortezi*, but these and other swordtails can be readily crossed in the laboratory. Both *X. pygmaeus* and *X. cortezi* inhabit thickly vegetated, shallow areas (approximately one meter in depth) of slow to moderate current. There is broad overlap in their use of space, and individuals of the two species are frequently observed within a few centimeters of one another—close enough for a mating opportunity. Could the loss of preference for large size in *X. pygmaeus* be associated with avoidance of *X. cortezi*?

We asked females to choose between *X. cortezi* and size-matched *X. nigrensis* males. Since *X. nigrensis* are in the sister group to *X. pygmaeus*, we controlled for phylogenetic distance by also comparing responses to *X. cortezi* versus size-matched *X. malinche*, sister to *X. cortezi* (Fig. 8.1). Responses paralleled preferences for body size: females from the *nacimiento* showed strong preferences for both *X. malinche* and *X. nigrensis* over the sympatric *X. cortezi*, while females from Y-Griega failed to show a preference in either case. This pattern appeared to be due to correlated responses in individual females: when we tested the same *X. nigrensis* male against a male *X. pygmaeus* and a male *X. cortezi*, the females that showed the strongest preference for *X. nigrensis* against *X. pygmaeus* showed the weakest for *X. nigrensis* over *X. cortezi*, and vice-versa. There is thus a negative correlation between species recognition—avoidance of heterospecifics—and a mating preference for large size. The loss of large size preference in *X. pygmaeus* females may thus be a correlated consequence of selection against mating with a closely related swordtail (Rosenthal & Ryan, in review).

This still leaves the mystery of the large males—how they are generated, and if there are any fitness consequences to mating with them. These large males are now appearing only sporadically in collections at the *nacimiento* and at Huichihuayán, so we may soon lose our chance to understand this phenomenon.

## Natural Selection on Sexually Dimorphic Traits

No account of the behavioral ecology of *Xiphophorus* would be complete without mention of the trait that gave the genus its name. Males in most swordtail species have the lower rays of the caudal fin extended into a conspicuous "sword" bearing high-contrast pigment patterns. The sword starts growing around the age of sexual maturity. Female *X. helleri* prefer to mate with males with longer swords (Basolo 1990). This preference is shared by females in species that diverged from swordtails prior to the appearance of the ornament (Basolo 1995), suggesting that swords arose in response to a preexisting bias on the part of females. This bias appears to reflect a more general preference for large body size: a swordless male enlarged to the same total length as a sworded male is equally effective at eliciting preference (Rosenthal & Evans 1998). Swords are metabolically inexpensive to produce, and males allocate proportionately more energy to swords when on a restricted diet (Basolo 1998). Given that swords are cheap to produce and that they exploit a shared ancestral mating preference, one should expect them to be ubiquitous throughout the swordtails.

Yet sword length varies considerably across our nine study species, ranging from the hypertrophied sword of *X. montezumae*, which exceeds the length of the rest of the body, to its close relative *X. continens*, which lacks the sword altogether (Fig. 8.1). Across populations of *X. multilineatus*, there is a threefold difference in mean sword length. Within populations of *X. multilineatus* and *X. nigrensis*, swords are highly polymorphic; one large male may have an elongate ornament, another none at all. What are the factors responsible for maintaining this variation?

Predation pressure is an obvious candidate. Signals that are more conspicuous to potential mates are often more conspicuous to predators as well (Endler 1980; Zuk & Kolluru 1998). Based on gut content analyses and total abundance, the major visual predator of swordtails is the Mexican tetra, *Astyanax mexicanus*, a widely distributed, omnivorous, toothy relation of the piranha (Rosenthal et al. in press). These and other small characid fish are infamous throughout Latin America as *pica-culo*, or ass-nippers, for their attacks on swimmers. They also nip the posteriors of swordtails; we commonly find otherwise healthy males with the sword or other portions of the caudal fin bitten off.

The visual preferences of *A. mexicanus* parallel those of female swordtails. Like females, they go preferentially to the large and intermediate morphs over small males, and yet fail to show a preference for large males over intermediate males. *A. mexicanus* prefer male *X. multilineatus* with long swords over males with artificially shortened swords. This preference appears to be ancient, rather than the result of coevolution with the prey; tetras from localities far outside the range of swordtails show the same preference (Rosenthal et al. in press).

Swordtails have responded to predation in two ways. With the exception of *X. malinche*, which lives in *Astyanax*-free waters, the swords of the Río Pánuco *Xiphophorus* have a strong component in the ultraviolet. The lenses of female swordtail eyes transmit ultraviolet light, while the lenses of *Astyanax* eyes, like ours, filter it out. Swords that appear black or clear to predators, both characid and scientist, may thus exhibit a private, high-contrast ultraviolet band to females (M. Cummings, G. Rosenthal, & M. Ryan, unpublished data).

Selection has also favored an overall reduction of sexually dimorphic traits in high-predation areas. Recall that body size is highly correlated with age at maturity: if males mature earlier, they are smaller throughout their lives. A basic prediction of life history theory is that predation pressure should drive down age at first reproduction (Stearns 1992; see also Ryan et al. 1992). Mean male body size is negatively correlated with predator density across swordtail populations; this may be due to the predators' visual preference for large males, but it may also simply reflect increased predation on juveniles (G. Rosenthal, M. Stephens, & M. Ryan, unpublished data).

Sword length is also negatively correlated with *A. mexicanus* density. Yet sword length is correlated with body size, and when we expressed sword length as a function of body size the relationship with predation disappeared. The allometry of sword length on body size—the slope of the relationship within each population—actually showed a *positive* relationship with predation! Swords were proportionately larger as a function of body size in high-predation regimes. This suggests to us that swords may represent yet another alternative mating tactic: they are inexpensive to produce and grow after sexual maturity, and they elicit the same female preference as an equivalent increment in body size. In high-predation regimes, it may thus benefit males to mature early and start growing a sword. Yet swords are themselves attractive to predators, and are easy to lose. Variation in sword length may thus reflect a compromise between accommodating lethal predation, by maturing early, and sublethal predation, by minimizing sword expression (G. Rosenthal, M. Stephens, & M. Ryan, unpublished data).

## Summary and Future Prospects

We were initially drawn to swordtails by the simple genetic mechanisms underlying variation in suites of conspicuous traits. Yet correlated traits cry out for dissection: What are the factors maintaining the correlation? Are all traits equally important from the point of view of female preference, or are females choosing males based on only a subset of these traits? We have begun to address these questions by playing back synthetic animations to females. We begin by using population parameters—the mean and standard deviation of a slew of behavioral and morphological measures—to construct three-dimensional models of large, courting males and small, noncourting males. We can then vary these parameters independently of one another and ask the females which ones are relevant. Is a large, courting male with a sword as effective against a small male when shrunk down to the same total length as that small male? Can a small, slender, swordless male be made attractive if he courts at a rate one standard deviation above the mean for large males? This approach will allow us to understand the precise nature of the female preferences maintaining the $P$ gene polymorphism in species like *X. nigrensis.* Moreover, it will allow us to elucidate the loss of preference for large size in *X. pygmaeus.* Have preferences for other traits, like body shape and courtship, been reduced along with the preference for large size? Are preferences for *X. cortezi*-like traits more reduced than others?

Using synthetic animations to break down correlated phenotypes is a powerful approach, yet it would be ideal if these correlated trait complexes could be dissected in nature—if there were wild populations with males recombinant for suites of sexually dimorphic traits. Yet despite the fact that swordtails hybridize readily in the laboratory, until now there have been no well-documented cases of hybridization in nature. In 1997, we stumbled on natural hybrids between *X. birchmanni* and *X. malinche* in the Río Calnali, in the state of Hidalgo. The two parental species are at opposite extremes of the signaling spectrum—*X. birchmanni* is vertically dimorphic, with a nuchal hump, a sail-like dorsal fin, a deep body, no sword, and parallel vertical bars, while *X. malinche* is a typical swordtail with a long sword, a more modest dorsal fin, a slender body, and irregular blotches on the flank. Most individuals in the hybrid zone are recombinant for these traits. We can now ask questions about female preferences for particular trait combinations found in nature, a task facilitated by the geographic structure of the hybrid zone, which shows an upstream-to-downstream gradient from *malinche* traits to *birchmanni* traits. Curiously, the sword, for which females show an ancestral preference, disappears earlier than other *malinche* traits, including allozyme markers. Perhaps novel trait combinations are more attractive to females than this ornament (Rosenthal et al., unpublished data).

## A View from the Field: 15 March 1999 above Calnali, Mexico

On this Saint Patrick's Day the steep relief and green hills are reminiscent of Ireland, but we have just hiked up to the ridge which separates the Río Calnali from the next valley, which you cannot reach by road. We get our first glimpse of the Río Pochula far below, a thin ribbon gleaming in the noonday sun. People in Calnali say there are deep pools with bright little fish—*poxtas de colores*. This may be a second, replicate contact zone between *X. birchmanni* and *X. malinche*. Will we see rampant hybridization like we do behind us, with the same distribution of traits? Will we climb all the way down the mountain only to find a stream that is swordtail-free? Or will these small treasures of the Sierra Madre once again surprise us with something totally unexpected? There's no alternative but to go and look— and no chance we'll make it back by nightfall.

## Acknowledgments

We are extremely grateful to Klaus Kallman and Don Morizot for introduction to the system; B. Causey, D. Hews, and W. Wagner for collaborating on early studies; M. Morris for her critical work on this system that is reviewed here; M. Cummings, T. Flores Martinez, F. García de León, and M. Stephens for collaboration; and to everyone who has provided invaluable assistance in the laboratory and in the field. We are indebted to Christopher Elmore for the line drawings in Figure 8.1. We appreciate the financial support from the National Science Foundation, the National Geographic Society, Dr. Lorraine Stengl, and the Department of Zoology, University of Texas at Austin. L. Gilbert, J. Crutchfield, and A. Alexander of the Brackenridge Field Laboratory, University of Texas at Austin, have provided facilities for stock maintenance and numerous experiments. We appreciate logistical support provided by the Laboratory of Zoology at the Technological Institute of Ciudad Victoria, Mexico. We are grateful to D. Hendrickson of the Texas Memorial Museum, Héctor Espinosa of the Mexican National Museum of Ichthyology, and A. Narvaez and the United States Embassy in Mexico for assistance with collecting permits. We thank the Mexican National Institute of Fisheries, National Institute of Ecology, and Foreign Ministry for allowing these studies. Finally, the people of the Huasteca region deserve special thanks for their warm hospitality and for their commitment to the rivers and streams they steward.

# References

Basolo AL, 1990. Female preference for male sword length in the green swordtail (Pisces: Poeciliidae). Anim Behav 40:332–338.

Basolo AL, 1995. Phylogenetic evidence for the role of a preexisting bias in sexual selection. Proc Roy Soc Lond B 259:307–311.

Basolo AL, 1998. Shift in investment between sexually-selected traits: tarnishing of the silver spoon. Anim Behav 55:665–671.

Borowsky RL, McClelland M, Cheng R, Welsh J, 1995. Arbitrarily primed DNA fingerprinting for phylogenetic reconstruction in vertebrates: the *Xiphophorus* model. Mol Biol Evol 12:1022–1032.

Dawkins R, 1980. Good strategy or evolutionarily stable strategy? In: Sociobiology: beyond nature/nurture? (Barlow GW, Silverberg J, eds). Boulder, CO: Westview Press.

DeBono M, Bargmann CI, 1998. Natural variation in a neuropeptide Y receptor homolog modifies social behavior and food response in *C. elegans*. Cell 94:679–689.

Dries L, Morris M, Ryan M, 2001. Why are some pygmy swordtails large? Copeia 2001:355–364.

Endler, JA, 1980. Natural selection on color patterns in *Poecilia reticulata*. Evolution 31:76–91.

Endler JA, 1983. Natural and sexual selection on color patterns in poeciliid fishes. Environ Biol Fishes 9:173–190.

Franck VD, 1970. Verhaltensgenetische Untersuchungen an Artbastarden der Gattung *Xiphophorus* (Pisces). Z Tierpsychol 27:1–34.

Hall JC, 1994. The mating of a fly. Science 264:1702–1714.

Holland B, Rice WR, 1998. Perspective: chase-away sexual selection: antagonistic seduction versus resistance. Evolution 52:1–7.

Howard RD, 1978. The evolution of mating strategies in bullfrogs, *Rana catesbeiana*. Evolution 32:850–871.

Hubbs CL, Gordon M, 1943. Studies of cyprinodont fishes. XIX. *Xiphophorus pygmaeus*, new species from Mexico. Copeia 1943:31–33.

Kallman KD, 1989. Genetic control of size at maturity in *Xiphophorus*. In: Ecology and evolution of livebearing fishes (Poeciliidae) (Meffe GK, Snelson FF, eds). Englewood Cliffs, NJ: Prentice-Hall; 163–185.

Lank DB, Smith CM, Hanotte O, Burke T, Cooke F, 1995. Genetic polymorphism for alternative mating behavior in lekking male ruff *Philomachus pugnax*. Nature 378:59–62

Mayr E, 1992. The growth of biological thought. Cambridge, MA: Harvard University Press.

Meyer A, Morrissey J, Schartl M, 1994. Recurrent origin of a sexually selected trait in *Xiphophorus* fishes inferred from a molecular phylogeny. Nature 368:539–542.

Morris MR, Batra P, Ryan MJ, 1992. Male-male competition and access to females in the swordtail *Xiphophorus nigrensis*. Copeia 1992:980–986.

Morris MR, De Queiroz K, Calhoun SW, Morizot DC, 2001:65–81. Phylogenetic relationships among the northern swordtails (*Xiphophorus*) as inferred from allozyme data. Copeia 2001:65–81.

Morris MR, Ryan MJ, 1990. Age at sexual maturity of male *Xiphophorus nigrensis* in nature. Copeia 1990:747–751.

Morris MR, Ryan MJ, 1995. Large body size in the pygmy swordtail *Xiphophorus pygmaeus*. Biol J Linn Soc 54:383–395.

Morris MR, Wagner WE, Ryan MJ, 1996. A negative correlation between trait and mate preference in *Xiphophorus pygmaeus*. Anim Behav 52:1193–1203.

Rauchenberger M, Kallman KD, Morizot DC, 1990. Monophyly and geography of the Rio Panuco Basin swordtails (Genus *Xiphophorus*) with descriptions of four new species. American Museum Novitates 2975:41.

Rosenthal GG, Evans CS, 1998. Female preference for swords in *Xiphophorus helleri* reflects a bias for large apparent size. Proc Natl Acad Sci USA 95:4431–4436.

Rosenthal GG, Evans CS, Miller WL, 1996. Female preference for a dynamic trait in the green swordtail, *Xiphophorus helleri*. Anim Behav 51:811–820.

Rosenthal GG, Flores Martinez TY, Garcia de Leon FJ, Ryan MJ, in press. Shared preferences by predators and females for male ornaments in swordtails. Amer Nat.

Rosenthal GG, Ryan MJ, in review. A negative association between species recognition and conspecific mating preferences in the swordtail *Xiphophorus pygmaeus*.

Ryan MJ, Causey BA, 1989. "Alternative" mating behavior in the swordtails *Xiphophorus nigrensis* and *Xiphophorus pygmaeus* (Pisces: Poeciliidae). Behav Ecol Sociobiol 24:341–348.

Ryan MJ, Hews DK, Wagner WE, 1990. Sexual selection on alleles that determine body size in the swordtail *Xiphophorus nigrensis*. Behav Ecol Sociobiol 26:231–237.

Ryan MJ, Pease CM, Morris MR, 1992. A genetic polymorphism in the swordtail *Xiphophorus nigrensis*: testing the prediction of equal fitnesses. Amer Natur 139:21–31.

Ryan MJ, Wagner WE, 1987. Asymmetries in mating preferences between species: female swordtails prefer heterospecific males. Science 236:595–597.

Shuster SK, Wade MJ, 1991. Equal mating success among male reproductive strategies in a marine isopod. Nature 350:608–610.

Stearns SC, 1992. The evolution of life histories. Oxford: Oxford University Press.

West-Eberhard MJ, 1989. Phenotypic plasticity and the origins of diversity. Ann Rev Ecol Syst 20:249–278.

Wilson EO, 1975. Sociobiology. Cambridge, MA: Harvard University Press.

Zuk M, Kolluru G, 1998. Exploitation of sexual signals by predators and parasitoids. Q Rev Biol 73:415–438.

# 9

# Learning from Lizards

Judy Stamps

From an early age, I was convinced that we share our planet with "aliens," creatures who speak their own languages and who view the world from perspectives quite different than our own. Many children probably begin life this way, but are later trained to view animals as inferior beings (and hence unworthy of serious attention) or provided such a surfeit of pets that they eventually outgrow their early interest in animals, and focus instead on members of their own species. Neither of these things happened to me: my parents respected and encouraged my interest in animals while never quite understanding it, and due to several factors, I never met enough animals when I was growing up. One reason is that my natal habitat (San Francisco) is low with respect to biodiversity, another was that as a child I had allergies to a long list of substances, including fur and feathers. As a result, the only pets allowed in the house were lizards and fish, and observations of homeotherms were restricted to the occasional urban birds that alighted in the backyard, or birds and mammals glimpsed during vacations and field trips.

In the days before Jane Goodall and other behavioral behaviorists were presented on TV as role models, it was not apparent to anyone in my family that one could make a living observing animals, and there were many occasions when I was caught staring fixedly at fish in my aquarium, and was told to stop wasting time and start doing my homework. This state of affairs lasted until I reached college at the University of California at Berkeley, and discovered a course called "animal behavior." The professor, Dr. George Barlow, not only revealed that many interesting intellectual puzzles could be addressed, if not solved, by watching animals, but also let drop in one of his lectures an offhand comment that "lizards might be good experimental subjects for studying stereotyped motor patterns," based on earlier work by Carpenter and his colleagues on the headbob patterns many lizards use in communication. Based on my limited experience with pet *Anolis carolinensis* lizards, and filled with a confidence and naivete characteristic of sophomores, I showed up at his next office hours and announced that I would like to do such a project. To his eternal credit, he did not laugh and send me away to get more experience, but accepted me as a student. Dr. Barlow played a pivotal role in my career, first as an undergraduate, and later as a graduate student. Knowledgeable in both classical ethology and the more recent innovations in behavioral ecology and sociobiology that were all the rage in the early 1970s, Dr. Barlow trained his students to simultaneously approach behavioral questions from a proximate and an ultimate perspective, and repeatedly emphasized the importance of choosing study animals that could be easily observed and manipulated in both the laboratory and the field.

In my case, the choice of study species was partly a matter of serendipity. Two floors up from Dr. Barlow's office, Dr. George Gorman had several room-sized cages filled with *Anolis* lizards that he had recently brought back from the West Indies. He kindly volunteered two species, *Anolis trinitatus* and *Anolis aeneus*, for my experiments on lizard communication signals. The next step was to decide which species would be most suitable for the purpose. Finally, the many hours spent observing pet *Anolis carolinensis* lizards as a child paid off, as it was quickly apparent that *A. trinitatus* males were not comfortable in the experimental apparatus, for example, they turned brown, plastered themselves against their perches, and refused to headbob at their image in a mirror. In contrast, *A. aeneus* exhibited behavior that I had earlier associated with relaxed, confident male anoles, including upright postures, bright color, spontaneous production of one type of head-bob patterns, and prolonged bouts of other types of headbob patterns when presented with a mirror-image. Hence, *A. aeneus* adult males were selected for the initial study of stereotyped headbob patterns in lizards (Stamps & Barlow 1973).

From that point on, my research proceeded by a series of predictable steps. After a few years of studying the behavior of adult male *A. aeneus* in the laboratory, I was encouraged to see how these displays were used in natural conditions, which led to the first of a series of field trips to the West Indian island of Grenada, where this species is endemic. The attractions of this part of the world were apparent as soon as I stepped off the plane and experienced the warm, balmy air, so I began making plans for subsequent field trips. Initial field observations showed that the headbob displays I had observed in the laboratory were intimately related to territoriality and dominance, and that every age and size class of this species spent appreciable amounts of time fighting, chasing, and displaying at one another. Indeed, even the tiniest hatchlings (0.2 g, 20 mm from tip of nose to vent) chased other juveniles out of tiny bits of real estate 0.5 m$^2$ in area, produced a wide array of head bobbing patterns, and exhibited social behavior every bit as interesting as that of the adults (Stamps 1978). There was something extremely attractive about the possibility of entering the world of vertebrates the size and weight of a postage stamp, animals so small that being hit by a raindrop is a dangerous experience, because it knocks them off their perches. There were also good practical reasons for focusing on juveniles, including the fact that little was known about the social behavior of any juvenile lizard, so that the background "natural history" required for any solid behavior study would be publishable in one journal or the other. More important, because their home ranges and territories were so small, entire "neighborhoods" of juveniles could be easily established and manipulated in the field, and observed in the laboratory under densities comparable to those preferred by free-living individuals.

Of course, by the time I had embarked on serious studies of juvenile *A.*

*aeneus*, a fascination with ideas had been added to my earlier fascination with animals. The animal behavior contingent at Berkeley in the 1970s was abuzz with excitement about the "new" discipline of behavioral ecology, and many of the graduate students were taken up with the challenge of testing theory generated during this heady period. Behavioral ecology as a discipline had inherited the ethologists' concerns with the effects of behavior on the survival and reproduction of organisms, and with the selective pressures that have contributed to the origin and evolution of behavior (see Wilson 1975; Krebs & Davies 1978 for contemporary reviews).

With respect to territorial behavior, the implicit goal of most studies was to understand the adaptive significance of territorial behavior, by formulating and testing models based on the presumed benefits accruing to territory owners, and the presumed costs of maintaining an established territory. Although earlier students of territoriality had discussed many potential benefits of territorial defense (especially for species with "all purpose" territories, e.g., see Hinde 1956), and had noted ways that territory owners might benefit from the presence of neighboring territory owners (Stamps 1988), during the late 1960s and early 1970s most theoreticians assumed that the primary function of territories was to confer priority access to either food or mates, and that social interactions between neighboring territory owners were entirely competitive or aversive (Stamps 1994). This was also a time when the distinction between retrospective and prospective approaches to the study of function were often blurred, so that most of us assumed that one could tell whether a particular factor had in the past contributed to the evolution of behavior in a given population by asking whether experimental manipulations of that factor generated changes in that behavior in present-day members of that same population. Thus, the notion that territories functioned in the defense of a food supply was tested by asking whether territory owners adjusted territory sizes in response to experimental manipulations of food supplies (review in Boutin 1990), or responded to changes in the density of competitors by altering territory size or overlap (e.g., see Stamps 1990).

Preliminary observations of juvenile *A. aeneus* seemed consistent with the notion that their territories might be related to the defense of a food supply. Juveniles hatch from eggs laid one at a time by females living in scrub-Acacia woodland habitats, then hatchlings emigrate to small clearings, where they establish small territories on small plants or piles of twigs near the ground. After growing to 30 mm snout-vent length, individuals move back into scrub-Acacia woodland habitats, where they continue to grow, and live the rest of their lives (Stamps 1983a). Since juvenile territories were far removed in both time and space from the territories defended by subadults or adults, it was unlikely that juvenile spacing behavior was directly related to access to mates or breeding resources. The only things juveniles did while living in these territories were grow and survive, and long-term studies monitoring food levels and juvenile growth rates indicated that juvenile growth rates were food-limited during at least a portion of the season when juveniles are present and growing (from August to April; Stamps & Tanaka 1981a). Thus, it seemed clear that juveniles defended "feeding territories" rather than nesting, mating, or all-purpose territories, so I dutifully spent the next few years recording or manipulating food supplies and population densities in the field and the laboratory, to determine the effects of these factors on juvenile aggressive behavior and spacing patterns (Stamps & Tanaka 1981b).

These early studies provided equivocal results. For instance, while it was clear that adding food supplements to particular areas attracted individuals from surrounding areas, individuals did not change their territory sizes in response to natural or artificial changes in food levels (Stamps & Tanaka 1981b; Stamps, unpublished data). Similarly, juveniles vigorously defended territories even when provided with a superabundance of food and space in the field or the laboratory, suggesting that territory defense was not triggered by food shortfalls or competition with conspecifics. With benefit of hindsight, it is clear that I invested more time in this line of inquiry than was either necessary or desirable. However, eventually the accumulation of negative or equivocal results encouraged me to spend more time observing the animals and reexamining assumptions of current theory in light of these observations. At this point, the research began to take a more interesting turn.

Anyone who has worked with wild animals for any length of time gets a feeling about the types of habitats in which they are likely to encounter their species. Thus, observations of free-living A. aeneus suggested that they preferred to live in clearings, in areas with particular configurations of twigs, plants, and other structural features, which were shaded from the hot tropical sun from 11 to 15:30 hr. The big advantage of this particular species was that tentative hypotheses about the habitat features preferred by juveniles were directly testable under field conditions. By cutting a small tree, it was possible to create a new clearing in an area previously devoid of juveniles. Then, one could try all sorts of manipulations of habitat features to determine those which were most attractive to new hatchlings searching for homesites. By a lengthy process of trial and error, setting up potential territories and determining whether (and how quickly) they were claimed and defended by free-living juveniles, one could ask subjects about the features they deemed most important when settling into novel habitats and establishing territories.

One insight from these manipulations was that food did not seem to be a major determinant of habitat and territory selection for the juveniles. Areas with an abundance of prey of the types eaten by juveniles remained unoccupied, until my students and I added perches suitable for juveniles, changed the sun-shade pattern by pruning a strategically placed tree, or removed branches harboring adult males of the same species. Conversely, juveniles refused to settle in wooded areas with insufficient light levels, even though many of those areas had an abundance of suitable prey and perches (Stamps 1983a). Similarly, even when provided with food ad libitum in the laboratory, juveniles preferred complex habitat configurations, and these preferences were more pronounced in the presence of predators (Stamps 1983b, 1984).

Eventually, this first set of studies led to the (belated) realization that the most important determinants of territory and habitat quality are those factors that lead to the acceptance or rejection of an area by every potential settler, not the features that generate minor variations in social and spacing patterns in habitats that every settler deems acceptable (Stamps 1994). The responses of juveniles to natural, artificial, and manipulated habitats in the field and laboratory suggested that predator-protection and suitable temperature regimes were highest on their priority list of features affecting habitat quality, and that food did not enter the equation until after their safety and thermal requirements had been met. Hence, it was not surprising that juveniles crowded into the few habitats that were acceptable with respect to structural and thermal considerations, and that after crowding into these areas, individuals occasionally competed with their neighbors for food. In turn, since habitat quality seemed primarily determined by nontrophic factors, it was not surprising that juveniles did not select habitats or territories based on food levels, nor dramatically alter their social and spacing patterns in response to perturbations in food supplies.

Another insight gained from watching free-living juveniles settle in these habitats was how difficult it was for them to gain accurate information about relevant social and environmental factors when settling into a novel habitat. From my lofty position as designer and observer of lizard neighborhoods, it was obvious which areas had already been claimed by territory owners and which were vacant, which areas overheated between the hours of 11 and 12 and which did not, and so forth. However, it was equally obvious that hatchlings entering these habitats for the first time lacked this information. Newcomers would often venture into a patch, be attacked by one resident and then another, and emigrate without ever discovering a prime, unclaimed territory in an opposite corner of the same patch. Similarly, some juveniles settled into potential homesites on overcast days, began the series of aggressive interactions required to establish a territory, and then were forced to abandon these areas when they overheated on subsequent, sunny days.

Aside from generating a feeling of omniscience unbecoming in a scientist, these observations suggested to me that juveniles might have difficulty measuring important environmental and social features while choosing habitats and territories. At the same time, observations of juveniles settling and living in territorial neighborhoods suggested that they were very interested in one another, paying close attention to the position, displays, and behavior of other juveniles. In combination, these two sets of observations implied that settlers in this and other species might use interactions with conspecifics to gain important information about an area and its inhabitants. For instance, the presence of satisfied territory owner in a patch indicates that the patch must be suitable for occupancy (e.g., it does not overheat at midday, is not regularly frequented by predators, and has adequate food levels for survival). The notion that juveniles might use the presence and behavior of conspecifics as cues to habitat quality led to field experiments designed to test whether juveniles pay attention to conspecifics when selecting territories and habitats.

An initial experiment hinged on the fact that juvenile *A. aeneus* rely on visual displays (pushups, headbobs, extensions of folds of skin under the throat [dewlaps]) when interacting with one another. In this type of species, animals often interact normally with one another when separated by transparent surfaces. In fact, free-living juveniles separated by transparent sheets of plastic initially seemed unaware of the fact, displaying at one another, approaching, and then lunging toward one another in an attempt to grab one another by the jaws. They seemed surprised when these lunges resulted in their bouncing off the "invisible force field" that separated them! Eventually they learned not to try and move past these transparent barriers, but they still enthusiastically exchanged visual displays with other juveniles on the other side of them.

An early experiment asked whether a territory previously inhabited by a conspecific would be more attractive to a settler than a previously vacant territory of equivalent quality (Stamps 1987a). By setting up pairs of territories surrounded by transparent plastic walls near one another in the same clearing, and then tossing a coin to decide which one would get a resident juvenile, I could offer juveniles living in the neighborhood views of two territories, one of which had a resident, and the other of which remained empty (Fig. 9.1). Then, after neighboring juveniles had observed both for a week or so, both sets of walls and the resident inside the walls were removed, and juveniles living in the vicinity were allowed free access to both territories. The results of this field experiment were clear: by a number of criteria, juveniles preferred the territory which previously harbored an occupant to the unused territory (Stamps 1987a). It was clear that observation of the previous resident inside the enclosure played an important role in this process, because juveniles who lived near the enclosure when it had a resident ("previous residents," see Fig. 9.1) quickly claimed the previously occupied territory when the walls were removed, whereas juveniles who arrived in the area after the walls and previous owner had been removed showed no such preference for the previously used territory (Stamps 1987b). This field experiment provided support for the hypothesis that juvenile *A. aeneus* use the behavior or presence of conspecifics to assess habitat quality.

I happily wrote up and submitted the results from this experiment, and then something happened during the review process that changed the course of my career. One of the submitted manuscripts from this project (Stamps

## EXPOSURE PERIOD

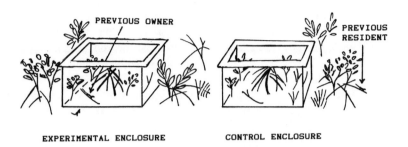

## ACCESS PERIOD

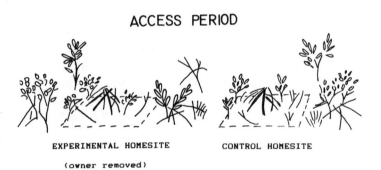

Figure 9.1. Experimental design used to determine whether juvenile *A. aeneus* lizards living in a clearing in the field preferred a previously used homesite (inside the experimental enclosure) to a comparable, unused homesite (inside the control enclosure). Free-living residents living in the clearing observed both enclosures (and the "owner" of the territory in the experimental enclosure) for a week, then the enclosures and the owner were removed, and residents were allowed free access to both homesites. From Stamps 1987a.

1987b) considered the role of familiarity with a neighborhood on success in territory acquisition, although it also rambled over a number of other topics. One of the reviewers of this manuscript was not impressed, commenting something along the lines of "these results are not interesting, because they are intuitively obvious." This comment got me to thinking about the role played by intuition in science, because in this particular case, the results may have been intuitively obvious, but I couldn't find any empirical studies indicating that this intuition was correct. The editor of the journal had graciously allowed me an opportunity to rewrite the paper in light of the reviewers' comments, so I rewrote the introduction, emphasizing the lack of any empirical data on the topic, and jettisoned results and discussion unrelated to the relationship between familiarity with a neighborhood and success in territory acquisition. Thereafter, I kept an eye out for ideas that were uncritically accepted because they were intuitively attractive, or conversely, were uncritically rejected because they failed to conform with contemporary notions about the way the world worked.

Field observations had suggested another context in which conspecifics might offer useful cues to habitat quality: when newcomers arrive and decide whether or not to settle in unfamiliar patches of habitat. This notion was tested using plastic-walled enclosures of a different design, in which a large enclosure was divided into two sides, each of which contained enough habitat to support four juveniles (Fig. 9.2). This time, arrangements of twigs and plants suitable for juvenile home ranges were set up around the outside of the enclosure. A coin toss was used to determine which side of the enclosure would be provided with juvenile "residents," and which side would remain empty, and then colonization by free-living lizards of the homesites around the edges of the enclosure was observed over a period of days. Again, the results were clear: based on a number of measures, juveniles preferred to settle next to conspecifics, arriving and settling first on the homesites that were directly outside the portion of the enclosure that already contained conspecific residents (e.g., Fig. 9.3). These and other field experiments indicated that newcomers to patches of suitable habitat were attracted to conspecifics while settling, a process termed "conspecific attraction" (Stamps 1988).

As is often the case, it wasn't until after I returned from the field and trotted off to the library that I discovered that the notion of conspecific attraction was not new: many authors over the years had suggested that animals might be attracted to conspecifics while settling, and several of them had suggested reasons why this behavior might be advantageous to newcomers (Stamps 1988, 1994). However, discussion of conspecific attraction by Nice, Lack, Barends, and other luminaries had been largely forgotten by the 1970s and 1980s, and most empirical studies on conspecific attraction during that period focused on benthic marine invertebrates, and were published in journals rarely read by animal behaviorists or ecologists working with terrestrial vertebrates. As a result, in the late 1980s, conspecific attraction was an unfamiliar notion for many biologists.

Since that time, my collaborators and I have worked on developing statistical techniques to allow researchers to study conspecific attraction using observational data, since many animals are not as amenable to field experiments as juvenile *A. aeneus* lizards. For instance, it is hard to imagine anyone manipulating owl population densities or habitat structure in order to determine if young owls prefer to establish territories next to those of pre-

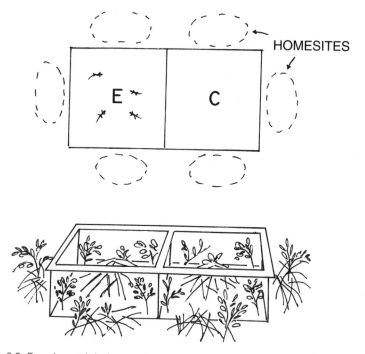

Figure 9.2. Experimental design used to determine whether free-living juveniles preferred to settle next to previous residents (on homesites near E, the side of the enclosure that contained four resident juveniles) or on comparable homesites near C, the side of the enclosure that lacked residents. From Stamps 1988.

vious territory owners. After casting about for a suitable study animal, we discovered a lovely monograph on territorial behavior in house wrens published by Kendeigh in 1941; this monograph contained more detail about the settlement of hundreds of individually marked birds than any editor could justify publishing today. The data, coupled with some new statistics developed to study habitat preferences, demonstrated that male house wrens breeding for the first time preferred to establish territories next to those of previous settlers in the same habitat, after controlling for variation in territory quality (Muller et al. 1997; Fig. 9.4). At this point, conspecific attraction has been demonstrated for a wide range of animals (Danchin & Wagner 1997; Stamps, in press), suggesting that it may be time to pay closer attention to the possible adaptive significance of this behavior.

At this point, hypotheses about the adaptive significance of conspecific attraction can be divided into two nonexclusive categories, one dealing with processes that occur when animals are searching for and settling into patches of unfamiliar habitat, and the other dealing with processes that occur after animals have settled into patches of suitable habitat (Stamps, in press). Conspecific cueing, mentioned earlier, belongs in the first category. This hypoth-

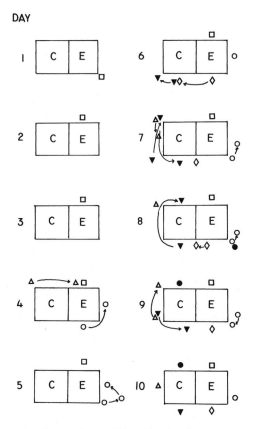

Figure 9.3. Settlement patterns for one trial of the conspecific attraction experiment in Figure 9.2. Each diagram represents a successive day, each symbol represents a different individual, and arrows indicate situations in which an individual changed position during the same day. The first juvenile (square) arrived at the E (experimental) side of the enclosure and eventually settled on the E side, the other homesites next to the E side of the enclosure filled next (by day 6), and eventually (by day 10) juveniles had settled on all of the homesites around both sides of the enclosure. From Stamps 1998.

esis suggests that prospective settlers use established conspecifics to reduce their costs of searching for a high-quality habitat, or to reduce their chances of erroneously choosing and settling in a poor-quality habitat. Recent variations on this theme are that newcomers might use the reproductive success, as well as the presence, of previous residents to assess habitat quality (Danchin et al. 1998), that newcomers might use the presence of heterospecifics as an indication of habitat quality (Monkkonen et al. 1997), or that dispersers improve their chances of locating any suitable habitat by tracking stimuli produced by individuals who have already settled in a habitat (Stamps, in press).

An obvious alternative hypothesis is that conspecific attraction occurs be-

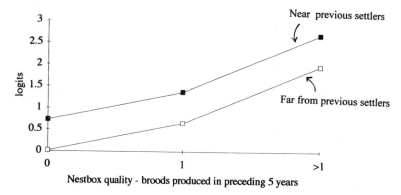

Figure 9.4. Factors affect the preference of naive house wrens for nestboxes, measured as those males arrived and settled in a territorial neighborhood. Preference scores (in logits) were computed based on each male's choice, in relation to the nestboxes that were still available to that male when he arrived in the study area. Naive males preferred nestboxes that were successful (produced broods) in previous years to nestboxes that were not successful, and preferred nestboxes near to previous settlers to nestboxes of the same quality that were far from previous settlers. From Muller et al. 1997.

cause individuals benefit from the presence of conspecifics after settling in a new habitat. This point of view was advanced by Allee in the early part of this century (1951); perhaps not coincidentally, Allee was a Quaker. However, Allee's ideas, and studies of the potential benefits conspecifics confer on one another, more or less disappeared during the middle of this century, except for situations in which it was patently obvious that conspecifics actively formed aggregations (e.g., colonial species, or animals forming schools, flocks, herds, or other "groups"). In part, the neglect of positive interactions among conspecifics was due to the fact that most behavioral biologists and ecologists in the middle of the century focused on competition rather than cooperation. However, now the pendulum seems to be swinging the other way, with renewed interest among theoreticians and empiricists in cooperation and mutualism (e.g., see *Ecology* 78[7]; Dugatkin 1997). Ecologists are rediscovering Allee effects, and are beginning to explore their consequences for extinction, metapopulation dynamics, and a variety of applied problems in conservation biology, an effort that has been supported and encouraged by many behavioral studies of the ways that individuals benefit one another when living in groups or neighborhoods.

These days, my colleagues and I continue to be interested in the significance of conspecific attraction for individual settlers, and the consequences of conspecific attraction for population-level phenomena. At the individual level, the issue is how to determine if conspecific attraction improves settlement efficiency in a given species. While it is relatively easy to test for the existence of Allee effects after settlement (e.g., see Aviles & Tufino 1998), it is not as obvious how to determine whether individuals who rely on con-specific cues while settling obtain higher-quality habitats, or are more likely to find suitable habitats (i.e., as opposed to dying en route), than individuals who ignore or avoid conspecifics while settling. Juvenile *A. aeneus* lizards are hardly ideal for studying this question, because they are difficult to find when traveling between suitable habitats in the field, so we are currently in the market for other suitable model species. At the population level, we are exploring the potential effects of conspecific attraction on the distribution of animals in patches of suitable habitat (Greene & Stamps, in review). The issue is especially pressing because so many populations and species have been reduced to locally small numbers (small neighborhood sizes) as a con-sequence of habitat fragmentation and other human activities. Regardless of the adaptive significance of conspecific attraction, this phenomenon implies that individuals will not be attracted to patches lacking a minimum number of residents, or patches of habitat (no matter how suitable) large enough to support individual neighborhoods, but too small to support a neighborhood. It is now clear that many animals are unable to persist in small fragments of habitat (Bender et al. 1998), but this pattern is typically attributed to demo-graphic stochasticity, or reductions in habitat quality associated with small patch size. However, if the members of a species are reluctant to settle in patches too small to support a critical number of conspecifics, then habitat enhancement (predator or parasite culling, food addition, etc.) is unlikely to do much to improve recruitment of that species to small patches of habitat. This is but one example of a situation in which more information about the proximal behavioral bases of habitat selection might play a useful role in conservation efforts.

By now, it should be clear that early studies of settlement patterns in the juvenile lizards led in directions hardly anticipated while I was sitting in the field, enjoying the Zen state of mind that I associate with observing animals and their behavior. Another area that began to attract my interest while watching juvenile lizards was the role that aggressive behavior (displays, fights, chases, etc.) played in their lives. Juveniles frequently interact with one another, especially when they first establish home ranges and territories, and early studies indicated that juveniles are, if anything, more aggressive than adults of the same species (Stamps 1978). Some of these fights are spectacular, as when two individuals exchange a bewildering array of head-bobbing patterns, tail lifts, dewlap extensions, and other displays for a half hour or more, gradually move closer and closer to one another, and then finally grab one another by the jaws and pound each other repeatedly against a branch.

One thing that became clear while observing these animals was that their interactions while settling in vacant habitats seemed to affect their social relationships and space use patterns later on, after they were living together in these habitats. Having learned my lesson while studying conspecific attraction, this time around I went to the literature before looking at the settlement process in these animals, and was surprised to find that although literally thousands of scientists have studied the behavior of territorial animals after they have established their territories, very little is known about the processes by which animals obtain territories in the first place (Maynard Smith 1982; Stamps 1994; Stamps & Krishnan 1995). Along the same lines, although many workers have investigated aggressive behavior when territorial animals compete for indivisible space (e.g., small tanks or arenas in the laboratory, or single territories embedded within established neighborhoods in the field), very little is known about the processes by which free-living animals compete for divisible space, that is, space that can be divided or shared (Stamps & Krishnan 1999). More dangerously, many workers have bridged this gap in our knowledge by assuming that processes that are important when animals compete for indivisible space, or when individuals defend territories in established neighborhoods, are also important when animals acquire territories in patches of vacant, divisible habitat.

For example, a widespread assumption in the territorial literature is that territorial animals gain space by winning aggressive contests (Stamps & Krishnan 1995, 1997). This assumption makes intuitive sense if animals are competing for indivisible space: presumably, the winner would end up with the space, while the loser would leave. However, it is less obvious how this process might work if animals are competing for divisible space, as occurs when prospective settlers establish territories next to one another within larger patches of suitable habitat. Hence, it seemed reasonable to use the lizards as a model system to determine whether and how winning contests influenced space acquisition when juveniles settle in territorial neighborhoods. This was a relatively easy proposition in this system: unenclosed patches of suitable habitat large enough to support a neighborhood were created within natural habitats, juvenile settlers were sprinkled into those patches, and then the space use and social interactions of those settlers were recorded on a continuous basis until they had either established stable home ranges or territories, or had emigrated from the patch (Stamps 1990, 1992).

We tested the assumption that juveniles gained space by winning contests in every conceivable way: by looking at individual space use over the short term (Stamps & Krishnan 1994b) or over the entire settlement period (Stamps & Krishnan 1998), by looking at the space use of interacting dyads over the short term (Stamps & Krishnan 1994b), or by looking at situations in which two individuals competed for the same space (Stamps & Krishnan 1995, 1997). However, in no case did our results support the hypothesis that juveniles gained space by winning contests (review in Stamps & Krishnan 1998). Instead, the type, rather than the outcome, of aggressive interactions turned out to be the best predictor of success in space competition for individuals and dyads. Thus, pairs whose first encounter was an escalated fight (regardless of outcome) had fewer subsequent interactions than pairs whose first encounter was a chase (Stamps & Krishnan 1994b). Individuals eventually achieved high social status and exclusive territories if they fought and chased other individuals over the course of the settlement period, but the outcome of those fights was unrelated to social status, territory size, or exclusivity (Stamps & Krishnan 1998). Similarly, pairs of opponents who resolved their dispute with a final fight were very likely to have mutually exclusive home ranges at the end of the settlement period, but the outcome of these last fights was irrelevant, and in any case, most of them ended in a draw (Stamps & Krishnan 1997).

Rather than supporting the hypothesis that winning contests was important to territory acquisition, these results suggested an alternative hypothesis, namely, that aggressive interactions during the settlement period act as punishment, discouraging opponents from returning to particular areas in that habitat. The notion that aggressive behavior acts as punishment is familiar to comparative psychologists (e.g., Blanchard & Blanchard 1986), and recently this idea has also made its way into the behavioral ecology literature (Clutton-Brock & Parker 1995). In the context of territory establishment, the high costs of an escalated fight to both opponents might discourage them from using space where they are likely to encounter that opponent in the future, thus encouraging the formation of mutually exclusive home ranges and mutually respected territory boundaries. In that case, the "goal" of a contest would be to deliver a lot of punishment to the opponent, not necessarily to "win" the contest. Conversely, other types of aggressive interactions (e.g., chases) are probably less costly than fights on a per encounter basis to both opponents, implying that repeated chases might be required to discourage the chaser and the chasee from encountering one another in the future. Interestingly, when two *A. aeneus* competed for the same space, they typically did so via a series of chases, and the contested space was eventually used by the individual that persisted longest in engaging in chases in that area, not by the individual who did the chasing (Stamps & Krishnan 1995).

Musings on the role of learning processes in territory acquisition led to extended hours in the library trying to decipher and understand the jargon of the learning literature, and visits to patient and supportive comparative psychologists for help in identifying the concepts and references in the learning literature most likely to shed light on the role of learning in territory acquisition. Eventually, this process led to the development of a simple, learning-based model of territory establishment in divisible space (Stamps & Krishnan 1999).

This initial model assumes that mobile animals assemble home ranges or territories out of a number of contiguous, smaller areas; that these areas are heterogeneous with respect to the location and type of important habitat features (but not with respect to habitat quality); and that individuals arrive and settle in patches large enough to support many residents. Because habitats are structurally heterogeneous, we assume that individuals benefit by learning what is in each area and where important sites are in relation to one another. Another potential benefit of familiarity is that animals might have an opportunity to practice motor skills that allow them to use that area efficiently, a point developed in more detail in Stamps (1995). As a result of these processes, we assume that the attractiveness of a given area increases exponentially as a function of the amount of time previously spent in that area. Conversely, if two individuals venture into the same area at the same time, we assume they have a fight ending in a draw, with equal effects on both individuals. Engaging in fights in an area is assumed to reduce the attractiveness of that area, with a negative exponential relationship between number of aggressive interactions in an area and the subsequent attractiveness of an area. Finally, the chances that an individual will return to a particular area is assumed to depend on the attractiveness of that area, relative to the attractiveness of other areas that are available to that same individual.

Even though the simulated settlers in our simulation models are far less sophisticated than even juvenile lizards (e.g., the simulated settlers always have fights ending in a draw, only detect opponents if they bump into them, are incapable of distinguishing one individual from another, and neither produce nor respond to territory advertisement signals), the spacing and social behavior generated by these simple models are in many ways reminiscent of that of real territorial animals (Stamps & Krishnan, in press). For instance, each individual eventually establishes a stable home range in a predictable location within a larger patch of suitable habitat, and aggressive interactions between settlers lead to larger territory sizes and more exclusive home ranges, as compared to models in which settlers ignore one another when they meet. The model has also been used to explore interactions between different types of settlers, as in cases in which some individuals are allowed to settle for a "week" (residents) and then new individuals (newcomers) are released into the patch. In this situation, the model generates a strong "prior residency advantage": when a resident and a newcomer have an encounter in an area familiar to the resident, the resident typically ends up using the area, while the newcomer usually abandons it. This last result is interesting because current theory assumes that the prior residency advantage is a result of residents winning contests with intruders (e.g., Enquist & Leimar 1987; Nuyts 1994), but in our simulations, a strong prior residency advantage was generated even though every fight ended in a draw, with equal effects on both opponents.

We were heartened by the fact that such a simple model can produce social behavior and spacing patterns reminiscent of those observed in juvenile *A aeneus* and other territorial animals, and are currently exploring other questions about the effects of aggressive interactions on territorial behavior in variants of the initial model. Thus, we are combining optimality approaches with assumptions of the learning model, to ask how residents on feeding territories ought to behave when attacking intruders, if attacks on intruders reduce the chances of their returning to the territory at a later time (Switzer et al., in preparation). Another current effort asks what happens to settlement and space use patterns if settlers vary with respect to the amount of punishment they deliver when fighting with their opponents, or if punishment varies as a function of the relative competitive ability of different opponents (Stamps & Krishnan, in review). Eventually, of course, we plan to return to the field to test basic assumptions and predictions of these models using real animals.

By now, the observant reader may have noticed several themes that have run through my research career. One is an emphasis on testing assumptions, rather than taking them at face value. This was a lesson learned early on, for example, when I wasted several years studying juvenile territoriality under the premise that food was the most important determinant of habitat selection and territorial behavior in this species. Because I assumed that food was the most important variable to measure in animals with "feeding territories," it took a long time for me to notice that juveniles only ate the equivalent of five to ten *Drosophila* per day, that arthropods this small were abundant in all of the habitats used by juveniles, that temporal fluctuations in arthropod abundance were far more important than spatial variation, and that virtually all of the microhabitats satisfying the juveniles' (long list of) other habitat requirements had sufficient food to support reasonable growth rates. Later examples of assumptions I have viewed with skepticism are that all interactions between territorial animals are competitive, that (as a result), territorial animals should avoid one another while settling, or that winning contests is the most important determinant of success when territorial animals compete for divisible space. This same streak of skepticism with respect to untested assumptions has also been a theme in other lines of research conducted in my laboratory over the years, for example, studies of parent-offspring interactions in budgerigars that revealed behavior more sophisticated than envisioned in contemporary models or empirical studies (Stamps et al. 1985, 1987).

Generally speaking, I suspect that overly enthusiastic, uncritical acceptance of untested assumptions is most likely to occur when those assumptions are intuitively attractive. As was noted above, the fact that an idea is intuitively attractive does not necessarily mean that anyone has bothered to verify it. This can be a major problem, since intuition about animal behavior is often based on our own prior experiences, not on scientific evidence. For instance, I wonder whether the notion that animals acquire territories by winning fights has been so widely accepted, despite the scarcity of support from field studies of territorial settlement, because most behavioral ecologists live in cultures in which success is measured in terms of "winning" competitive interactions. Would we would be more receptive to ideas about the importance of draws and standoffs in territorial systems if we lived in a society in which sports fans preferred games ending in a draw (indicative of two worthy, well-matched opponents) to games with clear winners and losers? Along the same lines, if competition is a lesson drummed into us from an early age, it may be difficult to look at territorial animals (who are obviously fighting and aggressive) and notice situations in which neighbors might actually (*mirabile dictu!*) benefit from one another's presence when settling or living in territorial neighborhoods.

Regardless of the reasons, it seems clear that many critical assumptions about territorial behavior and habitat selection have been widely accepted without close scrutiny, or on the basis of scanty evidence. Rather than continuing to build theoretical superstructures based on these underlying assumptions, or continuing to incorporate these assumptions into the design of basic or applied studies, it might be more worthwhile to take a "step back" and test them. The series of studies outlined in this essay provides several examples in which widely held assumptions were supported (e.g., aggression among settlers increases territory size and exclusivity in the simulation models) and several cases where they were not (e.g., winning contests is not critical for successful territory acquisition for the juvenile *A. aeneus*, or for the simulated settlers in the territorial models).

Of course, territorial behavior is hardly the only topic in which some widely accepted assumptions have not received the scrutiny they deserve. I encourage my own students not to take anything on faith when learning about a topic, and instead make it a point of examining the supporting evidence for assumptions that are critical to the problem that has attracted their interest. I also encourage them to do a complete literature search, rather than relying on on-line searches which only yield studies published over the past few years. This is because assumptions we take for granted these days may have been viewed differently in the past, as is illustrated by the ebb and flow of interest in, and discussion of, conspecific attraction, Allee effects, and related topics over the past seventy years. In some cases, a comprehensive literature review may reveal that a key assumption has never received the scrutiny it deserves. If future investigation reveals that major changes in theory or experimental design would ensue if the assumptions did not turn out to be correct, then the student has identified a promising research problem, whose results will be of interest to a wide audience no matter how they turn out.

Clearly, *Anolis aeneus* lizards have proven to be a very rewarding study species over the years. This is partly a consequence of the fact that they are small and diurnal, rely on visual communication signals, and are uninhibited by the presence of human observers, and partly a result of the fact that their habitat can be so readily manipulated in the field, thus revealing the features they consider important when selecting habitats and territories. As a result, these animals have been useful not only for testing existing hypotheses, but (more important) for suggesting new ones. In fact, some of the more interesting ideas that emerged from the research were suggested by the animals themselves, as they insisted in repeatedly engaging in behavior that made no sense in terms of current theory. Sometimes it turned out that this behavior was consistent with observations made years ago by early ecologists or behavioral biologists, or with processes studied by scientists working in other disciplines (e.g., comparative psychology). In many cases the behavior observed in this species has not turned out to be typical or unusual: once it was described and put into an appropriate theoretical context, investigators working with birds, fish, insects, and many other species have mentioned observing similar behavior in "their" species. This, of course, should be the hallmark of any model species, which by virtue of their ordinariness can yield insights applicable to many organisms other than themselves.

## Acknowledgments

I feel privileged to have been provided with the financial, logistical, and moral support that has allowed me to "enter the world" of these animals, and bring back some ideas that may prove useful or interesting to basic and applied scientists working with other animals. This was hardly a career I envisioned as a child, but thanks to the American taxpayers (through grants from NSF), a number of mentors, colleagues, and bemused but supportive laypersons in Grenada, it has turned out to be an extremely interesting and exciting one.

## References

Allee WC, 1951. The social life of animals. Boston, MA: Beacon Press.

Aviles L, Tufino P, 1998. Colony size and individual fitness in the social spider *Anelosimus eximius*. Am Nat 152:403–418.

Bender DJ, Contreras TA, Fahrig L, 1998. Habitat loss and population decline: a meta-analysis of the patch size effect. Ecology 79:517–533.

Blanchard DC, Blanchard RJ, 1986. Punishment and aggression: a critical reexamination. In: Advances in the study of aggression (Blanchard RJ, eds). Orlando: Academic Press; 121–164.

Boutin S, 1990. Food supplementation experiments with terrestrial vertebrates: patterns, problems, and the future. Can J Zool 68:203–220.

Clutton-Brock TH, Parker GA, 1995. Punishment in animal societies. Nature 373:209–216.

Danchin E, Boulinier T, Massot M, 1998. Conspecific reproductive success and breeding habitat selection: implications for the study of coloniality. Ecology 79:2415–2428.

Danchin E, Wagner RH, 1997. The evolution of coloniality: the emergence of new perspectives. Trends Ecol Evol 12:324–347.

Dugatkin AL, 1997. Cooperation among animals: an evolutionary perspective. New York: Oxford University Press.

Enquist M, Leimar O, 1987. Evolution of fighting behaviour: the effect of variation in resource value. J Theor Biol 127:187–205.

Hinde RA, 1956. The biological significance of territories of birds. Ibis 98:340–369.

Kendeigh SC, 1941. Territorial and mating behaviour of the house wren. Urbana: University of Illinois Press.

Krebs JR, Davies NB, 1978. Behavioural ecology. Sunderland, MA: Sinauer.

Maynard Smith J, 1982. Evolution and the theory of games. Cambridge: Cambridge University Press.

Monkkonen M, Helle P, Niemi GJ, Montgomery K, 1997. Heterospecific attraction affects community structure and migrant abundances in northern breeding bird communities. Can J Zool 75:2077–2083.

Muller KL, Stamps JA, Krishna, VV, Willits NH, 1997. The effects of conspecific attraction and habitat quality on territorial birds (*Troglodytes aedon*). Am Nat 150:650–661.

Nuyts E, 1994. Testing for the asymmetric war of attrition when only roles and fight durations are known. J Theor Biol 169:1–13.

Stamps JA, 1978. A field study of the ontogeny of social behavior in the lizard *Anolis aeneus*. Behaviour 66:1–31.

Stamps JA, 1983a. The relationship between ontogenetic habitat shifts, competition and predator avoidance in a juvenile lizard (*Anolis aeneus*). Behav Ecol Sociobiol 12:19–33.

Stamps JA, 1983b. Territoriality and the defence of predator-refuges in juvenile lizards. Anim Behav 31:857–870.

Stamps JA, 1984. Rank-dependent compromises between growth and predator protection in lizard dominance hierarchies. Anim Behav 32:1101–1107.

Stamps JA, 1987a. Conspecifics as cues to territory quality: a preference of juvenile lizards (*Anolis aeneus*) for previously used territories. Am Nat 129:629–642.

Stamps JA, 1987b. The effect of familiarity with a neighborhood on territory acquisition. Behav Ecol Sociobiol 21:273–277.

Stamps JA, 1988. Conspecific attraction and aggregation in territorial species. Am Nat 131:329–347.

Stamps JA, 1990. The effect of contender pressure on territory size and overlap in seasonally territorial species. Am Nat 135:614–632.

Stamps JA, 1992. Simultaneous versus sequential settlement in territorial species. Am Nat 139:1070–1088.

Stamps JA, 1994. Territorial behavior: testing the assumptions. Adv Study Behav 23:173–232.

Stamps JA, 1995. Motor learning and the value of familiar space. Am Nat 146:41–58.

Stamps JA, Barlow GN, 1973. Variation in the displays of *Anolis aeneus* (Sauria: Iguanidae). Behaviour 47:67–94.

Stamps JA, Ens B, in press. Habitat selection by dispersers: integrating proximate and ultimate approaches. In: Causes, consequences and mechanisms of dispersal at the individual, population and community level (Danchin E, Dhondt A, Nichols J, Clobert J, eds). Oxford: Oxford University Press.

Stamps JA, Clarke A, Arrowood P, Kus B, 1985. Parent-offspring conflict in budgerigars. Behaviour 94:1–40.

Stamps JA, Clarke A, Arrowood P, Kus B, 1987. The effects of parent and offspring gender on food allocation in budgerigars. Behaviour 101:177–199.

Stamps JA, Krishnan VV, 1990. The effect of settlement tactics on territory sizes. Am Nat 135:527–546.

Stamps JA, Krishnan VV, 1994a. Territory acquisition in lizards: I. First encounters. Anim Behav 47:1375–1385.

Stamps JA, Krishnan VV, 1994b. Territory acquisition in lizards: II. Establishing social and spatial relationships. Anim Behav 47:1387–1400.

Stamps JA, Krishnan VV, 1995. Territory acquisition in lizards: III. Competing for space. Anim Behav 49:679–693.

Stamps JA, Krishnan VV, 1997. Functions of fights in territorial establishment. Am Nat 150:393–405.

Stamps JA, Krishnan VV, 1998. Territory acquisition in lizards. IV. Obtaining high status and exclusive home ranges. Anim Behav 54:461–472.

Stamps JA, Krishnan VV, 1999. A learning-based model of territory establishment. Q Rev Biol 74:291–318.

Stamps JA, Krishnan VV, in review. How territorial animals compete for divisible space: a learning-based model with unequal competitors. Am Nat.

Stamps JA, Tanaka SK, 1981a. The influence of food and water on growth rates in a tropical lizard (*Anolis aeneus*). Ecology 62:33–40.

Stamps JA, Tanaka SK, 1981b. The relationship between food and social behavior in juvenile lizards (*Anolis aeneus*). Copeia 1981:422–434.

Wilson EO, 1975. Sociobiology: the new synthesis. Cambridge, MA: Harvard University Press.

# 10 Acoustic Communication in Frogs and Toads

H. Carl Gerhardt

Although pleased to be considered a behavioral ecologist, I am also intensely interested in the mechanisms underlying evolutionarily important behaviors. I believe that information about mechanisms underlying communication provides insights and hypotheses concerning the evolution of communication. The interaction of these two approaches is reciprocal: knowing the functional significance of signals and receiver selectivity frames and focuses research aimed at understanding how sensory systems work in the real world.

## How and Why I Became Interested in Acoustic Communication in Frogs

My youth was spent in Savannah, Georgia, where my interest in amphibians developed after hearing the calls of male frogs and toads and then tracking them down with a flashlight. I was fascinated by being able to recognize different species by their calls alone, especially when as many as a dozen species called from the same ponds after heavy summer rains. I could also easily approach and watch calling males of most species, and, under favorable conditions, I found dozens of mated pairs.

As an undergraduate student at the University of Georgia, I was first an English major; I committed to a career in biology in my senior year. My professors in the Department of Zoology impressed me with the advantages of conducting quantitative studies requiring large sample sizes, and given my earlier interest in calling frogs, I knew I had a winning combination: animals whose behavior fascinated me and which could be collected in very large numbers. My mentor at the University of Georgia was the late Bernard S. Martof, who was a pioneer in the study of anuran communication. He was one of the first scientists to show that playbacks of calls alone were sufficient to attract gravid females. Thus, visual, olfactory, and tactile cues are at most supplementary cues for mate identification (Martof 1961). Martof also conducted the first playback experiments that used artificially generated sounds (Martof & Thompson 1964). Although I knew Martof from my high school days when he had encouraged my interest in herpetology, by the time I knew the direction of my research interests, Martof had abandoned his studies of anuran communication. My only formal study with him resulted in a paper on the taxonomy of a salamander (Martof & Gerhardt 1965), and only later did I fully appreciate his early work with frogs.

When it was time to choose a graduate school, the most active group studying vocal behavior in frogs was in W. Frank Blair's laboratory at the University of Texas. Although Blair himself and many of his students focused on the role of frog calls in species isolation, other students such as William F. Martin were intrigued with mechanisms of sound production and hearing. Indeed, my interactions with Martin undoubtedly influenced my later decision to conduct postdoctoral research at Cornell University with Robert R. Capranica, who was a pioneer in using behavioral experiments to frame questions about the neural basis of signal recognition. I also met Murray J. Littlejohn, who was visiting Austin from the University of Melbourne the year I arrived, and from that initial contact I developed an interest in studies of hybrid zones and patterns of geographical variation in frog calls, which can provide insights concerning mechanisms of speciation. Meeting Murray ultimately resulted in several research leaves in Australia, which offers unique chances to study interesting patterns of geographical variation.

My initial graduate research proposal aimed at studying interspecific communication and territorial defense in ranid frogs and was inspired by my early experience with most of the seven sympatric species occurring near Savannah. However, a summer's project testing the phonotactic selectivity of females of the green treefrog (*Hyla cinerea*) and barking treefrog (*Hyla gratiosa*) convinced me that these and other treefrogs would be the main subjects of my research on acoustic communication. The animals were so easy to study that I accumulated sufficient data to write a master's thesis after just three months. For my doctoral research, I extended the work with *H. cinerea* and *H. gratiosa* (Oldham & Gerhardt 1975) and added projects involving six additional hylid species (Gerhardt 1973, 1974a, b). I was also lucky enough to find and record three kinds of interspecific hybrids. Tests of females of the parental species showed that these animals not only discriminated against the calls of other species but also against the acoustically more similar calls of hybrids (Gerhardt 1974c).

The experience in collecting data for the thesis reinforced my conviction that treefrogs had an array of important characteristics that make them ideal subjects for the study of both the evolution and neurobiology of acoustic communication. Moreover, the fieldwork, mainly done around Savannah, gave me a tremendous advantage for all of my future work. I learned which species reliably showed phonotactic behavior, and when and where to collect large numbers of responsive females. This basic knowledge is an essential component of successfully conducted quantitative behavioral studies. My experience at Texas and Cornell was invaluable because of the generous guidance and encouragement I received from fellow graduate students, postdoctoral associates, and faculty mentors. These individuals helped me see the important theoretical questions—about both proximate mechanisms and evolutionary processes—that could be addressed by studying animal communication in general, and acoustic communication in particular.

# Mechanisms and Evolution of Acoustic Communication

Studies of acoustic communication in anurans provide insights concerning both proximate mechanisms underlying behavior and the evolutionary significance of behavior. The anuran auditory system is a well-established model system for studying the biophysical and neural bases of sound pattern recognition (Fritzsch et al. 1988). We know a great deal about the response properties of the anuran's tympanic membrane and inner ear organs and how auditory neurons in various nuclei in the ascending auditory pathway respond to acoustic properties of known pertinence (reviews in Hall 1994; Lewis & Narins 1999). Furthermore, behavioral studies of sound localization suggested the current hypothesis that small anurans, like many insects and unlike higher vertebrates, use a pressure-difference mechanism to generate the binaural cues (e.g., differences in eardrum displacement between the two ears) needed to locate a sound source (Rheinlaender et al. 1979). Subsequently, anuran pressure-difference systems have been modeled (Eggermont 1988), and laser vibrometry and neurophysiological recordings have been used to characterize the directional patterns of the eardrum and to assess other potential inputs of sound to its inner surface (Jørgensen 1991; Narins et al. 1988). With regard to evolutionary questions, studies of acoustic communication in anurans have contributed empirical data bearing on theoretical models and controversies concerning sexual selection, mating systems theory, alternative mating tactics, and speciation (Gerhardt 1994a, b; Littlejohn 1993; Lucas et al. 1996; Perrill et al. 1982; Ryan & Rand 1993; Sullivan et al. 1995).

In the next few paragraphs, I outline a common first approach to studying both mechanisms and evolution of acoustic communication. This approach generates a body of basic information that serves as a framework for posing and refining questions that can be answered with a combination of behavioral, neurophysiological, and field experiments. I then provide some examples drawn from my research.

## Acoustic Criteria for Signal Identification and Mate Choice

A first step in studying sound pattern recognition, sexual selection, and speciation is to describe the physical properties of acoustic signals and their statistical variation within individuals and populations, and in different parts of the range of a species. Properties that show relatively little variation at one or more of these levels have often been the focus of studies of neural mechanisms, whereas both stereotyped and variable properties are relevant to studies of evolutionary processes. The next step, which has formed the core of my research program, is to conduct behavioral studies to learn which of the acoustic properties of communication signals mediate the identification of signals, male-male competition, and female choice. As many classical ethological studies have shown (e.g., Tinbergen 1951), not all features of animal signals or behaviors, even highly invariant ones that allow humans to identify a species or to predict its behavior, are actually used by the animals themselves.

Characterizing the sets of acoustic criteria that frogs and toads use for signal identification and in male-male interactions and mate choice requires synthesizing acoustic stimuli that reliably elicit normal behaviors. Such acoustic models should be just as effective as typical, natural signals because using marginally attractive signals can lead to erroneous conclusions about an animal's selectivity (Schmitt et al. 1993). For the species of anurans I have studied, the equivalence of synthetic and natural, pre-recorded calls has been confirmed in simultaneous choice tests (e.g., Doherty & Gerhardt 1984; Gerhardt 1974a, 1978a, 1981a, b). Such synthetic signals can be considered "standard" calls, the acoustic properties of which are then varied in a systematic fashion. Male or female frogs are tested in either single-stimulus or multiple-stimulus (choice) playback experiments. Single-stimulus experiments are often used to identify the acoustic properties that a stimulus must possess in order to reliably elicit responses. Measures such as the speed of responses can be used to estimate the optimum values of quantitatively varying properties (Schul et al., in preparation), even though choice tests are likely to be more efficient in this respect. If quantitative measures of phonotaxis are unavailable, and the investigator simply determines whether females respond or not, then single-stimulus tests can underestimate receiver selectivity. This point is forcefully illustrated by the fact that females of some species of insects and anurans readily respond to playbacks of heterospecific calls if they are given no choice, but seldom, if ever, respond to the same signals if conspecific calls are available at the same time (Gerhardt 1982; Schul 1998).

Multiple-stimulus playback experiments can also be used to estimate the strength of preferences based on particular acoustic properties. Strong preferences are based on relatively small differences in the values of pertinent acoustic properties. Larger differences are likely to elicit preferences that are at least moderately independent of relative intensity. That is, the playback level of a stimulus with the preferred value can be lowered relative to that of an alternative stimulus without abolishing the preference. Strong preferences can be maintained in face of reductions in the sound pressure level of a stimulus with the preferred value by 12 to 18 dB (sound pressure ratios of 4 to 8) relative to that of a stimulus with a nonpreferred value (e.g., Diekamp & Gerhardt 1995; Gerhardt 1982). The intensity-independence of a preference can indicate the costs that a female might be willing to incur to mate with a male producing calls with preferred values because sound pressure level roughly translates into relative distance (e.g., Gerhardt et al. 1996). Thus, a female is likely to choose a distant male with calls having optimum values of an important property rather than a nearby male with calls having suboptimum values. However, even weak, intensity-dependent preferences, which require relatively large differences in the value of an acoustic property, can have important evolutionary consequences over long spans of time.

In studies of proximate mechanisms, we want to learn how each pertinent acoustic property is transduced, encoded, processed, and ultimately decoded in order to elicit an appropriate behavioral response. Transduction is the conversion of the mechanical energy in sound waves to bioelectrical activity. Encoding is the representation of the value of an acoustic property in the firing patterns of single auditory neurons or groups of neurons. Processing involves transformations in encoding schemes in the ascending auditory pathway. One example is the transformation of a synchronization code to a rate code in the processing of pulse-repetition rate. In the first scheme, a neuron responds with a single spike or short burst of spikes to each sound pulse in a stimulus. In the second scheme, a neuron's firing rate is highest when a sound has a certain pulse rate and then drops off if the rate is increased, decreased, or both (Hall 1994). Decoding, or recognition, might be inferred from the responses of a highly selective set of neurons that respond maximally when a stimulus possesses behaviorally optimum values of one or more acoustic properties. One of the great challenges for neurobiology is to find out how the results of sensory processing by selective neurons are transformed into appropriate motor responses.

With regard to evolutionary questions about communication, the behavioral specifications for selective phonotaxis provide estimates of the form and intensity of sexual selection on call structure (e.g., Gerhardt 1991, 1994a; Ryan et al. 1992). For example, comparing the optimum range of values for a given property with the variation in that property among males indicates which subset of males should have higher mating success. Equivalently, playback experiments can be used to assess the extent to which differential mating success based on some acoustic property, or, more often, some morphological correlate such as body size might be explained by female preferences. Males that produce highly attractive signals might also possess phenotypic or genotypic traits that benefit the females choosing them directly, indirectly, or both (Kirkpatrick & Ryan 1991). Such benefits might serve to maintain the preference in the population or species and also could have contributed to its establishment. That is, if females initially benefit by responding to a "new" signal, both the preference and the signal are more likely to become established than if females receive no benefit (Dawkins & Guilford 1996).

Whereas the range of optimum signal variants is usually restricted to ranges of values typical of conspecific signals, systematic variation of acoustic properties sometimes reveals preferences for values outside the range of variation. Moreover, studies using the calls of other closely related species that are not found in the range of distribution of the target species can also reveal unexpected biases ("hidden preferences") for properties of hetero-specific calls (Arak & Enquist 1993). The best-known anuran examples are two species of neotropical frogs in the *Physalaemus pustulosus* species group, in which conspecific whines are enhanced by appending acoustic suffixes and prefixes typical of the calls of other species but not found in conspecific calls (Ryan & Rand 1993). These observations indicate that signal structure and receiver selectivity are unlikely to be perfectly matched at few, if any points during the course of their evolution. One reason may be that the initial propensity to respond to new signals must often evolve in a different behavioral context (Ryan & Rand 1993).

## MECHANISMS UNDERLYING SPECTRAL AND TEMPORAL SELECTIVITY IN TREEFROGS

Two examples of the interplay of behavioral and neurophysiological approaches to underlying mechanisms include preferences based on spectral patterns in green treefrogs (*Hyla cinerea*), and preferences based on temporal patterns in two species of gray treefrogs (*H. chrysoscelis* and *H. versicolor*). Both of these properties are static (low within-male variation) and mediate moderate to strong preferences that play a role in mate choice (Gerhardt 1987, 1991, 1994a, b).

In the green treefrog, males produce a noisy call with many frequency components (Fig. 10.1A, inset). Energy is concentrated, however, in two frequency bands: a low-frequency band, usually consisting of a single component around 1 kHz, and a high-frequency band, usually consisting of two to four components centered around 3 kHz. Behavioral studies show that females respond reliably to synthetic calls consisting of either band alone, but females do so at much lower playback levels when presented with a low-frequency component (900 Hz at about 48 dB SPL [sound pressure level in decibels re 20 μPa]) than when presented with a high-frequency component (3,000 Hz), which needs to be played back at about 90 dB SPL (Fig. 10.1; Gerhardt 1976). To put this difference into perspective, a 40-dB difference corresponds to a hundredfold difference in sound pressure. Offered a simultaneous choice, however, females prefer signals with both low- and high-frequency peaks (900 + 3000 Hz) to a signal with just the low-frequency peak (900 Hz) at playback levels as low as 55 dB SPL (Gerhardt 1981a). Females do not show such a preference at a lower level (48 dB SPL) that is still above the phonotactic threshold for the low-frequency peak alone. These results suggest that high-frequency energy becomes behaviorally relevant to the female *when it is combined with low-frequency energy* as in conspecific calls. A second requirement is that a signal's intensity has to be about 7 dB above the behavioral threshold for signals having only low-frequency energy.

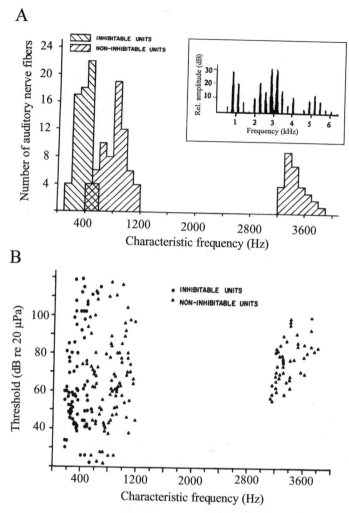

Figure 10.1. Spectral properties and auditory tuning in the green treefrog. (A) Distribution of characteristic frequencies (frequency to which the neuron has its lowest threshold) of auditory nerve fibers recorded in the green treefrog. The two populations tuned to low frequencies innervate the amphibian papilla, one of the two main auditory inner ear organs in frogs, and the population tuned to high frequencies innervate the basilar papilla, the other main auditory organ. The low-frequency-tuned population of neurons (high-to-low diagonal cross-hatching) has the property of tone-on-tone inhibition: presentation of a second tone of higher frequency can suppress the response of these neurons to a tone at their characteristic frequency. (B) Scatter plot of the characteristic frequencies and absolute thresholds of the auditory nerve fibers whose distribution is shown in (A). Notice that the most sensitive high-frequency-tuned neurons have thresholds that are significantly higher than those of the most sensitive low-frequency-tuned neurons. Modified from Capranica and Moffat (1983). *Inset*: Power spectrum of a typical advertisement call of a male *Hyla cinerea*. Modified from Gerhardt (1974a).

Neurophysiological studies show that low-frequency sounds are mainly transduced by one of the inner ear organs of the frog, the amphibian papilla, and high-frequency sounds by another inner ear organ, the basilar papilla (Lewis & Narins 1999). The absolute thresholds of the most sensitive auditory neurons that innervate the amphibian papilla and convey information to the central nervous system are about 20 to 30 dB lower than those of the most sensitive neurons that innervate the basilar papilla (Fig. 10.1; Capranica & Moffat 1983). Indeed, the most sensitive basilar-papilla neurons have absolute thresholds of about 50 to 55 dB SPL, the very level at which high-frequency acoustic energy begins to influence the preference for signals with both spectral peaks. Thus, in terms of mechanisms, the neural correlates of the preference for the combination of frequencies must be sought at some level in the ascending pathway where the inputs from these two auditory organs converge.

Paralleling the behavioral preference for signals with both spectral peaks, neurons on which information from the two papillae converge should have the property of responding much better to the combination of frequencies than to either frequency presented in isolation. Just such a property was found by Karen Mudry, working in Capranica's laboratory, at the level of the auditory thalamus (Mudry & Capranica 1987). Evoked potentials (summed responses of groups of neurons) in response to a combination of low- and high-frequency tones were much greater in amplitude than the linear sum of responses to the same tones presented in isolation. Single auditory neurons having an analogous property—firing only in response to combinations of appropriate tones and not to the same tones presented in isolation—were found commonly in the thalamus and rarely in the next lower station in the midbrain, the torus semicircularis (inferior colliculus of mammals) (Fuzessery 1988). Although some new studies indicate the thalamus is unlikely to mediate rapid decisions about the relative attractiveness of different acoustic signals (Walkowiak et al., unpublished data), this structure is still likely to be important for controlling motivation for phonotaxis by integrating sensory information with information about the animal's reproductive status (e.g., Wilczynski et al. 1993).

In two cryptic species of gray treefrogs, the spectral properties of advertisement calls are very similar, and the main species differences are in pulse rate, pulse duration, and pulse shape (Gerhardt & Doherty 1988). Females of the diploid species, *H. chrysoscelis*, show strong preferences based on pulse rate alone, and strongly reject pulse rates that are lower than those of conspecific males calling at about the same temperature as that of the female (Gerhardt 1994a). Females of its sibling tetraploid species, *H. versicolor*, not only prefer slow to high pulse rates but also prefer pulses with the shape (rise-time) typical of conspecific calls to pulses with the shape (rise-time) typical of the calls of *H. chrysoscelis* (Diekamp & Gerhardt 1995; Gerhardt & Schul 1999). In both species, changing the temperature of the female changes her preference for pulse rate in a fashion parallel to the change in pulse rate in the calls of conspecific males, at least over the normal range of breeding temperatures (Gerhardt & Doherty 1988; Gerhardt & Schul, in preparation). These parallel changes, called temperature coupling, have also been described in acoustic insects, fireflies, and weakly electric fish (citations in Gerhardt 1978a).

The neural processing of pulse rate in gray treefrogs provides a good example of the transformation of the encoding of relevant acoustic properties. Neurons in the auditory nerve and in the dorsolateral nucleus (the first station in the ascending auditory pathway) are *unselective* for pulse rate, pulse duration, or pulse rise-times typical of conspecific calls (Rose et al. 1985; Schul & Gerhardt, in preparation). These neurons simply synchronize their firing to sound pulses, and the synchronization is just as precise when the pulse rate has heterospecific values as when pulse rate is the same as in conspecific calls (Rose et al. 1985). At the level of the torus semicircularis (two stations higher in the pathway), however, many auditory neurons respond selectively to certain ranges of pulse rate, even though they synchronize only poorly to the pulses in the sound. One class of temporally selective neurons—bandpass neurons—fire most strongly to a relatively narrow range of pulse rates; their firing rates decrease as the pulse rate is changed to lower or higher values. In both species of gray treefrogs, a substantial proportion of the population of bandpass neurons is "tuned" to pulse rates typical of those in the calls of conspecific males (Rose et al. 1985), and in *H. versicolor*, the proportion of bandpass neurons is higher when the pulses in the stimulus also have shapes that are similar to pulses within conspecific calls (Diekamp & Gerhardt 1995). As expected from behavioral studies, the pulse rate tuning of bandpass neurons is temperature-dependent in the same way (at least qualitatively) as the behavioral preferences of females (Rose et al. 1985). However, many bandpass neurons that are tuned to conspecific pulse rates would not be likely to respond especially well to conspecific calls because these neurons are most sensitive to frequencies that are not emphasized in conspecific calls (Diekamp & Gerhardt 1995; Walkowiak 1984). In general, pattern recognition is likely to be based on the activity of an array of moderately selective neurons that are distributed throughout the auditory system rather than on the activity of a few, highly selective neurons found in a single locus in the central auditory system (Walkowiak & Luksch 1994). Support for this view derives from the discovery of parallel processing of temporal and spectral patterns by different parts of the brain in the leopard frog, *Rana pipiens* (review in Hall 1994).

## EVOLUTION OF COMMUNICATION:
## FEMALE PREFERENCES AND MALE SIGNALS

Preferences based on differences in carrier frequency, pulse rate, call duration, and call rate (or a combination of these last two: calling effort or call duty cycle) have been documented in more than a dozen species of frogs and toads (Gerhardt 1988, 1994b). These studies show a general trend for preferences based on acoustic properties that are highly invariant within bouts of calling by males (static properties such as carrier frequency, pulse rate, and call duration in some species) to be stabilizing or weakly directional (Fig. 10.2A; Castellano & Giacoma 1998; Gerhardt 1991; Grafe 1997). In stabilizing selection, values at or near the mean are preferred to lower and higher values representative of the two ends of the range of variation in male calls. In weakly directional selection, high values are preferred to low values but not to values at or near the mean.

Properties that vary significantly within bouts of calling (dynamic properties such as call rate and call duration in many species) often mediate strongly directional preferences, in which females prefer extreme values representative of one end of the distribution of the property in male calls to values near the mean (Fig. 10.2B). In some species, females even prefer values beyond the range of variation in male calls (Gerhardt 1991; Ryan & Keddy-Hector 1992). Similar patterns have recently been described for Hawaiian crickets (Shaw & Herlihy 2000).

Both weakly and strongly directional preferences have the potential to cause significant evolutionary change in a call property, depending on the heritability of the property and whether or not there is counterselection. For example, directional preferences on carrier frequency, which is usually negatively correlated with body size, could be countered by selection against large body size. Directional preferences for long calls or calls produced at a high rate are likely to be opposed by the high energetic costs of producing such signals (e.g., Wells & Taigen 1986). Males that produce costly calls might not be able to attend multiple choruses, and chorus attendance is almost universally correlated with mating success in anurans (Halliday & Tejedo 1995). Moreover, high rates of calling or long-duration calls might also increase predation risks (e.g., Grafe 1997; Ryan 1985).

Stabilizing patterns of female preferences might be a common explanation for the low within- and between-male variation in pulse rate, which is often used for species identification. Besides the advertisement call, males often produce calls in aggressive contexts that have a different pulse rate; indeed, sometimes males produce a continuum of aggressive calls that are graded by differences in pulse rate (Grafe 1995; Wells 1988). Thus, male frogs are unlikely, in principle, to be constrained by their motor system to produce calls with a single, narrow range of pulse rate.

By contrast, female preferences, even when stabilizing, are unlikely to be the main cause of low variability in carrier frequency. As mentioned above, carrier frequency is usually constrained by morphology (body size and the size and mass of laryngeal structures). Indeed, in both intraspecific and interspecific comparisons, a large proportion of the variation in carrier frequency is explained by variation in body size (Gerhardt 1991; Ryan 1988). Prefer-

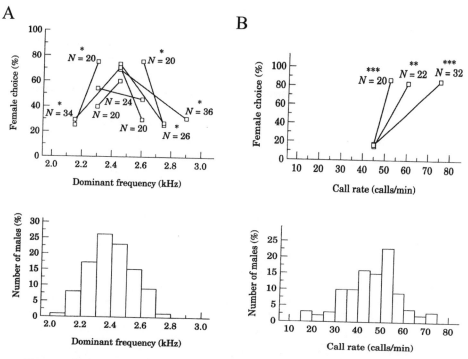

Figure 10.2. (A) Top panel: choices of females of the African reed frog (*Hyperolius marmoratus*) given choices between synthetic calls that differed in dominant frequency. *Bottom panel*: distribution of dominant frequency in the advertisement calls of males in the same population. (B) *Top panel*: choices of females given choices between synthetic calls that differed in call rate. *Bottom panel*: distribution of call rate in the advertisement calls of males. Each line connects points showing the percentages of females choosing each of the two alternatives in a two-speaker choice test. Asterisks indicate statistically significant preferences (* < 0.05; ** < 0.01; *** < 0.001).

ences based on carrier frequency can thus exert indirect selection on body size, which, however, is also likely to be the target of completely different selective forces. Thus species differences in carrier frequency could also be interpreted as an incidental effect of direct selection on body size. Some interesting exceptions, which may indicate direct selection on carrier frequency, include a few species that have calls with much lower frequencies than expected by their body size (e.g., the *Physalaemus pustulosus* species group and some African toads) (Martin 1972; Ryan 1988; Ryan & Drewes 1990). The trick used by these species is to add structures that load (add mass to) the vocal cords, hence lowering its fundamental frequency. Body size also influences auditory tuning within and between some species (Márquez & Bosch 1997a; Ryan et al. 1992; Zakon & Wilczynski 1988). Thus, at the population level, size-dependent variation in carrier frequency in males might be roughly matched by size-dependent tuning in females (see below; Robertson 1990).

Although a reasonably good match exists between the tuning of the anuran auditory system and the frequencies emphasized in conspecific calls in most species, differences of 20 percent or more between the mean values of tuning (best excitatory frequencies) and the mean carrier frequency in the population are often reported (review in Gerhardt & Schwartz 2001). These mismatches, if reflected in female preferences, can therefore serve as a source of directional selection on carrier frequency (e.g., Ryan et al. 1990, 1992).

Most support for female choice based on carrier frequency comes from demonstrations of large-male mating advantages or size-assortative mating (Robertson 1990; Ryan 1985). However, body size has also been found to be correlated with intensity (Arak 1988; Gerhardt 1975) and call rate (Green 1990). Indeed, experiments with the natterjack toad (*Bufo calamita*) showed that the large-male mating advantage was based on preferences involving call intensity and not frequency (Arak 1988). Selectivity based on carrier frequency and inferred from size-dependent mating success could be masked or underestimated if small males obtain matings by searching or satellite male tactics (Sullivan et al. 1995). Preferences based on carrier frequency could be overestimated if large males can physically monopolize females or resources required by females.

Some frogs and toads modify carrier frequency, although usually by less than 5 to 10 percent (e.g., Wagner 1989). These alterations are observed in male-male interactions or in response to playbacks and not as a response to detecting a female. In three of four species in which frequency alteration has been reported (*Acris crepitans*, *Bufo americanus*, and *Rana clamitans*), lowering carrier frequency appears to result in a drop in call amplitude (Bee & Perrill 1996; Howard & Young 1998; Wagner, personal communication). One hypothesis is that lowering frequency is a form of bluffing that is most likely to be performed by small males in response to large males (Bee et al. 2000).

Relatively few studies have directly examined the relationship between variation in call properties and mating success. No study, for example, has found a significant relationship between carrier frequency and mating success, although small sample sizes have usually limited the possibility of detecting anything but large correlations (Gerhardt 1994b). By comparison, most studies that have directly correlated call rate with mating success have a found significant positive relationship. One of the best such studies simultaneously monitored calling by groups of males at a time during periods before some of the males attracted females (Passmore et al. 1992). Females usually chose males whose call rate was above average in the sample of monitored males as well as the average estimate for the whole population.

## EVOLUTION OF PREFERENCES: WHAT MIGHT FEMALES GAIN FROM CHOOSING CERTAIN SIGNALS?

The reasons that females choose males with particular acoustic properties are diverse, and interpretations of the benefits, if any, of mate choice are often controversial. In the Australian toadlet *Uperoleia laevigata*, females appear to use differences in dominant frequency, at least in part, to pick a male with a body mass that is approximately 70 percent of their own (Gerhardt 1988; Robertson 1990). Doing so increases the proportion of their eggs that are fertilized because large males fertilize more eggs than small males; mating with still larger males, however, subjects the female to the danger of drowning. In the túngara frog, *Physalaemus pustulosus*, females increase fertilization success by mating with large rather than small males; large males add "chucks" to the "whine" part of their calls that have lower-frequency harmonics than chucks produced by small males (Ryan 1985). Ryan et al. (1990) argue, however, that the benefits of mating with large males is a consequence rather than a cause of the female preference for lower-than-average calls and at best serves to maintain the preference in contemporary populations. These authors use a comparative phylogenetic analysis to support their hypothesis that the preference for low-frequency chucks evolved earlier than the chuck itself and is thus a preexisting bias.

The most controversial explanation for female preferences is that females might receive indirect genetic benefits ("good genes") that result in more viable offspring of both sexes. Some frogs are excellent subjects for testing this idea because they offer no direct benefits, such as increased fertilization success, territorial resources, or parental care. Females receive only gametes. Because frogs fertilize eggs externally, clutches can be divided and some eggs fertilized with the sperm of males with certain call characteristics and other eggs, with the sperm of males with other call characteristics. Comparisons of various measures of fitness of the offspring can reveal if there is any difference in the genetic contributions of the male parents. One study of the gray treefrog (*Hyla versicolor*) finds that the offspring of males producing long calls are fitter in several respects than the offspring of males producing short calls (Welch et al. 1998). Clutches of eggs fertilized by males producing long calls did not have greater hatching success than eggs fertilized by short callers (Krenz et al., unpublished data).

Finally, I emphasize that females in nature must usually assess two or more pertinent properties of signals before making a choice. In the gray treefrog (*Hyla chrysoscelis*), for example, females also prefer long calls to short calls, and preferences based on pulse rate alone are even stronger than those in the sibling species *H. versicolor* (Gerhardt 1994a). In populations where *H. versicolor* is absent, about 50 to 60 percent of the females prefer short calls with the local, conspecific pulse rate rather than long calls with a lower, pulse rate (more similar to the calls of *H. versicolor*). In populations where *H. versicolor* also occurs, 85 to 100 percent of the females choose the short call with "correct" pulse rate. One interpretation is that the risk of mating with a genetically incompatible male (offspring of matings of *H. chrysoscelis* and *H. versicolor* are sterile triploids) in sympatric areas has selected for females that weigh pulse rate, which identifies the species, more strongly than call duration, which might indicate differences in quality among conspecific males (Gerhardt 1994a, b). Geographical differences in preference strength in other anurans have been explained in a similar way (Márquez & Bosch 1997b; Pfennig 2000).

## General Lessons for Developing a System for Studying Behavior and Communication

In my view, there is no substitute for exploring the suitability of a particular taxonomic group or particular species for testing in experiments designed to gain knowledge about how and why individuals respond selectively to particular communication signals. Two main steps are (1) observing the behavior of the animals in their natural habitat; and (2) conducting pilot experiments to see if the animals behave normally in laboratory or seminatural conditions. The first step is facilitated if the animals are common and can be readily approached in order to observe their behavior closely. Rare species might offer unique opportunities for comparative studies, but collecting sufficient data is likely to be difficult. The main reason for studying the animal in nature is to get a feel for its environment and the problems it faces. What are its predators, and how do the predators find and capture their prey? Does it compete with other species for resources? How does it compete with conspecifics for mates or resources? What are the encounter rates of predators, competitors, and mates? By what modality or modalities does it communicate with conspecifics? Are mutually effective signals used in interspecific interactions? How does the communication channel or channels distort, attenuate, or mask intraspecific communication? As I mentioned in my introduction, the abundance of several species of treefrogs and the ease with which they could be approached without disturbing them was a large part of my reason for choosing to study them. The fact that males often signal in huge aggregations and in complex, mixed species choruses enhanced my attraction to the system by posing challenging questions about how treefrogs might solve the communication problems arising from such environments.

The second step—assessing the suitability of the animal for manipulative experiments—is critical because a researcher ultimately needs to simplify, if not to control, the context of behavior in order to disentangle the multiple factors that influence behavior in general and communication in particular. My first playback experiments with green treefrogs were so exciting because of the reliability with which gravid females responded repeatedly to playbacks in the laboratory. I knew I would be able to easily collect extensive data about their selectivity with respect to variation in conspecific calls as well as their ability to reject the signals of other species commonly occurring in their environment.

For questions about evolutionary mechanisms, the possibility of conducting breeding experiments that produce large numbers of offspring is a significant advantage. These experiments can test hypotheses about the genetic consequences of mating decisions and provide estimates of heritable variation in behaviors such as signaling and selective phonotaxis. Frog and fish systems are particularly favorable because fertilization is external. Paternity is thus known or can be controlled. In species with internal fertilization and multiple mating, paternity is far less certain and females often have the ability to exercise postcopulatory choice (e.g., Choe & Crespi 1997).

Another important consideration for studying evolution is the existence of wide-ranging species showing geographical variation in behavior, the accessibility of a series of closely related species, or both. Geographical variation and the availability of an array of different species in the same taxonomic group provide opportunities to apply comparative approaches that can provide insights or generate hypotheses about broad-scale patterns of evolution and perhaps even the evolutionary origins of and patterns of change in particular behaviors.

For questions about proximate mechanisms, the choice of a system is often dictated by the availability of a body of existing literature and established methodology. For neurophysiological studies of communication, for example, a new researcher needs to know about the anatomy and physiology of sensory receptors and the ascending pathways. What loci in the central nervous system are most likely to contain neurons that are selective for biologically important signals? What kinds of microelectrodes are best suited to record from single neurons or groups of neurons in these areas of the brain? What kinds of anesthesia can be used safely and effectively with the study species? For me, anuran acoustic communication was attractive because a great deal of basic information about the frog auditory system was already available (reviews in Fritzsch et al. 1988) and many methodological details had been worked out.

# Future Directions

## Proximate Mechanisms

Most neurophysiological studies of sensory processing have been conducted with anesthetized or immobilized preparations. Moreover, in most studies the reproductive state of the animal is not even mentioned much less controlled. In treefrogs, for example, most females do not show phonotactic behavior unless they have ovulated and but not yet laid their eggs. Females stop responding after they have laid their eggs. We have no idea if immobilizing animals or recording from males or from females that are not in a state of phonotactic readiness can alter the patterns of selectivity (or lack thereof) of auditory neurons. Thus, it will be important to develop ways of recording from single neurons or groups of neurons in freely moving, behaving animals. Only by doing so can we learn if the filter properties—selectivity for particular values of behaviorally relevant acoustic properties—of auditory neurons are maintained, or perhaps even sharpened, during the times when the animal is evaluating and choosing among alternative signals. For anurans, higher centers, such as the auditory thalamus, might not be involved in the moment-to-moment evaluation of signals that leads to behavioral decisions (Walkowiak et al., in preparation) but rather serve as areas that integrate sensory information with inputs from hormone receptors that are evaluating the animal's reproductive state. Indeed, Casseday and Covey (1996) argue convincingly that sensory processing leading to rapid decisions in all vertebrates is mainly accomplished in a lower center—the inferior colliculus (torus semicircularis of anurans).

The biophysics and neural mechanisms underlying sound localization in anurans are still poorly understood. Because the putative internal pathways between the eardrums and other inputs to the inner ear organs can probably be modified easily by the behaving anuran, the development of a freely moving preparation will also be a prerequisite to further progress in understanding how the system works when the animal is trying to find a sound source. Finally, the interaction between sound pattern recognition and sound localization needs further study. We know, for example, that separation of two loudspeakers can improve the attractiveness of overlapping signals (Schwartz & Gerhardt 1995), and thus directional hearing can help to reduce masking interference in large choruses. Could directional hearing also impose selection on signal design? Is, for example, the attractiveness of a signal enhanced if it is easier to locate?

## EVOLUTION OF ACOUSTIC COMMUNICATION

For most species, we still lack robust preference functions that relate the strength of female preferences to variation in particular properties of communication signals that vary among males in the same population or species (Wagner 1995). Such functions require determining the smallest difference in a relevant property that elicits a preference, and then determining how the strength of the preference varies as that difference is systematically increased. It is also vital to explore how a particular difference mediates preferences as the absolute values of alternative stimuli are varied. In other words, is preference strength a constant for a given difference or is it stronger if the alternative stimuli are representative of one part of the range of variation as opposed to another part (e.g., Basolo 1990; Gerhardt et al., in press)?

Methods of estimating preference strength serve a dual purpose if they incorporate variables that might represent costs of assessment. I have already suggested that varying the relative intensity of alternatives can indicate the propensity of females to move extra distances to pick males whose calls are especially attractive. Another method of introducing potential costs involves varying the amount of cover between the female's release point and the sources of the attractive and unattractive signals, as in one study of mate choice in crickets (Hedrick & Dill 1993). Assessment costs are a key feature of handicap models of the evolution of female preferences, and we would expect that the costs females are willing to incur to mate with males with particular signals is related to the benefits of mating with these males. We are also only beginning to assess the effects of varying more than a single acoustic property at a time.

Finally, comparative studies such as those showcased by Ryan and his colleagues are likely to continue to provide hypotheses and data about large-scale patterns of signal and preference evolution (Ryan & Rand 1993). The data required to finely resolve phylogenetic relationships and to posit the direction of evolution of character states is slowly but surely becoming available for some groups. One promising study on a smaller scale involves comparisons of species in the diploid-tetraploid complex of gray treefrogs. Because the tetraploid (*H. versicolor*) has almost certainly arisen multiple times from the diploid species by autopolyploidy (Ptacek et al. 1994), we can be rather sure that signals and preferences in *H. chrysoscelis* represent the ancestral state with respect to those in *H. versicolor*. Contrary to the situation in the *Physalaemus pustulosus* group of frogs, sensory biases in the form of frequency preferences are not conserved in *H. versicolor*. Instead, females have different frequency preferences than females of the diploid species even though the frequency structure of the advertisement calls has been conserved (Gerhardt et al., in preparation). The neural bases for this difference and its probable effects on other aspects of call recognition in the tetraploids are currently being studied extensively in my laboratory.

# References

Arak A, 1988. Female mate selection in the natterjack toad: active choice or passive attraction? Behav Ecol Sociobiol 22:317–327.

Arak A, Enquist M, 1993. Hidden preferences and the evolution of signals. Phil Trans R Soc Lond B 340:207–213.

Basolo AL, 1990. Female preference for male sword length in the green swordtail. Anim Behav 40:332–338.

Bee MA, Perrill SA, 1996. Responses to conspecific advertisement calls in the green frog (*Rana clamitans*) and their role in male-male competition. Behaviour 133:283–301.

Bee MA, Perrill SA, Owen PC, 2000. Male green frogs lower the pitch of acoustic signals in defense of territories: a possible dishonest signal of size? Behav Ecol 11:169–177.

Capranica RR, Moffat AJM, 1983. Neurobehavioral correlates of sound communication in anurans. In: Advances in vertebrate neuroethology (Ewert JP, Capranica RR, Ingle DJ, eds). New York: Plenum Press; 701–730.

Casseday JH, Covey E, 1996. A neuroethological theory of the operation of the inferior colliculus. Brain Behav Evolution 47:311–336.

Castellano S, Giacoma C, 1998. Stabilizing and directional female choice for male calls in the European green toad. Anim Behav 56:275–287.

Choe JC, Crespi BJ, 1997. Mating systems in insects and arachnids. Cambridge: Cambridge University Press.

Dawkins MS, Guilford T, 1996. Sensory bias and the adaptiveness of female choice. Am Natur 148:937–942.

Diekamp BM, Gerhardt HC, 1995. Selective phonotaxis to advertisement calls in the gray treefrog *Hyla versicolor*: behavioral experiments and neurophysiological correlates. J Comp Physiol A 177:173–190.

Doherty JA, Gerhardt HC, 1984. Evolutionary and neurobiological implications of selective phonotaxis in the spring peeper (*Hyla crucifer*). Anim Behav 32:875–881.

Eggermont JJ, 1988. Mechanisms of sound localization in anurans. In: The evolution of the amphibian auditory system (Fritzsch B, Hetherington T, Ryan MJ, Wilczynski W, Walkowiak W, eds). New York: John Wiley & Sons; 307–336.

Fritzsch B, Hetherington T, Ryan MJ, Wilczynski W, Walkowiak, W (eds), (1988). The evolution of the amphibian auditory system. New York: John Wiley & Sons.

Fuzessery ZM, 1988. Frequency tuning in the anuran central auditory system. In: The evolution of the amphibian auditory system (Fritzsch B, Hetherington T, Ryan MJ, Wilczynski W, Walkowiak W, eds). New York: John Wiley & Sons; 253–273.

Gerhardt HC, 1973. Reproductive interactions between *Hyla crucifer* and *Pseudacris ornata* (Anura: Hylidae). Am Midl Natur 89:81–88.

Gerhardt HC, 1974a. The significance of some spectral features in mating call recognition in the green treefrog *Hyla cinerea*. J Exp Biol 61:229–241.

Gerhardt HC, 1974b. Behavioral isolation of the treefrog *Hyla cinerea* and *Hyla andersonii*. Am Midl Natur 91:424–433.

Gerhardt HC, 1974c. Vocalizations of some hybrid treefrogs: acoustic and behavioral analyses. Behaviour 49:130–151.

Gerhardt HC, 1975. Sound pressure levels and radiation patterns of the vocalizations of some North American frogs and toads. J Comp Physiol A 102:1–12.

Gerhardt HC, 1976. Significance of two frequency bands in long distance vocal communication in the green treefrog. Nature (London) 261:692–694.

Gerhardt HC, 1978a. Temperature coupling in the vocal communication system of the gray treefrog *Hyla versicolor*. Science 199:992–994.

Gerhardt HC, 1978b. Mating call recognition in the green treefrog (*Hyla cinerea*): the significance of some fine-temporal properties. J Exp Biol 74:59–73.

Gerhardt HC, 1981a. Mating call recognition in the green treefrog (*Hyla cinerea*):

importance of two frequency bands as a function of sound pressure level. J Comp Physiol 144:9–16.

Gerhardt HC, 1981b. Mating call recognition in the barking treefrog (*Hyla gratiosa*): responses to synthetic calls and comparisons with the green treefrog (*Hyla cinerea*). J Comp Physiol 144:17–25.

Gerhardt HC, 1982. Sound pattern recognition in some North American treefrogs (Anura: Hylidae): implications for mate choice. Am Zool 22:581–595.

Gerhardt HC, 1987. Evolutionary and neurobiological implications of selective phonotaxis in the green treefrog (*Hyla cinerea*). Anim Behav 35:1479–1489.

Gerhardt HC, 1988. Acoustic properties used in call recognition by frogs and toads. In: The evolution of the amphibian auditory system (Fritzsch B, Hetherington T, Ryan MJ, Wilczynski W, Walkowiak W, eds). New York: John Wiley & Sons; 455–483.

Gerhardt HC, 1991. Female mate choice in treefrogs: static and dynamic acoustic criteria. Anim Behav 42:615–635.

Gerhardt HC, 1994a. Reproductive character displacement on female mate choice in the grey treefrog *H. chrysoscelis*. Anim Behav 47:959–969.

Gerhardt HC, 1994b. The evolution of vocalization in frogs and toads. Annu Rev Ecol Syst 25:293–324.

Gerhardt HC, Doherty JA, 1988. Acoustic communication in the gray treefrog, *Hyla versicolor*: evolutionary and neurobiological implications. J Comp Physiol A 162:261–278.

Gerhardt HC, Dyson ML, Tanner SD, 1996. Dynamic acoustic properties of the advertisement calls of gray treefrogs: patterns of variability and female choice. Behav Ecol 7:7–18.

Gerhardt HC, Klump GM, 1988. Masking of acoustic signals by the chorus background noise in the green treefrog: a limitation on mate choice. Anim Behav 36:1247–1249.

Gerhardt HC, Schul J, 1999. A quantitative analysis of behavioral selectivity for pulse rise-time in the gray treefrog, *Hyla versicolor*. J Comp Physiol A 185:33–40.

Gerhardt HC, Schwartz JJ, 1995. Interspecific interactions in anuran courtship. In Amphibian biology, vol. 2: social behaviour (Heatwole H, Sullivan BK, eds). Chipping Norton, NSW: Surrey Beatty and Sons; 603–632.

Gerhardt HC, Schwartz JJ, 2001. Auditory tuning and frequency preferences in frogs and toads. In: Anuran communication (Ryan MJ, ed). Washington: Smithsonian Institution Press.

Gerhardt HC, Tanner SD, Corrigan CM, Walton HC, in press. Female preference functions based on call duration in the gray treefrog (*Hyla versicolor*). Behav Ecol.

Grafe TU, 1995. Graded aggressive calls in the African painted reed frog *Hyperolius marmoratus* (Hyperoliidae). Ethology 101:67–81.

Grafe TU, 1997. Costs and benefits of mate choice in the lek–breeding reed frog, *Hyperolius marmoratus*. Anim Behav 53:1103–1117.

Green AJ, 1990. Determinants of chorus participation and the effects of size, weight and competition on advertisement calling in the tungara frog, *Physalaemus pustulosus* (Leptodactylidae). Anim Behav 39:620–638.

Hall JC, 1994. Central processing of communication sounds in the anuran auditory system. Am Zool 34:670–684.

Halliday TR, Tejedo M, 1995. Intrasexual selection and alternative mating behaviour.

In: Amphibian biology: vol. 2: social behaviour (Heatwole H, Sullivan BK, eds). Chipping Norton, NSW: Surrey Beatty and Sons; 419–468.

Hedrick AV, Dill LM, 1993. Mate choice by female crickets is influenced by predation risks. Anim Behav 46:193–196.

Jørgensen MB, 1991. Comparative studies of the biophysics of directional hearing in anurans. J Comp Physiol A 169:591–598.

Kirkpatrick M, Ryan MJ, 1991. The evolution of mating preferences and the paradox of the lek. Nature (London) 350:33–38.

Lewis ER, Narins PM, 1999. The acoustic periphery of amphibians: anatomy and physiology. In: Comparative hearing: fishes and amphibians (Fay RR, Popper AN, eds). New York: Springer Verlag; 101–154.

Littlejohn MJ, 1993. Homogamy and speciation: a reappraisal. In: Oxford surveys of evolutionary biology (Futuyma D, Antonovics J, eds). Oxford: Oxford University Press; 135–164.

Lucas JR, Howard RD, Palmer JG, 1996. Callers and satellites: chorus behaviour in anurans as a stochastic dynamic game. Anim Behav 51:501–518.

Márquez R, Bosch J, 1997a. Female preference in complex acoustical environments in the midwife toads *Alytes obstetricans* and *Alytes cisternasii*. Behav Ecol 8:588–594.

Márquez R, Bosch J, 1997b. Male advertisement call and female preference in sympatric and allopatric midwife toads. Anim Behav 54:1333–1345.

Martin WF, 1972. Evolution of vocalization in the genus *Bufo*. In: Evolution in the genus *Bufo* (Blair WF, ed). Austin: University of Texas Press; 279–309.

Martof BS, 1961. Vocalization as an isolating mechanism in frogs. Am Midl Natur 71:198–209.

Martof BS, Thompson EF, 1964. A behavioral analysis of the mating call of the chorus frog, *Pseudacris triseriata*. Am Midl Natur 71:198–209.

Martof BS, Gerhardt HC, 1965. Observations on geographical variation in *Ambystoma cingulatum*. Copeia 1965:342–346.

Mudry KM, Capranica RR, 1987. Correlation between auditory thalamic area evoked potentials and species-specific call characteristics II. *Hyla cinerea* (Anura: Hylidae). J Comp Physiol A 161:407–416.

Narins PM, Ehret G, Tautz T, 1988. Accessory pathway for sound transfer in a neotropical frog. Proc Natl Acad Sci USA 85:1508–1512.

Oldham RS, Gerhardt HC, 1975. Behavioral isolation of the treefrogs *Hyla cinerea* and *Hyla gratiosa*. Copeia 1975:223–231.

Passmore NI, Bishop PJ, Caithness N, 1992. Calling behaviour influences mating success in male painted reed frogs, *Hyperolius marmoratus*. Ethology 92:227–241.

Perrill SA, Gerhardt HC, Daniel RE, 1982. Mating strategy shifts in male green treefrogs (*Hyla cinerea*): an experimental study. Anim Behav 30:43–48.

Pfennig, K. S. 2000. Female spadefoot toads compromise on mate quality to ensure conspecific matings. Behav Ecol 11:220–227.

Ptacek MB, Gerhardt HC, Sage RD, 1994. Speciation by polyploidy in treefrogs: multiple origins of the tetraploid, *Hyla versicolor*. Evolution 48:898–908.

Rheinlaender J, Gerhardt HC, Yager D, Capranica RR, 1979. Accuracy of phonotaxis in the green treefrog (*Hyla cinerea*). J Comp Physiol 133:247–255.

Robertson JGM, 1990. Female choice increases fertilisation success in the Australian frog, *Uperoleia laevigata*. Anim Behav 39:639–645.

Rose GJ, Brenowitz EA, Capranica RR, 1985. Species specificity and temperature dependency of temporal processing by the auditory midbrain of two species of treefrogs. J Comp Physiol 157:763–769.

Ryan MJ, 1985. The túngara frog: a study in sexual selection and communication. Chicago: University of Chicago Press.

Ryan MJ, 1988. Constraints and patterns in the evolution of anuran acoustic communication. The evolution of the amphibian auditory system (Fritzsch B, Hetherington T, Ryan MJ, Wilczynski W, Walkowiak W, eds). New York: John Wiley; 637–677.

Ryan MJ, Drewes RC, 1990. Vocal morphology of the *Physalaemus pustulosus* species group (Leptodactylidae): morphological response to sexual selection for complex calls. Biol J Linnean Soc 40:37–52.

Ryan MJ, Fox JH, Wilczynski W, Rand AS, 1990. Sexual selection for sensory exploitation in the frog *Physalaemus pustulosus*. Nature, Lond 343:66–67.

Ryan, MJ, Keddy-Hector A, 1992. Directional patterns of female mate choice and the role of sensory biases. Am Natur 139:S4–S35.

Ryan MJ, Perrill SA, Wilczynski W, 1992. Auditory tuning and call frequency predict population-based mating preferences in the cricket frog, *Acris crepitans*. Am Natur 139:1370–1383.

Ryan MJ, Rand AS, 1993. Sexual selection and signal evolution: the ghost of biases past. Phil Trans R Soc Lond B 340:187–195.

Ryan MJ, Rand AS, Weigt LA, 1996. Allozyme and advertisement call variation in the túngara frog, *Physalaemus pustulosus*. Evolution 50:2435–2453.

Ryan MJ, Wilczynski W, 1991. Evolution of intraspecific variation in the advertisement call of a cricket frog (*Acris crepitans*, Hylidae). Biol J Linnean Soc 44:249–271.

Schmitt A, Friedel T, Barth FG, 1993. Importance of pause between spider courtship vibrations and general problems using synthetic stimuli in behavioural studies. J Comp Physiol A 172:707–714.

Schul J, 1998. Song recognition by temporal cues in a group of closely related bush-cricket species (genus *Tettigonia*). J Comp Physiol A 183:401–410.

Schwartz JJ, Gerhardt HC, 1989. Spatially mediated release from auditory masking in an anuran amphibian. J Comp Physiol A 166:37–41.

Schwartz JJ, Gerhardt HC, 1995. Directionality of the auditory system and call pattern recognition during acoustic interference in the gray treefrog, *Hyla versicolor*. Auditory Neurosci 1:195–206.

Shaw KL, Herlihy D, 2000. Acoustic preference functions and song variability in the Hawaiian cricket *Laupala cerasina*. Proc R Soc Lond B 267:577–584.

Sullivan BK, Ryan MJ, Verrell PA, 1995. Female choice and mating system structure. In: Amphibian biology: Vol. 2: Social behaviour (Heatwole H, Sullivan BK, eds). Chipping Norton, NSW: Surrey Beatty and Sons; 469–517.

Tinbergen N 1951. The study of instinct. Oxford: Oxford University Press.

Wagner WE, Jr, 1989. Fighting, assessment, and frequency alternation in Blanchard's cricket frog. Behav Ecol Sociobiol 25:429–436.

Wagner WE, Jr, 1995. Measuring female mating preferences. Anim Behav 55:1029–1042.

Walkowiak W, 1984. Neuronal correlates of the recognition of pulsed sound signals in the grass frog. J Comp Physiol 155:57–66.

Walkowiak W, Luksch H, 1994. Sensory motor interfacing in acoustic behavior of anurans. Am Zool 34:685–695.

Welch AM, Semlitsch RD, Gerhardt HC, 1998. Call duration as an indicator of genetic quality in male gray tree frogs. Science 280:1928–1930.

Wells KD, 1988. The effects of social interactions on anuran vocal behavior. In: The evolution of the amphibian auditory system (Fritszch, B, Wilczynski, W, Ryan, MJ, Hetherington, T, Walkowiak, W, eds). New York: John Wiley & Sons; 433–454.

Wells KD, Taigen TL, 1986. The effect of social interactions on calling energetics in the gray treefrog (*Hyla versicolor*). Behav Ecol Sociobiol 19:9–18.

Wilczynski W, Allison JD, Marler CA, 1993. Sensory pathways linking social and environmental cues to endocrine control regions of amphibian forebrains. Brain Behav Evolution 42:252–264.

Zakon HH, Wilczynski W, 1988. The physiology of the anuran eighth nerve. In: The evolution of the amphibian auditory system (Fritzsch B, Hetherington T, Ryan MJ, Wilczynski W, Walkowiak W, eds). New York: John Wiley; 125–155.

# 11 Selection in Local Neighborhoods, the Social Environment, and Ecology of Alternative Strategies

Barry Sinervo

> It may metaphorically be said that natural selection is daily and hourly scrutinizing, throughout the world, the slightest variations; rejecting those that are bad, preserving and adding up all that are good; silently and insensibly working, whenever and wherever opportunity offers, at the improvement of each organic being in relation to its organic and inorganic conditions of life. We see nothing of these slow changes in progress, until the hand of time has marked the lapse of ages.
> Charles Darwin (1859, 133)

Reading through the *Origin* as a graduate student, I was struck by the contradiction raised by the "Bard of Evolutionary Biology" with regards to the time scale for natural selection. If natural selection "is daily and hourly scrutinizing," why should it take "the hand of time" to mark "the lapse of ages" before we can see any change? Fortunately for my graduate research, the study of natural selection was undergoing a renaissance. Gould and Lewontin's (1979) challenge of the "adaptationist's programme" was the floodgate that precipitated most of the change. Their skepticism of adaptation as the primary principle underlying organic evolution sparked a flurry of papers in defense of adaptational analysis. Notably, the study of adaptation shifted from pattern-based explanations, in which comparisons are made between populations found in different selective environments, to process-based explanations, in which the fitness of individuals is followed and differences between individuals are compared. Analysis of adaptational patterns became couched in a rigorous statistical framework of selective mechanism. Bock (1980), Lande and Arnold (1983), Arnold (1983), and Arnold and Wade (1984) made important conceptual contributions to quantitative methods for analyzing natural and sexual selection, while West-Eberhard (1979, 1983) synthesized perspectives on behavior and social selection.

We can now routinely view natural selection in the act of evolving populations. Observing direct evidence of natural selection in the wild is still difficult. Large-scale and long-term studies are still required to gain a rare glimpse into the process that Charles Darwin viewed as too slow to record. While many processes still remain far too glacial for us to perceive in our short lifetime, other processes can be captured much like a stop-motion camera captures images. From early field studies (Boag & Grant 1981; Endler 1986; Grant 1986), it became apparent that natural selection was very strong and the evolutionary response to selection could be extremely rapid. Selection on traits closely related to fitness, such as alternative reproductive strategies, would be expected to lead to quite rapid evolutionary change, provided that heritable variation is maintained during selection. It is also now apparent that rapid evolutionary change can be driven by cyclical genetic games (Sinervo & Lively 1996), which provides a model system for understanding how social environments can drive selection on behavioral traits.

Here, I examine ecological aspects related to social environment and social selection (West-Eberhard 1979, 1983) that promote rapid evolution in alternative behaviors. I couch arguments in terms of general predictions regarding phenotypic evolution in variable environments. Selection is more likely to produce genetically distinct phenotypes in coarse- compared to fine-grained environments (Levins 1962a, b). The social environment of an individual is composed of the ensemble of nearest neighbors with which it interacts. Individuals in species that maintain the same territory or have high site fidelity throughout their life span will experience coarse-grained social environments. Individuals in such species are likely to experience only a handful of neighbors that vary in competitive ability. If individuals vary in genetically based social behaviors (e.g., territorial behavior), then individuals in a population are likely to experience vastly different social neighbors owing to sampling effects that arise during the formation of neighborhoods. Sampling effects lead to "coarse" social neighborhoods and concomitantly produce variation in the frequency of neighboring behavioral phenotypes that surround individuals. I also extend the arguments to situations in which developmentally plastic alternative phenotypes are likely to evolve. In this case, the temporal environment of an individual is likely to be coarse- or fine-grained over its lifetime depending on the social conditions experienced during ontogeny and plastic alternative phenotypes will be favored. While arguments are directed at alternative male mating strategies, they will be generally applicable to adults and juveniles of both sexes.

## Frequency-Dependent Selection and Male-Male Competition

Natural selection results from the differential survival and reproduction of organisms as a function of their phenotypic attributes. Because of their phenotypes, which are due to the amalgam of traits that make up individuals, some individuals do better than others. The concept of selection is central to Darwin's theory of evolution and forms the cornerstone of many theories in the field of animal behavior (Tinbergen 1963). Selection is defined as a functional relationship between fitness and phenotype that leads to evolutionary or genetically based changes in phenotype distribution (Endler 1986). Phenotypic selection can be quantified by three functional relationships (Lande & Arnold 1983): directional, stabilizing, and disruptive (Fig. 11.1A). In each case, the fitness of an organism depends on the phenotypic traits of that organism. However, in these three simple cases, the fitness surface does not necessarily depend on the distribution of phenotypes in the population. When the phenotype distribution of the population changes (e.g., mean or variance evolves among years), the fitness surface remains relatively constant or invariant over evolutionary time. The selection surface may be a property of the abiotic environment, given that the surface remains static over evolutionary time. In contrast, understanding selection on behavioral traits requires an entirely different perspective because feedback between the phenotypes in the population and the fitness surfaces can promote rapid changes in the phenotype frequency, with ensuing changes in fitness surfaces.

Selection on social behaviors involves an interaction between two or more individuals. The outcome could be beneficial for both, deleterious for one and beneficial for the other, neutral for one but beneficial for the other, neutral for one but deleterious for the other, or neutral for both. The cumulative fitness outcome of many such behavioral interactions in a population leads to selection on behavioral traits. Understanding selection in such cases requires a game theoretic perspective. Von Neumann and Morgenstern (1953) developed game theory to explain human behaviors in conflict situations such as economic exchanges that lead to an economic payoff. Payoffs in economic games or social conflicts are measured by an abstract quantity called expected utility (Maynard Smith & Price 1973; Maynard Smith 1982). In biology, the payoff from a game of life is measured in terms of fitness (e.g., survival and reproductive success). Constructing a payoff matrix is a necessary step in analyzing fitness consequences of behaviors. If fitness of one individual in the contest is tabulated against phenotype of another individual, we can easily derive the payoff matrix for two-player games (e.g., Hawk-Dove). Nature is rarely so simple and social systems usually involve multiplayer games (Sinervo & Lively 1996; Henson & Warner 1997). To capture this complexity, we can plot the fitness of the focal individual against the number of players (or frequency of players) of a given type (Fig. 11.1B, C, D). Such "social" regressions describe how an individual's fitness is dependent on frequency of phenotypes with which it interacts.

Sexual selection arising from either male-male competition or female choice is a form of frequency-dependent selection. Darwin defined sexual selection as variation in mating success. He realized that sexual selection was a special case of natural selection, but he did not clearly elucidate the underlying process. Ronald Fisher (1930) realized that forms of sexual selec-

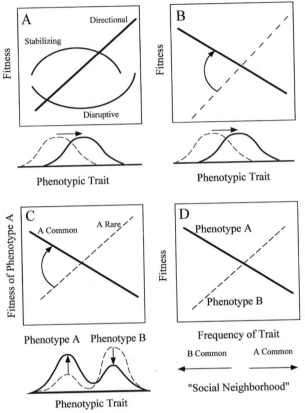

Figure 11.1. (A) Curves for non-frequency-dependent selection are assumed to remain invariant as the population evolves (Lande & Arnold 1983). The dashed line is the phenotype distribution in the first generation, and the solid line is the phenotype distribution in the next generation. Three forms of natural selection are depicted (directional, stabilizing, and disruptive). Regardless of the shape or mean of the phenotype distribution, the form (and position of the curve) of natural selection remains constant. (B) However, in the case of frequency-dependent selection, the curve describing the strength (and direction) of selection can change as the frequency distribution of the trait changes. When the phenotype is normally distributed, selection favoring large size should increase the mean of the population from one generation to the next. However, as the population mean changes, directional selection changes and now favors small size. (C) When competing phenotypes have a discrete distribution, two forms are easy to identify. Two curves that describe fitness of each morph can be measured in the same generation. As the population evolves from a preponderance of phenotype B (dashed line) to a preponderance of phenotype A (solid line), the direction of selection changes. (D) Another way to plot phenotype and its effect on fitness, particularly when morphs are discrete, is to plot fitness of each morph (A and B) as a function of the frequency or proportion of each morph. In this example, each morph has a fitness advantage when rare or the phenotypes are under negative frequency-dependent selection. For example, directional selection favors the small size when the small morph is rare, while directional selection favors large size when the large morph is rare. Negative frequency-dependent selection is not a special property of discrete morphs, and such effects may be operating on normally distributed traits (e.g., panel C). Morphs merely provide a useful tool to analyze such frequency dependence.

tion involving female choice can lead to a runaway process. The theoretical details of runaway sexual selection were only recently formalized, when O'Donald (1980), Kirkpatrick (1982), and Lande (1981) derived a population genetic model of this process. Female choice involves powerful frequency-dependent selection that promotes a reinforcing genetic correlation between genes for female preference and the preferred male trait. The strength of selection on female choice depends on the genetic correlation and on the strength of frequency-dependent selection that arises from the association between female choice and the male trait (e.g., assortative mating). The other form of sexual selection, male-male competition, likewise involves frequency-dependent selection. Frequency-dependent selection is not unique to sexual selection, but sexual selection invariably involves frequency-dependent selection because interactions between organisms are an implicit part of the biology of sex. Forms of natural selection that involve competition between species (Schluter & McPhail 1993; Schluter 1994) or competition among individuals in the same species (Smith 1987, 1993; Pfennig 1992a) likewise entail frequency-dependent selection.

In male-male competition, male fitness depends on the frequency of rival male types. The nature of frequency-dependent selection determines who will win and who will lose in the long run and whether or not all male types are present in a population. If we are considering discrete alternative male morphs, then "negative frequency-dependent selection" refers to a system of mating strategies in which each male morph has a fitness advantage when rare (Henson & Warner 1997) (Fig. 11.1). Game theory solves for the stability conditions resulting from such rare fitness advantage (Maynard Smith & Price 1973; Maynard Smith 1982). For a strategy to be considered an evolutionarily stable strategy or an ESS, one of two conditions must be met. First, when a strategy becomes common, the ESS is itself uninvadable by all other rare mutant strategies. A more stringent kind of ESS is referred to as a CSS or continuously stable strategy. For a strategy to be a CSS, when the strategy enters the population as a rare mutant it must be able to invade and spread against all possible strategies, for the jth "rare" strategy (Lessard 1990). Thus, an ESS is uninvadable, while a CSS is uninvadable and also able to invade all other strategies. A key condition that would prevent a unique ESS (e.g., only one male type wins) would be a rare advantage for each strategy in the population. In such situations, the population would contain a diversity of strategies. In our quest for explaining diversity in the biological world, we are rarely interested in demonstrating that a single strategy is an evolutionarily stable strategy, rather we are more often interested in explaining why systems of strategies in a single population exist in an evolutionarily stable state.

This chapter addresses ESS analysis in nature. How do we translate this esoteric formalism to the real world of male-male conflict or female choice? Many authors have erroneously oversimplified ESS analysis into hypotheses of equal versus unequal male fitness (Austad 1984). As demonstrated in the discussion, statistical analysis of equal male fitness is inherently flawed. Applying ESS analysis to nature requires an explicit analysis of the frequency dependence of male strategies. I develop methods for ESS analysis that are based on local neighborhoods of behavioral interaction. The "territory" or high site fidelity generates a finite set of local neighbors by virtue of geometry and boundaries. Territories are often stable across the lifetime of an organism, or at least over a single reproductive season. The spatial arrangement of territories can be used to derive the impact of neighbors on a focal individual's fitness. Moreover, territorial neighbors are invariably quite different among individuals in a population. Variation in "social environments" among neighborhoods provides the statistical leverage that is required to compute the payoff matrix.

As stated earlier, variation among neighborhoods also results in a "coarse-grained social environment," the raw material for frequency-dependent selection. I extend arguments on local neighborhoods to address the likelihood of alternative strategies evolving in a population and posit the following hypothesis. The coarse social environments of some mating systems generate selective conditions conducive to the evolution of genetically based alternative male strategies. A basic result from evolutionary analysis of phenotypes in variable environments is that genetic polymorphisms are much more likely to evolve in coarse-grained environments than fine-grained environments (Levins 1962a, b). In fine-grained environments, selection is more likely to lead to evolution of "plastic behaviors" or alternative phenotypes whose development is triggered by social cues. To develop these arguments, I first consider qualities of an ideal system for the analysis of behavioral evolution. In my search for a system, I learned many lessons that others may likewise find profitable in their quest for organisms with similar "model system" qualities.

# Model Systems for Evolutionary Analysis of Alternative Behaviors

Having read the early studies of the 1980s on natural selection, it was apparent that evolution could be quite rapid in nature. I embarked on a search to find an organism, a "model system," that could be used to study processes of selection. Stephen Adolph, a graduate-student colleague at the University of Washington, listened to my criteria for an ideal study organism to study natural and sexual selection:

1) A student of selection must be able to follow individuals during their entire life span. It is generally agreed that measures of lifetime reproductive success are the definitive measure of fitness (Endler 1986; Clutton-Brock

1988; de Jong 1994). Success is measured by the number of progeny that an individual recruits into the next generation.

2) A cohort of such "individuals" must be followed from birth to successful reproduction, tallying up along each phase of the life history those unfortunates that did not make the selective cut. The life history is divided up into episodes of selection, and differential survival of these different episodes is the hallmark of natural selection (Arnold & Wade 1984).

3) The student must be able to track the progeny of individuals that are fortunate enough to survive to maturity and reproduce. Thus, the study maps lifetime reproductive success of the parental generation onto lifetime reproductive success of the progeny generation. The mapping refers to those traits that contribute to an individual's success in recruiting progeny to the next generation. Mapping traits in the progeny cohort back onto traits in the original parental cohort is accomplished through parent-offspring regression, which also allows one to determine the heritability of traits, a requisite for evolutionary change.

4) A more pragmatic requirement is dictated by the life span of a graduate student dissertation: the organism must have a life span shorter than the dissertation to allow a study of as many such generations as possible. Evolutionary change is revealed when the traits change across several generations under the force of selection.

5) The final and most important criterion is related to the study of behavioral phenotypes and really only became apparent to me after years of natural history observations. It must be possible to observe behavioral interactions between individuals and map the outcome of such behaviors onto fitness. I was not interested in a generic study of natural selection. I was interested in a study of frequency-dependent natural selection. Effects of one phenotype on another are fundamental to behavioral interaction and also form the basis of frequency-dependent selection. A pragmatic requirement that is dictated by such "fitness mapping" is that the interactions cannot be so fluid over time as to defy measurement. Intuitively, I realized that in a territorial species the behavioral interactions are going to be stable across a reproductive season, which would allow me to map social interactions between individuals and then derive the relationship between social behaviors and fitness.

After listening to criteria 1–4, Steve suggested that I come down to southern California and work on lizards. I followed up on Steve's offer, and he set out to train me in the basic skills of herpetological research. I then spent the next decade searching for the ideal site to study natural selection, but I didn't actually get to observe evolution for another decade. My searches of the western United States carried me to the San Gabriel Mountains of southern California, the baking heat of the Mojave Desert, the chilling slopes of 12,000-foot Telescope Peak that overlook the depths of Death Valley, the high mountain passes and low canyons in the Sierra Nevada Range, the tranquil banks of the Deschutes River in eastern Oregon, the emerald green foothills of the Coast Range, and dozens of sites in between. The first two organisms that I chose were rejected after six years of deliberation. The western fence lizard, *Sceloporus occidentalis*, matures in one year; however, all populations have overlapping generations. The effect of age on behavioral interactions is profoundly interesting but not genetic. I was interested in systems where genetic variation is obvious. Age would only confound my study of genetic variation. The second species, the sage-brush lizard, *Sceloporus gracious*, is even more long-lived in its habitat on the tops of mountains where predators are few. Moreover, the habitat was too thick to simultaneously follow the behavior of many animals, particularly in sample sizes required to detect selection.

I needed an organism without overlapping generations and that occurred at high densities. Byron Wilson, another graduate colleague at UW, turned me on to the side-blotched lizard, *Uta stansburiana*. The side-blotched lizard lives life in the fast lane. An annual species, it matures in one year and most adults are dead after the first year. Collaborations with Byron also provided another bit of information on the suitability of the side-blotched lizard as an organism of study. We had just reared seven different populations in a large laboratory growth study of hatchlings. Hatchlings were obtained from eggs that field-collected gravid females had laid in the lab. We kept the adults in the lab as an afterthought. Adults from one population kept breeding in the laboratory. Adults from all six other populations shut down reproduction, presumably because the stress and density of captivity was not conducive to their reproduction. The "lab-rat" lizard population was from Corral Hollow in the Coast Range of California. An organism that breeds well in the laboratory is ideal because studies can be conducted in the laboratory under controlled conditions in tandem with field studies. I suspected that the ability of side-blotched lizards from the Coast Range of California to breed in the laboratory was related to high density in nature. Armed with this natural history information, I headed down to the University of California at Berkeley for a postdoctoral fellowship in Paul Licht's laboratory with the aim of studying the endocrine basis of life history and behavioral traits. Paul Licht was one of the experts on lizard endocrinology, and I realized that this training would become important in unraveling the genetic basis of behavioral traits related to reproduction. Fortunately, Corral Hollow was a short fifty-minute drive from UC Berkeley, admirably close for concurrent field studies of natural and sexual selection.

While Corral Hollow was attractive by virtue of proximity, I also visited two other more distant locations in the Coast Range, Los Baños Grandes and Del Puerto Canyon, in search of a study site that was ideal for demographic analyses that are necessary for studying selection. We studied the lizards at all three sites for three years and found one that was just right for selection analyses. Corral Hollow was just too dense. Hatchling densities were astronomically high, reaching into the thousands per hectare plot. Del Puerto had modest densities, but the presence of larger-bodied western fence lizards confounded my analysis of natural selection arising from behavioral interactions of a single species. Los Baños Grandes had moderate to high densities of lizards (hundreds to thousands of hatchlings per hectare), and the Los Baños site consisted of a monotypic stand of side-blotched lizards. Moreover, habitat for lizards at Los Baños Grandes consisted of rock-outcropping islands, ideal adult habitat for a saxicolous or rock-dwelling lizard. Populations are separated by a "sea of grass," unsuitable habitat for adult utas. Populations are separated by 0.25 to one kilometer and population size varies from ten to three hundred per rock outcropping. Each population could be used as an independent replicate to provide baseline data on natural selection. Moreover, replicate populations could be experimentally manipulated to address the causes of selection (Mitchell-Olds & Shaw 1987; Wade & Kalisz 1990; Sinervo & Basolo 1996; Sinervo 1998, 1999).

The discreteness of subpopulations at the Los Baños site has become critical in our recent analyses of natural selection. When fitness is not independent of other interacting individuals, as is the case for sexual selection or any selection mediated by social causes, statistical estimates of selection based on "individuals" are inappropriate (Hulbert 1984; Wade & Kalisz 1990; Svensson & Sinervo 2000). Side-blotched lizards are ideally suited for overcoming this problem. During the past decade, we have developed experimental and statistical protocols (Sinervo & Lively 1996; Sinervo & Svensson, submitted; Svensson & Sinervo 2000) for analyzing selection when fitness is dependent on interactions between conspecifics as in mate choice or male-male competition (Sinervo & Basolo 1996). To avoid pseudoreplication (Hulbert 1984), fitness can be estimated on replicate plots where social agents of selection are manipulated (Sinervo & Basolo 1996). In side-blotched lizards, I had found an organism with ideal properties for the evolutionary analysis of behavior in rigorous replicated field experiments.

Moreover, within the first few days on the Los Baños site I noticed that males exhibited a strange panoply of throat colors. I had not seen such diversity anywhere in the species range. Males possessed bright orange, bright blue, or yellow on their throats. Physical capabilities of orange-throated males were particularly striking. A 10 gram orange male was capable of defending a territory 100 $m^2$ in area compared to 40 $m^2$ for blue males. Naively, I thought that orange males with such robust capabilities would always have higher fitness than the staid territorial blue males. The yellow males remained entirely enigmatic. I set out to understand the "puzzle of the male morphs" as a side project while most of my energy was focused on studies of natural selection on offspring size (Sinervo et al. 1992; Sinervo & DeNardo 1996; Sinervo & Doughty 1996; Svensson & Sinervo 2000). A study of life history traits is a necessary first step in the analysis of selection on behavior. Life history traits describe processes of birth and death rate, and are central to understanding selection on behavior. Life history traits equate to fitness. Studying how juvenile and adult behaviors influence life history traits provides the mapping between behavior and natural selection (Arnold 1983).

## The Rock-Paper-Scissors Game of Alternative Male Strategies

An unusual game was being played out in the coast range of California. I kept careful track of male territories, scored throats in elaborate detail, and noted all exceptional natural history observations arising from territorial altercations between males. From these formative observations, I distilled the following set of "behavioral rules" for male engagement.

The three male morphs of side-blotched lizards, *Uta stansburiana*, evolved according to a variant on the "rock-paper-scissors" children's game (Maynard Smith 1982; Sinervo & Lively 1996). Males with bright orange on their throats embody the metaphorical "rock." Orange-throated males are ultradominant lizards and attack other males that make incursions onto their territory. If an orange is challenged by another orange, there is likely to be a stalemate because orange-throated lizards are equally matched in aggression. However, orange-throated males are very quick to attack neighboring blue-throated males. Blue-throated males embody the metaphorical "scissors." Blue-throated males have lower stamina than oranges. Orange males, by virtue of their greater resource holding power (Parker & Rubenstein 1981; Maynard Smith & Harper 1988), will eventually invade a blue-throated male's territory and copulate with females on the blue-throated male's turf. Rock beats scissors.

Success of orange-throated males does not remain unchecked for long. Yellow-throated males lurk in cracks, crevices, and interstices of orange male territories, and they embody the metaphorical "paper." Whereas blue and orange males fight for territory, yellow-throated males rarely fight. When cornered, yellow males either flee or elicit a "female rejection display," which is used by female lizards to indicate that they are not ready for copulation. Orange males appear to fall for the charade of female-mimicking yellow-throated males and leave them alone. Orange males also leave their females unguarded as they roam their large territories or force their way onto neighboring territories in search of additional females to court. At these times, yellow sneaker males obtain secretive copulations with females on an orange male's territory. Paper beats rock.

When yellow-throated males are common, blue-throated males gain a mating advantage that is not possible in the presence of orange-throated males. Blue-throated males are diligent at mate-guarding their females, particularly from sneaky yellow-throated males. When surrounded on all sides by sneaking yellows, blues are safe from ultradominant oranges. Scissors beats paper.

These are post hoc analyses of my natural history observations. Male strategies only became clear after game theoretic analysis. First hints of a dynamic game were evident after I plotted male morph frequency over the years. A cycle was apparent. Since the original study (1990–95, Sinervo & Lively 1996), a second rock-paper-scissors cycle has run its course (1996–99). To understand the dynamic cyclical game, we had to figure out a way to calculate the fitness of a male as a function of his neighbors, from which we could derive the payoff matrix. The real insight came when we (Sinervo & Lively 1996) were applying the natural history rules of male engagement to ESS models. After writing the model, it was clear that we had re-derived a variant on the rock-paper-scissor game, which Maynard Smith (1982) had written about in *Evolution and the Theory of Games*. As in the rock-paper-scissors game where rock beats scissors, paper beats rock, and scissors beat paper, the ultradominant orange-throated strategy beats blue, the sneaker strategy of yellow males beats orange, and the mate-guarding strategy of blue-throated males beats yellow. Orange beats blue to complete the cyclical dynamic.

## Measuring Frequency-Dependent Selection: Analysis of Local Neighbors

In lizard land, we construct elaborate home range maps (DeNardo & Sinervo 1994a, b; Sinervo & DeNardo 1996; Sinervo & Lively 1996) that describe the location of male territories relative to those of neighboring males (Fig. 11.2A, B) and females (Fig. 11.2C). We often observe neighboring males interacting, but cannot observe every interaction. We infer that males with overlapping home ranges are more likely to influence one another's fitness through behavioral interactions than more distant individuals. Fitness can be estimated any number of ways. In the past (Sinervo & Lively 1996), we estimated fitness from male control over female territories. If two or more males share a female territory, then the value of the female is distributed to each male in proportion to the number of visits he made to her territory (Fig. 11.2C). More recently we have used DNA paternity analysis to estimate fitness from the number of progeny in a female's clutch that a male sires (Zamudio & Sinervo, submitted). In the case of paternity data, clutches are analyzed for the frequency of co-siring (e.g., multiple paternity) as a function of social neighbors. In the case of territory data, fitness is estimated from the total number of female territories that each male monopolizes plus the fraction of shared female territories. The number of females acquired is then regressed as a function of the number of orange, blue, and yellow neighbors on a male's territory.

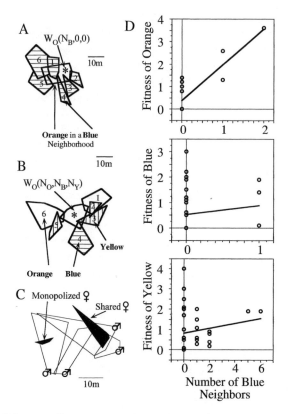

Figure 11.2. Polygons reflect space use by individual male side-blotched lizards. (A) A focal orange male (*) is surrounded by three blue neighbors (1–3). In this "idealized neighborhood," common blue males surround a "rare" orange. The focal orange male's fitness, $W_O$, is a function of number of blue males ($N_B = 6$) (e.g., $W_O [N_B]$). The male's fitness is also dependent on absence of orange ($N_B = 0$) and yellow ($NY = 0$) and thus the function is given by $W_O(N_B, N_B, N_Y) = W_O(N_B, 0, 0)$. (B) Few males in the population experience idealized neighborhoods. Another focal orange male (*) is surrounded by six males representing three morphs: blue ($N_B = 1$), orange ($N_o = 2$), and yellow ($N_y = 3$). The focal male's fitness depends on all three kinds of morphs (e.g., $W_O [N_B, N_B, N_y]$). (C) Males vary in their ability to control female territories. A male monopolizes the territory of the female on the left, while three males share the territory of the female on the right. Territory overlap provides an index of potential male success (e.g., control of female territories) as a function of number of male neighbors. More recently, we have used DNA paternity to estimate male success (Zamudio & Sinervo, unpublished data). (D) Fitness of each male type as a function of the number of surrounding blue neighbors (see text). Fitness is derived from a male's share of females (e.g., C).

A single regression equation describes success of one morph as a function of the local density of one kind of neighbor. For example, orange males gain access to females as the number of blue neighbors increases. Significance tests for a single regression equation that describes each male morph as a function of neighbors (Fig. 11.2D) are not as relevant as significance tests for a difference in slopes (e.g., success between pairs of morphs as a function of blue neighbors or an ANCOVA). We are interested in how one male type does relative to another, and the difference in slope indicates whether a morph is gaining or losing fitness relative to other morphs. For example, orange males also have a significantly greater slope than blue or yellow males (ANCOVA) and thus orange males gain fitness from having lots of blue neighbors (Fig. 11.2D). Yellows do not gain or lose fitness to blue neighbors (the slope between yellow males and blue males is not significantly different). A full ANCOVA model for each year of territory data included terms for Morph, $N_B$, $N_O$, $N_Y$, Morph $\times N_B$, Morph $\times N_O$, and Morph $\times N_Y$. The resulting multivariate equation constitutes the basic gain and loss function that describes effect of male social environments on morph fitness. Expressing male fitness in terms of neighborhood density of morphs allows us to compute idealized neighborhoods (Fig. 11.2A). The analysis of male neighborhoods does not eliminate the problem of pseudoreplication discussed above, but rather the multivariate regression explicitly explains spatial autocorrelation in fitness between adjacent males (Koenig 1998). To eliminate pseudoreplication it would be necessary to consider larger male neighborhoods that are entirely independent of other male neighborhoods (e.g., separated by unsuitable habitat, Sinervo et al., submitted).

## Generating a Payoff Matrix from Neighborhoods

Converting the multivariate gain and loss functions to an idealized payoff matrix completes the neighborhood analysis. The payoff of each strategy when rare, is compared to each common strategy. Each gain and loss equation (Fig. 11.3) is solved for "boundary conditions" that yield idealized neighborhoods in which a male has only one kind of neighbor (Fig. 11.2A). Data from six separate years of the male rock-paper-scissors cycle was used to compute the six relevant histograms of the payoff matrix. In practice, pooling variation among years does not enhance power, as years are quite different in the overall frequency of males. Different male morph frequencies across years (center, Fig. 11.3) resulted in good statistical power to detect frequency-dependent selection on one morph in any given year (e.g., rare morph) but not others (e.g., common morphs). Six years of territory data were required to detect the six relevant gain/loss functions that are needed in the construction of a payoff matrix from field data.

Each year, we standardized the fitness payoffs of a rare morph playing against the common morph (e.g., $W_{OB}$) by the fitness of a common morph playing against other common morphs (e.g., $W_{BB}$). In practice this also standardizes for any differences in overall fitness among years (e.g., relative fitness in each bar of the histogram in the payoff matrix, Fig. 11.4, is derived from a single year of data). When males compete against the same morph (e.g., blue versus blue), they do not gain or lose (e.g., slope of $W_B$ against $N_B$

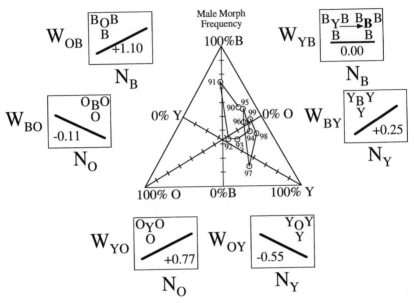

Figure 11.3. Changes in the frequency of male morphs from 1990 to 1999. The DeFiniti diagram (triangle in center) graphs frequency of the three male morphs as follows: 0 to 100 percent blue from base to apex, 0 to 100 percent orange from right side to left vertex, and 0 to 100 percent yellow from left side to right vertex. The two cycles observed to date have an average period of 4.5 years. A summary of all six functions describes fitness of each morph competing against each type of competitor. The number within each panel refers to morph difference in slope derived from ANCOVA analysis (e.g., number of females gained or lost per neighbor of a given type). The rock-scissors game is evident if we read the panels counterclockwise starting from the top left. A rare orange gains fitness from blue neighbors, while a rare blue loses fitness to his orange neighbors. A rare yellow gains fitness from orange neighbors, while a rare orange loses fitness to his yellow neighbors. Finally, a rare blue gains fitness from yellow neighbors, while a rare yellow does not lose fitness to his blue neighbors because yellow can transform to blue (e.g., when no orange males are present). Patterns are from six separate years of the cycle.

is zero, Fig. 11.2D) because the contest is "symmetric" (Parker & Rubenstein 1981). Thus, when rare and common morphs are the same (e.g., rare B in common B world), their relative fitness, $w_{BB} = W_{BB} / W_{BB} = 1.0$.

The number of players that a given male plays against is an important rule in multiplayer games because this parameter determines the neighborhood size for a "contestant." For simplicity, we assume that number of players varies for each morph, but that this number is constant regardless of the kind of neighborhood (e.g., common orange, blue, yellow). How many neighbors surround each male morph? Based on our neighborhood analyses of overlapping territories (1990–95) orange males play the rock-paper-scissors game with an average of 2.95 neighbors, blue males play with 2.35 neighbors, and yellow males play with 5.03 neighbors (Sinervo & Lively 1996). Using multivariate equations for male fitness as a function of morph and the number of neighboring morphs (e.g., the full ANCOVA model for each year), we can compute the expected fitness of each male type when rare, and when playing against a single common morph. I refer to such neighborhoods as "idealized dyadic neighborhoods" (dyads = two-way interactions) in which a single male of a given morph plays against male morphs that are of one type. For example, in a common blue world, rare orange plays against 2.95 blue neighbors, zero orange neighbors, and zero yellow neighbors. The orange male would gain 1.01 females (Fig. 11.3) from every blue or $2.95 \times 1.01 = 2.98$ females from a neighborhood of blues in addition to his baseline fitness (e.g., intercept for orange males, not shown). His more common blue competitors play against 2.35 blue neighbors, zero orange neighbors, and zero yellow neighbors. Note that only a few blue males see the rare orange, most blue males in a common blue world only see blue neighbors. While neighborhoods are finite, the entire world of blue male neighborhoods is infinite and the average effect of one orange on the population of blues is assumed to be inconsequential. Similar calculations yield fitness of all rare and common pair-wise combinations. Fitness of each rare morph when playing against each other type when common is also known as the payoff matrix. I graph the payoff matrix as a series of histograms to allow easy comparison of each rare morph in the three relevant common worlds (Fig. 11.4).

A rare orange competing in a neighborhood of blue males obtains 4.3 times the fitness of common blue (e.g., $W_{OB} > W_{BB}$). A rare blue competing in a neighborhood of orange obtains 0.2 times the fitness of common orange (e.g., $W_{BO} > W_{OO}$). A rare yellow-throated male competing in a neighborhood of orange males (e.g., $W_{YO} > W_{OO}$) obtains 1.8 times the fitness of common orange. A rare orange in a neighborhood of yellow males obtains 0.4 times the fitness of common yellow (e.g., $W_{OY} < W_{YY}$). A rare blue competing in a neighbor of yellow males obtains 1.8 times the fitness of common yellow (e.g., $W_{BY} > W_{YY}$). Finally, a rare yellow competing in a neighborhood of blue males obtains the same fitness as common blue (e.g., $W_{YB} \approx W_{BB}$). Each rare morph has high fitness against one common morph, but low fitness against another common morph. Each morph can be invaded when common and all morphs can invade when rare. Thus, no morph is an ESS and they will persist indefinitely.

## Calculating Fitness Consequences across Generations

We know the maternity of all offspring on the study site at Los Baños because eggs are obtained from uniquely marked females. We release uniquely marked offspring back on site when they hatch (Sinervo et al. 1992). We

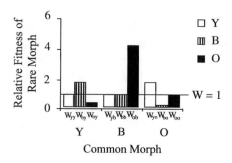

Figure 11.4. The payoff matrix for each throat-color morph of side-blotched when rare and when common. Data are derived from nature and represent "idealized neighbor-hoods" in which a rare male must only compete against a single type of competitor, the common morph. The home range used by each rare male type would be entirely sur-rounded by the home ranges of males with the common throat color (see text for details; from Sinervo & Lively 1996).

scored surviving male offspring at maturity for throat color. We estimated the putative sire for offspring from the probability of paternity that is based on territory data. We assumed that if a male monopolizes a dam (e.g., the only male that we observed on a dam's home range), he is the sire of all offspring. For shared dams, we computed an average color score for sires assuming probability of paternity is given by the number of visits that a male made to a dam's home range. Regression of offspring throat and putative sire's throat score yields a slope of 0.48 (P = 0.006) or $h^2$ = 0.96 (Falconer & MacKay 1996). More recent DNA paternity analysis yields a heritability for throat color of 0.86 (Zamudio & Sinervo, submitted).

From such behavioral estimates of putative sires, we can also calculate the number of progeny that each morph recruits into the next year (Fig. 11.5). Analysis of variance indicates that morphs recruit significantly different numbers of progeny depending on year (e.g., Year × Morph interaction is significant, P < 0.05, Fig. 11.5). Over the course of a single year (e.g., 1991–92), the population of side-blotched lizards changes dramatically as the more successful orange-throated males increase in frequency and rapidly drive blue males down in frequency (Fig. 11.5). The success of orange-throated males does not remain unchecked for long. Yellow-throated males have high fitness across the next generation (Fig. 11.5, 1992–93) and yellow males increase in frequency and rapidly drive orange males down in frequency. Finally, two seasons later, blue-throated males recruit more progeny into the next generation (Fig. 11.5, 1994–95), and blue males increase in frequency and rapidly drive yellow males down in frequency. The cycle is renewed in 1995, when orange males recruit far more progeny into the next generation (i.e., 1996, not shown). We have verified results based on behavioral analyses of male success with DNA paternity (Zamudio & Sinervo,

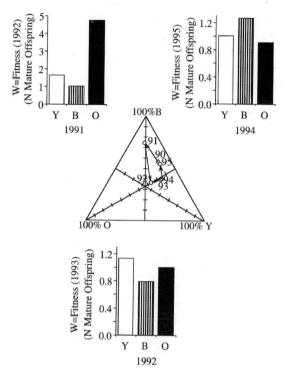

Figure 11.5. Changes in the frequencies of three color-morphs of side-blotched lizards, *Uta stansburiana*, during the years 1990–95 (DeFiniti diagram, axes as in Fig. 11.3). The cycle in frequency from 1990 to 1995 changed from common blue (1991) to common orange (1992) to common yellow (1993–94) and back again to common blue (1995). From 1991 to 1992, the frequency of orange nearly doubled while the frequency of blue was cut in half. This is because orange males sired nearly five progeny for every progeny sired by blue (histogram). However, the advantage of orange males disappeared when they became common. Yellow males sired more progeny in 1992. In 1994, blue males sired the most progeny. Each morph wins at some point in the cycle, and each morph loses (redrawn from data in Sinervo & Lively 1996).

submitted). The payoff matrix based on DNA paternity of nine nuclear microsatellites indicates a tremendous fitness advantage to each morph when rare (Sinervo et al., submitted). Moreover, the offspring produced by putative sires (Fig. 11.5) would be expected to have their fathers' throat color and the composition of the population should change between years in response to selection. We observed a change in the frequency of morphs from one year to the next (Fig. 11.3) because strong frequency-dependent selection (Figs. 11.3, 11.4, 11.5) drives a dynamic genetic game among the three players. Variation among neighborhoods in any given year invariably leads to a strong advantage to the rare strategy. The average neighborhood composition in any given year determines the direction of frequency change between years (e.g., arrows in Fig. 11.5).

Describing the fitness advantage using behavioral estimates of success is an essential step in the analysis of behavioral interactions. Males interact with other male morphs at the edges of their home ranges, and this generates the opportunity for frequency-dependent selection, especially among males in different neighborhoods. The realization of this selective opportunity occurs with successful copulation. Multiple paternity was found in over 80 percent of all clutches—remarkable given their internal fertilization (Zamudio & Sinervo, submitted). Modal clutch size was four to five eggs and the modal number of sires is two. In one case at least five sires inseminated a single female. Because females store sperm, male competition is not restricted to copulation success. Yellow-throated males posthumously sire more offspring than blue- or orange-throated males, indicating that sperm competition may be a novel aspect of their strategy. Results from DNA paternity confirm the behavioral analyses of fitness because probability of siring progeny is directly proportional to a male's territory area ($r = 0.24$, $P = 0.008$, $N = 126$ males). While variation in success explained by territory area remains a small but significant fraction of the total (e.g., $r^2 = 0.06$), tremendous variation is also explained by frequency-dependent interactions among males (e.g., DNA-paternity payoff matrix is comparable to Fig. 11.4, Sinervo et al., submitted).

## Discussion

### DEVELOPMENTAL ORIGIN AND EVOLUTION OF ALTERNATIVE TACTICS

An understanding of the origins and maintenance of alternative male strategies entails an understanding of ontogenetic (e.g., plastic) and genetic mechanisms that give rise to alternative phenotypes. Caro and Bateson (1986) provide a classification scheme for representing the diversity of alternative tactics that might be found in a population (Fig. 11.6). While their scheme only shows two possible strategies, the system can be expanded to encompass three or more discrete morphs in a population. The classification also simplifies strategies as discrete, while many animals show continuous variation in behavior. Because I am treating the evolution of genetically fixed alternatives, I will ignore systems with continuous variation. Finally, Caro and Bateson also consider ontogenetic changes in strategies from juvenile to adult. Their scheme can readily be generalized to organisms where the adult phase has two or more ontogenetic phases.

For example, protogynous hermaphroditic blue-headed wrasse, *Thalassoma bifasciatum*, are reproductive females during early life, intermediate phase (IP) males in midlife, and terminal phase (TP) males during late life

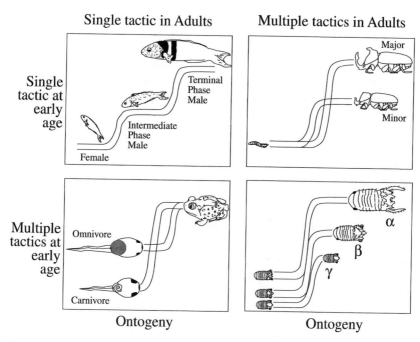

Figure 11.6. A classification scheme to describe the diversity of alternative male tactics that vary as a function of age. I have drawn examples of each of the four types from the literature, including sex change in protoynous hermaphroditic reef fish (*Thalassoma bifasciatum*), major and minor male morphs of horned beetles (*Podischnus agenor*), carnivore and omnivore morphs of spadefoot toad tadpoles (*Scaphiopus multiplicatus*), and a, b, and g male morphs of marine isopods (*Paracercis sculpta*). The y-axis reflects a change in any trait (gender, size, behavior, morphology) while the x-axis reflects age. The "ontogenetic trajectory" or developmental change that an individual can undergo is depicted by the lines. Trajectories redrawn from Caro and Bateson (1986.) Drawings of organisms by Sinervo were redrawn based on pictures in Eberhard (1980—beetles), Pfennig (1992—tadpoles), and Shuster (1989—isopods).

when they grow to a sufficiently large size that allows for nest defense (Warner & Hoffman 1980; Warner 1984; chapter 12, this volume). All individuals can be male or female, but social conditions dictate the change in behavior and morphology. Protogynous reef fish typify category 1.

In category 2, a single tactic is seen in juveniles, but two tactics are present in the adult phase. Horned beetles, *Podischnus agenor*, all pass through the same larval stage as they burrow through dung piles and gorge on food, but the nutritional state at metamorphosis dictates whether a male will develop with elaborate horns or small horns as an adult (Eberhard 1980). In principle, all individual male horned beetles are capable of one form or the other and environmental conditions (e.g., size of their dung nest when a larva) triggers development along alternative pathways.

In category 3, two tactics are possible in juveniles while only a single tactic is possible in adults. For example, larval spade-foot toads, *Scaphiopus multiplicatus*, can begin life by developing into either an omnivore or a carnivore (Pfennig 1992a, b). The alternative behaviors and morphology of the tadpole morphs are triggered by an environmental circumstance: the presence or absence of shrimp in their natal pond. Upon metamorphosing to the adult form, differences between individuals that were either carnivores or omnivores as juvenile tadpoles disappear.

In category 4, two tactics are present in the juvenile and two tactics are present in the adult. The differences in trajectories depend on genetic differences between morphs such as those found in the marine isopod *Paracercis sculpta* in which alpha, beta, and gamma males have a strong Mendelian genetic basis (Shuster & Wade 1992; Shuster & Sassaman 1996).

Throat color morphs in side-blotched lizards likewise have a strong genetic basis. However, one morph of side-blotched lizards is also capable of plastic transformation in the adult phase. Some yellow-throated males can transform and acquire a small blue patch on their throat and then begin defending territories (Sinervo, unpublished data). The environmental cue for yellow-transformation is death of a territorial orange male. Thus, it is entirely possible for some organisms to span two or more categories of Caro and Bateson's scheme and side-blotched lizards exhibit categories 3 and 4.

Caro and Bateson's scheme (1986) places emphasis on the development of tactics and on the control mechanisms underlying the development of alternative strategies. However, genes (e.g., isopods), environment (e.g., spade-foot toads or horned beetles), or interactions between genes and environment (e.g., lizards), could govern strategies in the four categories. In addition to the proximate source of variation, we must also consider whether or not the changes are reversible. Many short-term strategies based purely on behavior are reversible in that a male can adopt a variety of behaviors depending on context, such as male wasps that can adopt aggressive male or female-mimicking behavior on a short time scale (Field & Keller 1993). Most of the changes in morphology that occur during maturation are typically irreversible and once adopted the individual is stuck with that form for life (Gross 1984).

Finally, the developmental mechanisms governing the evolution of alternative strategies are undoubtedly related to the selective environment. Given that all of the examples described above involve behavioral interactions among individuals, it is likely that frequency-dependent selection shapes the alternative strategies. Couching the fitness of the morphs in terms of the environmental conditions that lead to the frequency dependence of the male strategies encapsulates selective mechanisms that govern evolution of the morphs in a succinct form. What are the conditions promoting plastic or genetically fixed alternative strategies?

## Coarse- versus Fine-Grained Social Environments

Classic work on the evolution of phenotypes in changing environments (Levins 1962a, b) and later theoretical studies (see references below) make some definitive predictions regarding the likelihood of evolving genetically based alternative phenotypes as opposed to developmentally plastic phenotypes. This theory provides the selective framework for understanding the developmental outcomes in Caro and Bateson's (1986) scheme. The dichotomy between genetically fixed or developmentally labile alternative phenotypes is also couched in terms of the evolution of genotype environment interactions or phenotypic plasticity. Genotype × environment interactions have been treated with diverse theoretical approaches in recent years, including gametic models (Levins 1962a. b; see Scheiner [1993]), game theory (Lively 1986), quantitative genetics (Via & Lande 1985; Gillespie & Turelli 1989; Gomulkiewicz & Kirkpatrick 1992), and single-locus models (Pigliucci 1992). One basic theoretical result depends on whether spatial or temporal environments are fine- or coarse-grained (Levins 1962a, b). The graininess of a spatial environment determines whether or not individuals are likely to encounter a single type of environment or several types of environments during their life span. In extremely coarse-grained environments, an individual is only likely to experience a single extreme environment and the match between phenotype and the environment will be honed by selection depending on which environment is encountered. When environments are patchy, progeny dispersal to an environment suitable for its genotype is also under strong selection. Coarse-grained environments promote genetically fixed alternatives.

In fine-grained environments selection is likely to refine the norm of reaction of the genotypes such that plastic responses are favored. The ability of a genotype to plastically change morphology or physiology is under selection in such situations. This allows the organism to match the optimum phenotype across the range of different environments that might be encountered.

Models have also considered competition between plastic morphs and genetically distinct morphs (Lively 1986). In cases where the environment is fine-grained, a single plastic genotype will often prevail. Genetically fixed morphs that do well only in a single environment will be eliminated in fine-grained environments. Some "medium-grained" environments permit both genetic and plastic morphs to coexist. Competition between plastic and genetically fixed morphs critically depends on the costs of plasticity and the strength of cues that trigger the transformation between phenotypes. If costs of plasticity are high (e.g., selective costs) then genetically fixed morphs may coexist with plastic forms. The reliability of environmental cues does have an effect on the outcome of competition between genetically fixed and phenotypically plastic morphs, but the cues can be relatively weak if the cost of plasticity is low.

### Ecology of Neighborhoods, Alternative Strategies, and Origin of Social Grain

Generalizations regarding evolution of plastic and genetically fixed morphs can also be applied to alternative systems of mating. Under what conditions might we expect to see alternative reproductive strategies, and under what conditions might we expect strategies to be genetic versus plastic in their developmental determination? Coarse-grained environments are more likely to support genetically fixed morphs while fine-grained environments are more likely to support plastic morphs. The coarse- and fine-grained dichotomy predicts the evolution of genetically fixed alternative male strategies in some species and plastic or condition-dependent strategies in others. The relative graininess of social environments in species (or populations) will determine the likelihood of evolving discrete morphs and thus whether genetically fixed alternative male strategists evolve, or whether plastic strategists evolve. Finally, frequency-dependent selection arises from environmental graininess per se (e.g., small neighborhoods in side-blotched lizards) and thereby determines the relative height of histograms in the payoff matrix or the strength of selection (e.g., Fig. 11.4). In situations where the environment is extremely coarse-grained, there is also large variation among social environments and this generates extreme frequency-dependent selection that acts on phenotypic variation. Thus, strength of frequency-dependent selection is driven by the degree to which frequencies of morphs vary among neighborhoods—the graininess of the social environment. In situations where neighborhoods are small, such as lizards, the graininess can be dramatic. One blue might experience three orange neighbors while another blue might experience three yellows (Fig. 11.2). Stochastic effects that arise from sampling in small neighborhoods can accentuate the strength of frequency-dependent selection.

Extreme frequency-dependent selection serves to continually refine each morph. Aspects of morphology that favors success within each class of alternative strategies is also under frequency-dependent selection. Males with a slight beneficial variation in alternative morphology are favored over other males within the same morphological class that are not so well endowed. In this fashion, correlated selection on morphological and physiological traits will be continually refined and we would expect correlated selection on a large number of genetic loci governing such traits. Because frequency-dependent selection in such systems is chronic, selection can promote extreme linkage disequilibrium between loci controlling morph development and loci governing functional, morphological, or physiological traits, even if these loci are unlinked. Selection is chronic even in situations when an equilibrium is reached because sampling effects will lead to variation among small neighborhoods that promotes frequency-dependent selection.

The origin of grain in the social environment invariably arises from variation in the physical environment. The variation in grain in the physical environment interacts with interindividual variation in resource holding potential (RHP) that is genetically based. RHP allows some individuals to monopolize resources (Emlen & Oring 1977). In situations where resources are concentrated or the mating system concentrates females, a despotic male that can defend the resources will be at a tremendous advantage. Orange-throated males are despotic in their ability to control rich thermal territories where females aggregate. When such despotic male morphs arise as a novel rare phenotype in a population (e.g., mutant), the neighborhoods around these males become coarse-grained and select for alternative behavioral phenotypes that cannot exploit resources by physical ability, but rather use alternative strategies such as sneaking behavior. Differences in RHP of the despotic male morph must remain relatively constant during an organism's life span because changes in RHP are likely to lead to temporally fine-grained social neighborhoods if the entire life span of an individual is considered. In order for the spatially coarse-grained social environments to promote genetically fixed alternatives, it is critical that the temporal social environment be coarse-grained. The temporal environment will be coarse-grained in an annual species such as *Uta stansburiana* because a maturing male faces a very short breeding season with a narrow window of reproductive opportunities. When the social environment experienced by an individual is temporally fine-grained (e.g., longer-lived species) it is likely that phenotypic plasticity will be favored and the effect of spatially coarse-grained social environments that promote genetically fixed alternatives will be minimized.

For example, if the despotic morph has a high probability of experiencing benefits that accrue with age or growth such as large size then the temporal social environment is less likely to fine-grained. The temporal environment is fine-grained when an individual's social opportunities change with age and a single individual can experience in a variety of social environments over its life span. For example, in blue-headed wrasse discussed above, large TP males control nesting sites where females lay eggs. However, early in life the same individuals could have been small female or IP males. Likewise, male bluegill sunfish come in three size morphs (Gross 1984), one of which is a large parental male. Territorial parental males actively court females and defend a nest into which females deposit eggs. The ability of despots to control resources generates coarseness in the distribution of females. This coarseness provides a selective opportunity for the evolution of an alternative strategy that can exploit any weaknesses of the despotic male morphs. In blue-gilled sunfish, a medium-sized satellite male mimics females and gains fertilization success by interrupting a territorial male while he courts a female. Often this interruption results in the satellite male squirting his sperm onto the eggs, and mixing it with the sperm of the territorial male. Finally, sneaker males are very small and they can dive in between a territorial male and the female that he is courting. When the female begins to lay the eggs, the sneaker quickly squirts ejaculate in a "dive bombing raid." An important component of the three behavioral strategies is that the male types mature at different ages. The territorial males typically mature at six to seven years, whereas the other two male types mature in two to three years. The frequency of morphs appears to be a function of the density and condition of males at maturity (Ehlinger et al. 1997). The fine-grained nature of ontogeny promotes the evolution of plasticity or "condition-dependent" strategies; any individual might experience the spectrum of social environments as it grows in size. While temporal graininess favors plasticity, spatial graininess favors fixed strategies owing to the despotic behavior of terminal phase males that defend nests. In the case of blue-gill sunfish, the prematurational environment of individuals is fine-grained in that any individual in the population might experience conditions that promote slow growth, which favors transformation into sneakers, or more rapid growth that favors transformation into the larger territorial male.

A theoretical treatment of the interaction between spatial and temporal graininess is clearly warranted. Nevertheless, it is possible for us to generate a hypothesis that integrates the interaction between temporal and spatial environmental grain and the evolution of genetically fixed or phenotypically plastic morphs (Table 11.1). Genetically based morphs are favored when the spatial and temporal environments are both coarse-grained. Plastic morphs are favored when the spatial social environment is coarse-grained while the temporal social environment is fine-grained (or vice versa). Finally, when both the temporal and spatial environments are fine-grained, plasticity is also expected to be favored. Thus, three of the four pair-wise conditions favor plasticity while only a single condition favors genetically fixed morphs. The hypothesis does not consider all conditions that limit evolution of genetically based and plastic morphs or those that favor a monomorphic species with no alternative strategies. The conditions regarding temporally and spatially graininess in social environments are not a sufficient condition for the evolution of genetically based alternative strategies, but rather a necessary condition.

Neighborhoods vary in the concentration of resources and some male strategies may reap a windfall in some neighborhoods, while other males of the same morph do poorly in other neighborhoods. The importance of local neighborhoods in driving the evolution of alternative strategies has been considered in a number of other systems. I illustrate the principles of local neighborhood analysis and the coarse-grained nature of social environments with a variety of systems with alternative male strategies, including fish, isopods, birds, and my own work on lizards. I do not explicitly consider the dichotomy between coarse- and fine-grained environments because this will require additional comparative studies in species or populations with and without alternative male strategies. It is also clear that initial differences in physical environment will promote coarse grain in social environments and

Table 11.1
Interactions between Spatial versus Temporal Coarse- and Fine-grained Social
Environments Promote the Likelihood of Evolving Either Plastic versus Genetically
Fixed Alternative Strategies

| | | Spatial Environment | |
| --- | --- | --- | --- |
| | | Coarse grain | Fine grain |
| Temporal | Coarse grain | Genetically fixed | Plasticity |
| Environment | Fine grain | Plasticity | Plasticity |

such variation in the physical environment is expected to be a ubiquitous feature of most organisms and their ecology.

## TEMPORALLY FINE-GRAINED BUT SPATIALLY COARSE-GRAINED SOCIAL ENVIRONMENTS

Warner and Hoffman (1980; chapter 12, this volume) studied sex change in the blue-headed wrasse, *Thalassoma bifasciatum*, a protogynous hermaphroditic reef fish. As described above, some fish skip the usual female phase and mature as an IP male, while other fish begin life as a female. Each of these two types can then mature into a TP male later in life. Why not skip both phases and become a TP male straight away? The selective advantages of a sex transformation in reef fish is due to the fitness advantages of large size. Theory predicts that females should change to the TP male phase when female success as a function of size equals male success as a function of size (Charnov 1982). A small juvenile is nowhere near the size required to be a territory-holding TP male, so they should start out life as a female or perhaps an IP male. The size threshold or switch occurs when TP males are large enough to spawn about 1.5 times per day, compared to one spawning per day for females. TP males grow to be quite large and they can spawn up to forty times per day.

Why do some juveniles mature into IP males while others mature into females? We have to consider the relative success of IP males in terms of their resource, the TP males. Because IP males are sneakers, they parasitize copulations with females that are spawning on a territory defended by TP males. If too many females turn into sneaky IP males, there won't be enough females in the harems for the sneaker males to mate with. Conversely, if too few females transform into sneaker males, then the harems of TP males are somewhat underexploited by the now rare sneakers. Thus, the frequency of IP males should be a function of the proportion of TP males and the average number of females spawning on a TP male's territory. Empirical estimates of the equilibrium frequency of IP males are in close agreement with the predictions of ESS theory for two species of fish (Warner & Hoffman 1980;

Charnov 1982). In this example, frequency-dependent selection is rampant: frequency of IP males depends on number of TP males, and number of females on a TP male's territory. The local neighborhoods consist of isolated reefs, which vary in population density. Thus, the local density of IP males on a single reef drives the frequency of morphs (Warner & Hoffman 1980) and the transitions between morphs. Plastic strategies are favored under these conditions even though the reefs are spatially grainy because an individual may experience the complete spectrum of social environments during its long life span as it grows.

### RUFFS IN COARSE-GRAINED LEKS THAT VARY IN SIZE

Male ruffs, *Philomacus pugnax*, congregate in leks to attract females. Ruffs use the puffy feathers on their neck in the "ruff-display," which attracts females and intimidates rivals (van Rhijn 1991). Males vigorously defend a "postage-stamp" piece of turf (approximately 1.5 $m^2$) on the leks. Males vary in their ability to defend even this small space. Male ruffs are commonly found in two genetically based plumage morphs (Lank et al. 1995). Dark-colored morphs are "resident" males that defend a territory. Light-colored morphs are "satellite" males that do not defend a territory, but rather move between leks (Lank et al. 1995). A third morph is called a nape-necked male because it lacks the ruff altogether (van Rhijn 1991). The success of nape-necked males is currently unknown, because the overall frequency of nape-necked males is extremely low.

Residents gain success by displaying for females. Once a female makes a choice, the female copulates with her "pick" male. Satellites gain quick and furtive copulations when resident ruffs are engaged in territorial disputes. Most birds do not have an intromittent organ, and fertilization is achieved with a "cloacal kiss." The female lifts her tail (lordosis) and a male flies up against the female to briefly engage their cloacas. Resident ruffs can easily drive off satellite males. Yet often the residents will allow the satellites a certain measure of peace. Why should a resident male tolerate any cuckoldry by satellite males? Höglund et al. (1993) suggested that the number of males on a lek enhances the attractiveness of the lek to passing females. Resident males gain fitness on larger leks.

It is interesting to consider how neighborhood size or lek size allows the two common morphs to persist. I reanalyzed data from Höglund et al. (1993). The unit of analysis remains the lek, a neighborhood of males, but I used a repeated measures analysis of covariance to analyze the effect of lek size on fitness (Fig. 11.7). A repeated measures ANCOVA compares satellites and residents on the same leks (cf. ANCOVA models for male lizards, above). Moreover, I included a quadratic term to test for differences in curvilinearity of the gain functions between residents and satellites (not consid-

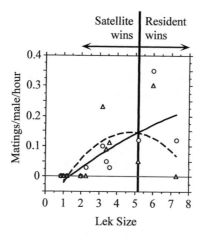

Figure 11.7. Reanalysis of data on effect of lek size (number of males) on number of copulations per male per hour in the ruff, *Philomacus pugnax* (data collected by Höglund et al. 1993). In reanalyzing their data, I assumed a quadratic relationship between fitness gains and lek size. Difference in curvature of fitness response between satellites and residents is significant in a repeated measures ANCOVA (see text), implying a fitness advantage to satellite males in leks with two to five males, but a fitness advantage to resident males in leks with more than five males.

ered by Höglund et al. [1993]). The interaction between morph and the quadratic term was significant ($F_{1,8} = 7.11$, $P < 0.03$), which indicates that the optima for lek size are significantly different between the two morphs. Graphical analysis indicates that variation in lek size promotes high fitness in each morph. Satellites benefit from being in leks with two to five males. Residents benefit from being in leks with more than five males.

Satellites should continuously be on the move, trying to find a lek that is small enough to gain high success. However, other satellites are also searching for and joining small leks, and the system is dynamically unstable on a very short time scale. The presence of additional satellites may increase the attractiveness of a lek to passing females (Fig. 11.7). However, leks may be stable over the lifetime of an individual resident male, who may revisit the same lek year after year. While satellites "parasitize" copulation opportunities from the residents, residents must allow the satellites onto the lek to enhance the number of females visiting their lek (Höglund et al. 1993). The raw variation in lek size and female preference for large leks leads to a coarse-grained environment. Satellites are free to choose among leks of different sizes, while resident ruffs must remain at a single lek and defend their territories. This coarse-grained social system generated by variation in lek size or the density-dependence of leks, contributes to the evolution and maintenance of alternative male strategies in the ruff. Substantial variation in fitness among individuals arises from variation in social neighborhoods on small versus large leks. Undoubtedly, the variation in the frequency of satellite males and resident males at leks contributes to selection, but the interaction effects of morph frequency on a lek (e.g., number of residents versus satellites) and density (e.g., lek size) is not known at this time and merits further study. In addition, the females may also appear to have a strategy of mate (or lek) choice that further interacts with the male strategies.

## COARSE-GRAINED NEIGHBORHOODS OF MARINE ISOPODS IN SPONGOCOEL CAVITIES

The final example of graininess in social environments is related to males of the marine isopod species *Paracercis sculpta* that come in three discrete size morphs (Shuster 1989; Shuster & Wade 1992; Shuster & Sassaman 1996). Male morphs and females reside inside spongocoels, which are hollow chambers found in sponges. The morphs have a simple genetic basis and the three types of male morphs are controlled by one locus with three alternative alleles. Alpha males are much larger than females and possess elaborate horns or uropods protruding out of their tail. Large alpha males evict other males out of the sponge if detected. Alphas wield the long posterior facing uropods that are wielded like "horns" to wrestle males and defend their "female harem." Beta males are about the same size and morphology as a female, and they lurk among females in the harems, attempting to mimic the appearance of females. The gamma males are much smaller than females and beta males. Shuster (personal communication) has described gamma males as little sperm bombs, consisting mainly of spermatheca. They lurk around the sponge waiting for an opportunity to dive bomb a pair in copulo.

Shuster and Wade (1992) addressed the stability of the three morphs. In the long run, does one morph tend to dominate and win because it has higher fitness? The null hypothesis in this case is that all three morphs have equal fitness. If the morphs have the same fitness, then no morph has an advantage in the long term and all will persist. The alternative hypothesis is that one morph has higher fitness, and the morph with high fitness would tend to increase in frequency until it became the only male type present in the population. Shuster and Wade (1992) found that they could not reject the null hypothesis. They did not detect any statistically significant differences in the mean number of progeny each male type was likely to produce (Fig. 11.8).

Sexual selection is defined as variation in mating success. Does equal fitness among morphs imply that there is no sexual selection on isopods? Shuster and Wade (1992) did not find variation in average mating success among the morphs (Fig. 11.8). However, they did find substantial mating success among males within each mating type, which results in tremendous opportunity for selection (Arnold & Wade 1984). There may be strong selection favoring some alphas over other alphas (e.g., large uropod size), or for

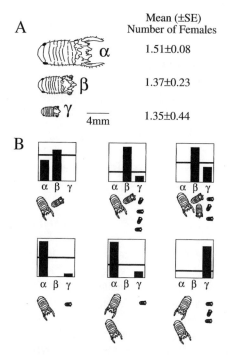

Figure 11.8. (A) Three alternative male morphologies in marine isopods have a strong genetic basis. Mean fitness (number of mates) is not significantly different among the morphs, suggesting that all three may persist over long evolutionary time scales. (B) While average fitness of morphs is similar, variation among the composition of spongocoel neighborhoods generates extreme frequency-dependent selection. Six of many possible neighborhoods are shown, along with relative fitness (histograms) of each morph (Shuster & Wade 1992). The heavy horizontal line indicates relative fitness = 1. For example, an alpha male with two b and three g neighbors (B) will have markedly different success (relative fitness = 0) compared to an alpha with one g neighbor (relative fitness = 1.84). A complete payoff (not shown) would require consideration of all possible neighborhoods, or regression equations of fitness as a function of morph and morph frequency. Derived from mating rules in Shuster and Wade (1992). (Drawings of morphs by Sinervo based on Shuster [1989].)

betas to be "good at being sneaky" (e.g., stabilizing selection of female form), or for gammas to have exceptionally large sperm storage for their body size (directional selection on gonad volume). This would be selection that serves to refine the phenotypes of the three morphs. However, the most obvious reason for tremendous variation in mating success within each morph class is the kind of neighborhoods in which an individual morph is found. Different spongocoels contain a variety of neighborhoods (Fig.11. 8B, C). Shuster and Wade explicitly calculated "mating rules" that determine each male's success, given the number and kind of males present in isolated spongocoels. Their mating rules are an elaborate set of equations that define the frequency dependence of the three male strategies. The individual chambers of spongocoels generate a very coarse-grained environment that refines the phenotypes of genetically distinct alternative male strategies, much in the same way that territorial neighborhoods of lizards refine genetically distinct morphs. Parallels between lizards and isopods are particularly striking in that both systems have three morphs. Moreover, isopods appear to cycle in frequency (Shuster, personal communication) like side-blotched lizards.

## Null Hypotheses, Equal Fitness, and Process-Based Measures of Selection

The philosopher Karl Popper (1962) provided scientists with a guiding rule for scientific inquiry. It is not possible to prove something is true beyond a shadow of a doubt, but we can be certain that some hypotheses are false. Falsification is the only truth that we can know. However, if we fail to reject the null hypothesis, are we sure it is true? No. This is because any pattern, any process, might be explained by many alternative processes, not merely the null and alternative hypotheses considered by the researcher. We can couch Popper's ideas into the notion of a statistical test. We can never really be sure that the alternative hypothesis is true. Nor can we be sure that the null hypothesis is true. We can only be sure that hypotheses are false. If we collect data that refutes a hypothesis, we can safely (probabilistically speaking) reject that hypothesis.

The null hypothesis for mating systems with alternative male strategies is often couched in terms of equal fitness arguments. With Popper's ideas in mind, we can address the suitability of equal male fitness as an explanation for the existence of alternative male types. Arguing that male types have equal fitness is unusual with regards to how scientists traditionally view acceptance of the null versus alternative hypotheses (Austad 1984). The alternative hypothesis is that the morphs do not have equal fitness and one morph will ultimately prevail. We often know a priori that there are different morphs, so the goal is really to collect enough data to accept the null hypothesis. This is contrary to the way we traditionally construct null versus alternative hypotheses. Moreover, it is contradictory to the principles of ESS analysis to define the persistence of a morph in terms of a fitness advantage when rare (e.g., payoff matrix), and then analyze fitness with the aim of showing no significant difference in fitness among morphs (e.g., equality of male fitness). This is only true if the system is at equilibrium, which is rarely the case in nature. Moreover, any given social neighborhood will be wildly out of "equilibrium" from the population average by virtue of the stochastic effects of sampling and the effect this has in generating frequency-dependent selection (see above for ruffs, reef fish, isopods, and lizards).

We usually want to collect enough data to reject the null hypothesis and accept the alternative hypothesis. If we fail to reject the null hypothesis we had better collect a large volume of data to convince ourselves that the null hypothesis is likely to be true. The power of the test must be sufficient to reject the alternative hypothesis. Shuster and Wade (1992), in their treatment of isopods, collected a convincing amount of data ($N > 300$) to show that the males had equal male fitness. The case for equal male fitness gets stronger and stronger, but according to the Popperian view of science, we can never know for sure. In fact, the demonstration of significant differences in fitness among male morphs in different spongocoel neighborhoods (e.g., opportunity for selection is high within male morphs but not between morphs, see above) is a more powerful argument for the stability of the isopod mating system. In the isopod example, Shuster and Wade calculated the fitness of alpha, beta, and gamma males in a variety of neighborhoods. In each neighborhood, they established that the morphs differed in relative fitness and the likelihood of siring progeny. They then used this "process-based" information on individuals in neighbors in their computation of equal male fitness. In the coarse-grained world of isolated isopod spongocoels fitness is never equal, but strikingly frequency-dependent.

Process-based arguments avoid the statistical ambiguity of "equal-male-fitness" arguments by identifying mechanisms that lead to morphs being preserved when they become rare. In the case of male strategies, social environments generate variation in fitness. Describing a fitness advantage for a rare morph yields a powerful mechanism for preserving variation. In this case, the alternative hypothesis is that a male gains or loses fitness as a function of the kind of males in his neighborhood. The null hypothesis is that males have equal fitness. In the rock-paper-scissors game we rejected the null hypothesis (e.g., equal fitness) in favor of alternative hypotheses (e.g., unequal fitness among years [Fig. 11.5] or among males found in different neighborhoods [Figs. 11.3, 11.4]). The discrepancy between Shuster and Wade's acceptance of the null hypothesis and our refutation of the null hypothesis lies in the temporal scale of analysis. Shuster averaged results across several seasons, while we calculated fitness from different years. Frequency-dependent selection is best studied as a selective process within a single generation that promotes evolutionary change (or stasis via stabilizing selection) across generations. Lumping or pooling data among years may hide an interesting dynamic. Likewise, pooling variation among disparate neighborhoods obfuscates interesting spatial variation among social environments that drives sexual selection.

## FUTURE ANALYSIS OF SOCIAL GRAININESS AND LOCAL NEIGHBORHOODS

Future studies should address the frequency dependence of alternative mating strategies rather than the red-herring equal male fitness. Moreover, comparative studies should address the degree to which graininess (temporal or spatial) in the social environment contributes to the likelihood of evolving alternative strategies in the face of frequency-dependent selection. Finally, a challenge for developing any new model system with alternative male strategies entails determining the relevant spatial scale for analyzing frequency-dependent selection. Natural history observations should guide us in choosing a natural spatial scale that depends on the degree to which individual neighborhoods (e.g., ensembles of competitors) are independent of other neighborhoods.

Moreover, future studies need to be focused on the variation in social environments at a variety of spatial scales. Variation in social environments among individuals is the underlying process that drives sexual selection and promotes discrete alternative strategies within a population. Variation in abiotic and social environments among populations may explain why some species (or populations) have alternative morphs while others contain only a single morph. The arguments discussed above implicate the relative graininess of social environments as a potent environmental effect that drives the evolution of fixed versus developmentally plastic alternatives. A population that experiences a fine-grained environment would be expected to evolve plastic morphs while a population that experiences coarse-grained environments (both temporal and spatial) would be more likely to evolve genetically fixed strategies. The interaction between spatial versus temporal graininess in social environments may also explain the evolution of genetically fixed versus plastic studies. Additional theoretical and empirical studies will be required to determine how the two forms of heterogeneity contribute to the evolution of alternative strategies.

Side-blotched lizards (*Uta stansburiana*) (Sinervo, unpublished) and tree lizards (*Urosaurus ornatus*) (Moore & Thompson 1990; Thompson & Moore 1992; Carpenter 1995) exhibit tremendous among-population variation in the presence and number of alternative morphs. For example, side-blotched lizards are described as being monogamous in Texas (Tinkle 1967), but extremely polygynous in the coast range of California (Sinervo & Lively 1996). Perhaps this among-population variation is driven by variation in the graininess of the social environments that arises from a consistent abiotic cause (e.g., concentrated resources would favor despots). The Texas utas are found in small, isolated bushes that only support single females and males (e.g., homogeneous and fine-grained), while the California lizards are found in large heterogeneous rock outcroppings that support high densities of females in a single male territory but lower densities in other territories (e.g., coarse-grained distribution of females as a resource for males). In this regard lizards may provide fruitful model systems to extend the ideas presented in this chapter. Similarly, sunfish exhibit variation in alternative male strategies found in different lakes and in different species (pumpkin seed versus blue gill) and salmon vary in frequency of alternative males among streams (Gross 1984, 1985, 1991). This variation may likewise be due to abiotic factors that promote more or less grain in the social environment. Additional comparative studies elucidating the role of "social grain" in promoting the evolution of alternative male strategies are clearly warranted.

## Acknowledgments

I thank S. Alonzo, A. Chaine, R. Calsbeek, T. Comendant, C. Both, B. Lyon, C. Lively, S. Shuster, R. Dukas, D. Lank, and E. Svensson for stimulating discussions on frequency- and density-dependent selection. Supported by NSF grant IBN 9631757.

## References

Arnold SJ, 1983. Morphology, performance and fitness. Am Zool 23:347–361.

Arnold SJ, Wade MJ, 1984. On the measurement of natural and sexual selection: theory. Evolution 38:709–719.

Austad SN, 1984. A classification of alternative reproductive behaviors and methods for field-testing of ESS models. Am Zool 24:308–309.

Boag PT, Grant PR, 1981. Intense natural selection in a population of Darwin's finches (Geospizinae) in the Galapagos. Science 214:82–85.

Bock WJ, 1980. The definition and recognition of biological adaptation. Amer Zool 20:217–227.

Caro GM, Bateson P, 1986. Organization and ontogeny of alternative tactics. Anim Behav 34:1483–1499.

Carpenter GC, 1995. The ontogeny of a variable social badge: throat color development in tree lizards (*Urosaurus ornatus*). J Herp 29:7–13.

Charnov EL, 1982. The theory of sex allocation. Princeton: Princeton University Press.

Clutton-Brock TH, 1988. Reproductive success: studies of individual variation in contrasting breeding systems. Chicago: University of Chicago Press.

Darwin C, 1859. The origin of species by means of natural selection. London: J. Murray (Penguin Press facsimile).

de Jong G, 1994. The fitness of fitness concepts and the description of natural selection. Q Rev Biol 69:3–29.

DeNardo DF, Sinervo B, 1994a. Effects of corticosterone on activity and territory size of free-ranging male lizards. Horm Behav 28:53–65.

DeNardo DF, Sinervo B, 1994b. Effects of steroid hormone interaction on territorial behavior of male lizards. Horm Behav 28:273–287.

Eberhard WG, 1980. Horned beetles. Sci Am March:124–131.

Ehlinger TJ, Gross MR, Phillip DP, 1997. Morphological and growth rate differences between bluegill males of alternative reproductive life histories. N Am J of Fish Manag 17:533–542.

Emlen S, Oring L, 1977. Ecology, sexual selection, and the evolution of mating systems. Science 197:215–222.

Endler JA, 1986. Natural selection in the wild. Princeton: Princeton University Press.

Falconer DS, MacKay TFC, 1996. Introduction to quantitative genetics. Essex: Longman.

Field SA, Keller MA, 1993. Alternative mating tactics and female mimicry as postcopulatory mate-guarding behaviour in parasitic wasp *Cotesia rubecula*. Anim Behav 46:1183–1189.

Fisher R, 1930. The genetical theory of natural selection. London: Clarendon Press.

Gillespie JH, Turelli M, 1989. Genotype-environment interaction and the maintenance of polygenic variation. Genetics 121:129–138.

Gomulkiewicz R, Kirkpatrick M, 1992. Quantitative genetics and the evolution of reaction norms. Evolution 46:390–411.

Gould SJ, Lewontin RC, 1979. The spandrels of San Marco and the Panglossian paradigm: a critique of the adaptationist programme. Proc R Soc Lond B 205:581–598.

Grant PR, 1986. Ecology and evolution of Darwin's finches. Princeton: Princeton University Press.

Gross MR, 1984. Sunfish, salmon, and the evolution of alternative reproductive strategies and tactics in fishes. In: Fish reproduction: strategies and tactics (Wooton R, Potts G, eds). London: Academic Press; 55–75.

Gross M, 1985. Disruptive selection for alternative life histories in salmon. Nature 313:47–48.

Gross MR, 1991. Evolution of alternative reproductive strategies: frequency dependent sexual selection in male bluegill sunfish. Phil Trans R Soc Lond B 332:59–66.

Henson SA, Warner RR, 1997. Male and female alternative reproductive behaviors: a new approach using intersexual dynamics. Ann Rev Ecol Syst 28:571–592.

Höglund JR, Motgomerie R, Widemo F, 1993. Costs and consequences of variation in the size of ruff leks. Behav Ecol Sociobiol 32:31–39.

Hulbert SH, 1984. Pseudoreplication and the design of ecological field experiments. Ecological Monographs 54:187–211.

Kirkpatrick M, 1982. Sexual selection and the evolution of female choice. Evolution 36:1–12.

Koenig WD, 1998. Spatial autocorrelation of ecological phenomena. Trends Ecol Evol 13:24–28.

Lande R, 1981. Models of speciation by sexual selection on polygenic traits. PNAS 78:3721–3725.

Lande R, Arnold SJ, 1983. The measurement of selection on correlated characters. Evolution 37:1210–1226.

Lank DB, Smith CM, Hanotte O, Burke T, Cooke F, 1995. Genetic polymorphism for alternative mating behaviour in lekking male ruff *Philomachus pugnax*. Nature 378:59–62.

Lessard S, 1990. Evolutionary stability: one concept, many meanings. Theo Pop Biol 37:154–170.

Levins R, 1962a. Theory of fitness in a heterogeneous environment. I. The fitness set and adaptive function. Amer Natur 96:361–373.

Levins R ,1962b. Theory of fitness in a heterogeneous environment. II. Developmental flexibility and niche selection. Amer Natur 97:74–90.

Lively CM, 1986. Canalization versus developmental conversion in a spatially variable environment. Amer Natur 128:561–572.

Maynard Smith J, 1982. Evolution and the theory of games. Cambridge: Cambridge University Press.

Maynard Smith J, Harper DGC, 1988. The evolution of aggression: can selection generate variability? Phil Trans R Soc B 319:557–570.

Maynard Smith J, Price GR, 1973. The logic of animal conflict. Nature 246:15–18.

Mitchell-Olds T, Shaw RG, 1987. Regression analysis of natural selection: statistical and biological interpretation. Evolution 41:1149–1161.

Moore MC, Thompson CW, 1990. Field endocrinology of alternative male reproductive tactics. In: Progress in clinical and biological research (Epple A, Scanes CG, Stetson MH, eds). New York: Wiley-Liss; 342:685–690.

O'Donald, P. 1980. Genetic models of sexual selection. Cambridge: Cambridge University Press.

Parker GA, Rubenstein DE, 1981. Role assessment, reserve strategy, and acquisition of information in asymmetric animal conflicts. Anim Behav 29:221–240.

Pfennig DW, 1992a. Polyphenism in spadefoot toad tadpoles as a locally adjusted evolutionarily stable strategy. Evolution 46:1408–1420.

Pfennig DW, 1992b. Proximate and functional causes of polyphenism in an anuran tadpole. Funct Ecol 6:167–174.

Pigliucci M, 1992. Modelling phenotypic plasticity. I. Linear and higher-order effects of dominance, drift, environmental frequency and selection on a one-locus, two-allele model. J Genet 71:135–150.

Popper K, 1962. Conjectures and refutations: the growth of scientific knowledge. New York: Basic Books.

Scheiner SM, 1993. Genetics and evolution of phenotypic plasticity. Annu Rev Ecol Syst 24:35–68.

Schluter D, 1994. Experimental evidence that competition promotes divergence in adaptive radiation. Science 266:798–801.

Schluter D, McPhail JD, 1993. Character displacement and replicate adaptive radiation. Trends Ecol Evol 8:197–200.

Shuster M, Wade MJ, 1992. Equal mating success among male reproductive strategies in a marine isopod. Nature 350:606–661.

Shuster SM, 1989. The reproductive behavior of α-, β-, γ-male morphs in *Paracerceis sculpta*. Evolution 43:1683–1698.

Shuster SM, Sassaman C, 1996. Genetic interaction between male strategy and sex ratio in a marine isopod. Nature 388:373–377.

Sinervo B, 1998. Mechanistic analysis of natural selection and a refinement of Lack's and William's principles. Amer Natur 154 SUPPL:S26–S42.

Sinervo B, 1999. Adaptation, natural selection, and optimal life history allocation in the face of genetically-based trade-offs. In: Adaptive genetic variation in the wild (Mousseau T, Sinervo B, Endler JA, eds). Oxford: Oxford University Press; 41–64.

Sinervo B, Basolo AL, 1996. Testing adaptation using phenotypic manipulations. In: Adaption (Rose MR, Lauder GV, eds). New York: Academic Press; 149–185.

Sinervo B, DeNardo DF, 1996. Costs of reproduction in the wild: path analysis of natural selection and experimental tests of causation. Evolution 50:1299–1313.

Sinervo B, Doughty P, 1996. Interactive effects of offspring size and timing of reproduction on offspring reproduction: experimental, maternal, and quantitative genetic aspects. Evolution 50:1314–1327.

Sinervo B, Doughty P, Burghardt G, 1992. Allometric engineering: a causal analysis of natural selection on offspring size. Science 258:1927–1930.

Sinervo B, Lively CM, 1996. The rock-paper-scissors game and the evolution of alternative male reproductive strategies. Nature 380:240–243.

Sinervo B, Svensson E, submitted. Frequency dependent selection on egg size and a cyclical offspring-quantity-quality game of r- and K-strategists. Nature.

Sinervo B, Zamudio K, Corrigan GN, Rollo D, 2000. Evolutionary cycle of alternative male strategies and the rock-paper-scissors game. Nature 406:985–988.

Smith TB, 1987. Bill size polymorphism and intraspecific niche utilization in an African finch. Nature 329:717–719.

Smith TB, 1993. Disruptive selection and the genetic basis of bill size polymorphisms in the African Finch *Pyrenestes*. Nature 363:618–620.

Svensson E, Sinervo B. 2000. Experimental excursions on fitness landscapes: density-dependent selection on egg size. Evolution 54:1396–1403.

Thompson CW, Moore MC, 1992. Behavioral and hormonal correlates of alternative reproductive strategies in a polygynous lizard: test of the relative plasticity hypothesis and challenge hypothesis. Hormones and Behavior 26:568–585.

Tinbergen N, 1963. On aims and methods of ethology. Z Tierpsychol 20:410–433.

Tinkle DW, 1967. The life and demography of the side-blotched lizard, *Uta stansburiana*. Misc Publ Mus Zool University of Michigan 132.

van Rhijn JG, 1991. The ruff. London: T. & A. D. Poyser.

Via S, Lande R, 1985. Genotype-environment interaction and the evolution of phenotypic plasticity. Evolution 39:505–522.

von Neumann J, Morgenstern O. 1953. Theory of games and economic behavior. Princeton: Princeton University Press.

Wade M, Kalisz JS, 1990. The causes of natural selection. Evolution 44:1947–1955.

Warner R, 1984. Mating behavior and hermaphroditism in coral reef fishes. Am Sci 72:128–136.

Warner RR, Hoffman SG, 1980. Local population size as a determinant of mating system and sexual composition in two tropical marine fishes (*Thalassoma* spp.). Evolution 34:508–518.

West-Eberhard MJ, 1979. Sexual selection, social competition and evolution. Proc Am Phil Soc 123:222–234.

West-Eberhard MJ, 1983. Sexual selection, social competition and speciation. Q Rev Biol 58:155–183.

Zamudio K, Sinervo B, submitted. Polygyny, mate-guarding, and posthumous fertilizations as alternative male strategies. PNAS.

# 12 Synthesis: Environment, Mating Systems, and Life History Allocations in the Bluehead Wrasse

Robert R. Warner

## Beginnings

I began working with the bluehead wrasse (*Thalassoma bifasciatum*) because I wanted to test an idea I'd developed in my Ph.D. thesis. The subject of the thesis was why fish change sex, and the idea was simple: certain mating systems lead to situations where size-specific reproductive success differs between males and females. When that occurs, it pays to be whatever sex has the greater reproductive payoff at a particular size (Warner 1975). For example, if the mating system is such that large males monopolize reproduction, then individuals would profit from being a female while small and change sex when they are large enough to compete successfully.

While many people find sex change inherently bizarre, the topic of sex allocation forms an important link between behavioral ecology and evolutionary ecology. At the time of my first interest in the topic, Trivers' seminal paper on sexual selection (Trivers 1972) was just beginning to be appreciated; as far as I was concerned, I was working on life history evolution. However, the clear link between sexual selection (arising out of the mating system) and a striking life history shift (sex change) would eventually help to point out that behavioral ecology isn't just animal behavior, but a study of important traits that can shape many other aspects of the organism (Warner 1980). But before I went much further into this area, I needed to find out more about behavior. The idea was there, but the data was not. I knew fish changed sex, but I did not know whether they had the predicted mating system.

A perfect opportunity presented itself in the form of a postdoctoral fellowship at the Smithsonian Tropical Research Institute in Panama. I had been working at the Scripps Institution of Oceanography in California. Fish in temperate waters are hard to watch: the water is often murky, and reproduction is seasonal. Coral reef fish, on the other hand, live in clear, warm water, and many of them reproduce year-round. Even better, many of the reef-dwelling fish families (wrasses, parrotfish, groupers, damselfish) were known to change sex. So to begin all I really needed was to choose a species.

For a behavioral ecologist, coral reef fish are nearly perfect subjects: they are as colorful and active as birds, with the added advantage that the observer can "fly." Bluehead wrasses are one of the most common fish on Caribbean reefs, and most individuals live in shallow water where they can be observed via snorkeling. They are small fish, never hunted by man, so they show no fear or avoidance. In fact, they are insatiably curious, and readily approach a diver on the off chance that such an ungainly creature might dislodge a food item in the course of disturbing the reef. Inexperienced divers often have a cloud of bluehead wrasses following them across a reef. In contrast, a snorkeler hanging quietly in the water above the reef is quickly ignored by the wrasses.

Having a common and fearless species was not sufficient. We also needed to observe reproductive behavior, and we needed to establish the pattern of sex change. On both these counts, blueheads proved to be ideal. Mating takes place year-round at specific sites on the reef, and on any particular day all mating occurs in a predictable two-hour midafternoon period. Individual females mate nearly every day, and these females tend to be highly faithful to particular mating sites (Warner 1986). So it is a simple matter for a team of observers to witness all the matings occurring in a local population, and to observe them day after day. Thus questions that might take some researchers many mating seasons to resolve can be answered quickly with blueheads, and sample sizes can be very large.

Even more important is the fact that blueheads are extremely easy to catch unharmed. As mentioned above, these fish are attracted to disturbance and are very opportunistic feeders. When frightened, they dive straight down. So using a simple lift-net baited with a sea urchin, a pair of workers can easily capture fifty or more fish in one attempt, and repeat this over and over. The technique is almost species-specific, with well over 99 percent of the individuals caught being blueheads. Thus population samples can be taken quickly and easily. Beyond that, it is easy to transplant whole segments of populations (for example, all females, or all individuals above a certain size) from one reef to another.

While local populations are genetically mixed due to extensive dispersal in the pelagic larval stage, after settlement juveniles and adults do not leave the reef on which they find themselves. Thus a series of patch reefs is a set of discrete local populations over which experiments can be replicated and controlled. It is a bit like having replicate aquaria in a laboratory, without the inconvenience of feeding or cleaning.

## Synthesis: Environment, Behavior, and Life History

My early work on this species helped to establish the idea that some aspects of life history (such as sex change) can be best explained in the context of behavioral ecology (such as the mating system). I have long been interested in this general theme (Warner 1980). Over the past twenty years, I have tried to build a picture of how adaptations to the physical environment in turn can generate selection for other aspects of life history and behavior. The simultaneous consideration of population biology, evolutionary ecology, and behavioral ecology leads to an enriched understanding of all three fields.

As an illustration of this approach, I describe what we know about the bluehead wrasse as a series of steps. First, I provide the physical context in which the fish finds itself, and how the mating system reflects responses to this context. Then, given the mating system, I describe a cascading series of allocations that show just how pervasive the influence of the mating system is on the rest of life history. To me, understanding allocations and how they change under different conditions is the key to understanding natural selection on adaptive traits.

## BACKGROUND

### POPULATION STRUCTURE

After a pelagic larval phase of about fifty days (Victor 1982), blueheads settle out onto reefs and metamorphose into small (< 10 mm standard length, or SL) juveniles. The fish mature at about 35 mm SL, at three to four months of age. All small adult fish are in the initial color phase (IP; see vignette in Fig. 12.1). If an individual lives long enough, it transforms into the distinctive terminal color phase (TP) at about 75 mm SL. All TP individuals are males, and thus there are two sources for these individuals: they may result from simple color change from an IP male, or they may be the result of both color and sex change of an IP female (Fig. 12.1a). Thus the basic polymorphism in this species is not between males and females, but between protogynous hermaphrodites (operating as either females or secondary males) and non-sex-changing (primary) males. Smaller TP males range over the reef as bachelors or "floaters," but larger individuals maintain permanent territories around mating sites.

### THE ENVIRONMENTAL CONTEXT

*Reefs, currents, and migration*: Blueheads live in relatively shallow water over coral reefs and other hard-bottom substrates throughout the tropical western Atlantic. Their primary diet is zooplankton, and thus most individuals can be found feeding in schools on the upcurrent ends of reefs for much of the day. Because the eggs produced by females are pelagic and there is no parental care, the eggs themselves immediately become part of the plankton. Not surprisingly, mating and egg release occur at the *downcurrent* ends of reefs (Warner 1988a), and individual females must migrate from the upcurrent feeding areas to the downcurrent spawning sites. In many cases, this is only a short distance and takes only a few minutes; however, certain reef

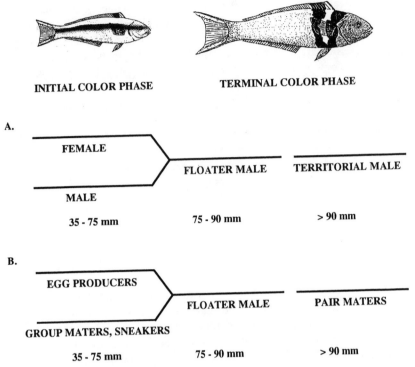

Figure 12.1. A. Outline of the size-specific expression of sex and coloration in the blue-head wrasse. Note both Initial Phase males and Initial Phase females can become large Terminal Phase males. B. Outline of how mating modes map onto size and coloration in blueheads. Floater TP males generally do not engage in reproduction, while IP males and TP males are in competition for the eggs produced by females. *Sources*: Warner et al. 1975; Warner & Hoffman 1980b.

configurations result in daily spawning migrations that can comprise more than 10 percent of the daily activity (Warner 1995).

*Recruitment and demography*: As is true of most marine fish, the arrival of young on the reef is sporadic and episodic. Recruits tend to arrive near the dark of the moon, but the actual magnitude of the recruitment pulse is highly variable (Victor 1986). Thus the demographic environment in which a particular individual finds itself is unpredictable. This is important because many aspects of fitness in this species are frequency-dependent and may also depend on relative position within a size structure. Given the timing and magnitude of recruitment, an individual of a particular size may be relatively small or large, and (for example) this can dramatically affect the probability of ever attaining successful territorial male status. Because demography is so variable and yet so important to fitness, individuals should respond to local size structure in their behavioral and life history allocations. This phenotypic plasticity permits us to engage in a variety of experimental field manipulations with a reasonable expectation of success (Warner 1991).

*Costs and benefits of a tropical existence*: *T. bifasciatum* is a tropical species. This is a double-edged sword in terms of behavioral ecology. Environmental conditions are relatively constant over the course of the year, and one result of this is that blueheads engage in continuous reproduction. On any day of the year, there are blueheads mating throughout the Caribbean. The implication is that larval survival does not differ dramatically over the course of the year (thus there is no reason to store energy for reproduction at a better time). Unfortunately, it also means that there is no "better" time, in that tropical production is poor and larval survival is uniformly low. Blueheads were one of the first species to be shown to be recruitment-limited (Victor 1983); that is, the supply of recruits to any particular location is generally so low that local adult populations are not limited by density-dependent mortality. This means that the magnitude and size structure of any local population may fluctuate wildly, but there should be abundant food for any settled individual. Thus we expect allocations to be responses to demography and mating opportunities, but not to food supply.

All fish must die, and blueheads are subject to predation from a wide variety of larger species. While mortality rates are enormous just after settlement (Victor 1986; Caselle 1999), tagging studies indicate that mortality is relatively constant over the adult size range (Warner 1984a). Individuals rarely live beyond the age of three (Victor 1986), but they pack an enormous amount of reproduction into this short time, and this is one of the keys to understanding the life history of this species.

## FEMALE RESPONSE AND THE BASIS FOR THE MATING SYSTEM

As outlined above, the physical environment results in abundant food and continuous reproduction in the bluehead wrasse. Females may expect to reproduce approximately 480 times as egg-producers in the first two years of their lives, and, if they survive to become successful terminal phase individuals, another 2,300 matings as a TP male (Warner 1984a). Given the measured mortality rates in this species (Warner 1984a), each reproductive event as a female accounts for approximately 0.25 percent of the total expected lifetime reproductive success. In species in which each reproductive event comprises a relatively small proportion of that individual's lifetime reproductive success, any activity that might increase fitness in the short term may not convey increased lifetime reproductive success: even slightly higher risks of mortality can heavily discount future reproduction. In other words, for females that reproduce many times over their lifetime, fitness lies in survival rather than in marginal increases in current reproductive output. Stated more formally, life history theory suggests that repeated reproduction (iteroparity) discounts the relative importance of any particular reproductive event, and places a premium on survival in order to reproduce in subsequent periods (Mertz 1971; Stearns 1976, 1992; Warner 1998).

Survival, in turn, entails avoidance of risk. I suggest that the bluehead mating system may be explained through a consideration of just how much a female might be willing to risk in any one reproductive event. Highly iteroparous females will avoid mortality risks and thus be very conservative in their behavior, and this will have a profound influence on mating system structure. For example, bluehead females will be expected to seek assurance that any current reproductive activity is safe (Warner & Dill, unpublished). By the same token, these same females will not be expected to engage in mate assessment or mate searching to the same degree as less iteroparous species, if these activities involve increased risk of mortality. Field experiments have shown that females are relatively more cautious than other blueheads (Warner 1998).

The mating system reflects conservative female behavior. Downcurrent mating sites consist of upward projections from the edge of a reef, and serve to minimize the distance a female must travel in open water to release her pelagic eggs near the surface (Robertson & Hoffman 1977; Warner 1988a). Tagging studies have shown that once a mating site is chosen by a female, she is highly faithful to that particular site, migrating to spawn at it by the same specific route (Warner 1986, 1995). Wholesale removals and replacements of populations showed that mating site location itself is determined by tradition, involving social learning passed on through females; the same mating sites are used over many successive generations (Warner 1988a). Successive whole population replacements showed that females are in fact capable of mating site resource assessment, but under normal circumstances they avoid individual assessment and only copy other females (Warner 1990a). Females do not engage in intensive mate assessment or searches for mates. Experiments shifting males from one mating site to another indicate that sites, rather than males, are the objects of female choice (Warner 1987). Finally, by replacing only males or females on a reef it was shown that females determine mating site location without reference to male behavior (Warner 1990b). Thus female mating behavior appears to operate without much reference to males (Emlen & Oring 1977).

MALE RESPONSE TO FEMALE BEHAVIOR
AND THE RESULTING MATING SYSTEM

The fact that females pay little attention to their mates does not mean that mating is random, or that sexual selection is weak. In fact, the highly predictable site-specific mating behavior of females sets the stage for intense male-male competition for mating sites and determines the relative strength of the two components of sexual selection. Experiments in which we replaced all of the TP males on a reef allowed us to track the degree to which variance in male mating success was determined by winning a mating territory, defending that territory, or attracting new females to the territory. Females do exercise minimal mate choice (they occasionally switch sites), but overall most of the variance in male mating success results from competition between males to acquire and defend mating sites (Warner & Schultz 1992). That is, conservative female behavior both lowers the contribution from female mate choice (because females will not bear the cost of careful mate assessment) and strengthens the contribution from intrasexual competition (because consistent female spatial behavior intensifies competition between males for control of mating sites).

Once a TP male successfully occupies a mating site, he can have enormous reproductive success: average *daily* mating success ranges between 15 in St. Croix (Warner & Schultz 1992) to over 25 in the San Blas Islands of Panama (Warner & Hoffman 1980a; Hoffman et al. 1985). The current record is at a site on reef 13, in Tague Bay of St. Croix, where a male accrued an average of 163 matings a day (Warner, unpublished). TP males own a mating site until death; the average tenure at a site is about 90 days (Warner 1984a). There is no usurpation of mating sites by roving TP males (Dunham et al. 1995), but roving TP males and neighboring territorial males will contend for territorial ownership when a site becomes vacant.

Control of mating sites is not always the prerogative of large males. TP males spend most of their time in defense against small IP males who are trying to contribute their sperm to the pelagic matings. About 3 percent of the matings of territorial TP males are subject to sperm competition from IP males. Interestingly, the persistence of the alternative IP male mating type (small males that sneak-mate or group-spawn) in this species is fostered by the tendency of females to be indifferent to the type of male with whom they mate. Field experiments that alter local population density show that when the relative number of IP males is high, the most successful mating sites become economically undefendable (Warner & Hoffman 1980a). TP males simply do not have enough time to both defend a site and mate with arriving females, and these areas become group-spawn sites (Warner & Hoffman 1980a). The fact that a mating site is occupied by a group-spawning IP male aggregation does not appear to affect a female's fidelity to that site (Warner 1985, 1986), and the most successful sites on larger reefs are occupied by large groups of small IP males.

Thus the sexual polymorphism in blueheads (Fig. 12.1a) is paralleled by a mating polymorphism (Fig. 12.1b). Small IP males and large TP males are in competition for the eggs of the females, and use very different means to achieve fertilizations.

## ALLOCATION RESPONSES TO THE MATING SYSTEM

Given that environmental effects on female mating behavior establish the basis for the mating system, we can now turn to the major theme: how mating systems affect allocations. We explore allocations on several levels.

SEX ALLOCATION

The basics of the size-advantage model for sex change were outlined in the introduction. Basically, we expect that animals would benefit from sex change from female to male in species where large males can achieve relatively high reproductive success. This is certainly true for the bluehead wrasse. Observations of tagged individuals indicate that females and initial phase males average about one mating a day, while territorial TP males average over 25 times that rate (Fig. 12.2d; Warner et al. 1975; Warner 1985; Hoffman et al. 1985). The mating rate for IP males was estimated as "pair-spawning equivalents," calculated by adding up a particular male's proportional share of multiple-male matings (e.g., participating in a five-male, one-female group gives the subject male a fifth of a spawn).

So blueheads appear to change sex in the right direction, but when should the change actually occur for any particular individual? This depends on an individual's future prospects as a male versus as a female, discounted by the probability of being alive at future sizes and ages. This is calculated as reproductive value (RV; see Stearns 1992). It turns out that the curve for male RV crosses that of female RV at about 75 mm SL, which is the average size at sex change (Warner 1988b). Further experiments showed that sex change is in fact socially controlled in blueheads (Warner & Swearer 1991). That is, an individual female is acutely aware of her immediate prospects as a male or a female, and will initiate sex change almost immediately upon the removal of a locally dominant TP male if she is then the largest remaining individual in the area. She behaves as a male within minutes, and will court as a male and mate with other females on the day of TP male removal. She can be producing sperm as few as five days after the initiation of sex change. This has led to a series of fascinating experiments investigating the relative roles of brain and gonad in controlling sexual behavior (e.g., Godwin et al. 1996).

The fact that large males can accrue very high mating success explains why females might change sex when they achieve large size, but it does not explain the presence of primary males. In fact, many sex-changing species have only young females and older secondary males (Warner 1984b). The key to understanding the existence of small males in blueheads lies in the polymorphic mating system (Fig. 12.1b). Large males can control all the mating sites on small reefs, because mating densities are low. However, on larger reefs the most successful mating sites are economically undefendable (Warner & Hoffman 1980a), and most of the matings go to IP males in group-spawnings (Warner & Hoffman 1980b; Warner 1984b). Thus the relative fitness of the sex-changers and the primary males changes with reef size (Warner et al. 1975), and the polymorphism of sexual types may be explained by the polymorphism of reef sizes. Since the male types are in competition with each other for fertilizations, the proportion of IP males in the population (call this $P$) is frequency-dependent and should be determined by

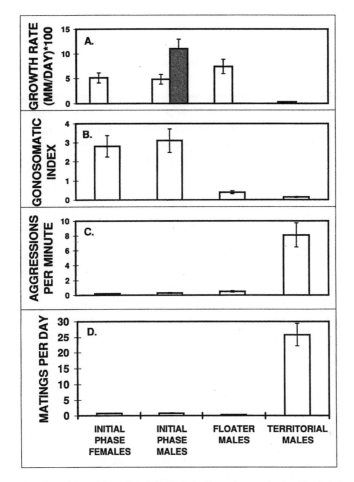

Figure 12.2. Allocation patterns for the color/size/sex classes in the bluehead wrasse. Error bars are 95 percent confidence intervals for the mean. A. Growth rates of marked individuals. The gray bar represents IP males from small reefs, where they are unable to mate. *Sources*: Warner 1984a; Schultz & Warner 1990. B. Gonosomatic indices (percent of body weight devoted to gonad). *Sources*: Warner et al. 1975; Warner & Robertson 1978. C. Intraspecific aggression rates from focal-individual data. *Source*: Warner & Hoffman 1980a. D. Average number of matings per day. For IP males, mating success was estimated as pair-spawn equivalents, the summed share of successive matings in which an individual participated. *Sources*: Warner et al. 1975; Warner & Hoffman 1980a, b; Warner 1984a, 1985; Hoffman et al. 1985; Warner & Schultz 1992.

their share of the total matings ($S$), by the formula $S = P/(1-P)$ (Warner & Hoffman 1980b; Charnov 1982). Given the distribution of reef sizes in the San Blas Archipelago of Panama, we estimated $S$ as 0.53, and thus predicted $P$ as 0.34. The actual value of $P$ was 0.32 (Warner & Hoffman 1980b). The remarkably close agreement suggests that the sexual diversity seen in blueheads is driven by physical features (namely, reef sizes) that affect mating systems.

## SEX-SPECIFIC LIFE HISTORY ALLOCATIONS

If the two sexual types in blueheads are in equilibrium, as suggested above, then the life history trajectories of IP males and females should be similar. This is because they both can become successful TP males (Warner et al. 1975). We have already seen that the mating success of these two types is similar, and they devote approximately the same amount of tissue to gonad (Fig. 12.2b; Warner et al. 1975). Growth rates of small males and females are also very similar, averaging about 0.05 mm per day (Fig. 12.2a), and mortality rates are identical as well (Warner 1984a; Schultz & Warner 1990). The only exception to this pattern is the growth rate of IP males on small reefs, where TP males exclude them from reproduction; males in those circumstances grow twice as fast as any other IP individuals (gray bar in Fig. 12.2a; Warner 1984a). Transplants of IP males between large and small reefs showed that individuals were fully capable of adjusting their mating rate and growth rate to a new situation (Warner 1984a), so this allocation pattern is plastic and a good demonstration of the cost of reproduction.

The other striking pattern in the growth versus reproduction profiles is that of the floater TP males. These individuals essentially abandon reproduction, rapidly shrinking their gonad and not engaging in mating (Fig. 12.2b, d). At first glance, this seems counter to the size-advantage model, since mating success *declines* after sex change (Hoffman et al. 1985; Warner 1988b, c). In fact, these individuals have reached the point where RV is maximized by trading current reproduction for growth (Warner 1988b). The probability of reaching successful territorial TP status increases as an individual grows, of course, and eventually the very large rewards associated with territorial ownership comprise a large part of RV. The large size is reached more quickly by shifting allocations to growth, and relative size is the prime determinant of winning a contest for a vacant territory (Warner & Schultz 1992). This reallocation is particularly convincing because sex-changers abandon guaranteed reproduction as females to enter a nonreproductive phase.

## REPRODUCTIVE ALLOCATION: MATE GUARDING
## VERSUS GAMETE PRODUCTION

In many species, males can compete for fertilizations in two basic ways. A male can prevent other males from gaining access to females (mate guarding), or he can allow access to females and attempt to create situations in which his sperm has a higher likelihood of fertilization (sperm competition). The allocation pattern for territorial TP males is dramatically different from other types within the species (Fig. 12.2). They do not grow at all, and have extremely small testes relative to other males. However, their rate of aggression is much higher than any other type (Warner & Hoffman 1980a). As was suggested above, relative size is critical in determining direct contests between individuals, and only the largest males can successfully defend a mating territory. Successful territorial defense is reflected in the low rate of interference by IP males (only 2.8 percent of 26,878 territorial pair-spawns observed have had an interfering IP male).

IP males, on the other hand, are essentially always in sperm competition, either in group-spawns or in attempts to join a pair-spawning TP male; it is they, and not TP males, that show mate choice (van den Berghe & Warner 1989). When fertilization is external, gamete release is nearly simultaneous, and the principal way for a male to increase fertilizations relative to others is simply to release more sperm (Parker 1984). The testes of IP males are enormous (much bigger on an absolute scale than those of TP males: Warner & Robertson, 1978; Fig. 12.2b), and these IP males are estimated to release more than 50 million sperm each time they mate (Shapiro et al. 1994). This is far in excess of the sperm numbers needed for maximal fertilization of the eggs (Warner et al. 1995; Marconato et al. 1997). In contrast, TP males release 3.3 million sperm per spawn on average, and show no signs of sperm depletion over the course of many matings (Warner et al. 1995; Petersen et al., submitted). This amount is just sufficient to achieve high fertilization (Warner et al. 1995).

In summary, relative size and the nature of the mating system interact to effect a radical shift in male allocations. Small males divide their energy between growth and gamete production, engaging only in sperm competition. Large males, who can win contests, shift almost all of their energy to aggression and mate guarding, with minimal gamete production. Since there is no territory usurpation, there is no requirement for further growth. Note that mate guarding *allows* low allocation to gamete production in this species.

GAMETE ALLOCATION

The fact that TP males expend most of their energy on aggression and have small testes implies that careful allocation of sperm output may be of considerable advantage (Warner 1997a). Douglas Shapiro has termed such allocations "sperm economy," where males will release the amount of sperm necessary to effect a relatively high, but not maximum fertilization rate (Shapiro et al. 1994; Marconato et al. 1995; Warner et al. 1995). In these cases, a male may be economizing on sperm in any particular mating because he is faced with the possibility of many future matings. In other words, there may be a level of "adaptive infertility" in each mating such that overall, the number of eggs fertilized by a particular male is maximized (Ball & Parker 1996).

With recent advances in techniques to measure sperm output and fertilization rates (Shapiro et al. 1994; Marconato et al. 1995), we have been able to observe and manipulate gamete allocation. The results have been exciting and challenging. Shapiro and his colleagues found that TP males regulate the amount of sperm they release in relation to the size of the female with whom they are mating (Shapiro et al. 1994). The increased amounts of sperm may be needed to raise fertilization rates (Petersen et al. 1992; Marconato et al. 1997), or to counter the possibility of sneaker male interference (Warner 1997a).

We have also found that males with very high mating success (over fifty matings a day) appear to reduce their sperm output per mating to a point where the fertilization rate is lowered by about 5 percent (Warner et al. 1995; Petersen & Warner 1998). By conserving the number of sperm per mating, highly successful males are able to both defend their females and provide a fairly high fertilization rate to each mate. While males can clearly benefit from this allocation strategy by avoiding sperm depletion, each female that mates with these successful males suffers a lower fertilization rate. Females do not appear to benefit from this sperm economy, and yet they do not change mating sites. A parallel result indicates that the fertilization rate achieved by females in group-spawns with IP males is marginally but significantly higher than that obtained from pair-spawns with average-success TP males (Marconato et al. 1997). Again, females show no preference for these situations where higher fertilizations might be obtained (Warner 1985). In this case, males may be exploiting the conservative nature of females, in the sense that females may be reticent to change mating sites.

Further experiments have shown a remarkable ability of TP males to adjust the amount of sperm released per spawn to compensate for changes in expected daily mating success. For example, when a male is provided with twice as many mates as is usual at his site, he suffers sperm depletion and no fertilization in spawns in the latter half of the mating period. By the next day, however, the male will release less sperm per spawn for every mating of the period. In this way, the same total amount of sperm produced is metered out over all spawns, resulting in a moderately high fertilization rate and no depletion (Shapiro, Petersen, & Warner, unpublished).

Remember, this sperm economy results from a prior allocation of most of the available energy to mate defense. Mate defense, in turn, is profitable only for relatively large males in situations where females are highly faithful to particular mating sites. Adaptations at one level resonate through other levels, and a broad perspective helps to set the context for the action of natural and sexual selection.

## And So . . .? How Is This Useful for Someone Beginning a Career in Behavioral Ecology?

Here I was able to put together past work to give a fairly complete picture of the mating system and life history of this species as a reflection of adaptations to particular problems posed by the environment. But it is important to realize that granting agencies are extremely unlikely to fund extensive studies in natural history. Each step of the way, over twenty-five years, I proposed to use the bluehead system to answer specific questions of general interest in behavioral and evolutionary ecology. Blueheads are an attractive system for experimentation because the animals are very plastic and can be easily manipulated, and intraspecific variability can be used in experimental tests of hypotheses in evolutionary ecology (Warner 1991). But at each iteration, it is a specific question that must drive the research. If it had ever appeared that we were simply trying to uncover another detail of the mating system of the bluehead wrasse, funding would have stopped.

So the key is not the particular species, it is the question you are asking. Read extensively, go to meetings, talk with colleagues, teach the subject. Once the questions become clear in your mind, ask two things: First, am I interested in this topic, and second, can I use the system I already know to answer the question? Interest, clarity, and enthusiasm are what make a proposal exciting to read. Many people have good ideas. Once a reviewer is convinced that the question is worth asking, the next step is to ask whether the research can answer the question. This is precisely where a detailed knowledge of a single system can tip the balance in your favor. If you can say that you know the experiments will not founder in logistics, that you can get started quickly and efficiently because the groundwork is there, you have gone a long way toward securing funding for future research. The zeitgeist produces the questions, and a detailed knowledge of a particular species allows one to sort through those questions to come up with a feasible plan.

## Future Directions

New tools provide the ability to answer old questions, and new approaches provide another way of asking those questions. In birds and mammals, for example, molecular techniques have shown unexpected patterns of paternity and reinvigorated the study of female mate choice. For externally fertilizing species like most fish, multiple paternity can often occurs because males can have simultaneous access to eggs released by the female. In accord with this, alternative male mating strategies are common in fish (Taborsky 1998). While standard DNA-based paternity analyses provide data on the outcome of male-male competition, up until now we have had little information on the costs involved with achieving a certain level of paternity. As was seen above, sperm is by no means cheap, and sperm allocation can be a major aspect of a male's mating strategy. In my laboratory, we have recently developed a technique that allows us to assess directly the relative amounts of sperm that different males contribute to a mixed mating, independent of paternity analysis (Wooninck et al., in press). We found that relative sperm contributions are not necessarily reflected in the distribution of paternity: that is, sperm competition is not a fair raffle (cf. Parker 1990a, b). Sperm competition and allocation remains a very active field in behavioral ecology (Petersen & Warner 1998), and this new technique should have application far beyond the bluehead wrasse.

Equally, our view of the evolution of mating systems has been hindered by a lack of proper tools, but in this case the missing element was a rigorous theoretical approach. In the past, most treatments of mating systems considered optimal behavior for only one class of players against a static background (e.g., Emlen & Oring 1977, for dominant males). Recently, Suzanne Henson Alonzo has pioneered an approach in which all players respond to each other in an interactive dynamic game (Henson & Warner 1997). Using this method, conflict becomes obvious as a difference in optima between the different players (e.g., between dominant and alternative males, or between females and dominant males), and the eventual equilibrium reached represents the mating system itself. The modeling approach thus allows us to view mating systems as the resolution of intra- and intersexual conflict, and predicts patterns that cannot be generated from single-player considerations (Alonzo & Warner 1999, 2000, in press). Again, I anticipate that this new direction will extend well beyond the mating system of a particular fish.

Finally, I think the future of behavioral ecology, in marine systems at least, will be greatly improved by a consideration of the geographical scale of adaptation. My past research has stressed the fact that many marine species occur in a wide variety of environments, and the resulting intraspecific variability and phenotypic plasticity are the key to successful experiments in behavioral and evolutionary ecology (Warner 1991). However, not all traits are plastic in the bluehead wrasse. A problem faced in many models in evolutionary ecology is that gene flow complicates the development of local adaptations. This problem is particularly acute in many marine organisms because they have distinct bipartite life cycles in which the larval stage is highly dispersive and adults are quite sedentary. Larvae are small and often planktonic, and large-scale oceanographic processes may be important in determining the patterns of settlement. This has several fundamental effects on the ability of organisms to adapt to local environments.

A plastic response to a highly variable physical or demographic environment may be adaptive, but some environmental factors simply cannot be predicted or detected by adults. For such cases fixed responses are the best solution, averaged over the dispersed range of the species. The problem for behavioral ecology is that such large-scale fixed responses may be maladaptive at many local levels. A naive adaptationist approach that assumes the life histories and behaviors of organisms to be adaptations to their current local environment will often fail when dealing with these higher-level adaptations. Experiments will not evoke a response, and there often will not be a good match between a particular trait and the local environment. The validity of these traits cannot be interpreted in studies that restrict themselves to a few localities.

A fascinating area for future research for marine species, at least, would be to identify the relative roles of local genetic differentiation, phenotypic plasticity, and large-scale fixed responses in determining behavior, life history, and morphology (Warner 1997b). This should be a function of the selective regime, the detection abilities of the settled individuals, and the degree of larval dispersal (and hence the gene flow among populations). It would be a major step forward in the attempt to integrate behavioral and evolutionary ecology into management and conservation biology.

The degree of dispersal and connection between populations is a major focus of current research in marine population biology, and this information will be of tremendous use in determining the proper scale of study for behavioral ecologists. Information for the bluehead wrasse is just beginning to appear (Caselle & Warner 1996; Caselle 1997; Swearer et al. 1999; Warner et al. 2000), but it is one of the first marine species for which we are even able to guess at the extent to which local production results in local recruitment. This is an area of tremendous promise, and a place where interdisciplinary studies will have the most impact.

## Acknowledgments

I am deeply grateful to the U.S. National Foundation for funding since 1976. Graduate students and research assistants too numerous to be named helped me in all phases of this work.

## References

Alonzo SH, Warner RR, 1999. A tradeoff generated by sexual conflict: in a Mediterranean wrasse, males refuse present mates to reduce competition in the future. Behav Ecol 10:105–111.

Alonzo SH, Warner RR, 2000. Female choice, conflict between the sexes, and the evolution of male alternative reproductive behaviors. Evol Ecol Res 2:149–170.

Alonzo SH, Warner RR, in press. Dynamic games and field experiments examining intra- and inter-sexual conflict: explaining counter-intuitive mating behavior in a Mediterranean wrasse, *Symphodus ocellatus*. Behav Ecol.

Ball M, Parker GA, 1996. Sperm competition games: external fertilization and 'adaptive' infertility. J Theor Biol 180:141–150.

Caselle JE, 1997. Small-scale spatial variation in early life history characteristics of a coral reef fish: implications for dispersal hypotheses. Proc 8th Int Coral Reef Symp 2:1161–1166.

Caselle JE, 1999. Early post-settlement mortality in a coral reef fish and its effects on local population size. Ecol Monog 69:177–194.

Caselle JE, Warner RR, 1996. Variability in recruitment in coral reef fishes: importance of habitat at large and small spatial scales. Ecology 77:2488–2504.

Charnov EL, 1982. The theory of sex allocation. Princeton: Princeton University Press.

Dunham ML, Warner RR, Lawson J, 1995. The dynamics of territory acquisition: a model of two coexisting strategies. Theoret Pop Biol 47:347–364.

Emlen ST, Oring LW, 1977. Ecology, sexual selection, and the evolution of mating systems. Science 197:215–223.

Godwin J, Crews D, Warner RR, 1996. Behavioural sex change in the absence of gonads in a coral-reef fish. Proc Roy Soc B 263:1683–1688.

Henson S, Warner RR, 1997. Male and female alternative reproductive behaviors in fishes: a new approach using intersexual dynamics. Ann Rev Ecol Sys 28:571–592.

Hoffman SG, Schildhauer MP, Warner RR, 1985. The costs of changing sex and the ontogeny of males under contest competition for mates. Evolution 39:915–927.

Marconato A, Tessari V, Marin G, 1995. The mating system of *Xyrichthys novacula*: sperm economy and fertilization success. J Fish Biol 47:292–301.

Marconato A, Shapiro DY, Petersen CW, Warner RR, Yoshikawa T, 1997. Methodological analysis of fertilization rate in the bluehead wrasse, *Thalassoma bifasciatum*: pair versus group spawns. Mar Ecol Progr Ser 161:61–70.

Mertz DB, 1971. The mathematical demography of the California condor population. Amer Natur 105:437–453.

Parker GA, 1984. Sperm competition and the evolution of animal mating strategies. In: Sperm competition and the evolution of animal mating systems (Smith RL, ed). Orlando: Academic Press; 1–60.

Parker GA, 1990a. Sperm competition games: raffles and roles. Proc Roy Soc Lond B 242:120–126.

Parker GA, 1990b. Sperm competition games: sneaks and extra-pair copulations. Proc Roy Soc Lond B 242:127–133.

Petersen CW, Warner RR, 1998. Sperm competition and sexual selection in fishes. In: Sperm competition and sexual selection (Birkhead TR, Møller AP, eds). New York: Academic Press; 435–463.

Petersen CW, Warner RR, Cohen S, Hess HC, Sewell AT, 1992. Variation in pelagic fertilization success: implications for production estimates, mate choice, and the spatial and temporal distribution of spawning. Ecology 73:391–401.

Robertson DR, Hoffman SG, 1977. The roles of female mate choice and predation in the mating systems of some tropical labroid fishes. Z Tierpsychol 45:298–320.

Schultz ET, Warner RR, 1990. Phenotypic plasticity in life-history traits of female *Thalassoma bifasciatum* (Pisces: Labridae) II. Correlation of fecundity and growth rate in comparative studies. Env Biol Fish 30:333–344.

Shapiro DY, Marconato A, Yoshikawa T, 1994. Sperm economy in a coral reef fish, *Thalassoma bifasciatum*. Ecology 75:1334–1344.

Stearns SC, 1976. Life-history tactics: a review of the ideas. Q Rev Biol 51:3–47.

Stearns SC, 1992. The evolution of life histories. Oxford: Oxford University Press.

Swearer SE, Caselle JE, Lea DW, Warner RR, 1999. Larval retention and recruitment in an island population of a coral-reef fish. Nature 402:799–802.

Taborsky M, 1998. Sperm competition in fish: 'Bourgeois' males and parasitic spawning. Trends Ecol Evol 13:222–227.

Trivers R, 1972. Parental investment and sexual selection. In: Sexual selection and the descent of man 1871–1971 (Campbell B, ed). Chicago: Aldine Press; 136–179.

van den Berghe EP, Warner RR, 1989. The effects of mating system on male mate choice in a coral reef fish. Behav Ecol Sociobiol 24:409–415.

Victor BC, 1982. Daily otolith increments and recruitment in two coral-reef wrasses, *Thalassoma bifasciatum* and *Halichoeres bivittatus*. Mar Biol 71:203–208.

Victor BC, 1983. Recruitment and population dynamics of a coral reef fish. Science 219:419–420.

Victor BC, 1986. Larval settlement and juvenile mortality in a recruitment-limited coral reef fish population. Ecol Monog 56:145–160.

Warner RR, 1975. The adaptive significance of sequential hermaphroditism in animals. Amer Natur 109:61–84.

Warner RR, 1980. The coevolution of behavioral and life history characteristics. In: Sociobiology: beyond nature–nurture? (Barlow GW, Silverberg J, eds). Boulder: Westview Press; 151–188.

Warner RR, 1984a. Deferred reproduction as a response to sexual selection in a coral reef fish: a test of the life historical consequences. Evolution 38:148–162.

Warner RR, 1984b. Mating systems and hermaphroditism in coral reef fish. Amer Sci 72:128–136.

Warner RR, 1985. Alternative mating behaviors in a coral reef fish: a life-history analysis. Proc Fifth Int Coral Reef Cong 4:145–150.

Warner RR, 1986. The environmental correlates of female infidelity in a coral reef fish In: Indo-Pacific fish biology: Proceedings of the Second International Conference on Indo-Pacific Fishes (Uyeno T, Arai R, Taniuchi T, Matsuura K, eds). Tokyo: Ichthyological Society of Japan; 803–810.

Warner RR, 1987. Female choice of sites versus males in a coral reef fish, *Thalassoma bifasciatum*. Anim Behav 35:1470–1478.

Warner RR, 1988a. Traditionality of mating-site preferences in a coral-reef fish. Nature 335:719–721.

Warner RR, 1988b. Sex change and the size advantage model. Trends Ecol Evol 3:133–136.

Warner RR, 1988c. Sex change in fishes: hypotheses, evidence, and objections. Env Biol Fish 22:81–90.

Warner RR, 1990a. Resource assessment vs. traditionality in mating site determination. Amer Natur 135:205–217.

Warner RR, 1990b. Male vs. female influences on mating site determination. Anim Behav 39:540–548.

Warner RR, 1991. The use of phenotypic plasticity in coral reef fishes as tests of theory in evolutionary ecology. In: The ecology of coral reef fishes (Sale P, ed). New York: Academic Press; 387–398.

Warner RR, 1995. Large mating aggregations and daily long-distance spawning migrations in the bluehead wrasse, *Thalassoma bifasciatum*. Env Biol Fish 44:337–345.

Warner RR, 1997a. Sperm allocation in coral reef fishes. BioScience 47:561–564.

Warner RR, 1997b. Evolutionary ecology: how to reconcile pelagic dispersal with local adaptation. Coral Reefs 16S:115–128.

Warner RR, 1998. The role of extreme iteroparity and risk-avoidance in the evolution of mating systems. J Fish Biol 53(Suppl A):82–93.

Warner RR, Hoffman SG, 1980a. Population density and the economics of territorial defense in a coral reef fish. Ecology 61:772–780.

Warner RR, Hoffman SG, 1980b. Local population size as a determinant of mating system and sexual composition in two tropical marine fishes (*Thalassoma* spp.) Evolution 34:508–518.

Warner RR, Marconato A, Shapiro DY, Petersen CW, 1995. Sexual conflict: males with highest mating success convey the lowest fertilization benefits to females. Proc Roy Soc B 262:135–139.

Warner RR, Robertson DR, 1978. Sexual patterns in the labroid fishes of the Western Caribbean. I. The Wrasses (Labridae). Smithsonian Cont Zool 254:1–27.

Warner RR, Robertson DR, Leigh EG, Jr, 1975. Sex change and sexual selection. Science 190:633–638.

Warner RR, Schultz ET, 1992. Sexual selection and male characteristics in the bluehead wrasse, *Thalassoma bifasciatum:* mating site acquisition, mating site defense, and female choice. Evolution 46:1421–1442.

Warner RR, Swearer, SE, 1991. Social control of sex change in the bluehead wrasse, *Thalassoma bifasciatum* (Pisces: Labridae). Biol Bull 181:199–201.

Warner RR, Swearer SE, Caselle JE, 2000. Larval accumulation and retention: implications for the design of marine reserves and essential fish habitat. Bull Mar Sci. 66.

Wooninck L, Fleischer R, Warner RR, in press. Assessing relative sperm contributions using quantitative PCR. Mol Ecol.

# 13 The Economics of Sequential Mate Choice in Sticklebacks

Manfred Milinski

## Why Sticklebacks?

Although Konrad Lorenz's books had some flavor of old-fashioned group selectionist thinking in several places, they were fascinating to read and convinced me that investigating why animals and humans behave the way they do would be immensely valuable. I decided to become an ethologist and to accomplish this I studied both biology and mathematics, the latter because I admired the simple elegance of a mathematical proof. I had no idea that algebra would become the most important tool I would learn for building hypotheses about the function of behavior. It was a stroke of luck that exactly at that time the field of behavioral ecology was born with the insight that natural selection should favor individuals that decide economically in every situation of their life, for example, how to choose a diet (MacArthur & Pianka 1966; Emlen 1966), when to leave a food patch (Charnov 1976), how to deal with competitors (Fretwell & Lucas 1970; Parker 1970), whom to help, while still maximizing inclusive fitness (Hamilton 1964; Trivers 1971), and how to be an optimal parent (Trivers 1974; Dawkins & Carlisle 1976; Parker & MacNair 1979). All these concepts were firmly based on elegant algebraic models. Sexual selection was not yet a hot topic in those days, although provocative ideas had been floating around since the 1930s (Fisher 1930; Zahavi 1975).

The concepts I have worked with come from combining evolutionary thinking with economics. For example, in order to understand the economics of foraging, researchers expected that an animal would select the "diet" that maximizes its fitness by maximizing its net rate of energy intake. They always begin with a mathematical model—a theory—to generate quantitative predictions (MacArthur & Pianka 1966; Emlen 1966), which can be tested empirically with animals that have evolved in the kind of ecology that the model assumes (Milinski 1997). The first experimental tests of optimal diet theory confirmed its predictions to a convincing extent (Werner & Hall 1974; Krebs et al. 1977) and "optimal foraging" became fashion.

At first sight schools, flocks, and herds of prey seem to be a paradise for a hungry predator. It often pays single prey to join such groups because of the risk dilution effect ("selfish herd" effect, Hamilton 1971), which, if anything, also helps the predator. However, the predator may also suffer from the "confusion effect" (Neill & Cullen 1974; Ohguchi 1981; see Pitcher & Parrish 1993 for review) when it tries to track a schooling target. It was tempting to try to predict the optimal foraging strategy of a predator that has coevolved with swarming prey. To do so for my Ph.D. thesis at the University of Bochum, I had to find a system that was suitable for controlled lab experiments with many independent replicates to avoid what is now known as "pseudo-replication" (Hurlbert 1984). Small fish predators and *Daphnia* that can form plankton swarms seemed to be ideal for such work and although I was surrounded by ornithologists (Curio et al. 1978), I tested several species of cyprinid fish and found that shoals became quickly accustomed to experimental tanks. When tested singly, however, these fish were delicate. It would have been too time-consuming to get several hundred single fish accustomed to the test tank. When I talked about my problems with some of my fellow-students in a pub, one of them, his name was Udo Hallau, said, "Why don't you try the sticklebacks from the pond behind the lab?"

In the end, I decided to focus on three-spined sticklebacks (*Gasterosteus aculeatus*) for several reasons: (1) they were plentiful in the ponds around my university; (2) much of their behavior was known via Tinbergen's classical papers (e.g., Pelwijk & Tinbergen 1937; Hoogland et al. 1957); (3) they are found everywhere in the Northern Hemisphere from Canada to Switzerland (Wootton 1976, 1984), under a broad range of ecological conditions, so that many questions can potentially be answered with them (no expensive and adventurous fieldwork in nice parts of the world is necessary, which I regret now and then); (4) they are easy to breed in large numbers in the lab; (5) they are only a few centimeters long, and thus do not need much lab space; (6) they have spines that are ideal for individual marking (Milinski 1984a, 1994); (7) they become quickly accustomed to tanks also when kept singly; and (8) they are extremely interesting animals—it is still a pleasure to work with them. Sticklebacks also pose a big disadvantage in that it is often difficult to explain to nonscientists why research on sticklebacks is worth public money; they are neither of commercial interest such as salmon, nor do they potentially kill people as do lions, nor are they vectors for human pests such as are rats or ticks. If you work with sticklebacks you need a really interesting question to convince people that your research is worth their taxes.

Optimal foraging on swarming prey turned out to be beyond the scope of the theory available at the time I initiated my work in this area (Milinski 1977a, b). New concepts and a modified theory that included a hunger state-dependent balancing of the conflicting demands of feeding and avoiding predation were yet to be developed. During my early work, I learned of the importance of collaborations between mathematicians and empiricists. I joined forces with Rolf Heller, a "pure" mathematician, who became fascinated by optimality problems and could handle sophisticated mathematical tools such as "Pontryagin's maximum principle" (Milinski & Heller 1978; Heller & Milinski 1979; Milinski 1984b; see Milinski 1993 for a review). I lined up with many other researchers who had come to the conclusion that three-spined sticklebacks are suitable for testing predictions generated by different theories such as "ideal free distributions" (e.g., Milinski 1979, 1984a, 1994; review in Milinski & Parker 1991), reciprocal cooperation (e.g., Milinski 1987; Milinski et al. 1990, 1997), and sexual selection (Milinski & Bakker 1990, 1992; Bakker & Milinski 1991; Bakker 1993).

## Hamilton and Zuk, and Mate Choice in Sticklebacks

After my time in Bochum I spent half a year in Oxford working next door to Bill Hamilton, who was enthusiastic about the role of parasites in shaping sexual preferences: I became somewhat imprinted by Bill's ideas. Hamilton was not only a most ingenious evolutionary theorist but also a biologist who knew everything about his animals. Around that time I had the privilege to take part in a Berlin Dahlem Workshop on sexual selection (Bradbury & Andersson 1987). Although such workshops rarely generate new ideas, they both describe completely the state of the art of a field of research and work out unsolved problems and open questions. Equipped with this knowledge and stimulated by Hamilton's fascination, I went to Bern in Switzerland to build up a new department for behavioral ecology. If you have to start from scratch, it is the right moment to switch to a new field of research, such as sexual selection.

Males of many species are handicapped by luxurious sexual ornaments, for example, bright colors, long tail feathers, songs, and display behavior. Darwin (1871) assumed that despite their lower survival, such "handicapped" males have more offspring because they are preferred as mates by females. There are two major hypotheses to explain how such a female preference has evolved. (1) Fisher's (1930) runaway hypothesis assumes a strong genetic covariance between male trait and female preference in both sons and daughters. Under certain conditions, a runaway process could start that pushes both the trait and the preference until mating advantage and viability costs are equal or the population eventually goes extinct. (2) For parasite-driven sexual selection (Hamilton & Zuk 1982), luxurious ornaments are assumed to reveal a male's resistance against currently predominant parasites. Females select healthy fathers for their offspring when they prefer ornamented males. Until I moved to Bern, no experimental test of the Hamilton–Zuk model had been published. I had not worked on sexual selection in Bochum, but I was used to demonstrating Tinbergen's famous experiments on the male stickleback's zigzag dance, its red breeding coloration, and the females' head-up display to students in practicals. Sticklebacks actually seemed to be ideal for testing the Hamilton–Zuk hypothesis.

In springtime, a male three-spined stickleback builds a nest to which it attracts up to twenty ripe females within a few days (Wootton 1976). The males develop a bright red breeding coloration from carotenoids at the beginning at the breeding season, and it has been shown that the females' visual sensitivity for red coloration increases at the beginning of the reproductive season and reaches a higher level than that of males (Cronly-Dillon & Sharma 1968). Carotenoid-based signals have become an integral part of the theory of sexual signaling (Olson & Owens 1998; Møller et al. 2000).

In order to test whether females actually take the intensity of the sexual coloration into account when they choose a mate, one has to change the coloration experimentally, otherwise the evidence would be only correlational and not decisive: redder males may be bigger, for example, and females may prefer redder males because they are bigger. It is difficult to use dye on a wet fish and one might produce artifacts by doing so. I found the solution to this problem when my wife and I saw Stravinsky's ballet *The Fire Bird* in the opera house in Bern. The dancers appeared in beautiful colors, which changed during dancing. When the lights of the hall were illuminated for applauding, all dancers had white dresses. The beautiful colors had been produced by colored light! Then it was easy to think of neutralizing differences in redness among male sticklebacks by illuminating the tanks by green light. I developed the design for mate choice experiments for my first courses that I had to teach in Bern. Thereafter, I was lucky enough to attract Theo Bakker, a geneticist from Leiden, where the ethologist Piet Sevenster had followed Tinbergen's tradition of stickleback research. Together Theo and I had twenty-five years of experience with sticklebacks, which is a good prerequisite for meeting the animals' ecological requirements when handling them in experiments, providing some insurance against producing artifacts.

## Does Male Breeding Coloration Reveal Condition?

To examine whether male breeding coloration reveals condition, we (Milinski & Bakker 1990) placed male sticklebacks with developed breeding coloration individually into small tanks. After all males had completed nests and courted females vigorously, students were asked to score each male's intensity of red coloration on a 10-point scale. Thereafter we measured each male's condition factor (Bolger & Connolly 1989). Figure 13.1a shows that the intensity of his red breeding coloration revealed a male stickleback's condition, suggesting that female sticklebacks may be able to use such information in their mate choice decisions.

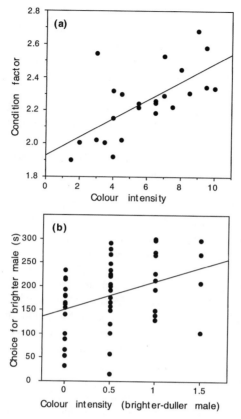

Figure 13.1. (a) Correlation between the intensity of red breeding coloration and the condition factor of 24 reproductive male sticklebacks, $r^2 = 0.44$, $F = 17.27$, $p < 0.001$; (b) correlation between the difference in color intensity of 12 pairs (brighter–duller male) and active female choice for the brighter male, $r^2 = 0.14$, $F = 7.62$, $p < 0.01$ (from Milinski & Bakker 1990).

## Do Female Sticklebacks Prefer Brighter Males?

To examine whether female sticklebacks prefer brighter males we arranged tanks in a line starting with male 1, which had the lowest average color rank, and ending with male 24, which had the highest rank. Neighboring males were defined as pairs for simultaneous presentation to ripe females. In a separate tank positioned centrally in front of each pair of males, a cell containing a single gravid female was placed and her choice between the two males was videorecorded during a five-minute period. Choosing females maintain the head-up courtship posture while pointing toward the chosen male. This is a great advantage of sticklebacks because they reveal accurately when they are choosing a mate instead of approaching a conspecific for other reasons. There was a significant preference for the brighter male in each pair, this preference being intensified as the difference in coloration increased (Fig. 13.1b). It follows from the previous correlation that by preferring the brighter male, the females actually preferred the male that was in better condition.

From this observation we could not decide whether female preference was ultimately based on coloration or on some correlated character, for example, directly on the condition factor. We repeated the same protocol with new males. Females were asked to choose between males under white light, and others—thanks to the Bern ballet—under green light, where they were unable to judge differences in the intensity of red coloration. Under white light, the females preferred again the redder males, whereas this preference disappeared under green light (Milinski & Bakker 1990). Thus female sticklebacks take the intensity of the males' red coloration into account when they choose a mate. Parasitization caused deterioration in the males' condition and a decrease in the intensity of their red coloration. Tests under both lighting conditions revealed that the females recognize the formerly parasitized males by the lower intensity of their breeding coloration (Milinski & Bakker 1990), which is a necessary condition for coloration to be judged as a revealing handicap. Our results support the parasite hypothesis of female choice (Hamilton & Zuk 1982). In our experiments females could always see two males simultaneously, which is a simplification of the natural situation. In nature, the males' territories may be separated by several meters, and the females have to visit males sequentially—presumably a more difficult task than simultaneous choice. We might have produced artifacts with our unnatural simultaneous choice situation. There are two ways to approach this problem: (1) A quick and simple (not dirty) way: we present each female with only one male at a time and measure whether she displays "head-up" longer in front of redder males. (2) We look for a theory for sequential choice and test specific hypotheses of this theory under the assumption that sticklebacks have evolved economic rules for sequential choice.

## The Concept of Sequential Choice

In the following I concentrate on the concept of the "job choice" or "marriage problem" from economics, and how it applies to sequential mate choice in sticklebacks and other animals. Imagine you need a job. Let's assume for simplicity's sake that neither your qualifications nor those of your potential competitors are important: you get the job that you choose from the ones that are offered. Of course, you'd like to have a top job with a top salary. Unfortunately, such jobs are rarely offered. Usually you have to choose between several suboptimal jobs and you may decide to wait until a job is offered that is close to your wishes. In such a scenario simultaneous choice among top jobs is probably rare. Better jobs are offered sequentially and disappear quickly. One problem you may face is that sometimes a job is offered that may be acceptable if there would be no better option to come, but how can you determine what options the future holds? This job-search problem has been treated as a dynamic programming problem in the economics literature (e.g., McNamara & Collins 1990; Collins & McNamara 1993; see Real 1990 for review). An analogous situation is sequential mate choice. Theorists have investigated the rules that a member of the choosy sex (usually the female) could adopt when she tries to find a top mate (Janetos 1980; Janetos & Cole 1981; Parker 1983; Wittenberger 1983; McNamara & Collins 1990; Real 1990; Collins & McNamara 1993; Dombrovsky & Perrin 1994; Getty 1995; Luttbeg 1996; Mazalov et al. 1996; Wiegmann et al. 1996; Johnstone 1997).

## THEORETICAL SOLUTIONS TO THE SEQUENTIAL CHOICE PROBLEM

The next step would be to describe the theory of sequential choice and to generate quantitative predictions that can be tested with a biological system, for example, with sticklebacks. Although this would be far beyond the scope of this essay, I will not drop this step completely but give instead some flavor of the kind of approach that would be taken.

Let's assume that a female has no problem in ranking available males, for example, by the redness of their breeding coloration. Given some distribution of mate qualities in a population, and some time series of encounters with potential mates, how should unmated individuals shape their preferences and selection of mates? Janetos (1980) examined two rules. The first is the best-of-n males. A female may take the better of the first two males she encounters, or the best of the first three she encounters, or in the general case, the best of n males. This strategy demands some capacity for remembering the qualities and locations of previously encountered males, and the female must be free to return to a previously encountered male if need be (this assumption is referred to as "complete recall"). How large should n be? Janetos calculated that the best of n + 1 is not much better than the best of n when n ≥ 5. One therefore expects females using this strategy to inspect a small number of males, somewhat in the vicinity of five or six. The expected gain from inspecting more males is small.

Janetos's second strategy is called the one-step decision rule. In such a model there are two possible outcomes of any encounter with a male: the female may mate and end her search, or she may reject the male and go on to the next male. A female should only mate and end her search if the male on hand is better than she could expect to find by continuing her search. The optimal decision strategy for such a problem begins by starting at the end (step n) and working backward. In fact, the criterion for mating becomes more and more stringent as one goes back to step one (see also McNamara & Collins 1990). Thus one has a picture of a female starting out as very picky indeed, but becoming less so as she runs out of time, until her last chance, when she will mate with any conspecific male (Janetos 1980).

Which of the above rules is the better choice? The best-of-n strategy dominates others when searching and sampling are not built into the model (Real 1990). The costs of sampling mates are real and probably in many cases large, and when potential mates are rejected they are often no longer available for future matings.

Once costs are added to the mate-choice problem, the relative performance of the various strategies is quite different (Parker 1983; Real 1990). A sequential-search optimal-threshold model now emerges as the best solution. Females must determine the threshold value that equates the cost of search with the expected fitness benefit of sampling an additional potential mate (Real 1990). Increasing search cost would decrease the optimal threshold value: females become less selective and lower-quality males are acceptable. There are scenarios, however, under which various rules generate different predictions (Real 1990; Wiegmann et al. 1996). That being said, decreasing selectivity with increasing costs of search and sampling is a general prediction of economic models of sequential search, which can be used for a test for economic mate choice in any species with that problem.

Janetos's (1980), Parker's (1983), and Real's (1990) models assume that searching females have complete knowledge of the distribution of qualities among males in a population prior to search. This assumption, made for analytical purpose, may be unrealistic (Dombrovsky & Perrin 1994; Mazalov et al. 1996). If the female has no a priori expectation of the distribution of mate qualities she has to do some sampling. After sampling some males the female then needs to determine a critical value of acceptable mates. If the next male that she encounters is above that value she accepts it, otherwise she continues to search (Dombrovsky & Perrin 1994; Mazalov et al. 1996). So we need to test whether females rate the very first male higher if it is a high-quality male or, alternatively, whether the sequence of males that a female has just encountered influences her decision whether to accept the male at hand. In the latter case a given male should be rated higher when preceded by a low-quality male than by a high-quality male. Now we need to design experiments for testing these predictions.

## Do Female Sticklebacks Prefer Brighter Males When Encountered Sequentially?

We presented females with males of different color intensity sequentially. During each presentation of two minutes, we measured how long the female displayed her head-up posture in front of either a dull, a medium, or a bright male (Bakker & Milinski 1991). In our experimental design, each female saw each male twice, and the position of each type of male in order of presentation was the same (when averaged over all females.) Therefore, we could compare the duration of the head-up display that each type of male received per presentation to a female. The duration of the female's head-up posture (which correlates with her probability to spawn with that male [McLennan & McPhail 1990]) increased significantly with the intensity of the male's brightness (Fig. 13.2a), demonstrating that female sticklebacks can use red coloration as a cue, even in sequential choice problems. This is the quick and simple approach. But how do they do it? Do they know a distribution of male qualities a priori?

To address the above question, half of the females saw a dull male first and the other half were presented with a bright male first. They rated the bright male significantly higher than the dull male (significant difference between relative height of first column of Fig. 13.2b and that of Fig. 13.2c). The females thus have an internal representation of a distribution of male qualities, which supports the assumption of various models of sequential choice (Janetos 1980; Parker 1983; Real 1990; Wiegmann et al. 1996). But this result raises a new question: Do female sticklebacks not learn the actual distribution of male qualities?

Since half the females were presented to the three males in the order "bright, medium, bright, pause of thirty minutes, dull, medium dull" and the other half saw the males in the reverse order (see Fig. 13.2b and 13.2c), we can test whether bright males are rated higher when they are seen after medium males than before them. Similarly, we can address whether a dull male is rated lower after a medium male than before it as would be expected if the females learned about the actual distribution of male qualities presented. The combined probability of the six possible comparisons supports the hypothesis that the females' mate choice is attuned to the quality of previously encountered males ("previous male effect," Bakker & Milinski 1991; see also Downhower & Lank 1994; Collins 1995 for similar results). This attuning demonstrates that searching females learn something about the distribution of qualities of potential mates and adapt their behavior accordingly. This seems to be adaptive: a female that takes no account of the local distribution of male qualities would often end up with a male of relatively low quality in a population of high-quality males. Adaptive search may thus function as a fine-tuning mechanism of an internal expectation: although female sticklebacks have an a priori estimate of the distribution of male qualities, they modify this estimate on the basis of experience.

## Costs Influence Sequential Mate Choice in Sticklebacks

Models of sequential search predict that with increasing costs sampling selectivity for preferred males should decline (Parker 1983; Real 1990). Costs

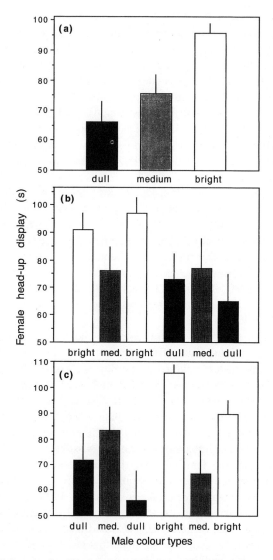

Figure 13.2. (a) Average (+ SE) duration of female head-up display directed at each of three different male color types (i.e., dull, medium, bright) by 28 female sticklebacks; p < 0.0001 after one-tailed Page test for ordered alternatives; (b) average (+ SE) duration of the female head-up display directed at each of 3 different male color types (i.e., dull, medium, bright) in the specified sequence by the 14 females starting with a bright male; (c) by the 14 females starting with a dull male (from Bakker & Milinski 1991)

may consist of long distances between territorial males, so that a female has to invest time and energy for sampling. For example, female pied flycatchers in nest box groups choose males with high-quality nest boxes, whereas females presented with a more costly choice (single boxes) do not make this distinction (Alatalo et al. 1988). Energy demand and competition for mates were suggested as potential costs of searching in this species (Slagsvold et al. 1988; Dale et al. 1992). Costs of mate choice influence the predictions of a number of models of sexual selection (e.g., Pomiankowski 1987). We have experimentally investigated whether time and energy costs of moving between males reduce a female stickleback's selectivity so that a dull male can become attractive even when the female has seen a bright male before (Milinski & Bakker 1992).

A bright male and a dull male were each placed with its nest in an individual "experience tank," with yet another dull male in a "test tank" (Fig. 13.3). Several hours before the start of an experiment, females were selected and placed in individual tanks. A partition consisting of a frame with a green net was placed in the test tank and in each of the two experience tanks (Fig. 13.3).

A female was placed in the large compartment of the experience tank with the dull male. After three minutes she was transferred to the test tank and released behind the plants. After about one minute she usually swam past the plants and approached the male, who typically started displaying at this point. We then measured the time she remained in his vicinity (Fig. 13.3). After five minutes she was transferred to her individual tank, where she stayed for about one hour until the whole procedure was repeated starting with transferring her to the experience tank with the bright male. In an additional round, females were forced to hold their position in a water current: before being transferred to the test tank, the female was placed into a ring channel with a water current. By this procedure we tried to simulate time and energy costs of moving from the bright to the dull male. The sequence of treatments was varied among females.

The females spent significantly more time with the dull test male when they had come from the dull male than when they had seen the bright male before (Fig. 13.4). However, when the females had spent ten minutes in the water current after they had seen the bright male, they displayed significantly longer near the dull test male than when they had come from the bright male directly (Fig. 13.4). This suggests that females take time and/or energy invested on the way from one male to another into account when they decide whether to accept the present male. Thus, when costs of searching and sampling are increased female sticklebacks become less selective, as predicted. In further experiments we could distinguish experimentally between costs of time and costs of energy. The results demonstrated that female sticklebacks become less selective in their sequential mate choice when either only energy or time costs of searching and sampling are increased (Milinski & Bakker 1992).

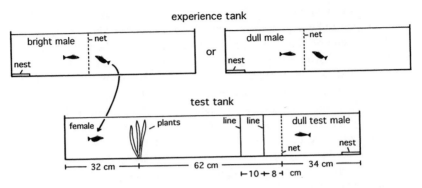

Figure 13.3. Experimental setup. A female spent three minutes in the experience tank either with a bright red or a dull (only experiments 1) male. Then she was transferred to the test tank either directly or after a specified time which she spent either in still or in running water (from Milinski & Bakker 1992). See text for further explanations.

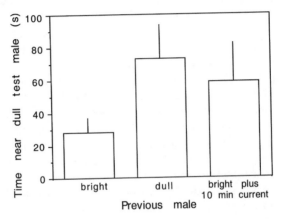

Figure 13.4. Time (average + SE) the females spent near the dull test male after they had come directly either from a bright male or a dull male (p = 0.005, directed paired t-test), or after they had come from the bright male and then had to maintain their position in a water current for 10 minutes before the test (p < 0.05, directed paired t-test), n = 12 females (from Milinski & Bakker 1992).

There is both a theory of sexual selection and a theory of economic choice. Without these theories it would have been difficult to even speculate on what would be optimal solutions to the problems and thus good candidates for being close to evolved strategies (see McNamara & Collins 1990). We have verified various assumptions of the theory of sexual selection in sticklebacks: costly sexual ornaments (i.e., the red breeding coloration) of males reveals their physical condition and state of parasitization and female sticklebacks prefer brighter males (and hence avoid weak and parasitized ones) under both simultaneous and sequential choice conditions. We have also verified two assumptions of the theory of sequential choice that seemed to be extremes: female sticklebacks have both an a priori expectation of the distribution of male qualities and they learn the actual distribution. This combination seems to be optimal as they can accept the very first male if it is of high quality, and yet they are able to modify this expectation on the basis of experience. Increasing costs of sampling males (in terms of either time or energy) reduces the females' selectivity: they accept low-quality males that they reject otherwise. The strength of sexual selection therefore depends on the spatial structure of the population.

## Where I Would Like to Go in the Future

Several empirical studies (either observational or experimental) have investigated the sequential choice problem in various species (see Gibson & Langen 1996 for a review). However, as in other fields of research, theorists have made enormous progress and offered refined and more complex solutions to the problem of sequential choice. For example, Parker (1983) looks at the case where both males and females are actively choosing on the basis of their partner's fitness, with a mating pair being formed if both animals choose the other (see also McNamara & Collins 1990). Dombrovsky and Perrin (1994) and Mazalov et al. (1996) drop the common assumption that the distribution of the quality of potential mates is known a priori. Using stochastic dynamic programming, they develop a model that includes the possibility for searching individuals to learn about the distribution, and in particular to update mean and variance during the search. They use optimal-stopping theory to predict sequential choice. Getty (1995) uses signal detection theory (SDT) to include imperfect discrimination of mate quality type. SDT is an optimal Bayesian approach to probabilistic categorization. Luttbeg (1996) presents a comparative Bayes model of information gathering and mate choice, which relaxes the assumption of random encounters and perfect information. Wiegmann et al. (1996) develop analytical predictions about how females should behave when either a (infinite time horizon) sequential search tactic or best-of-n rule is employed to search for mates. Lastly, Johnstone (1997) models mutual mate choice as a dynamic game, which yields predictions about mating behavior under the influence of time constraints, choice costs, and competition for mates. Thus, refined concepts and predictive theory are there. Empiricists should now take their turn. Young behavioral ecologists who are at the start of their career can profit from working through this large body of theory to do the decisive experiment.

I am tempted to try to understand my model system to a greater extent by testing predictions of these refined new mathematical models that meet the ecology of sticklebacks. However, the ultimate goal would be to measure the fitness consequences of mate choice and, last but not least, of sexual reproduction. What is the advantage of sex (Maynard Smith 1978; Stearns 1987; Hamilton et al. 1990)? Why do three-spined sticklebacks reproduce sexually?

Even if you like to study different problems, it is an enormous advantage to stay with the same system. This advantage grows with each additional year you have spent with it. I share my enthusiasm for this little clever fish with many other researchers from all over the Northern Hemisphere, whom I can meet every five years on an international symposium on stickleback behavior. I have changed places again and moved from Bern to Ploen, where I have another chance to start from scratch. It is again the right moment to switch to a new field of research, host-parasite coevolution. The first new building for the new department contains just aquaria rooms, each one tailored for working with little fish, and the many lakes around Ploen are full of three-spined sticklebacks.

## Acknowledgments

I thank all my collaborators who shared my enthusiasm for sticklebacks and Lee Dugatkin for his patience. I also wish to thank the German Science Foundation (DFG) and the Swiss National Foundation for support.

## References

Alatalo RV, Carlson A, Lundberg A, 1988. The search cost in mate choice of the pied flycatcher. Anim Behav 36:289–291.

Bakker TCM, 1993. Positive genetic correlation between female preference and preferred male ornament in sticklebacks. Nature 363:255–257.

Bakker TCM, Milinski M, 1991. Sequential female choice and the previous male effect in sticklebacks. Behav Ecol Sociobiol 29:205–210.

Bolger T, Connolly PL, 1989. The selection of suitable indices for the measurement and analysis of fish condition. J Fish Biol 34:171–182.

Bradbury JW, Andersson MB (eds), 1987. Sexual selection: testing the alternatives. Chichester: John Wiley & Sons.

Charnov EL, 1976. Optimal foraging: the marginal value theorem. Theor Popul Biol 9:129–136.

Collins EJ, McNamara JM, 1993. The job-search problem with competition: an evolutionarily stable dynamic strategy. Adv Appl Prob 25:314–333.

Collins SA, 1995. The effect of recent experience on female choice in zebra finches. Anim Behav 49:479–486.

Cronly-Dillon J, Sharma SC, 1968. Effect of season and sex on the photopic spectral sensitivity of the three-spined stickleback. J Exp Biol 49:679–687.

Curio E, Augst H-J, Böcking H, Milinski M, Ohguchi O, 1978. Wie Singvögel auf Feindrufe hassen lernen. J Orn 119:231–233.

Dale S, Rinden H, Slagsvold T, 1992. Competition for a mate restricts mate search of female pied flycatchers. Behav Ecol Sociobiol 30:165–176.

Darwin C, 1871. The descent of man, and selection in relation to sex. London: John Murray.

Dawkins R, Carlisle TR, 1976. Parental investment, mate desertion and a fallacy. Nature 262:131–132.

Downhower JF, Lank DB, 1994. Effect of previous experience on mate choice by female mottled sculpins. Anim Behav 47:369–372.

Emlen JM, 1966. The role of time and energy in food preference. Am Nat 100:611–617.

Fisher RA, 1930. The genetical theory of natural selection. New York: Dover.

Fretwell SD, Lucas HL, Jr, 1970. On territorial behavior and other factors influencing habitat distribution in birds. Acta Biotheor 19:16–36.

Getty T, 1995. Search, discrimination, and selection: mate choice by pied flycatchers. Am Nat 145:146–154.

Gibson RM, Langen TL, 1996. How do animals choose their mates? Trends Ecol Evol 11:468–470.

Hamilton WD, 1964. The genetical evolution of social behaviour. J Theor Biol 7:1–52.

Hamilton WD, 1971. Geometry for the selfish herd. J Theor Biol 31:295–311.

Hamilton WD, Axelrod R, Tenese R, 1990. Sexual reproduction as an adaptation to resist parasites (a review). Proc Natl Acad Sci USA 87:3566–3573.

Hamilton WD, Zuk M, 1982. Heritable true fitness and bright birds: a role for parasites? Science 218:384–387.

Heller R, Milinski M, 1979. Optimal foraging of sticklebacks on swarming prey. Anim Behav 27:1127–1141.

Hoogland R, Morris D, Tinbergen N, 1957. The spines of sticklebacks (*Gasterosteus* and *Pygosteus*) as a means of defense against predators (*Perca* and *Esox*). Behaviour 10:205–236.

Hurlbert SH, 1984. Pseudoreplication and the design of ecological field experiments. Ecol Monogr 54:187–211.

Janatos AC, 1980. Strategies of female mate choice: a theoretical analysis. Behav Ecol Sociobiol 7:107–112.

Janetos AC, Cole BJ, 1981. Imperfectly optimal animals. Behav Ecol Sociobiol 9:203–209.

Johnstone RA, 1997. The tactics of mutual mate choice and competitive search. Behav Ecol Sociobiol 40:51–59.

Krebs JR, Erichsen JT, Webber MI, Charnov EL, 1977. Optimal prey selection in the great tit (*Parus major*). Anim Behav 25:30–38.

Luttberg B, 1996. A comparative Bayes tactic for mate assessment and choice. Behav Ecol 7:451–460.

MacArthur RH, Pianka ER, 1966. On optimal use of a patchy environment. Am Nat 100:603–609.

Maynard Smith J, 1978. The evolution of sex. Cambridge: Cambridge University Press.

Mazalov V, Perrin N, Dombrovsky Y, 1996. Adaptive search and information updating in sequential mate choice. Am Nat 148:123–137.

McLennan DA, McPhail JD, 1990. Experimental investigations of the evolutionary significance of sexually dimorphic nuptial colouration in *Gasterosteus aculeatus* (L.): the relationsh between male colour and female behaviour. Can J Zool 68:482–492.

McNamara JM, Collins EJ, 1990. The job search problem as an employer-candidate game. J Appl Prob 28:815–827.

Milinski M, 1977a. Do all members of a swarm suffer the same predation? Z Tierpsychol 45:373–388.

Milinski M, 1977b. Experiments on the selection by predators against spatial oddity of their prey. Z Tierpsychol 43:311–325.

Milinski M, 1979. An evolutionarily stable feeding strategy in sticklebacks. Z Tierpsychol 51:36–40.

Milinski M, 1984a. Competitive resource sharing: an experimental test of a learning rule for ESSs. Anim Behav 32:233–242.

Milinski M, 1984b. A predator's costs of overcoming the confusion-effect of swarming prey. Anim Behav 32:1157–1162.

Milinski M, 1987. TIT FOR TAT in sticklebacks and the evolution of cooperation. Nature 325:433–435.

Milinski M, 1993. Predation risk and feeding behaviour. In: The behaviour of teleost fishes, 2nd ed. (Pitcher TJ, ed). London: Chapman & Hall; 285–305.

Milinski M, 1994. Long-term memory for food patches and implications for ideal free distributions in sticklebacks. Ecology 75:1150–1156.

Milinski M, 1997. How to avoid seven deadly sins in the study of behavior. Adv Study Behav 26:159–180.

Milinski M, Bakker TCM, 1990. Female sticklebacks use male coloration in mate choice and hence avoid parasitized males. Nature 344:330–333.

Milinski M, Bakker TCM, 1992. Costs influence sequential mate choice in sticklebacks, *Gasterosteus aculeatus*. Proc R Soc Lond B 250:229–233.

Milinski M, Heller R, 1978. Influence of a predator on the optimal foraging behaviour of sticklebacks (*Gasterosteus aculeatus* L.). Nature 275:642–644.

Milinski M, Külling D, Kettler R, 1990. TIT FOR TAT: sticklebacks "trusting" a cooperating partner. Behav Ecol 1:7–11.

Milinski M, Lühti JH, Eggler R, Parker GA 1997. Cooperation under predation risk: experiments on costs and benefits. Proc R Soc Lond B 264: 831–837.

Milinski M, Parker GA, 1991. Competition for resources. In: Behavioural ecology: an evolutionary approach, 3rd ed. (Krebs JR, Davies NB, eds). Oxford: Blackwell Scientific Publications; 137–168.

Møller AP, Biard C, Blount JD, Houston DC, Ninni P, Saino N, Surai PF, 2000. Carotenoid-dependent signals: indicators of foraging efficiency, immunocompetence or detoxification ability? Poultry Avian Biol Rev 11:137–159.

Neill SRStJ, Cullen JM, 1974. Experiments on whether schooling by their prey affects the hunting success of cephalopod and fish predators. J Zool Lond 172:549–569.

Ohguchi O, 1981. Prey density and selection against oddity by three-spined sticklebacks. Adv Ethol 23:1–79.

Olson VA, Owens IPF, 1998. Costly sexual signals: are carotenoids rare, risky or required? Trends Ecol Evol 13:510–514.

Parker GA, 1970. The reproductive behaviour and the nature of sexual selection in *Scatophaga stercoraria* L. II. The fertilization rate and the spatial and temporal relationships of each sex around the site of mating and oviposition. J Anim Ecol 39:205–228.

Parker GA, 1983. Mate quality and mating decisions. In: Mate choice (Bateson P, ed). Cambridge: Cambridge University Press; 141–166.

Parker GA, MacNair MR, 1979. Models of parent-offspring conflict: II. Suppression: evolutionary retaliation by the parent. Anim Behav 27:1210–1235.

Pelkwijk JJter, Tinbergen N, 1937. Eine reizbiologische Analyse einiger Verhaltensweisen von Gasterosteus aculeatus L. Z Tierpsychol 1:193–200.

Perrin N, Dombrovsky Y, 1994. On adaptive search and optimal stopping in sequential mate choice. Am Nat 144:355–361.

Pitcher TJ, Parrish JK, 1993. Functions of shoaling behaviour in teleosts. In: Behaviour of teleost fishes, 2nd ed. (Pitcher TJ, ed). London: Chapman & Hall; 363–439.

Pominiankowsi A, 1987. The costs of choice in sexual selection. J Theor Biol 128:195–218.

Real LA, 1990. Search theory and mate choice. I. Models of single-sex discrimination. Am Nat 136:376–405.

Slagsvold T, Lifjeld T, Stenmark G, Breiehagen T, 1988. On the cost of searching for a mate in female pied flycatchers Ficedula hypoleuca. Anim Behav 36:433–442.

Stearns SC (ed), 1987. The evolution of sex and its consequences. Basel: Birkhäuser.

Trivers RL, 1971. The evolution of reciprocal altruism. Q Rev Biol 46:35–57.

Trivers RL, 1974. Parent-offspring conflict. Amer Zool 14:249–265.

Werner EE, Hall DJ, 1974. Optimal foraging and the size selection of prey by the bluegill sunfish Lepomis machrochirus. Ecology 55:1042–1052.

Wiegmann DD, Real LA, Capone TA, Ellner S, 1996. Some distinguishing features of models of search behavior and mate choice. Am Nat 147:188–204.

Wittenberger JF, 1983. Tactics of mate choice. In: Mate choice (Bateson P, ed). Cambridge: Cambridge University Press; 433–447.

Wootton RJ, 1976. The biology of the sticklebacks. London: Academic Press.

Wootton RJ, 1984. A functional biology of sticklebacks. London: Croom Helm.

Zahavi A, 1975. Mate selection: a selection for a handicap. J Theor Biol 53:205–214.

# PART III       Bird Model Systems

# 14 Conversing with a Bird: Studies of Mating and Parental Behavior in Red-Winged Blackbirds

David F. Westneat

I began studying red-winged blackbirds (*Agelaius phoeniceus*) in order to understand what forces contribute to who mates with whom and how patterns of matings affect variation in parental behavior. My approach to both issues has been influenced by an interaction among the conceptual "wisdom" of my field at the time, my own biases and preconceptions, and what the birds did. This interaction has, in turn, led to new theory. The process could be likened to a long and sometimes frustrating conversation. My initial hypotheses were first formed from reading literature or discussing ideas with colleagues. I was then "told" by the birds that many of those ideas were at best only partially right. In struggling to interpret what the birds actually did, new theory has resulted, which in turn has suggested new questions to ask the birds. This interchange has gone beyond simply testing a hypothesis and getting a "yes" or "no" answer. It also has produced more inclusive and complex conceptual frameworks for addressing both of my original aims.

I have been both frustrated and rewarded by how my conversation with redwings has unfolded. The story I tell here is, of necessity, a personal one, with its own unique quirks. I suggest, however, that emerging out of the details of both my thought processes and what red-winged blackbirds do, comes some general lessons about modern behavioral ecology. Indeed, the value of multiple approaches, the central theme of this book, is one of these lessons. I found that learning new approaches in order to understand redwings better has been the most rewarding aspect of my studies of them.

I color-banded my first male red-winged blackbird, male "RGRX," on 30 March 1988. But, the study really began much earlier, through the development of the ideas and a survey of alternative study subjects. The ideas emerged slowly, starting many years earlier as part of my thesis research on indigo buntings (*Passerina cyanea*). I began the bunting project naively intending to test whether or not paternity affected paternal care. All I knew at that point was that paternal care in indigo buntings was variable (a few males in any population of this species feed fledglings, most do not), and that some researchers (e.g., Alexander 1974; Ridley 1978) had suggested that paternity ought to affect paternal care. The later seemed pretty obvious to me—paternity reduced the benefits of paternal care, so should not cuck-olded males reduce their care?

Haven Wiley, my Ph.D. advisor at the University of North Carolina, sug-gested to me that I would need to determine paternity and urged me to learn protein electrophoresis. And so I entered my first field season on the bun-tings with three goals: (1) to observe matings so I could estimate paternity, (2) to determine if muscle biopsies could be done on breeding buntings without harm and if so, to collect samples for protein analysis of paternity from as many families as possible, and (3) to observe paternal care at sam-pled nests to test if it varied with paternity. In the end, the results from goal 2 were more stunning than those from 1 or 3. The biopsies did work (West-neat et al. 1986) and so protein electrophoresis could be done. The results were astounding; about 15 percent of the nestlings had genotypes incompat-ible with the combination of parents (Westneat 1987a). Given that allele frequencies at the loci I was using could only detect 40 percent of any real extra-pair fertilizations (EPFs), this meant that over 35 percent of the off-spring must have come from EPFs (Westneat et al. 1987).

This level of extra-pair paternity certainly surprised me, raised many fas-cinating new questions about mating patterns, and is the most cited finding from my studies of buntings. However, it became less interesting to me than the results I obtained from goals 1 and 3 of my study. The behavioral obser-vations of matings were frustrating, as females could be seen for less than 40 percent of the time I was trying to watch them. When I could see what they were doing, there was no evidence that females initiated or otherwise en-couraged matings with males other than their pair-bonded mate (Westneat 1987b). Important questions about what rampant extra-pair matings might mean for our understanding of female reproductive behavior could not be addressed very effectively if females were difficult to watch and males often approached them aggressively when they could be seen.

Testing ideas about paternity and paternal behavior on the buntings was also frustrating. Male indigo buntings provide very little direct parental care to nestlings (only about 10 percent of males feed nestlings; 30 percent feed fledglings; Westneat 1988). Male defense of the nest is more common, yet it is highly variable from sample to sample within the same male, which makes for low power of tests for correlations with other variables. Finally, the pro-tein electrophoresis, with its power of detection of only 40 percent, undoubt-edly left undetected many of the offspring that were actually from EPFs. This meant that paternity measures for particular males were inaccurate. Thus it was disappointing, but not surprising, that I found no association between paternity and male provisioning. Young male indigo buntings in Michigan provided less care than older males, and were more likely to have reduced paternity as well (Westneat 1988). That interesting coincidence had an influence on my thoughts about the theory relating paternal care to pater-nity (see below). However, because alternative hypotheses for the result ex-isted (such as independent effects of experience on both paternity and paternal care), at the time I felt that I had not learned very much from the buntings about the relationship between paternity and paternal care.

In the fall of 1987, I started a post-doc at Cornell University with the primary aim of learning DNA techniques (e.g., Jeffreys et al. 1985), which had the promise of detecting all cases of extra-pair fertilizations. I also wanted to find a system in which I could observe how females behaved during encounters with extra-pair males and could measure paternal care more effectively than in the buntings. After considering several species and listening to advice from several colleagues at Cornell, I chose to work with red-winged blackbirds. Redwings are easy to see and to catch, males in eastern populations provide more food to nestlings than do male buntings (Searcy & Yasukawa 1995), and the circumstances were right—a population was present at a facility at Cornell not far from where I lived. Redwings are also socially polygynous (e.g., Searcy & Yasukawa 1995), and I was intrigued by the possibility of studying the interaction between polygyny and extra-pair matings. By the spring of 1988, I had learned how to use DNA fingerprinting to analyze some old indigo bunting samples (Westneat 1990), and so I was ready to study the blackbirds and reattack those two issues that had so frustrated me before.

## Sexual Conflict over Mating in Redwings

How should male and female red-winged blackbirds interact over copulation? Trivers (1972) predicted that males would be under selection to pursue a mixed mating strategy (pair with one female, and then attempt promiscuous matings with additional females) and to prevent other males from doing the same by guarding social mates. But how should females behave during matings, particularly extra-pair matings? Trivers did not make any direct predictions, but he did discuss an array of potential advantages for females of having particular males as mating partners. However, at the time I started with red-winged blackbirds, the data from birds on female behavior leading up to and during copulation was confusing. Most of the early studies of EPCs in birds had focused on waterfowl or colonial seabirds. Conspicuously aggressive extra-pair copulations initiated by males were the norm in ducks (e.g., McKinney et al. 1983) and some seabirds (e.g., *Uria aalge*, Birkhead et al. 1985). Males initiated most EPCs in white ibis (*Eudocimus albus*), but females mixed cooperation with resistance (Frederick 1987). In herring gulls (*Larus argentatus*), females successfully resisted all EPC attempts, typically initiated by males (Fitch & Shugart 1984). Hatch (1985) reported a few instances of females initiating EPCs in fulmars. My own observations of indigo buntings revealed only male-initiated and female-resisted EPCs. However, Susan Smith had found that female black-capped chickadees (*Parus atricapillus*) sometimes left their mate's territory, flew to a nearby male, and initiated a copulation (Smith 1988). Her sightings of this behavior were all at first light, emphasizing the possibility that females might have an active role in EPCs but that it might be difficult to see them behaving that way.

During the year leading up to starting the redwing project I began discussing female involvement in EPCs with Paul Sherman, a faculty member at Cornell. This discussion culminated with a paper on the causes of EPCs in birds (Westneat et al. 1990). In theory, the net benefits of EPCs to either sex could vary within or between populations. The types of benefits that could be obtained by females have been the subject of many subsequent reviews (e.g., Birkhead & Møller 1992; Petrie & Kempenaers 1997). However, for the redwings, I initially focused on what females did leading up to and during extra-pair encounters. Benefits might influence selection on these behaviors, but so too might the potential costs, and many of both of these could be difficult to measure. Heading into the redwing project, I was not sure what to expect from female redwings. On the one hand, females appeared to make decisions about with whom they paired based on habitat quality and not male quality, which are not correlated (Searcy & Yasukawa 1989; Beletsky & Orians 1996). Thus females might benefit from settling in one spot but copulating with another male. On the other hand, all territorial males are a biased set of the male population; they are older and capable of seizing an open territory and defending against neighbors and floaters. Hence the difference in quality between neighboring territory owners might be slight, reducing the benefits to females of initiating EPCs. Furthermore, males are half again as large as females. Resistance to an aggressive male approach might be costly to females, thereby potentially favoring accepting an EPC despite it being of no benefit or perhaps even costly. Some have argued that costs of EPCs to females should result in no EPFs, because in most birds, males lack an intromittent organ and therefore female cooperation is required for insemination (e.g., Fitch & Shugart 1984; Birkhead 1998). Redwings seemed like a species in which testing some of these overlapping and partially conflicting ideas might be feasible.

At the start of the study, I did not know for sure that redwings engaged in EPCs, but previous work indicating that mates of vasectomized males still produced fertile eggs (Bray et al. 1975; Roberts & Kennelly 1980), or reporting a few cases of EPC (e.g., Monnet et al. 1984) suggested that they did. Besides using DNA techniques to examine paternity, I also recorded focal female time budgets (Altmann 1974), which simultaneously gave me as much or more information about male time budgets because males were conspicuous and both sexes typically ranged over small areas.

## Red-Winged Blackbird Mating Interactions

The focal female observations I conducted over several seasons produced several important insights into female reproductive behavior. One was that females often did some boring stuff; during their fertilizable period (the several days leading up to egg-laying) they spent 95+ percent of their time foraging. This made for some long days of watching females forage, but perseverance had some rewards; extra-pair events did occur on occasion. Surprisingly, EP events were always male-initiated (in all 57 cases of extra-pair courtship, the extra-pair male approached the female first; Westneat 1992). Because of Smith's (1988) results on chickadees, I and my assistants conducted observations at all times of day (between 0500 and 2000), and never witnessed a female initiating an EPC in any way similar to what Smith had observed or Gray (1996) later found in a western population of redwings. However, this does not mean females never initiated EPCs. I calculated that if half of the approximately 40 females I observed each season initiated a single EPC during a fertilizable period, the probability of seeing even a single case in an entire season was about 0.5, given the average of 5 hours of observation per female out of the approximately 115 hours of daylight during a fertilizable period (days $-5$ to $+2$). Thus, while my observations suggest females are not often pursuing EPCs, they cannot eliminate the possibility that some do it infrequently.

Meanwhile, DNA fingerprinting revealed that 25 to 30 percent of the nestlings were extra-pair fertilizations (Westneat 1993a). Even if I was missing some female-initiated EPCs, I knew I was not missing enough to account for all of these EPFs. Some of the male-initiated EPCs we witnessed had to be inseminating the female and fertilizing the eggs. This conclusion was supported indirectly by data on paternity within territories in which more than one female was nesting. Paternity was unaffected by the presence of two fertilizable females at the same time (Westneat 1993a). However, when a female was fertilizable at the same time another female was settling, a different pattern emerged. When a new female is prospecting for a place to nest, a series of interesting behavioral interactions occur. Female redwings are aggressive toward each other (Searcy & Yasukawa 1995), and this is especially frequent when a new female first arrives. The male conspicuously intervenes in these interactions, and does so by guarding the new arrival and chasing the established female away from her. The result is the established female is herded to the margins of the territory and is subjected to more extra-pair chases as well (Westneat 1993a). Not surprisingly, the frequency of extra-pair fertilizations tends to be considerably higher than normal in the broods of such females (Westneat 1993a). These dynamics on polygynous territories suggested that male behavior greatly influences with whom a female copulates.

Male mate-guarding behavior therefore became very interesting to me. Males might guard mates for a variety of reasons (Birkhead & Møller 1992), but mate-guarding might increase paternity by either preventing other males from getting access to females, or by preventing females from meeting up with other males (Lifjeld et al. 1994). I was interested in experimenting with male mate-guarding to distinguish between those ideas and to explore the ways ecology might influence paternity via its effects on mate-guarding. Focal observations revealed that male-initiated EPCs were predominantly by nearby territory holders and occurred more often when the female's mate had left his territory (Westneat 1993b). Such absences declined during the period when females were most receptive (two days before egg-laying), but then increased again while the female was still fertilizable (during early egg-laying; Westneat 1993b). This pattern suggested that male presence on the territory prevented extra-pair events.

I thought the best way to test this idea was to remove the male temporarily. I knew, however, that removing a male would quickly lead to floater males attempting to claim that territory (reviewed by Beletsky & Orians 1996). Males leave naturally for a few minutes to as long as twenty minutes (rarely). So, I decided to capture the male and hold him for sixty minutes. This experiment resulted in a hundredfold increase in the rate of extra-pair events, and increased the frequency of EPFs in the broods on the removed male's territory if the male was removed just before egg-laying (Westneat 1994). All of the extra-pair events observed during these experiments were male-initiated, implying that mate-guarding prevented other males from gaining access to a female, and not that it prevented females from seeking out other males.

The experimental reduction in male mate-guarding simulated by short-term removal gives only part of the picture. A more powerful experiment would be to increase male mate-guarding and see an improvement in paternity. Not only would this further confirm the influence of male guarding on paternity, but it would also be a critical step in testing theory on conflicting demands. Optimization theory (e.g., Williams 1966; Trivers 1972; Krebs & McCleery 1984) predicts that trait values are at intermediate optima as a result of a balance of costs and benefits. Tests of such theory are more robust if they demonstrate: (1) that reducing the trait of interest reduces the benefits, (2) that reducing the costs of the trait increases the trait value, and (3) that an increase in trait value increases the benefits. The focal observations I had conducted on redwings provided a hint of how to alter the costs of mate-guarding. Many of the natural trips males made away from the female were into open fields nearby where we could see them foraging. This implied that males were faced with a conflicting demand; food supplies usable by the male may have been scarce on the territory where the fertilizable female resided. Perhaps increasing the food supply on a territory would cause males to leave less, thereby improving their ability to guard their mates and sire more offspring. In 1990 and 1991, I placed feeders containing corn, sunflower seeds, and mealworms out on a random draw of territories. The territory resident males stayed on their territories more, and their paternity was increased, over that on unsupplemented territories (Westneat 1994). Besides providing experimental evidence that ecology influenced paternity, this experiment confirmed the observational data on female behavior; altering male behavior altered paternity, whereas the experiment did not, as far as I could tell, alter female behavior.

Females thus did not seem to initiate EPCs, a somewhat disappointing result, because it seemed to limit the types of questions one could ask about female behavior. However, my conversation with the redwings about female mating tactics was not over. Slowly, after several seasons of regular focal observations and removal experiments, I began to realize that female redwings were telling me something much more interesting and ultimately more challenging.

Although I failed to see much variation in female initiation of EPCs, I did begin to observe a surprising range of female responses to male approaches (Table 14.1). Sometimes females reacted to extra-pair males by resisting (running or flying away from him and giving the normally aggressive "chatter" vocalization). At other times, females cooperated fully with the male by going into the characteristic premating posture of birds and allowing the male to complete a behaviorally complete copulation (his cloaca pressed to hers for two or more seconds). Such cooperation sometimes happened in response to the first approach I saw a male make during removal observations, sometimes it happened immediately after an aggressive chase of the female by an extra-pair male, and sometimes it happened after a series of approaches in which the female gradually resisted less strenuously. Occasionally, females would resist one extra-pair male but not another. I also began to see some unusual sequences. Females sometimes would give the precopulatory posture, allow the male to approach within a few inches, and then turn around suddenly and peck at the male, preventing the copulation. This type of behavior also occurred between social mates.

As I built up a series of these anecdotal sequences, many new questions began to emerge. Were females sometimes manipulating which males were sires through differential resistance? I had found that broods containing EPFs fledged more young than those without EPFs, suggesting a benefit of EPFs to females (Westneat 1992). However, the mechanisms by which this occurred remain unclear. Furthermore, much of what females did during EP events is consistent with the possibility that females assess the dangers of male aggressiveness and resist only when the costs of doing so appear to be low. Or, perhaps females were both manipulating and resisting, depending on the situation.

The data suggest that some new theory on mating interactions is needed. The interplay between males and females in mating is an ideal subject for game theory models, and indeed many have been constructed to examine other types of mating games (e.g., Lazarus 1990; Johnstone 1997). However, the interactions involved in extra-pair matings have some unique features. First, they occur in sequences: males and females must first encounter one another, then the encounter must lead to a copulation, and the copulation to a fertilization. Both players can do things at each step to influence the proba-

Table 14.1

Sequences of Male-Female Behaviors During Male-initiated Within-Pair and Extra-Pair Interactions

| Sequence of Events* | | | | | | Within-Pair N | Extra-Pair N |
|---|---|---|---|---|---|---|---|
| Male | Female | Male | Female | Male | Female | N | N |
| Approach | Vocalize | Ignore | | | | 9 | 20 |
| | | Court | Ignore | Ignore | | 0 | 1 |
| | | Attack | Resist | Chase | | 2 | 1 |
| | Resist | Ignore | | | | 1 | 11 |
| | | Chase | | | | 21 | 17 |
| | Crouch | Ignore | | | | 4 | 3 |
| | | Court | Ignore | Ignore | | 2 | 0 |
| | | | Resist | Ignore | | 2 | 0 |
| | | Mount | | | | 1 | 0 |
| | | | Resist | Ignore | | 1 | 0 |
| | | Interrupted | | | | 1 | 0 |
| Court | Ignore | Ignore | | | | 4 | 6 |
| | | Attack | Resist | Chase | | 1 | 0 |
| | Vocalize | Ignore | | | | 5 | 10 |
| | | Approach | Crouch | Mount | | 1 | 0 |
| | | | Resist | Ignore | | 0 | 1 |
| | Resist | Ignore | | | | 1 | 13 |
| | | Court | Resist | Ignore | | 0 | 1 |
| | | | Crouch | Mount | | 0 | 2 |
| | | | | Mount (x) | | 0 | 3 |
| | | | | Mount | Resist | 0 | 1 |
| | | | | Mount (x) | Resist | 0 | 6 |
| | | Mount | Resist | Ignore | | 0 | 1 |
| | | Chase | | | | 0 | 5 |
| | Crouch | Ignore | | | | 1 | 2 |
| | | Approach | Resist | Ignore | | 0 | 1 |
| | | Court | Resist | Ignore | | 0 | 4 |
| | | | | Chase | | 0 | 1 |
| | | Mount | | | | 2 | 3 |
| | | | Resist | Ignore | | 1 | 1 |
| | | Mount (x) | | | | 1 | 0 |
| | | | Resist | Ignore | | 3 | 0 |
| Attack | Vocalize | Mount | Resist | Chase | | 1 | 0 |
| | Resist | Ignore | | | | 4 | 3 |
| | | Chase | | | | 18 | 2 |
| Total Events | | | | | | 86 | 119 |

*Definitions of behaviors: Male: Approach = Movement of male toward female, Court = Wing flutter and head bowing display, Mount = Male hops on female's back, attempts to press cloaca to hers (x) = repeated mounts, Attack = Male flies quickly and directly at female and comes within 1 m; Female: Vocalize = Gives either a soft twitter or chatter call, Resist = Female moves away from male, raises bill, or pecks at male, Crouch = Female bends forward and raises tail; Both sexes: Ignore = No response to preceding event.

bility of reaching the next step in the sequence (Fig. 14.1). However, because the sequence must pass through each step, what happens at earlier steps influences the conditions under which later steps are played. For example, by avoiding males during the encounter step, those encounters that do occur may be with a biased set of males, so the optimal tactic in that circumstance might not be the same as in the case when a female does not attempt to avoid males. Figure 14.1 provides a conceptual outline of a possible model of this sequential game. I have not completed a full model, but I am convinced that it will lead to some novel ways of testing possible explanations for variable female behavior.

More generally, my conversation with red-winged blackbirds about mating dynamics has led me to a position on mating dynamics that is not widely emphasized by my colleagues. The majority of workers on extra-pair matings in birds are now focusing on the potential benefits females receive from engaging in such matings (e.g., Petrie & Kempenaers 1998; Møller 1998; Birkhead 1998). This focus is an exciting development, because it is drawing attention to factors influencing females, a much neglected topic until recently. However, I see two misleading assumptions being made. One is to assume that females are benefiting from EPCs if there are EPFs. My preliminary models of the concept of mating sequences illustrated in Figure 14.1 suggest that females might be favored to allow fertilization even if the mating interaction is costly. Hence any time there is a correlation between a male trait and EPF success (e.g., Graves et al. 1993; Sundberg & Dixon 1996; Hasselquist et al. 1996; Møller et al. 1997; Møller & Birkhead 1994), two hypotheses exist: (1) the male trait signals benefits that females might receive from mating with him, or (2) the male trait either directly, or indirectly via another variable, leads those males to be better than other males at gaining access to females or overcoming female reluctance to copulate. Uncertainty about whether (1) or (2) explains the correlation between male trait and EPFs exists for nearly every example of EPFs studied to date.

The data on mating in redwings also reveals an often ignored gap that exists for all studies of EPFs to date—we do not know what proportion of EPFs comes from a given type of extra-pair event. Thus even in populations for which there is some evidence of female pursuit (e.g., Kempanears et al. 1992; Gray 1996), we do not know that all EPFs (let alone what proportion) are gained via female pursuit. The existence of variation in male and female behavior within populations is incompatible with simple ideas about the types of benefits that females might receive. DNA techniques alone will not help us understand this variation—they will have to be combined with careful, persistent observations. At one level, the existence of variation in female behavior is bringing some badly needed attention to hypotheses about female behavior (Eberhard 1996). However, too much enthusiasm over the idea of female control could also lead to the glossing over of many challenging and intriguing questions about the interaction between male and female, which

## SEQUENCE OF EXTRA-PAIR EVENTS

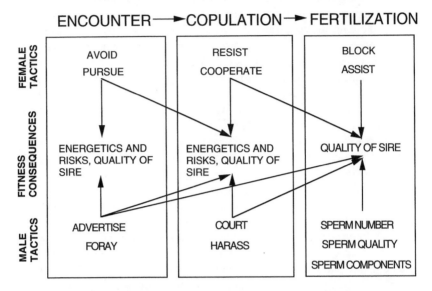

Figure 14.1. Conceptual model of some possible interactions between males and females over a mating sequence. Male or female tactics during each step (encounter, copulation, or fertilization) will influence the probability of reaching the next step. The tactics of both sexes will influence the fitness consequences to females of both events within a stage and those at subsequent stages (single-headed arrows).

necessarily occurs in every mating event. Eberhard (1997) used an appropriate analogy for this, but perhaps applied it incompletely. He argued that many aspects of female behavior and morphology could determine the topography of the "playing field" on which processes like sperm competition are played out. The data from redwings suggests to me that a more appropriate use of that analogy is that both males and females influence the topography of this "playing field" in sometimes indirect and complex ways. A key challenge currently facing workers in this area is to devise conceptual approaches to mating dynamics that fully account for this dual influence of the two sexes.

## Does Paternity Affect Paternal Care?

My tortuous conversation with red-winged blackbirds was not confined to mating behavior. My studies of paternity and paternal behavior followed a similar path of mixing theory with empirical studies. Male red-winged blackbirds in eastern populations often provision nestlings (e.g., Muldal et al. 1986). After being frustrated in my attempts to understand how paternity might affect paternal care in indigo buntings, I was eager to try anew with the blackbirds. My first goal was to collect correlational data on patterns of male provisioning from behavioral observations and paternity data from DNA fingerprinting in the first two seasons (1988 and 1989). The next two seasons (1990 and 1991) were devoted to the mate-guarding experiments (above). Although the male removals were not designed for testing effects on paternity, at the time I reasoned they might provide some useful experimental confirmation of the correlational data.

Simultaneously, Paul Sherman and I began to discuss just what the consequences of paternity ought to be. Reductions in paternity clearly diminish the benefits of care (e.g., Ridley 1978; Houston & Davies 1985; Winkler 1987). However, some models of paternal care found that paternity either did not reduce paternal care (e.g., Maynard Smith 1978; Grafen 1980) or was highly confounded with other variables (Werren et al. 1980). These models assumed that males had no means to assess variation in paternity among individual nests, so mean paternity was the relevant variable, and it did not vary among types of nests.

My own experience with measuring paternity and observing the behaviors that influenced it indicated to me that paternity did vary between nesting attempts, possibly in systematic ways, and that there were likely to be events that might serve as cues to males of their paternity. Paternity of male indigo buntings was highly variable between nesting attempts. Part of this variation was due to differences in the age of the male; males in their first breeding season sometimes had lower paternity than older males (Westneat 1988). Furthermore, because intrusions by neighboring males seemed to be the main route by which EPCs occurred in both buntings and redwings, I reasoned that males might be able to use the frequency of interactions with intruders and/or how those interactions ended to assess the likelihood EPCs had occurred. However, at the time I had no means to test these ideas on the buntings. Moreover, as hunches, they needed to be formalized to confirm that I was thinking clearly about the relationships between the variables involved.

Paul Sherman and I therefore began fiddling around with graphical analyses of life history models that explored the consequences of trade-offs among types of effort, including parental effort (e.g., Williams 1966; Gadgil & Bossert 1970; Sargent 1985). Paternity is a parameter in life history models, and so by speculating on the patterns of paternity and the ways males might assess their situation, we developed scenarios describing how paternity might influence paternal care (Westneat & Sherman 1993). One of these focused on the fact that paternity itself could change across the lifetime of males, and so could influence life history patterns of effort in a manner similar to the effects of aging, or senescence (Williams 1966; Schaffer 1974; Clutton-Brock 1984). That is, senescence alters the costs to parental effort (older males have a dimmer future regardless of their current level of effort, so the cost of their current parental effort is less). Changes in paternity with age also could change the costs of present effort: if paternity in future broods is likely to be low, then there is little cost to providing more parental effort now. Patterns of paternity with age thus could produce selection on life history patterns of parental care. Generally similar effects could also be generated if there were seasonal patterns of paternity.

Even if paternity differed from one nesting attempt to another in non-systematic ways, if males could assess their paternity in the current brood, they could adjust paternal effort assuming some different level of paternity in the future. In this case, theory confirms the idea that paternity might affect paternal behavior, but specifies that there must be some reliable cue of paternity that males can use. Polyandrous male dunnocks (*Prunella modularis*) provide an ideal example of the use of paternity cues; male provisioning of nestlings varies in proportion to male access to the female during the laying period (Davies 1992). Davies (1992) confirmed this with an experimental manipulation of mating access and the perceived timing of egg-laying.

Another consequence of incorporating paternity into life history models was that its effects depended directly on the relative shapes of curves relating parental effort to costs and benefits. I had been suspicious of the popular idea that if males provided valuable parental care benefiting females, then by engaging in EPCs, a female might risk losing such care (e.g., Birkhead & Møller 1992). However, male care is valuable because it dramatically improves offspring survival. Hence, it is also valuable to a male as long as paternity is not zero. If such care also had relatively little cost, then reduced paternity would have less effect than when males hardly benefit from parental care in the first place.

This idea was confirmed by Whittingham et al. (1992) and explored in detail by Houston (1995) in more formal models. These models showed explicitly that the shapes of the benefit and cost curves had major effects on the decline in paternal effort with declines in paternity. In some cases, major declines in paternity had only slight effects on paternal effort. In others, slight declines in paternity had a major impact on the expected decline in paternal effort.

These advances in theory have been well summarized elsewhere (Wright 1998; Whittingham & Dunn, in press). The complexity revealed by these models really began sinking in as I was preparing to analyze the data I had collected from the blackbirds (Westneat 1995).

## Paternity and Paternal Care in Blackbirds

Male red-winged blackbirds would seem to be a species in which reductions in paternity would have measurable effects on paternal care. Males in eastern populations provision nestlings in many nests, sometimes providing a substantial proportion of the total feeds. However, this care is variable, has some but not a huge effect on nestling survival, and is sensitive to variation in costs (Muldal et al. 1986; Yasukawa et al. 1990). Although the exact shape of the cost and benefit curves has not been estimated, the benefit curve does not appear to increase steeply compared to the cost curve. Paternity varies between 0 and 100 percent among nests (Gibbs et al. 1990; Westneat 1993a, b; Weatherhead & Boag 1995; Gray 1996).

Despite seemingly fitting a case where paternity should have an effect, provisioning behavior (measured on a standardized scale relative to other males) was not correlated with paternity (Westneat 1995). This held in bivariate tests among nests, in bivariate tests of total care and total paternity in all nests for a male within a season, as well as in multivariate analyses in which other potentially confounding factors (e.g., brood size, nest order, harem size, female age, settlement date, and mating activity) were included. No patterns of paternity or of paternal care with male age were evident. In addition, there were no effects of male removal during the fertilizable period on male parental care (Westneat 1995). In sum, paternity seemed to have no effect on paternal behavior in this population.

These correlative analyses provide an initial look at the possible effects of paternity on paternal care. They are, however, subject to many limitations, as has been pointed out by Kempenaers and Sheldon (1997). In lieu of experimental manipulations of paternity (which also must be interpreted with caution; Kempenaers & Sheldon 1997; Wright 1998), matched analyses of male responses to two broods differing in paternity provide better control for possible confounding factors (Dixon et al. 1994; Weatherhead et al. 1994).

Female red-winged blackbirds rarely produce two broods in a season, so any matched comparison in this species must be between social mates of polygynous males. Figure 14.2 plots the differences between these nests in paternity and in the relative level of male provisioning effort. Relative provisioning by males to each nest was not associated with relative paternity (Fig. 14.2).

Why do males with different levels of paternity between two simultaneous nests not provision them differently? One possibility is that the shapes of the benefit and cost curves are such that reduced paternity influences the optimal level of care only slightly (e.g., Fig. 14.3a). Other sources of variation thus could easily mask the difference in paternal behavior that should result from the differences in paternity.

However, such curves seem unlikely for red-winged blackbirds. No one has actually measured cost and benefit curves for paternal care, but, as mentioned above, the data we have suggest that the benefits of care are slight, and that the costs can, on occasion, be relatively large (e.g., Muldal et al. 1986). Figure 14.3b would seem to be the more likely scenario for male redwings, and those curves predict a larger relative effect of paternity on paternal care.

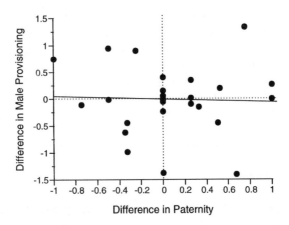

Figure 14.2. Plot of the difference in paternity (proportion of chicks sired by the paired male) between two red-winged blackbird nests found on the same territory (x-axis) by the difference in the standardized score of male provisioning (see Westneat 1995) to those two nests (y-axis). The relationship is not significant (R = 0.05, t = $-0.26$, P = 0.80, df = 26).

It is important to remember that the curves in Figure 14.3 depict only the potential selection pressures acting on males. In order to appropriately allocate parental care in regard to paternity, males must have a good estimate of their paternity in each nest. Thus, another possible explanation for the lack of response to paternity in redwings is that males are not able to assess their paternity. Interestingly, few behavioral events tied to EPC attempts in redwings actually correlate with paternity (Westneat 1993b). One possible cue is the number of times the male copulated with the female; the correlation between that and paternity is significantly greater than zero (Tau = 0.22, P < 0.035; Westneat 1993b). It, however, explains less than 5 percent of the variance in paternity. This low explanatory value might occur because the human observers collecting data on each male can only sample a portion of that bird's day, whereas the bird might know exactly how many times he copulated with each female. Nevertheless, it is possible that copulation rate is still a relatively poor predictor of paternity, because fertilization of any one egg is achieved by only one sperm on one day, and paternity depends as well on how often and when the female copulated with another male (Birkhead 1998). Without good cues of paternity, males might have no basis by which to adjust their paternal care in relation to their paternity.

In contrast to the Ithaca males, males in an Ontario population of red-winged blackbirds do respond to paternity by reducing their nest defense (Weatherhead et al. 1994). No one knows yet if the cost and benefit curves for nest defense are such that paternity reductions have greater effects than for provisioning behavior or if the situation in Ontario makes for better cues

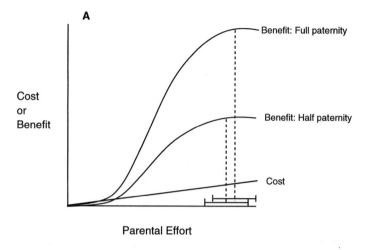

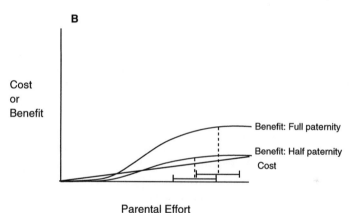

Figure 14.3. These curves illustrate how cost and benefit might influence the effect of paternity on paternal care. A. Steep benefit curves result from a high value of male parental care. If such care also has a shallow cost curve, then reducing paternity to one-half has a small effect on optimal parental effort. B. Shallow benefit curves result from increases in male care having minimal effect on the survival of young. Reduction in paternity by one-half results in reduction of optimal paternal effort. In both (A) and (B), horizontal error bars depict standard deviations in male parental behavior. A study would be more likely to detect the effect of paternity if the curves were like (B) than if they were like (A) (after Westneat & Sherman 1993).

of paternity. These uncertainties aside, the combination of the empirical re-sults from both populations and the insights gained by the models provides the context for a variety of more detailed comparative studies between popu-lations. Studies that attempt to identify the cues of paternity, and the effects of paternity on either nest defense or provisioning, could lead to interesting parallel experiments manipulating those cues and testing for appropriate responses.

## Summary

Red-winged blackbirds did not match my original expectations. Hence those ideas had to be reexamined and my assumptions rethought. In the end, the challenge of confronting a complex system was more rewarding to me, and I believe more instructive to the field, than would have been the case if the birds had conformed to those simple early ideas. The data obtained from eight years of study on the Ithaca red-winged blackbirds do not provide a simple picture of either the interactions between males and females over mating or the ways paternity influences paternal care. Many new questions have been stimulated by these results and those from other studies of reproductive behavior in birds. Most of these have to do with understanding a higher level of complexity in behavior than I originally imagined.

Indeed, it is this complexity of the redwing system that provides several lessons about doing modern behavioral ecology. A powerful message is that there are tremendous payoffs if a researcher's curiosity drives a "do-what-it takes" attitude about the study organism. Progress in the redwing system came from some unusual and unexpected directions, ones that forced me to learn new skills. An obvious example was that I needed to learn something about molecular techniques in order to study mating patterns more effectively. Today, using DNA techniques is almost routine, but for me it was a daunting undertaking. While I enjoy doing fieldwork on whole organisms in "natural" environments, eventually the long hours working with things I could not see (DNA molecules in small tubes) at a lab bench were worth it because the patterns on the gels held the key to knowing what had happened in the field. I feel the same way about modeling. By learning something about marginal value models, I eventually came to understand the complexity of how paternity might affect paternal care a bit better, and this has had a big influence on my ideas for doing appropriate field studies of paternal behavior. I am convinced as well of the potential utility of two other types of models: game theory models of sequences of male-female interactions and quantitative genetics models incorporating social selection (e.g., Wolf et al., in press) and indirect genetic effects (e.g., Moore et al. 1997). Game theory and quantitative genetic models of social selection both help define the influence of one individual on the fitness of others. Indirect genetic effects models portray the effects of a trait in one individual on the phenotype of another, which seems an especially valuable approach for describing the evolution of sequences of interactions. It seems possible these approaches will help us to devise a better framework for understanding the incredible diversity of mating patterns that now is apparent (Westneat 2000). The redwing system is therefore an excellent example of the theme of this book: that by combining into one research program conceptual ideas, empirical studies, and mathematical models, a richer set of results emerges.

My enthusiasm for this approach extends to the prospects for behavioral ecology in general. Behavioral ecology is, in my opinion, now the premier example of an integrative field, and it seems headed for even greater integration. Even just within the study of mating and parental behavior, interesting questions have arisen that demand an array of techniques and ideas from a diverse set of fields. For example:

1. If female redwings are indeed biasing matings toward certain males, what do they gain? One often-cited possibility is that males are somehow signaling underlying genetic quality that might benefit a female's offspring (Trivers 1972). A variety of hypotheses for what those "good" genes might be are stimulating investigations into the immune system from a behavioral-ecological perspective (e.g., Saino et al. 1997; Zuk et al. 1995), the physiology of carotenoid-dependent (e.g., Olson & Owens 1998) or UV reflecting plumage (e.g., Hunt et al. 1998; Johnsen et al. 1998), and learning and nutrition (e.g., Nowicki et al. 1998).

2. How might copulation by females with multiple males affect transmission of either detrimental or beneficial (Lombardo et al. 2000) microbes? Little is known about microbes in avian reproductive tracts (Sheldon 1993), but a couple of studies suggest that microbes could be transferred via copulation (Lombardo et al. 1996; Westneat & Rambo 2000). Further work in this area will require learning a great deal about techniques from microbiology.

3. How flexible is female mating behavior? Female preferences and the ways females bias mating interactions could be conditional (e.g., Jennions & Petrie 1997). The variation in redwing mating sequences certainly raises the possibility that female behavior is part of a conditional strategy based on learning. Females might be learning about males in a variety of ways, so future studies of mate choice might focus on patterns of learning. Social influences on mate preferences, such as forms of nonindependent mate choice (e.g., Pruett-Jones 1992), also draw attention to psychological approaches to understanding behavioral flexibility (e.g., Galef 1988; Heyes 1994; Zentall 1996). The evolutionary implications of such flexibility have been practically unstudied (but see Kirkpatrick & Dugatkin 1994), and yet there is quantitative genetic theory both on plasticity (e.g., Gomulkiewicz & Kirkpatrick 1992) and on social effects (indirect genetic effects; Moore et al. 1997) that might be used to model the evolutionary consequences of flexible preferences. Again, integration of these diverse perspectives seems likely to yield enormous benefits.

4. The most likely scenarios for paternity affecting paternal behavior require male assessment of relatedness. Mechanistic and behavioral-ecological studies of assessment and recognition are on the rise (e.g. Sherman et al. 1997), and might integrate sensory neurobiology with theory on assessment

(Reeve 1989). Studies of the hormonal basis of paternal behavior in birds are increasing (Ketterson & Nolan 1994). How sensory information might integrate with hormones to produce specific changes in paternal behavior with respect to paternity could be a fascinating area of study. Studies of dominance and aggression provide some interesting hints that there are interactions between sensory information and hormone levels (e.g., Archewaranon et al. 1991). Whether similar interactions might influence how parental males respond to variable paternity is unclear.

In sum, the study of mating and parental behavior is an integrative endeavor. Progress in this area now demands making use of all the tools available, be they conceptual, mathematical, or any of a rapidly expanding array of empirical methods. Our understanding of the redwing system has clearly benefited from combining approaches, and much more could yet be done. Although this means conversing with a study organism is becoming a bewildering and exhausting business, I think the insights that result will be more rewarding to us researchers and more relevant to a broader spectrum of biologists.

## Acknowledgments

My work on mating behavior has been supported by many people. My advisors, R. Haven Wiley, Chip F. Aquadro, and Paul W. Sherman, have provided a tremendous amount of their time and resources and have generously shared their wisdom. I am grateful to Bob Payne for providing me with my start on indigo buntings, and Nelson Hairston Jr. and Bob Johnson for helping me make the most of the blackbirds at Cornell University's Pond Units. Anne Clark has been an energetic and invaluable collaborator on the blackbirds. I wish to thank the dozens of field and lab assistants that have done much of the data collection on redwings. Finally, I am indebted to the National Science Foundation, the NSF-Kentucky EPSCoR program, and the University of Kentucky for their financial support of my research.

## References

Alexander RD, 1974. The evolution of social behavior. Ann Rev Ecol Syst 5:325–383.

Altmann J, 1974. Observational study of behavior: sampling methods. Behaviour 49:227–267.

Archewaranon M, Dove L, Wiley RH, 1991. Social inertia and hormonal control of aggression and dominance in white-throated sparrows. Behaviour 118:42–65.

Beletsky LD, Orians GH, 1996. Red-winged blackbirds: decision-making and reproductive success. Chicago: University of Chicago Press.

Birkhead TR, 1998. Sperm competition in birds: mechanisms and function. In: Sperm competition and sexual selection (Birkhead TR, Møller AP, eds). San Diego: Academic Press; 579–622.

Birkhead TR, Johnson SD, Nettleship DN, 1985. Extra-pair matings and mate guarding in the common murre *Uria aalge*. Anim Behav 33:608n–619.

Birkhead TR, Møller AP, 1992. Sperm competition in birds: evolutionary causes and consequences. London: Academic Press.

Bray OE, Kennelly JJ, Guarino JL, 1975. Fertility of eggs produced on territories of vasectomized red-winged blackbirds. Wilson Bull 87:187–195.

Clutton-Brock TH, 1984. Reproductive effort and terminal investment in iteroparous animals. Am Nat 123:212–229.

Davies NB, 1992. Dunnock behaviour and social evolution. Oxford: Oxford University Press.

Dixon A, Ross D, O'Malley SLC, Burke T, 1994. Paternal investment is inversely related to degree of extra-pair paternity in the reed bunting. Nature 371:698–700.

Eberhard WG, 1996. Female control: sexual selection by cryptic female choice. Princeton: Princeton University Press.

Eberhard WG, 1998. Female roles in sperm competition. In: Sperm competition and sexual selection (Birkhead, TR, Møller, AP, eds). San Diego: Academic Press; 91–116.

Fitch MA, Shugart GW, 1984. Requirements for a mixed reproductive strategy in avian species. Am Nat 124:116–126.

Frederick PC, 1987. Extrapair copulations in the mating system of white ibis (*Eudocimus albus*). Behaviour 100:170–201.

Gadgil M, Bossert WH, 1970. Life historical consequences of natural selection. Am Nat 104:1–24.

Galef BG, Jr, 1988. Imitation in animals: history, definition and interpretation of data from the psychological laboratory. In: Social learning: psychological and biological perspectives (Zentall TR, Galef BG, Jr, eds). Hillsdale, NJ: Erlbaum; 3–28.

Gibbs HL, Weatherhead PJ, Boag PT, White BN, Tabak LM, Hoysak DJ, 1990. Realized reproductive success of polygynous red-winged blackbirds revealed by DNA markers. Science 250:1394–1397.

Gomulkiewicz R, Kirkpatrick M, 1992. Quantitative genetics and the evolution of reaction norms. Evolution 46:390–411.

Grafen A, 1980. Opportunity cost, benefit and degree of relatedness. Anim Behav 28:967–968.

Graves J, Ortega-Ruano J, Slater PJB, 1993. Extra pair copulations and paternity in shags: do females choose better males? Proc Roy Soc Lond B 253:3–7.

Gray EM, 1996. Female control of offspring paternity in a western population of red-winged blackbirds (*Agelaius phoeniceus*). Behav Ecol Sociobiol 38:267–278.

Hasselquist D, Bensch S, von Schantz T, 1996. Correlation between male song repertoire, extra-pair fertilizations and offspring survival in the great reed warbler. Nature 381:229–232.

Hatch SA, 1985. Copulation and mate guarding in the northern fulmar. Auk 100:593–600.

Heyes CM, 1994. Social learning in animals: categories and mechanisms. Biol Rev 69:207–231.

Hoogland JL, Sherman PW, 1976. Advantages and disadvantages of bank swallow (*Riparia riparia*) coloniality. Ecol Mongr 46:33–58.

Houston AI, 1995. Parental effort and paternity. Anim Behav 50:1635–1644.

Houston AI, Davies NB, 1985. The evolution of cooperation and life history in the dunnock, *Prunella modularis*. In Behavioural ecology: the ecological consequences of adaptive behaviour (Sibly R, Smith R, eds). Oxford: Blackwell Scientific; 471–487.

Hunt S, Bennett ATD, Cuthill IC, Griffiths R, 1998. Blue tits are ultraviolet tits. Proc Roy Soc Lond B 265:451–455.

Hunter FM, Burke T, Watts SE, 1992. Frequent copulation as a method of paternity assurance in the northern fulmar. Anim Behav 44:149–156.

Jeffreys AJ, Wilson V, Thein SL, 1985. Hypervariable "minisatellite" regions in human DNA. Nature 314:67–73.

Jennions MD, Petrie M, 1997. Variation in mate choice and mating preferences: a review of causes and consequences. Biol Rev Camb Phil Soc 72:283–327.

Johnsen A, Andersson S, Örnborg J, Lifjeld JT, 1998. Ultraviolet plumage ornamentation affects social mate choice and sperm competition in bluethroats (Aves: *Luscinia s. svecica*): A field experiment. Proc Roy Soc Lond B 265:1313–1318.

Johnstone RA, 1997. The tactics of mutual mate choice and competitive search. Behav Ecol Sociobiol 40:51–59.

Kempenaers B, Sheldon BC, 1997. Studying paternity and paternal care: pitfalls and problems. Anim Behav 53:423–427.

Kempenaers B, Verheyen GR, de Broeck MV, Burke T, Broeckhoven CV, Dhondt AA, 1992. Extra-pair paternity results from female preference for high quality males in the blue tit. Nature 357:494–496.

Ketterson ED, Nolan V, Jr, 1994. Male parental behavior in birds. Ann Rev Ecol Syst 25:601–628.

Kirkpatrick M, Dugatkin LA, 1994. Sexual selection and the evolutionary effects of copying mate choice. Behav Ecol Sociobiol 34:443–449.

Krebs JR, McCleery RH, 1984. Optimization in behavioural ecology. In: Behavioural ecology: an evolutionary approach, 2nd ed (Krebs JR, Davies NB, eds), Sunderland, MA: Sinauer Associates; 91–121.

Lazarus J, 1990. The logic of mate desertion. Anim Behav 39:672–684.

Lifjeld JT, Dunn PO, Westneat DF, 1994. Sexual selection through sperm competition in birds: male-male competition or female choice? J Avian Biol 25:244–250.

Lombardo MP, Thorpe PA, Cichewicz R, Henshaw M, Millard C, Steen C, Zeller TK, 1996. Communities of cloacal bacteria in tree swallow families. Condor 98:167–172.

Lombardo MP, Thorpe PA, Power HW, 2000. The beneficial sexually transmitted microbe hypothesis of avian copulation. Behav Ecol 10:333–337.

Maynard Smith J, 1978. The evolution of sex. Cambridge: Cambridge University Press.

McKinney F, Cheng KM, Bruggers DJ, 1984. Sperm competition in apparently monogamous birds. In: Sperm competition and the evolution of animal mating systems (Smith RL, ed). Orlando: Academic Press; 523–545.

McKinney F, Derrickson SR, Mineau P, 1983. Forced copulation in waterfowl. Behaviour 86:250–294.

Møller AP, 1998. Sperm competition and sexual selection. In: Sperm competition and sexual selection (Birkhead TR, Møller AP, eds). San Diego: Academic Press; 55–90.

Møller AP, Birkhead TR, 1994. The evolution of plumage brightness in birds is related to extra-pair paternity. Evolution 48:1089–1100.

Møller AP, Saino N, Taramino G, Galeotti P, Ferrari S, 1997. Paternity and multiple signalling: effects of a secondary sexual character and song on paternity in the barn swallow. Am Nat 151:236–242.

Monnet C, Rotterman LM, Worlein C, Halupka K, 1984. Copulation patterns of red-winged blackbirds (*Agelaius phoeniceus*). Am Nat 124:757–764.

Moore AJ, Brodie ED, III, Wolf JB, 1997. Interacting phenotypes and the evolutionary process: I. Direct and indirect genetic effects of social interactions. Evolution 51:1352–1362.

Muldal AM, Moffatt JD, Robertson RJ, 1986. Parental care of nestlings by male red-winged blackbirds. Behav Ecol Sociobiol 19:105–114.

Nowicki S, Peters S, Podos J, 1998. Song learning, early nutrition and sexual selection in songbirds. Amer Zool 38:179–190.

Olson VA, Owens IPF, 1998. Costly sexual signals: are carotenoids rare, risky or required? Trends Ecol Evol 13:510–514.

Petrie M, Kempenaers B, 1998. Extra-pair paternity in birds: explaining variation between species and populations. Trends Ecol Evol 13:52–58.

Pruett-Jones S, 1992. Independent versus non independent mate choice: do females copy each other? Am Nat 140:1000–1009.

Reeve HK, 1989. The evolution of conspecific acceptance thresholds. Am Nat 133:407–435.

Ridley M, 1978. Paternal care. Anim Behav 26:904–932.

Roberts TA, Kennelly JJ, 1980. Variation in promiscuity among red-winged blackbirds. Wilson Bull 92:110–112.

Sargent RC, 1985. Territoriality and reproductive trade-offs in the threespine stickleback, *Gasterosteus aculeatus*. Behaviour 93:217–226.

Saino N, Bolzern AM, Møller AP, 1997. Immunocompetence, ornamentation, and viability of male barn swallows (*Hirundo rustica*). Proc Natn Acad Sci USA 94:549–552.

Schaffer WM, 1974. Selection for optimal life histories: the effects of age structure. Ecology 55:291–303.

Searcy WA, Yasukawa K, 1989. Alternative models of terrestrial polygyny in birds. Am Nat 134:323–343.

Searcy WA, Yasukawa K, 1995. Polygyny and sexual selection in red-winged blackbirds. Princeton: Princeton University Press.

Sheldon BC, 1993. Sexually transmitted disease in birds: occurrence and evolutionary significance. Phil Trans R Soc Lond B 339:491–497.

Sherman PW, Morton ML, 1988. Extra-pair fertilizations in mountain white-crowned sparrows. Behav Ecol Sociobiol 22:413–420.

Sherman PW, Reeve HK, Pfennig DW, 1997. Recognition systems. In: Behavioural ecology: an evolutionary approach, 4th ed. (Krebs JR, Davies NB, eds) Oxford: Blackwell Scientific; 69–96.

Smith, SM, 1988. Extra-pair copulations in black-capped chickadees: the role of the female. Behaviour 107:15–23.

Sundberg J, Dixon A, 1996. Old, colourful male yellowhammers, *Emberiza citrinella*, benefit from extra-pair copulations. Anim Behav 52:113–122.

Trivers RL, 1972. Parental investment and sexual selection. In: Sexual selection and the descent of man, 1871–1971 (Campbell B, ed). Chicago: Aldine; 136–179.

Weatherhead PJ, Boag PT, 1995. Pair and extra-pair mating success relative to male quality in red-winged blackbirds. Behav Ecol Sociobiol 37:81–91.

Weatherhead PJ, Montgomerie R, Gibbs HL, Boag PT, 1994.The cost of extra-pair fertilizations to female red-winged blackbirds. Proc Roy Soc Lond B 258:315–320.

Werren JH, Gross MR, Shine R, 1980. Paternity and the evolution of male parental care. J Theor Biol 82 619–631.

Westneat DF, 1987a. Extra-pair fertilizations in a predominantly monogamous bird: genetic evidence. Anim Behav 35:877–886.

Westneat DF, 1987b. Extra-pair copulations in a predominantly monogamous bird: observations of behaviour. Anim Behav 35:865–876.

Westneat DF, 1988. Parental care and extrapair copulations in the Indigo Bunting. Auk 105:149–160.

Westneat DF, 1990. Genetic parentage in Indigo Buntings: a study using DNA finger-printing. Behav Ecol Sociobiol 27:67–76.

Westneat DF, 1992. Do female red-winged blackbirds engage in a mixed mating strategy? Ethology 92:7–28.

Westneat DF, 1993a. Polygyny and extra-pair fertilizations in eastern red-winged blackbirds (*Agelaius phoeniceus*). Behav Ecol 4:49–60.

Westneat DF, 1993b. Temporal patterns of within-pair copulations, male mate guard-ing, and extra-pair events in the red-winged blackbird (*Agelaius phoeniceus*). Be-haviour 124:267–290.

Westneat DF, 1994. To guard mates or go forage: conflicting demands affect the paternity of male red-winged blackbirds. Am Nat 144:343–354.

Westneat DF, 1995. Paternity and paternal behaviour in the red-winged blackbird, *Agelaius phoeniceus*. Anim Behav 49:21–35.

Westneat DF, 2000. A retrospective and prospective look at the role of genetics in mating systems: toward a balanced view of the sexes. In: Vertebrate mating sys-tems (Appollonio M, Fiesta-Bianchet M, Mainardi D, eds). Singapore: World Sci-entific; 253–306.

Westneat DF, Clark AB, Rambo K, 1995. Within-brood patterns of paternity and parental care in red-winged blackbirds. Behav Ecol Sociobiol 37:349–356.

Westneat DF, Frederick PC, Wiley RH, 1987. The use of genetic markers to estimate the frequency of successful alternative mating tactics. Behav Ecol Sociobiol 21:35–45.

Westneat DF, Payne RB, Doehlert SM, 1986. Effects of muscle biopsy on survival and breeding success in Indigo Buntings. Condor 88:220–227.

Westneat DF, Rambo TB, 2000. Copulation exposes female red-winged blackbirds to bacteria in male semen. J Avian Biol 31:1–7

Westneat DF, Sherman PW, 1993. Parentage and the evolution of parental behavior. Behav Ecol 4:66–77.

Westneat DF, Sherman PW, Morton ML, 1990. The ecology and evolution of extra-pair copulations in birds. In: Current ornithology, vol. 7 (Power DM, ed). New York: Plenum Press; 331–369.

Whittingham LA, Dunn PO, 2001. Male parental care and paternity in birds. Current Ornithology 16:257–298.

Whittingham LA, Taylor PD, Robertson RJ, 1992. Confidence of paternity and male parental care. Am Nat 139:1115–1125.

Williams GC, 1966. Natural selection, costs of reproduction, and a refinement of Lack's principle. Am Nat 100:687–690.

Winkler DW, 1987. A general model for parental care. Am Nat 130:526–543.

Wolf JB, Brodie EB, III, Moore AJ, 1999. Interacting phenotypes and the evolutionary process II. Selection resulting from social interactions. Am Nat 153:254–266.

Wright J, 1998. Paternity and paternal care. In: Sperm competition and sexual selection (Birkhead TR, Møller AP, eds). San Diego: Academic Press; 117–146.

Yasukawa K, McClure JL, Boley RA, Zanocco J, 1990. Provisioning of nestlings by male and female red-winged blackbirds, *Agelaius phoeniceus*. Anim Behav 40:153–166.

Zentall TR, 1996. An analysis of imitative learning in animals. In: Social learning in animals: the roots of culture (Heyes CM, Galef BG, Jr, eds), San Diego: Academic Press; 221–244.

Zuk M, Johnsen TS, Maclarty T, 1995. Endocine-immune interactions, ornaments and mate choice in red jungle fowl. Proc Roy Soc Lond B 260:205–210.

# 15 The Evolution of Virtual Ecology

Alan C. Kamil and Alan B. Bond

The relationship between the perceptual and cognitive abilities of predatory birds and the appearance of their insect prey has long been of intense interest to evolutionary biologists. One classic example is crypsis, the correspondence between the appearance of prey species and of the substrates on which they rest which has long been considered a prime illustration of effects of natural selection, in this case operating against individuals that were more readily detected by predators (Poulton 1890; Wallace 1891). But the influences of predator psychology are broader, more complex, and more subtle than just pattern matching. Many cryptic prey, including the underwing moths (*Catocala*) that rest against tree bark in the daytime, are also polymorphic. This discontinuous variation in appearance is thought to make prey harder to find initially and harder to detect even after the predator has learned of their appearance (Poulton 1890).

In contrast to cryptic prey, other prey appear to have taken the opposite route, being quite conspicuous in appearance. The aposematic or warning coloration displayed by many distasteful or poisonous species appears to facilitate avoidance learning by predators (Guilford 1990; Schuler & Roper 1992). Batesian mimicry (Bates 1862), in which palatable prey evolve to imitate the appearance of aposematic species, appears to take advantage of the predator's tendency to stimulus generalization (Oaten et al. 1975). Even a century after many of these ideas were first proposed, however, much of the work in this area is still based on correlational data or on experiments that bear only indirectly on the issue.

The reason lies in part in the intractable nature of hypotheses about behavioral evolution. Behavior lacks a substantive fossil record, which means that any account of its origins must necessarily be inferential. Worse, behavioral evolution usually involves a dynamic interplay between the behavior and the environment that is difficult to reconstruct adequately under controlled conditions. As a result, even inferential investigations of the phenomenon generally bear only indirectly on the conditions under which the behavior actually evolved. To circumvent this constraint, we have developed an experimental method that allows realistic, repeatable simulation of the original processes involved in the evolution of color patterns in prey organisms. This "virtual ecology" technique provides an innovative approach to the experimental study of evolutionary dynamics and a quantum improvement in the ability to test evolutionary hypotheses. Our purpose in this chapter is to describe the development of the virtual ecology technique as the product of the cross-fertilization of ideas from a variety of disciplines, including operant psychology, behavioral ecology, population dynamics, and evolutionary computer algorithms. We have chosen to do this in a personal, autobiographical way in an attempt to communicate the flavor of what it has actually been like to experience these developments. In order to be historically accurate, we needed to emphasize the roles of coincidence and happenstance as well as of knowledge and synthesis in this process. We hope that the chapter will be fun to read and that it will provide useful insight into one example of how scientific research actually takes place.

## The Original Problem

Traditionally, psychologists have studied the sensory and cognitive capacities of animals in settings far removed from the natural world. Yet processes such as perception, attention, and memory are quite important in many natural systems, and the authors have long shared an interest in studying cognition within natural systems, in its original evolutionary context. One system that lends itself to this approach is the interaction between cryptic insects and predators that conduct a visual search for prey. The cognitive characteristics of such predators have influenced the evolution of both cryptic coloration and polymorphism, the occurrence of multiple cryptic forms of a single prey species. Cryptic polymorphic species are common in many groups of insects, particularly grasshoppers, leaf-hoppers, true bugs, and moths (Fig. 15.1).

The selective pressure for crypticity derives from the tendency of predators to feed most on prey that are easiest to find. To the degree that prey appearance is genetically based, one would expect selection to favor individuals that were harder to detect. Polymorphism as a response to predation is less obvious. A predator that has to hunt for several different prey types will, however, be less efficient than a predator that can search for just one. As a result, selection will promote the accumulation of novel color variants in these species. A new morph, arriving either through immigration or mutation, which differs sufficiently in appearance from prey with which the predators are already familiar should gain advantage, escaping predation long enough to become established as an element of the local population.

Even after establishment of a novel morph, cognitive features of the predators may continue to act on the prey population. If prey detection is frequency-dependent, in that more common prey types are more likely to be taken (what is often termed "apostatic selection"), then the prey population will tend to stabilize with relatively constant numbers of the different morphs. Apostatic selection may result from a number of different factors in the predator's behavior (Allen 1988), but the most interesting mechanism from a behavioral standpoint is hunting by searching image. Searching im-

Figure 15.1. Five specimens of *Catocala relicta*, demonstrating their forewing polymorphism. Note the variation in the patterning and brightness of the forewing, which is the only part of the wing seen when the moth is at rest.

age was developed as an ecological hypothesis by L. Tinbergen (1960), based on a theory originally proposed by von Uexküll (1934/1957). Tinbergen studied the densities of cryptic insect prey in the diet of small birds living in pine woodland. Prey were consumed less frequently than expected by chance when their numbers were relatively low and more often than expected by chance when their numbers were high. As a consequence, relatively small increases in prey density could result in large increases in predation, as the number of prey of a particular type appeared to cross some sort of threshold beyond which the predators "switched on" to that prey type.

Tinbergen (1960) proposed that the probability of detecting a prey item was a function of the recent previous experience of the predator. A series of successive detections of the same prey type, he reasoned, would serve to activate a representation of the prey item in the predator's nervous system. Subsequent search would then be guided by particular features of the representation, increasing the predator's efficiency at detecting that prey type. If predators possessed such a mechanism, then when the density of a particular prey type reached a high enough level relative to the density of other prey types, that prey type would be more likely to be encountered several times in succession and subsequent detections of it would be enhanced. As a result, searching image would result in a sudden increase in predation, resembling the observed pattern.

Although Tinbergen's hypothesis was fully consistent with the observed relationship between prey density and predation, it quickly became apparent that a number of other behavioral mechanisms could also contribute to the effect (Krebs 1973). For example, if different prey types live in different microhabitats, and if predators tend to search most intensively within those microhabitats that had proved most profitable in the recent past, a similar relationship would be obtained. Indeed, there seems little doubt that Tinbergen's findings probably reflected a number of different population processes, and did not constitute unequivocal evidence in favor of searching image. Nonetheless, the cognitive complexity and evolutionary implications of the searching image hypothesis have attracted many investigators over the years, including both of the authors of this chapter.

The searching image hypothesis makes sense in terms of many aspects of the biology of cryptic prey. For example, the natural distribution of cryptic organisms is often highly dispersed (Sargent 1976), which could function to limit the chances of a predator's acquiring a searching image by making successive encounters with the same prey type less likely. Searching image could also account for the high degree of polymorphism found in the appearance of many cryptic prey, through the selective advantage provided to novel, rare forms. The effects of hunting by searching image could, thus, be of substantial biological importance. In addition, the searching image concept appears to provide an immediate link between psychology and ecology. That is, although the idea of the searching image was based on biological and ecological phenomena, it is closely related to psychological concepts, such as selective attention and feature-based visual search (Bond 1983; Blough 1991). Thus, it provides an excellent context for multidisciplinary studies of cognitive mechanisms.

However, testing the searching image hypothesis presents substantial methodological challenges. In its most rigorous form, the hypothesis requires measuring changes in the ability to detect a cryptic prey as a function of recent experience with that type of prey, while controlling for all other factors that may influence the detectability. At minimum, therefore, it demands a procedure that allows complete experimental control over the sequence of prey encounters experienced by a predator. Furthermore, if, as Tinbergen (1960) proposed, searching image formation is only important when the prey are cryptic, the procedure would have to allow a fairly realistic level of difficulty.

## The Operant Analog to Predatory Search

Kamil was first introduced to Tinbergen's hypothesis and the challenges that it presents in the early 1970s by a colleague, Ted Sargent, in the context of his studies of underwing moths of the genus *Catocala* (Sargent 1976). Underwing moths have brilliantly colored hind-wings, but their fore-wings are complex patterns of muted tones that closely resemble the bark on which the insects rest during the day. One of the primary predators on these cryptic moths during the daylight hours is the blue jay (*Cyanocitta cristata*), and, by happy coincidence, Kamil happened to have a laboratory population of hand-raised blue jays from a research program on complex learning in Corvidae (Kamil & Mauldin 1987). The predator and the prey were available. And the question was most interesting: Would blue jays get better at detecting a particular type of cryptic *Catocala* moth if they encountered several of that type in succession? The problem was to come up with a technique for asking the question.

The issue was actively pondered and discussed for several months, until Kamil attended a departmental seminar about "concept formation" in pigeons (Herrnstein 1964, 1985). In this research, pigeons were presented with a large array of projected images, shown one at a time on a pecking key in an operant chamber. Some of the images contained an example of the "concept," for example, a tree, while others did not. The rule for the pigeon was that it received food if it pecked at the key when a tree was shown and no reward for pecking when no tree was present. The images were extremely variable in content and included many different trees of different species, in photos taken from many different angles and distances. Herrnstein found that pigeons learned this task well, pecking at images with trees with much higher probability then at images without trees. More important, they responded appropriately to new images, both with and without trees, which they had never seen before.

While concept learning was crucial to Herrnstein, the technique he used was the focus of Kamil's interest. The seminar gave rise to an astoundingly simple idea: substitute cryptic moths for trees and blue jays for pigeons, make a few changes in the responses required to more accurately reflect foraging behavior, and, *voila*, a technique that would allow direct and elegant testing of the searching image hypothesis. The use of projected images would allow visual detection of cryptic moths and make it easy to control the order of encounters with different prey types by controlling the sequence of the slides.

Pietrewicz and Kamil (1977) took a camera and a set of dead moths into the woods and made a series of photographs of trees, some with moths and some without. For each image that contained a moth, there was an identical image without one, and the location of the moth within the image was randomized across the set of images. These images were then shown to the jays, one at a time, in an operant chamber that had a large rectangular pecking key, a small round pecking key, and a food cup mounted on one wall (Fig. 15.2).

We modified Herrnstein's contingencies to more closely approximate foraging behavior. Each day, each blue jay was tested for sixty-four trials, thirty-two with moths, and thirty-two without moths, in random order. During each trial, one slide was projected on the rectangular key. The bird could either peck at the moth or peck at a small round key located to one side of the large key. The contingencies are summarized in the table below. The results of a peck at the large key depended on whether the trial was "positive," that is, whether a moth was present in the projected image.

| | "YES" RESPONSE (PECK AT LARGE KEY) | "NO" RESPONSE (PECK SMALL GIVING-UP KEY) |
|---|---|---|
| POSITIVE TRIAL (MOTH PRESENT) | FOOD (10 S) FOLLOWED BY 2-S DELAY (ITI) | 2-S DELAY (ITI) |
| NEGATIVE TRIAL (MOTH ABSENT) | 30-S DELAY (ITI) — (penalty for pecking an area without prey) | 2-S DELAY (ITI) |

Pecks to the large key on positive trials earned the bird a piece of mealworm. It was given ten seconds to eat the mealworm, followed by a two-second delay to the beginning of the next trial. If there was no moth in the image, pecking at the image produced just a thirty-second delay before the start of the next trial. The results of a peck at the small round key always produced a two-second delay followed by the start of the next trial. Thus, from the jay's point of view, each trial was like a patch to be searched for a moth. If a moth was found, a "Yes" response (pecking at the moth image) resulted in a food reward. If no moth was found, then the jay could move to the next patch by making a "No" response. Attacking an image without a moth resulted in wasted time, as does attacking nonprey items in natural foraging. Saying "No" when a moth was actually present produced no special result. In nature, when a predator fails to detect a prey item, there is no signal that informs the predator of its mistake.

Figure 15.2. The top panel shows a blue jay in the test apparatus. The bottom panel shows an example of a cryptic *C. relicta* on a white birch tree.

## Internal and External Validity

Any attempt to construct a controlled experiment to investigate a natural phenomenon inevitably raises the issues of external and internal validity (Cook & Campbell 1979; Kamil 1988). Internal validity refers to the extent to which a research design allows unambiguous attribution of cause and effect. That is, a study with high internal validity is well-controlled, so that if there is a difference between treatments, the difference can be ascribed to effects of the independent variable. External validity refers to the range of situations to which the results of the study can be generalized or applied. Obviously, any researcher would like each and every study to have both high internal and high external validity. Unfortunately, this is impossible. The steps required for high internal validity almost always lead to a reduction in external validity and vice versa (see Kamil 1988, for more detailed discussion).

The procedures developed by Pietrewicz and Kamil (1977) provide a good example of this trade-off. In order to control the appearance of the moths, we resorted to the use of photographic images. This allowed us to increase internal validity in several ways. For example, critical to testing Tinbergen's searching image hypothesis, it allowed control of the order of encounters with prey by controlling the order of slides in the projector. Thus, we could set up blocks of trials that were absolutely identical except for the order of the prey within the block. But this high level of internal validity came with a price. How could we know that the use of projected images (and all of the many other features of the operant procedures) had resulted in a situation in which the behavior of the jays was representative of behavior in natural situations?

The operant simulation appears logically reasonable. Consider a predator hunting for cryptic moths. The moths normally rest on tree trunks all day long, without moving, and press themselves down flat against the substrate, so a two-dimensional still image provides a reasonable visual approximation. Cryptic prey generally disperse, so that any given area is unlikely to contain more than one moth, and many areas will contain none. Thus the predator must go from patch to patch searching. In each patch it must search until it either finds a prey or decides to move on to the next patch. If it finds a moth it gets food; otherwise it merely expends time. If the predator attacks a nonfood item, it wastes additional time. These characteristics of the relationship between foraging behavior and its outcomes were present in our procedures. Thus there is a strong argument that we had constructed a reasonable simulation of hunting for highly dispersed, cryptic prey.

However, this is a very subjective judgment. Was there any way to empirically test the adequacy of these procedures, some kind of objective test of external validity? We accomplished this by testing hypotheses derived from detailed studies of the moths themselves (Sargent 1976). For example, each species of *Catocala* selects a specific substrate on which it prefers to rest. We reasoned that if our procedures were reasonable facsimiles of the real thing, the moths should be least detectable when presented on their species-specific substrates. So, Pietrewicz and Kamil (1977) took *Catocala* that differ in substrate preferences and tested them on several substrates. Each species tended to be least detectable on the type of tree on which they normally rest.

Another source of validation derives from an analysis of the type of search that the jays must use (Kamil et al. 1993), a type called exhaustive, self-terminating search. Consider the process of scanning a discrete area for the presence of a hard-to-find prey item. When a prey item is detected, it should be attacked immediately. However, it makes little sense to leave the patch until the whole area has been scanned. Thus, on the average, it should take longer to decide to leave an empty patch than to attack a prey item. This is also characteristic of the behavior of the jays in our simulation. Thus, we are reasonably confident that the operant procedures capture at least some of the important aspects of natural foraging for cryptic prey by visual predators. The next step was applying the operant procedures to the searching image hypothesis.

## Testing the Searching Image Hypothesis

Tinbergen specified that a specific experience, encountering a single, cryptic prey type several times in succession, should produce a specific effect, an improvement in the predator's ability to detect that prey type. In order to test the hypothesis properly, it is therefore necessary to create conditions that are identical in every respect except that in one condition the prey have been encountered in succession while in the other they have been encountered in random order. Furthermore, conditions of the experiment must be such that any change in the capture of prey must be reasonably thought to be due to a change in the predator's ability to detect the prey.

It is not easy to meet this pair of requirements, and many studies have failed to control for one or another factor (reviewed by Krebs 1973). For example, one common design is to expose the predator to just one type of prey for some time, then give it that type plus a new type. In these conditions the predator will usually take the type to which it was preexposed. But clearly this could be caused by many factors other than changes in detection. Simple familiarity could lead to a preference; the predator might detect all of the prey, but prefer to consume the familiar one. Detection of the prey needs to be dissociated from consumption. The operant technique of Pietrewicz and Kamil does, however, meet the requirements. It separates detection from consumption: the birds detect the moths in the images, but eat pieces of mealworm. It allows complete control over the order in which the prey types are presented to the predator. And it is clearly a detection task.

To test Tinbergen's (1960) hypothesis, Pietrewicz and Kamil (1979, 1981) tested each of a group of jays daily with a series of images, half of which contained moths. Two types of moths appeared equally often: *C. relicta* on birch and *C. retecta* on oak (as in 2). During each session, an experimental block of sixteen trials, half positive and half negative, in random order, occurred. One-third of these blocks were "nonruns," and included four instances each of *C. relicta* and *C. retecta*, in random order (and eight slides without moths). The other two-thirds were "runs" during which all eight of the moths in the set were either *C. relicta* or *C. retecta*. Thus the run and nonrun conditions were identical except for the order of encounter with the prey types. While nonruns exposed the jays to both prey types in random order, runs exposed them to eight consecutive encounters with a single prey type. The results were unequivocal (Fig. 15.3). Performance improved during run blocks but not during nonruns. This data provided the first methodologically sound test of the searching image hypothesis. Subsequent experiments with jays (e.g., Kono et al. 1998) and with pigeons searching for cryptic grains (Bond & Riley 1991; Langley 1996) have replicated these basic findings. These experiments demonstrate how operant techniques developed by psychologists can be used to evaluate biological hypotheses.

## Development of the Collaboration

The authors were both in the Psychology Department at Berkeley during the 1976–77 academic year. We got to know each other professionally, discovering many mutual research interests, particularly a general interest in studying cognition in meaningful ecological contexts as well as a more specific interest in searching image. Bond had come to searching image through a wholly different route. During a graduate zoology seminar in 1974, he realized that for apostatic selection actually to stabilize a prey population, it must be exhibited as a sustained bias in favor of more abundant prey types, even when the appearance of all potential prey was thoroughly familiar to the predator. Most of the experimental evidence that existed at that time did not effectively address this issue, since the designs generally involved presenting birds with two prey that differed in their familiarity. If learning to detect a novel prey were all that was involved, apostatic selection would require that predators rapidly forget the appearance of rarer prey items, something that seemed highly unlikely.

Bond put together a postdoctoral proposal to explore this issue using pigeons searching for cryptic grain. The choice of species was dictated by the decision to cross over from biology to psychology for this work: pigeons were what they had to work with in Donald Riley's comparative psychology

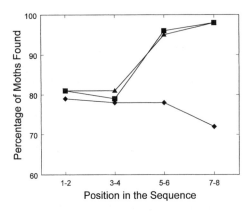

Figure 15.3. Detection of moths improved when several of the same type (*C. relicta* or *C. retecta*) were encountered in a row (run condition) than when they were presented in random order (nonrun condition). Redrawn from Pietrewicz and Kamil (1979).

laboratory. But the choice of organism, in turn, determined the experimental approach. Pigeons feed on the ground on scattered grains, and different types of grain are generally intermixed in a given area. Thus the birds were shown trays containing plastic-coated gravel and a random scattering of grains of two different types, beans and wheat. Unlike Kamil's design, which involved presenting the stimuli one at a time and regulating the bird's attentional state by modifying the sequence of presentation, the pigeon experiments presented all grain stimuli simultaneously, and the sequence of grains taken was a function of the bird's cognitive processes.

Pigeons showed a natural bias in favor of more abundant grain types when the food was presented on a cryptic background, even when the birds were thoroughly familiar with all grain types (Bond 1983). The bias was lower when the stimuli were harder to detect, and it was reversed when the grains were presented on a uniform gray background. The effect was large and robust, clearly implicating a process that would be important in the real-world ecology of the species. The problem with this natural grain preparation was that it was very difficult to manipulate the relationship between the background and the grains, and it was not feasible to record responses to individual grains to obtain direct measures of speed and accuracy of response.

In 1985, Bond and Riley returned to the issue of searching image in pigeons, using the Pietrewicz and Kamil (1979) operant approach. Photographs of cryptic grains were back-projected on a ground-glass pigeon key, and the birds were required to peck grains if they were present in the image. If the image was of gravel alone, the birds were trained to peck an advance key. The experiment proved very difficult to conduct. Pigeons are reluctant to peck a key that terminates the trial without a food reward and it proved necessary to provide food rewards for correct responses to nongrain stimuli. This, in turn, required careful titration of reward rates to avoid response biases. It is a good example of the dangers inherent in taking a design that works well for one species and porting it to another, ecologically unrelated organism. Although Bond and Riley (1991) did find a significant run effect, the operant approach to the detection of cryptic food was clearly problematic with pigeons. More recent operant investigations of searching image in pigeons have commonly used a different design, in which target stimuli are present in all displays and the searching image effect is detected solely in terms of response time (e.g., Blough 1991; Langley 1996). By 1989, Bond had been convinced that there was nothing more to be learned from such inflexible and complex experimental designs and had shifted his focus to social behavior. Aside from collaborating on one final attempt to improve on the natural grains preparation (Langley et al. 1996), he was effectively out of the searching image business.

Kamil and his lab continued to exploit the operant procedure with blue jays to investigate searching image (Pietrewicz & Kamil 1979, 1981; Kono et al. 1998), as well as a variety of other phenomena. One series of studies used the procedure to emulate foraging in patches of varying quality (Kamil et al. 1985; Kamil et al. 1988; Kamil et al. 1993). Another study (Olson et al., unpublished; Endler 1991) varied the level of prey crypticity by photographing moths on trees whose bark varied in complexity. As complexity increased, the match between the appearance of the bark and the appearance of the moth also tended to increase, which had several effects on the jays' behavior. With higher crypticity, the birds became less accurate, but they also modulated their speed of response, taking longer to decide when confronted with more complex trees.

Despite these successes, Kamil also found that the use of images of real prey items imposed severe methodological limitations. The presentation of photographic images to birds was technically cumbersome, and the array of possible stimulus displays necessarily limited. But the most restrictive feature of these existing techniques was that they were essentially static and unidirectional. The researcher determined the degree of crypticity, the proportions of different prey types, and the order of stimulus presentation, and the birds simply responded to the imposed treatments. The evolutionary interaction between predators and prey involves a dynamic, reciprocal interaction between prey appearance and abundance and predator perception that was wholly lacking in these studies. The question of how the psychology of predators might affect the evolution of their prey (Guilford 1990) appeared wholly out of reach.

And then, coincidentally, both authors ended up at the University of Nebraska. Given our previous contacts and separate work on searching images, we naturally began to explore working together on the detection of cryptic prey. The discussions centered around the lovely experiments that could be done if we could devise an artificial prey that emulated the complexity and crypticity of real moths and yet was amenable to systematic dimensional variation. But the discussions were merely idle speculation until Kamil returned from a vacation in England with the germ of the crucial idea. While visiting at Cambridge, Kamil had attended a dinner with friends, which led to an invitation to visit Nick Mackintosh's operant lab. The invitation was initially declined because Kamil's trip (with spouse) was strictly vacation. But Nick mentioned that one of his students, Kate Plaisted, was doing research on searching image in pigeons. A quick visit was arranged, in the course of which Kamil saw the stimuli that Plaisted had developed (Plaisted 1997; Plaisted & Mackintosh 1995). Plaisted's targets were black-and-white checkerboards displayed on a background of black-and-white squares. Crypticity was manipulated by varying the proportions of different sized patches of black in the background.

On returning to Nebraska, Kamil and his newfound enthusiasm for the checkerboard pattern led to a series of crucial brain-storming sessions. Plaisted's checkerboards appeared to lack the requisite dimensional complexity and were probably insufficiently cryptic to fool blue jays, but the basic idea of stimuli consisting of coarse arrangements of pixels was very attractive. Bond had extensive experience with digital graphics and multimedia programming, and he felt that it might be fairly simple to scan images of *Catocala*, convert them to a 64-level gray scale, and reduce them to coarsely digitized 16 × 16 pixel squares. The resulting digital prey were then presented on backgrounds of a complex texture that was similar in brightness and grain to the patterns on the moths. By varying the match between the distributions of gray-scale values in the moths and the background, the degree of crypticity could be controlled with great sensitivity (Fig. 15.4). Detecting these digital moths was comparably difficult, at least for human observers, to finding real moths in photographic images, but as was the case with the original use of photographic images (Pietrewicz & Kamil 1977), a formal test of external validity was required.

## Validation of the Digital Moth Approach

One purpose of our initial experiment, therefore, was to compare the behavior of jays searching for digital moths with that shown by jays searching photographic images for real moths (Bond & Kamil 1999). We trained the birds to detect three distinctive digital moths presented on backgrounds of varying crypticity. As the crypticity increased, detection accuracy decreased and response speed increased. We also found much longer latencies to say "No" on trials without moths than to say "Yes" on trials with moths, suggesting an exhaustive, self-terminating search of the display. Both of these features were also characteristic of the behavior of jays searching for real moths.

Jays tested with images of real moths also showed searching image effects, so we used a design very similar to the original Pietrewicz and Kamil (1979) study to examine the effects of changes in the trial sequence. Over the course of the experiment, each bird received 180 "treatment blocks" of

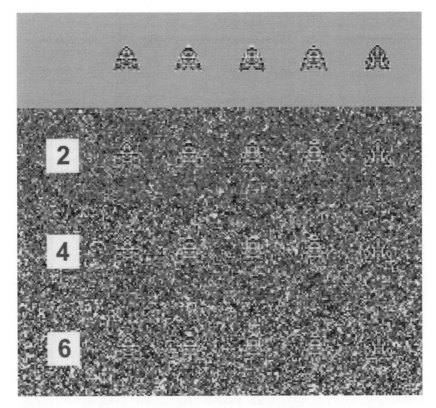

Figure 15.4. Five types of digital moths used in these experiments presented against backgrounds of three levels of crypticity to illustrate the difficulty of the detection task. When projected on the computer monitor, the moths were about 6 millimeters high.

16 trials, 8 containing digital moths and 8 without (in random order). Forty-five of these were control blocks, containing all 3 prey types in random order. The other 135 were run blocks, containing 8 prey of the same type (45 for type I, 45 for type II, and 45 for type III). The results were clear-cut: as with photographic images, a searching image effect was obtained.

The most important finding of this experiment, however, concerned one of the earliest predictions of searching image theory, the idea that a searching image for one prey type actively interferes with detection of prey of a disparate appearance (von Uexküll 1934/1957). The prediction is now generally interpreted in terms of selective attention. Because organisms have only a limited ability to process information, selective attention to one specific stimulus or modality has commonly been seen as requiring a reduction in attending to alternatives (Kahneman 1973). Thus, a bird with a searching image for one prey type would experience not only an improvement in the ability to detect that prey, but also a decreased ability to detect alternative prey items. The concept is attractive and easily understood, but demonstration of interference effects in searching image had proven very difficult. While the facilitative effects on detection of the target prey type have been demonstrated in a number of well-controlled studies (e.g., Blough 1989, 1991; Bond & Riley 1991; Kono et al. 1998; Pietrewicz & Kamil 1979; Plaisted & Mackintosh 1995), there was no good evidence for interference effects during cryptic prey detection. We therefore included another feature in our experimental design to test for interference effects. Each run or non-run treatment block was followed by a four-trial probe block, containing two positives of one of the three digital moths and two negatives. These were arranged so that each type of run block was followed equally often by each type of probe.

The issue under test was whether the jays were poorer at detecting moths in probes following a run of a different prey type than in probes following a nonrun. To examine this question, we measured the effect of the run in terms of the difference in accuracy between probe trials following runs and those following nonruns. Detection of the same prey type that was presented in the run was facilitated in the probes, a continuation of the searching image effect. The effects of runs of a different prey type on probe accuracy depended on the relative levels of crypticity, however. When the prey type in the probe trials was more conspicuous than that used in the previous run, there was no effect. But when the probe type was more cryptic than that used in the run, detection accuracy was significantly reduced, providing clear evidence of interference. Similarly asymmetrical interference effects have been seen in other studies, suggesting that it may be a general phenomenon (Blough 1989, 1992; Bond & Riley 1991; Reid & Shettleworth 1992; Lamb 1988).

Finally, we used the data from this experiment to test an alternative hypothesis, the suggestion that searching image effects are consequences, not of shifts in selective attention, but of selective forgetting of short-term memory traces (Plaisted 1997). Plaisted pointed out that the mean time interval between successive appearances of the same prey type was shorter during a run of a single prey type then during a random series of two or more prey types. Thus, she argued, increased forgetting during nonruns could account for the relatively higher levels of detection during runs. She then provided evidence that detection was improved when the amount of time since the last target of the same type was reduced, even in nonrun blocks (Plaisted 1997).

Although this was a tenable hypothesis, several considerations suggested that mnemonic effects were unlikely to provide a complete account of searching image. The time intervals Plaisted used were relatively short, generally well less than a minute, and experiments with pigeons feeding on natural grains had found no decrease in a response bias to the most abundant grain type over intervals less than three minutes (Langley et al. 1995). Furthermore, Plaisted's pigeons could not make a patch departure decision; they were not reinforced for pecking at displays without targets. We therefore thought it important to test Plaisted's hypothesis in our experimental paradigm. We examined the effects of the time interval between presentations of the same prey type, confining our analysis to cases in which the time interval was less than three minutes and tested to see whether it provided a superior account of the results. We found that the interval between successive presentations of the same prey type had no effect on the probability of detection and that response time actually decreased as the time interval increased. These results, combined with the interference effects we discovered, are clearly contrary to Plaisted's hypothesis.

## Virtual Ecology

Reassured that our digital moths replicated earlier work with images of real moths, we undertook a study that broke new ground, taking advantage of our increased ability to control and manipulate the stimuli. One strength of the digital moths is that they do not limit the number of different stimuli that can be presented to the birds. Novel backgrounds can be generated at will. The placement of the moths in the image need not be fixed. It can be determined randomly at the time of presentation. There was, thus, no concern that specific images might differ in their inherent difficulty level or that the birds might memorize the moth's location in particular images, and there was no need to control the particular sequence of stimuli being displayed. This opened the door to "virtual ecology," in which populations of digital moths are maintained in the computer as lists of individuals, and the impact of the predator's behavior on prey population dynamics is played out directly.

Our first virtual ecology study (Bond & Kamil 1998) tested the effects of apostatic selection on the maintenance of a prey polymorphism. We created a virtual prey population of 240 individual digital moths, 80 of each of 3 distinctive morphs, and exposed them to predation by 6 blue jays. The moths the jays detected were considered "killed" and were removed from the population. Those that were overlooked were allowed to breed (asexually), bringing the population back up to 240 for the following day. Each day thus constituted a generation, and our only experimental intervention was to set the initial numbers of the 3 morphs. No matter what initial conditions were imposed, however, the relative numbers of the 3 morphs rapidly approached a set of characteristic values. One morph was somewhat more cryptic than the others, so it stabilized at a higher population level. We perturbed this system several times, drawing down the abundance of the most common morph and raising that of one of the less abundant. Each time, the system returned to the same equilibrium levels (Fig. 15.5). Subsequent analysis demonstrated that the equilibrium was a result of negative feedback between prey abundance and prey detectability. The more common any of the prey types became, the more likely it was to be detected and removed from the population.

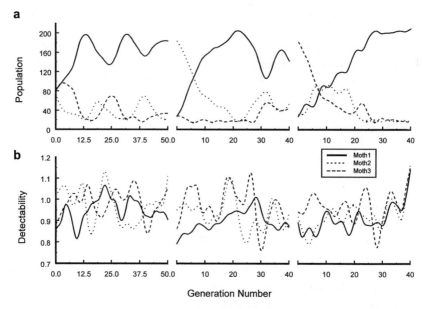

Figure 15.5. Population numbers of three species of digital moths in three successive replications of the virtual prey procedure. Curves were smoothed with weighted least squares, using an eight-generation window.

We then tested the effects of apostatic selection on novel morphs. In theory, since a new morph will initially be present in low numbers, and since predators will have had no prior experience with them, apostatic selection should favor rare novel forms. In our first introduction of a novel morph, the new prey type increased in abundance until the jays caught on to its presence and began to search for it in earnest. Its numbers subsequently declined until it reached an equilibrium level comparable to that of two of the original morphs. In the second iteration, the new type was extremely cryptic. Only two of the six jays learned to detect it, even after fifty generations. As a result, it quickly came to dominate the population, and the other morphs would have been driven to extinction if we had not brought their numbers up to a minimum (five) with each new generation. This result illustrates the power of selection in virtual ecology: a new morph introduced at very low levels can, in a short time, rise almost to fixation, driving out less cryptic variants.

## Virtual Genetics

The wing of a digital moth consists of seventy-five unique gray-scale pixels on one wing, replicated on the opposite wing. Given this relatively simple phenotype, Bond saw that it was feasible to eliminate the constraint of having uniform appearance within a moth type and endow digital moths with a virtual genetics that would determine the appearance of their wings. The genetic algorithm he developed was largely based on current understanding of the control of lepidopteran wing patterns (Robinson 1971; Nijhout 1991, 1996; Brakefield et al. 1996). We did not attempt to simulate lepidopteran genetics in full detail, but, rather, constructed a simpler system that nevertheless remains true to many of its salient features. The most significant simplification we adopted is that the moth phenotype is generated from a single, haploid chromosome, a string 117 bytes in length, in which each byte constitutes a gene. The relationship between gene and phenotype is not one-to-one, however. As in real moths, phenotypic characters are polygenic. The wing pattern of a digital moth is encoded by eighteen loci, each consisting of five genes which, taken together, define an elliptical wing patch. The location, eccentricity, orientation, extent, and peak intensity of each patch are each determined by the value of a separate gene within each locus. The intensity of each pixel in the phenotype is thus affected by several overlapping patches, which interact additively. One attractive feature of this design is that it incorporates redundancy in the genome that helps to maintain the level of genetic variance in the population, in spite of the absence of heterozygosity.

In this final stage of the development of our technique, our population of digital prey could now reproduce sexually. Their wing patterns are transmitted to their offspring, but because of recombination and mutation, the offspring are imperfect copies of their parents. And this enables us to ask questions, not just about the maintenance of prey crypticity and polymorphism, but about its evolutionary origins. Initial experiments on our virtual moth genome suggest that it is ideally suited for investigating natural selection. In particular, a random selection from the $2^{936}$ possible genotypes produces a population that is strikingly variable in appearance, but all of the phenotypes still appear reasonably moth-like (Fig. 15.6). This is essential, because the jays are trained to treat any of these objects as a potential prey item.

Equally essential, the genome is responsive to selective pressure. We demonstrated this by presenting blue jays with an initially random population of two hundred digital moths displayed on a background of moderate crypticity. Prey items that are overlooked during their exposure to the predator have three times as great a probability of being allowed to breed as those that are discovered. After twenty generations, two effects of selection by the jays can be seen. The population of moths became significantly darker. The moths also became more cryptic to the jays. Although this may not be obvious to human eyes looking at the population (Fig. 15.7), the moths in the twentieth generation take longer to detect than the moths in the first generation. These results confirm the role of predation in determining appearance of the moths. To extend this work, we are currently developing a series of experiments to test the effects of predator psychology on the evolution of cryptic prey polymorphism. The experiments will be unique in combining a simulated prey population (that nevertheless retains many of the critical features of real organisms) with a laboratory population of real predators. Although these

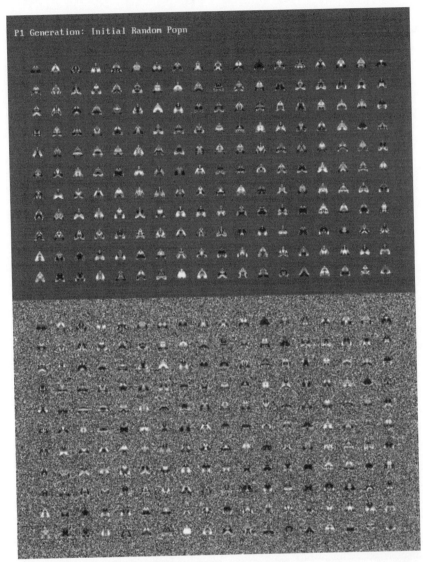

Figure 15.6. A population of digital moths created by randomly sampling the population of possible digital moth genomes 198 times. The moths are shown on both a plain gray background and a patterned background.

predators cannot be considered "evolutionarily naive" (Guilford 1990; Schuler & Roper 1992), our technique will enable us to explore these questions under conditions in which predator choice and prey appearance are dynamically related and, for the first time, to test directly theoretical predictions about the course of predator/prey evolution.

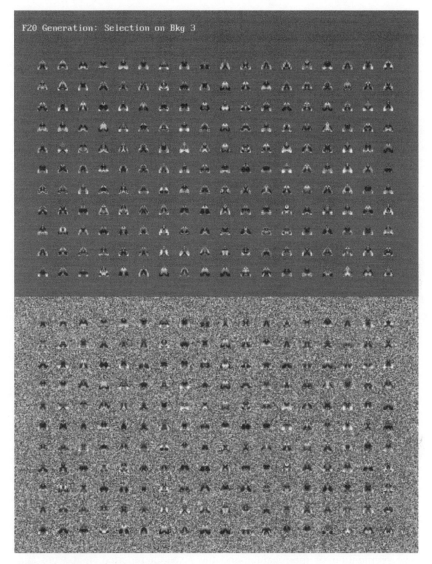

Figure 15.7. The same population of moths as shown in Figure 15.6, after twenty generations of selection by blue jays.

## Coincidence and Insight in the Evolution of Virtual Ecology

Virtual ecology blends ideas and methods from many disciplines, including operant psychology, behavioral ecology, population dynamics, and evolutionary computing. We could not have developed the technique if we had limited ourselves to just biology or psychology. This synthesis has resulted in an exciting, innovative procedure for the experimental study of evolutionary dynamics. Adaptationist stories are the bane of behavioral and evolutionary ecology, but in many cases the stories are actually hypotheses, hypotheses that have simply been inaccessible to empirical test. Virtual ecology provides a methodology that can be used to begin to explore these hypotheses under rigorous, realistic, and repeatable circumstances. As an example, consider the hypothesis that background heterogeneity may facilitate the evolution of polymorphisms in cryptic prey. Virtual ecology can be used to test this idea simply by subjecting different populations of moths to selective predation by jays under different conditions of background heterogeneity. We believe that virtual ecology can probably be adapted to investigate any circumstance in which the detection, selection, and/or choice among signals emanating from an organism affects the evolution of those signals. And it was only made feasible by the synthesis of ideas across traditional disciplinary boundaries.

The element of coincidence in the history of this research is characteristic of many developments in science. Experimental techniques evolve opportunistically, seldom being driven by the ideas of a single researcher working in isolation from the rest of the field. Would the procedure ever have developed to its current form if Herrnstein hadn't given that seminar or if Kamil hadn't had dinner with Mackintosh or if the authors had not both ended up at the same university? As is true of organic evolution, the evolution of an idea takes paths that are not readily repeatable. Chance plays a major role. Should this be taken to mean that an aspiring researcher should sit back and wait for lightning to strike? Absolutely not! There are two morals to draw: (1) Be alert and attentive to possibilities that arise. Chance favors the prepared mind. (2) Maximize your chances. Increase the probability of novel developments by exposing yourself to as broad a variety of ideas and facts as possible. Read, listen, and be prepared for the idea when it happens. And be sure not to allow disciplinary boundaries to constrain your thinking.

## Acknowledgments

The preparation of this chapter and the research reported in it were supported by grants from the National Science Foundation.

## References

Allen JA, 1988. Frequency-dependent selection by predators. Phil Trans R Soc Lond B 319:485–503.

Bates HW, 1862. Contributions to an insect fauna of the Amazon valley (Lepidoptera: Heliconidae). Trans Linn Soc Lond 23:495–566.

Blough PM, 1989. Attentional priming and search images in pigeons. J Exp Psychol: Anim Behav Proc 15:211–223.

Blough PM, 1991. Selective attention and search images in pigeons. J Exp Psychol: Anim Behav Proc 17:292–298.

Blough PM, 1992. Detectability and choice during visual search: joint effects of sequential priming and discriminability. Anim Learn Behav 20:293–300.

Bond AB, 1983. Visual search and selection of natural stimuli in the pigeon: the attention threshold hypothesis. J Exp Psychol: Anim Behav Proc 9:292–306.

Bond AB, Kamil AC, 1998. Apostatic selection by blue jays produces balanced polymorphism in virtual prey. Nature 395:594–596.

Bond AB, Kamil AC, in press. Searching image in blue jays: facilitation and interference in sequential priming. Anim Learn Behav.

Bond AB, Riley DA, 1991. Searching image in the pigeon: a test of three hypothetical mechanisms. Ethology 87:203–224.

Brakefield PM, Gates J, Keyes D, Kesbeke F, Wijngaarden PJ, Amonteiro A, French V, Carroll SB, 1996. Development, plasticity and evolution of butterfly eyespot patterns. Nature 384:236–242.

Cook, TD, Campbell, DT, 1979. Quasi-experimentation: design and analysis issues for field settings. Chicago: Rand McNally.

Darwin C, 1871. The descent of man and selection in relation to sex (2nd ed.). London: John Murray.

Endler JA, 1991. Interaction between predators and prey. In: Behavioural ecology: an evolutionary approach (Krebs JR, Davies NB, eds). Oxford: Blackwell Scientific; 169–196.

Guilford T, 1990. The evolution of aposematism. In: Insect defenses: adaptive mechanisms and strategies of prey and predators (Evans DL, Schmidt JO, eds). Albany: SUNY Press; 23–61.

Kahneman D, 1973. Attention and effort. Englewood Cliffs, NJ: Prentice-Hall.

Kamil, AC, 1988. A synthetic approach to the study of animal intelligence. In: Comparative perspectives in modern psychology: Nebraska Symposium on Motivation, Vol. 35 (Leger DW, ed). Lincoln: University of Nebraska Press; 230–257.

Kamil AC, Lindstrom F, Peters J, 1985. Foraging for cryptic prey by blue jays. I: The effects of travel time. Anim Behav 33:1068–1079.

Kamil AC, Mauldin JE, 1987. A comparative-ecological approach to the study of learning. In: Evolution and learning (Bolles RC, Beecher MD, eds). Hillsdale, NJ: Erlbaum; 117–133.

Kamil AC, Misthal RL, Stephens DW, 1993. Failure of simple optimal foraging models to predict residence time when patch quality is uncertain. Behav Ecol 4:350–363.

Kamil AC, Yoerg SI, Clements KC, 1988. Rules to leave by: patch departure in foraging blue jays. Anim Behav 36:843–853.

Kono H, Reid PJ, Kamil AC, 1998. The effect of background cuing on prey detection. Anim Behav 56:963–972.

Krebs JR, 1973. Behavioral aspects of predation In: Perspectives in ethology, vol. 1 (Bateson PPG, Klopfer PH, eds). New York: Plenum Press.

Lamb MR, 1988. Selective attention: effects of cuing on the processing of different types of compound stimuli. J Exp Psychol: Anim Behav Proc 14:96–104.

Langley CM, 1996. Search images: selective attention to specific visual features of prey. J Exp Psychol: Anim Behav Proc 22:152–163.

Langley CM, Riley DA, Bond AB, Goel N, 1995. Visual search for natural grains in pigeons: search images and selective attention. J Exp Psychol: Anim Behav Proc 22:139–151.

Nijhout HF, 1991. The development and evolution of butterfly wing patterns. Washington, DC: Smithsonian Institution.

Nijhout HF, 1996. Focus on butterfly eyespot development. Nature 384:209–210.

Oaten A, Pearce EM , Smyth MEB, 1975. Batesian mimicry and signal detection theory. Bull Math Biol 37:367–387.

Pietrewicz AT, Kamil AC, 1977. Visual detection of cryptic prey by blue jays (*Cyanocitta cristata*). Science 195:580–582.

Pietrewicz AT, Kamil AC, 1979. Search image formation in the blue jay (*Cyanocitta cristata*). Science 204:1332–1333.

Pietrewicz, AT, Kamil AC, 1981. Search images and the detection of cryptic prey: an operant approach. In: Foraging behavior: ecological, ethological, and psychological approaches (Kamil AC, Sargent TD, eds). New York: Garland Press.

Plaisted K, 1997. The effect of interstimulus interval on the discrimination of cryptic targets. J Exp Psychol: Anim Behav Proc 23:248–259.

Plaisted KC, Mackintosh NJ, 1995. Visual search for cryptic stimuli in pigeons: implications for the search image and search rate hypotheses. Anim Behav 50:1219–1232.

Poulton EB, 1890. The colours of animals: their meaning and use, especially considered in the case of insects. New York: Appleton.

Reid PJ, Shettleworth SJ, 1992. Detection of cryptic prey: search image or search rate? J Exp Psychol: Anim Behav Proc 18:273–286.

Robinson R, 1971. Lepidopteran genetics. Oxford: Pergamon.

Sargent TD, 1976. Legion of night: the underwing moths. Amherst: University of Massachusetts Press.

Schuler W, Roper TJ, 1992. Responses to warning coloration in avian predators. Adv Study Behav 21:111–146.

Tinbergen L, 1960. The natural control of insects in pine woods. I: Factors influencing the intensity of predation by songbirds Archives Néerlandaises de Zoologie 13:265–343.

von Uexküll J, 1957. A stroll through the worlds of animals and men In: Instinctive behavior (Schiller CH, ed). New York: International Universities Press; 5–80. (Original work published 1934).

Wallace AR, 1891. Darwinism: an exposition of the theory of natural selection with some of its applications. London: Macmillan.

# 16 Wood Ducks: A Model System for Investigating Conspecific Parasitism in Cavity-Nesting Birds

Paul W. Sherman

In the winter of 1983 I was awarded tenure at Cornell. Of course, I was ecstatic. But, surprisingly, the euphoria didn't last long. In fact, apprehension set in disturbingly quickly as questions swirled through my head: What would I do next? What were my new directions and aspirations? Could I stand being my own "boss"?

Since 1974, my third year in graduate school, I had studied the behavioral ecology of Belding's ground squirrels (*Spermophilus beldingi*) at Tioga Pass, California (e.g., Sherman 1977, 1981, 1985; Sherman & Morton 1984; Sherman & Holmes 1985). After a decade it was a well-known, productive project—some even referred to it as a "model system" for studying nepotism (assisting kin) in mammals. I had always assumed I would continue that work indefinitely.

But something was wrong. The research funding certainly had not run out and the site, near Yosemite National Park, was spectacular. However, my curiosity had begun to wane. I could tell I was getting "stale" because I found myself inventing clever, technically sophisticated ways to address the same basic questions that I had already answered. I had to make a choice: either move on or stagnate. Of course, I opted for the former.

But this was much easier said than done. Every graduate student knows the mixed emotions associated with initiating a new field project: excitement tempered by uncertainty, enthusiasm moderated by self-doubt. First, there is the question of approach: Do you start with a key theoretical issue and then identify an animal suitable for investigating it, or start studying a potentially interesting creature in hopes that something important will emerge? Either way you wonder if the project will be exciting and "career enhancing," and if someone else is already doing the same thing (and just about to publish the results!). Then there is the difficulty of obtaining research funding for a project with no published record of achievement and no guarantee of success. Inevitably there are also logistical problems, like obtaining governmental permission and landowners' consent, learning how to capture, mark, and observe the study animals, and figuring out where to stay, shop, bank, and so on. Finally, and most significantly, fieldwork invariably entails abandoning or uprooting loved ones; either way it creates resentments and leaves you feeling guilty for inconveniencing them while you seek high-sounding, intangible rewards like "knowledge."

Yes, tenure means job security. But it does not alleviate these concerns. In fact, responsibilities to my family (we now had a one-year-old daughter), my own graduate students, my department, and my research career weighed on me more heavily than ever in the winter of 1983–84. For weeks I had spent my free time in Cornell's research library, reading and fantasizing about potential projects, but things were not falling into place. I had considered and rejected many possibilities, and the days were getting longer. If I missed the spring field season, I would lose an entire year.

Then one afternoon I hit pay dirt. A new book with the unlikely title *Producers and Scroungers* contained a chapter on intraspecific parasitism by Malte Andersson (1984). He highlighted waterfowl (Anatidae) generally and wood ducks (*Aix sponsa*) particularly as birds that frequently lay eggs in each other's nests. Andersson also reminded readers that waterfowl are unusual because females are philopatric. This dispersal pattern, which is more characteristic of mammals than birds (Greenwood 1980), raises the possibility that "parasitism" actually represents nepotism. Kin selection (Hamilton 1964) and nepotism (Sherman 1980) were among my favorite theoretical and empirical topics. The idea of linking them to conspecific parasitism appealed to me as an exciting new way to approach an old problem. Eagerly, I plunged into the literature on parasitism, with my hopes soaring.

## Conspecific Parasitism

Parasitism occurs when a female lays her egg(s) in a nest that is incubated by another female, either a conspecific or a heterospecific. Interspecific parasitism has been documented in 122 species of birds, which is less than 0.02 of all species (Payne 1977; Yom-Tov 1980; Lyon & Eadie 1991). Conspecific parasitism is far more common. Currently it is known in 185 species (Rohwer & Freeman 1989; Eadie et al. 1998), and the list is expanding as molecular techniques for inferring maternity are applied to more species (MacWhirter 1989).

Most intraspecific parasites are precocial and parasitism is facultative (Eadie et al. 1988). The behavior is especially common among waterfowl. It occurs in at least 76 of 162 anatids (0.47) worldwide, including $\geq$ 33 of 55 (0.60) North American species (Beauchamp 1997). Whereas waterfowl comprise only about 0.02 of birds, $\geq$ 0.26 of known conspecific parasites are anatids. Conspecific parasitism also is overrepresented among birds that nest in cavities. Whereas 139 of 2,177 species (0.06) from North America, Europe, Australia, and southern Africa are obligate cavity-nesters (Newton 1994), 49 of the 185 known parasitic species (0.26) nest in cavities (Beauchamp 1997; Eadie et al. 1998).

Table 16.1
Hypotheses for the Ubiquity of Conspecific Parasitism Among Cavity-Nesting Waterfowl

| |
|---|
| 1. Bet Hedging |
| 2. Nest-Site Saturation |
| 3. Salvage Reproduction |
| 4. Cooperative Breeding |
| 5. Reproductive Enhancement |
| 6. Side Effect of Nest-Site Competition |

Among birds exhibiting *both* characteristics—waterfowl that nest in cavities—conspecific parasitism is universal. It occurs in 38 of 38 hole-nesting anatids worldwide, including 11 of 11 species from North America and the western Palearctic (Rohwer & Freeman 1989; Eadie et al. 1998). Moreover, proportions of nests that are parasitized consistently are higher among anatids that nest in cavities than species that nest in emergent wetland vegetation or grasslands (Eadie et al. 1998).

Six hypotheses (Table 16.1) have been proposed to explain the ubiquity of conspecific parasitism (henceforth simply "parasitism") among cavity-nesting waterfowl (Eadie et al. 1988; Sayler 1992; Sorenson 1998). All assume that having a few extra eggs in a nest does not jeopardize a female's own egg-hatching success or chick survival—which is reasonable because female waterfowl usually can incubate enlarged clutches and ducklings are precocial (Grice & Rogers 1965; Eadie et al. 1988). First, females may lay in multiple sites to spread predation risks and "hedge their bets" against reproductive failure (Payne 1977; Rubenstein 1982). Second, if all the suitable nest sites are fully occupied, females may have nowhere else to lay (Bellrose et al. 1964; Gowaty & Bridges 1991; Evans 1988). Third, if a female's partially completed clutch were preyed on or her nest cavity were destroyed, parasitism might enable her to salvage some reproduction (Haramis et al. 1983; Evans 1988; McRae 1997). Fourth, females might enhance their own genetic success by incubating eggs of close relatives, especially if suitable nest cavities were limited (i.e., Andersson 1984). Fifth, females might parasitize opportunistically, laying in unattended nests whenever they are encountered (Brown & Bomberger Brown 1989, 1996; McRae 1998); hole-nesting might facilitate this because cavity locations are predictable from year to year (Rohwer & Freeman 1989). Finally, several females may start laying in the same high-quality cavity, but only one will remain to incubate. In this case, parasitism is not favored directly; rather, "joint clutches occur because the donor is forced to abandon the eggs she has already laid in the nest rather than as a result of an intentional attempt to lay eggs 'parasitically'" (Eadie et al. 1988, 1713).

I planned to use this framework of alternative hypotheses to investigate the adaptive significance of parasitism in wood ducks, with special emphasis at the outset on the fourth (nepotism) hypothesis.

## Parasitism in Wood Ducks

The nesting biology of wood ducks was reviewed by Bellrose and Holm (1994) and Hepp and Bellrose (1995). Briefly, the birds breed throughout eastern North America (east of the Mississippi River) and in central California and Oregon. They nest in tree cavities created either by large woodpeckers or limb breakage and subsequent heart rot. Females form a nest bowl in the punky wood at the cavity bottom, and lay one egg per day, usually in the early morning. They spend the rest of the day foraging to meet the energetic demands of oogenesis (Drobney 1980, 1982)—a complete clutch of ten to twelve eggs (Semel & Sherman 1992) comprises > 0.75 of the female's body mass (Bellrose & Holm 1994).

During egg laying, the male stays with the female, mating and guarding her against sexual assaults. He departs when she begins incubating, which starts after the final egg is laid and lasts thirty days. Eggs hatch simultaneously and the next morning the female departs with the ducklings; any unhatched eggs are abandoned. The family forages together in wetlands for two to three months, during which time the mother molts and the ducklings reach flight stage. In late summer the birds migrate to wintering grounds along the Gulf Coast and in southern California. Pairs form during the winter and, in the spring, males follow their mate back to her natal area to breed.

Among the hole-nesting Anatidae, the wood duck, its congener the mandarin duck (*A. galericulata*: Davies & Baggott 1989), and the black-bellied whistling duck (*Dendrocygna autumnalis*: McCamant & Bolen 1979) stand out regarding the prevalence of parasitism. In wood ducks, the best studied of the 3 species, clutch sizes range from 7 to 31 eggs in natural cavities (Semel & Sherman 1986; Ryan et al. 1998) and from 5 to > 50 eggs in man-made nesting structures (Clawson et al. 1979; Haramis & Thompson 1985; Sayler 1992; Semel & Sherman 1995), and proportions of parasitized nests range from 0.26 to 0.33 in natural cavities and from 0.14 to 0.95 in artificial structures.

The birds' well-known breeding biology, wide geographic range, and high rates of parasitism, as well as their obvious aesthetic appeal (e.g., in 1997 a U.S. postage stamp featured a male wood duck) together seemed to make *A. sponsa* ideal for investigating ecological factors that enhance parasitism in cavity-nesting waterfowl. However, there was one major drawback: wood ducks favor dense stands of deciduous bottomland forests, and the birds' canopy-level nest holes are hard to find, dangerous to access, and difficult to observe (e.g., Weier 1966; Yetta et al. 1999). Luckily, wood ducks also readily accept artificial cavities, and over the past fifty years several hundred thousand nest boxes have been erected by wildlife managers and citizens' groups (Bellrose 1990). I therefore attempted to locate a suitable managed population for behavioral studies.

## Puzzling Behaviors in the Swamp

In the spring of 1984 a Cornell colleague put me in touch with Leigh H. Fredrickson at the University of Missouri. For many years, Fredrickson and his students had studied wood ducks at Duck Creek, a reservoir adjacent to Mingo National Wildlife Refuge in the "bootheel" of Missouri. They had marked hundreds of ducklings individually, so the adult population likely included many putative kin (hatched in the same clutch).

Fredrickson kindly granted me permission to observe his birds, and I spent the summer and fall preparing (i.e., obtaining permits, investigating how to catch, handle, and mark wood ducks, etc.). In the field I built an 80 meter long wooden catwalk that snaked through the swamp, connecting an ancient duck-hunting blind with twelve boxes I mounted on metal poles. I rigged each box with a remotely controlled shutter, enabling me to trap any bird that entered. The blind would serve as my observation post and the catwalk would provide quick access to the boxes (for capturing birds and counting eggs).

When the wood ducks returned to Missouri in March 1985 I was there waiting for them, accompanied by Brad Semel, a Cornell graduate and former field assistant on the ground squirrel project. Brad had just finished his master's degree at Purdue University (Semel & Anderson 1988), and he was looking for adventure while awaiting the outcome of job applications. We set up housekeeping in spare rooms at Fredrikson's field lab and went to work. Every day at 0400 we maneuvered a small motorboat around stumps and weed beds out to the rickety observation blind, crawled in, and waited. We caught many females, recorded their band numbers (or banded them), and attached a brightly colored, individually numbered nasal saddle to their bill to facilitate identification at a distance. We were confident that before long we would know if kinship and parasitism were linked. Our spirits and adrenaline levels were high.

And, sure enough, what we observed each dawn was astonishing—but not in the way we had hoped. Whenever a female entered or left her box, pairs of unmarked birds began hurriedly arriving from all directions. Groups quickly coalesced and interacted noisily as one after another of the females attempted to enter the box, while males struggled to guard their mates. Females often fought at the hole mouth: they jostled, beat each other with their wings, and shoved each other aside, each attempting to enter the box first. Those that made it in stayed only long enough to lay, then departed squealing loudly.

Clutches often increased explosively (e.g., 4 eggs in 21 minutes) and erratically (0 to 8 eggs per day); final clutch sizes ranged from one to 37 eggs.

In the melee, eggs were laid in inappropriate places, such as on the ground, on top of boxes, in the water, and even on the back of an incubating female! Nearly every active nest was parasitized (0.95) even though nearly half the boxes (0.46) were unoccupied. Parasitism was random with regard to relatedness. Heavily parasitized nests were abandoned, and even nests that were incubated were largely unsuccessful. In the biggest clutches eggs lay 10 to 15 centimeters deep and many were crushed, overwhelmed by bacterial infections, or incompletely incubated. The smell was horrendous.

Obviously it was not reasonable to try to evaluate evolutionary hypotheses for conspecific parasitism (Table 16.1) under such chaotic social circumstances. Thoroughly discouraged, Brad and I packed up in mid-season, drafted a short, "terminal" manuscript (Semel & Sherman 1986) documenting what we had seen, and slunk home. We figured that we were through with wood ducks—forever.

## An Illuminating Diversion

I was depressed for months. Literally. My first post-tenure project, my great experiment, my "perfect" setup, all had been a complete bust. At age 34 I felt as if I were over the hill.

For months I moped around, feeling low and replaying the mental tape "ducks in swamp = you failed," until it finally wore out—meaning I was regaining some perspective (aided by familial understanding and therapy). As mental health returned, so did scientific curiosity. What triggered those bizarre, apparently maladaptive social behaviors? How could dump nesting persist in the face of relentless natural selection?

I knew that when an organism's environment changes suddenly, formerly adaptive traits can become nonadaptive and can remain so until selection has had time to operate in the new environment. For example, certain currently maladaptive aspects of human physiology (e.g., cravings for fat: Nesse & Williams 1994) and behavior (e.g., child abuse: Daly & Wilson 1988; rape: Thornhill & Palmer 2000) undoubtedly are legacies of selection in different, antecedent environments. Could extraordinary parasitism and density strife in wood ducks similarly be due to some kind of habitat disturbance?

The one variable humans obviously have changed is the placement of nest cavities. Natural cavities used by wood ducks usually are widely dispersed and concealed in the forest canopy (Weier 1966; Soulliere 1988, 1990; Robb & Bookhout 1995), whereas wildlife managers typically attach boxes to poles over open water (to reduce mammalian predation) and cluster them (to facilitate maintenance). When a normally solitary species is encouraged to nest in close proximity to conspecifics, unusual competitive and reproductive behaviors sometimes materialize (e.g., Gowaty & Wagner 1988; Robertson & Rendell 1990; Barber et al. 1996; Purcell et al. 1997). In wood ducks extreme parasitism and density strife characterize populations managed by erecting large numbers of boxes in open marshes (e.g., Jones & Leopold 1967; Morse & Wight 1969; Heusmann et al. 1980; Bellrose & Holm 1994).

But why does this so-called dump nesting occur? Our observations suggested a possibility: when boxes are conspicuous it is relatively easy for would-be parasites to find active nests. Females normally behave surreptitiously around their nest site, but it is impossible to conceal one's comings and goings to boxes over open water. If females' tendencies to lay parasitically vary inversely with the time and energy it takes them to locate active nests, clustering boxes in the open enhances parasitism because it minimizes nest-searching costs of would-be parasites.

This hypothesis yields two predictions. First, females should locate active nests to parasitize by observing and following each other. This was supported by our behavioral observations (Semel & Sherman 1986) and those of Heusmann et al. (1980). It was confirmed by experiments in which stuffed decoys were placed near certain boxes and subsequent rates of box entry and parasitism were quantified (Wilson 1993).

Second, making boxes less conspicuous should reduce dump nesting. Brad and I decided to test this. By 1986, he was employed at the Max McGraw Wildlife Foundation in northeastern Illinois. Brad's employer also was interested in wood ducks because of their importance as game birds (they are second only to mallards in hunters' bags: Hepp & Bellrose 1995). Max McGraw and the nearby Moraine Hills State Park allowed us to erect two sets of boxes at each site: "experimental" boxes were widely dispersed in woodlands and attached high on tree trunks so as to mimic natural cavity locations, whereas "control" boxes were clustered on poles in open marshes to mimic traditional management practices. Experimental and control sites were adjacent (approximately 0.5 kilometers apart) to minimize the possibility that any differences we found could be attributed to differences between areas (e.g., in predators). Over the next seven nesting seasons we conducted four replicated studies comparing parasitism and productivity of these boxes.

Results of all four experiments were substantially similar (Semel et al. 1988, 1990; Semel & Sherman 1993, 1995). Briefly, we found that:

1. Dump nesting had a major effect on population dynamics. In visible, clustered boxes, clutch sizes were directly related to rates of parasitism (Fig. 16.1a) and parasitism was inversely related to egg-hatching success (Fig. 16.1b). In turn, egg-hatching success was inversely related to the population size (numbers of eggs laid) and rates of parasitism (Fig. 16.2a, c). As a result, at low population densities parasitism was infrequent and egg hatchability was high, leading to population increases; as population densities increased, so did dump nesting, which reduced hatchability and led to population declines. The effects occurred at our own study sites and many others (Fig. 16.2 b, d, e; Fig. 16.3).

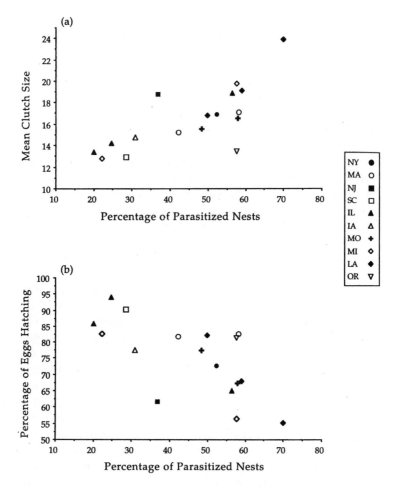

Figure 16.1. Relationships between the frequency of conspecific nest parasitism and (a) mean clutch size ($r_s$ = 0.76, P < 0.01), and (b) average hatching success of eggs ($r_s$ = −0.65, P < 0.01) of wood ducks from 17 studies in 10 states. The data are from Bellrose and Holm (1994), and the figure was modified from Eadie et al. (1998). Reprinted by permission of Oxford University Press.

Figure 16.2. (*Right*), Relationships between total number of eggs laid (an index of the size of the breeding population) and egg hatching success in visible, clustered boxes at the Montezuma National Wildlife Refuge in central New York (a, b) and the Max McGraw Wildlife Foundation in northeastern Illinois (c, d), and at ten national and state wildlife areas (e). Panels a and c illustrate that yearly egg numbers (*open circles*) and hatching success (*solid circles*) are virtually mirror images, and panels b, d, and e demonstrate the negative relationships between egg numbers and hatching success. Data in a and b is from Haramis and Thompson (1985), $r_s$ = −0.82, P < 0.05; data in c and d are from Semel et al. (1988), $r_s$ = −0.76, P < 0.02; data in e are from Semel et al. (1990) and Semel and Sherman (unpublished), $r_s$ = −0.65, P < 0.06. The figure was modified from Eadie et al. (1998), and reprinted by permission of Oxford University Press.

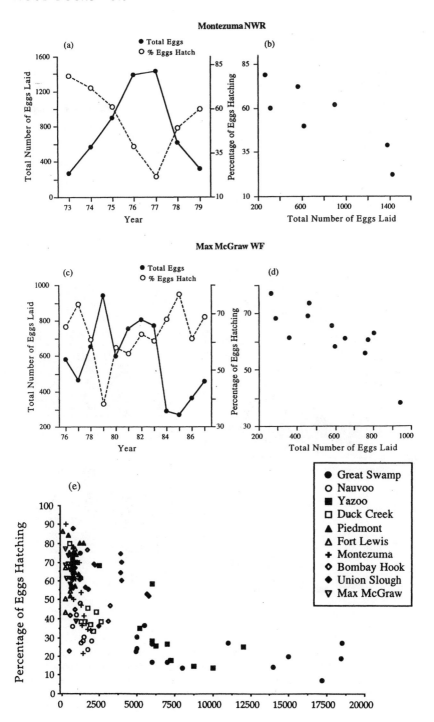

**Montezuma NWR**

(a)
● Total Eggs
○ % Eggs Hatch

(b)

**Max McGraw WF**

(c)
● Total Eggs
○ % Eggs Hatch

(d)

(e)

● Great Swamp
○ Nauvoo
■ Yazoo
□ Duck Creek
▲ Piedmont
△ Fort Lewis
+ Montezuma
◇ Bombay Hook
◆ Union Slough
▽ Max McGraw

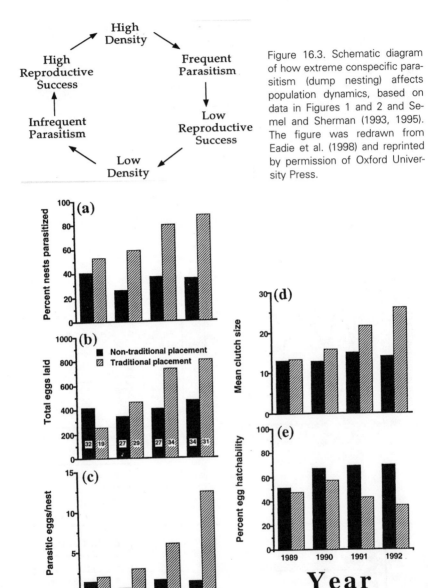

Figure 16.3. Schematic diagram of how extreme conspecific parasitism (dump nesting) affects population dynamics, based on data in Figures 1 and 2 and Semel and Sherman (1993, 1995). The figure was redrawn from Eadie et al. (1998) and reprinted by permission of Oxford University Press.

Figure 16.4. Comparison of reproduction by wood ducks nesting in "traditionally" placed nest boxes (pairs mounted back-to-back, on poles over open water; hatched bars) and nontraditionally placed boxes (dispersed in woodlands, attached 5 to 6 meters up on large tree trunks; dark bars) during 1989–92 at Moraine Hills State Park in northeastern Illinois: (a) total eggs laid, (b) frequency of parasitism, (c) parasitic eggs laid per nest, (d) mean clutch size, and (e) egg-hatching success. Parasitism was inferred when greater than or equal to two eggs appeared in a nest within twenty-four hours. The figure was modified from Semel and Sherman (1995) and reprinted by permission of The Wildlife Society.

2. Rates of parasitism and clutch sizes were considerably greater in visible, clustered boxes (traditional placement) than in hidden, dispersed boxes (nontraditional placement) (Fig. 16.4, a–d). The cyclic "boom-crash" dynamics that characterize many managed wood duck populations (e.g., Haramis & Thompson 1985; Semel et al. 1990; Semel & Sherman 1993) appears to be causally linked to parasitism and nest-box placement (Fig. 16.3).

3. Hatching success was significantly lower in visible, clustered boxes than in hidden, dispersed boxes (Fig. 16.4e). Although on average the same number of ducklings hatched per box from the two configurations, more than twice as many eggs were laid (parasitically) in each visible box. These unhatched eggs represent losses to individual reproductive success and population productivity.

4. Dump nesting was not due simply to scarcity of empty boxes. There was no relationship between weekly rates of parasitism and weekly proportions of occupied boxes in either the visible or hidden boxes (Semel & Sherman, in press). Parasitism was most frequent early in each nesting season (see also Clawson et al. 1979), when less than half the boxes were occupied.

5. Nests in hidden, dispersed boxes did not differ significantly from nests in natural tree cavities in any reproductive parameter, including nest initiation dates, frequencies of parasitism (0.25 to 0.35 of nests), parasitic eggs laid per nest (1 to 6 eggs), proportions of very large ($\geq 16$ eggs) "dump" clutches ($< 0.10$ of nests), probabilities that $\geq 1$ egg would hatch (i.e., that the nest was "successful") ($> 0.60$), probabilities that each egg would hatch ($> 0.50$), and numbers of unhatched eggs per nest ($< 4$ eggs).

Before our studies, it was assumed that dump nesting and associated reproductive interference, poor nesting success, and low egg hatchability characterized the breeding biology of wood ducks. This was because these phenomena occurred in nearly all managed populations, and no one had studied the birds' reproductive behaviors around natural cavities. Thus dump nesting aroused little suspicion or even concern in the wildlife community, especially since population productivity usually is gauged by the number of ducklings produced per box per year. Dump nesting does not necessarily reduce that, because the small fraction of eggs in huge clutches that sometimes do hatch equals the large fraction of eggs in unparasitized clutches that usually hatch. However, if one instead considers how many ducklings *could have been* produced from each box if most of the eggs hatched, the picture changes dramatically. Our data revealed a tremendous decrement in population growth relative to reproductive potential, and associated catastrophic declines in duckling production resulting in population crashes in heavily parasitized populations.

Replicated results showing that clustering nest boxes over open water could have serious negative consequences led us to recommend that managers begin hiding boxes at densities and in locations approximating those of natural cavities (e.g., Semel & Sherman 1993, 1995). The idea is to position boxes so the ducks are in a nesting environment more like their native habitat, where their behavioral mechanisms can function normally. In particular, to reduce dump nesting it is necessary to increase the costs in time and effort for females attempting to find active nests to parasitize.

Our recommendations have been widely publicized (e.g., Sherman & Semel 1989, 1992; Fellman 1993; Hope 1999) and, gradually, are being adopted nationwide (e.g., Bellrose & Holm 1994; Laboratory of Ornithology 1998). There is a growing realization that conservation and management programs must be developed in light of, rather than in spite of, the behavioral ecology of any target species, including wood ducks (see Caro 1998). Brad and I felt as if we had contributed to this emerging attitude, and we were proud of it. However, after seven years of study we still did not know why parasitism occurred under natural circumstances.

## Behavioral Ecology of Parasitism

Brad and I remained curious—and determined not to let wood ducks keep their private lives secret! In 1988 we had begun establishing a population at Moraine Hills State Park nesting in thirty-four appropriately located boxes—attached high on trunks of large trees ($\geq$ 6 meters off the ground) in oak-hickory woodlands at densities similar to natural cavities in northern hardwood forests (i.e., > 1/10–12ha: Ryan et al. 1998; Yetter et al. 1999). A summary of our results follow; the complete data are presented in Semel and Sherman (in press).

*Site fidelity.* Natal philopatry and fidelity to breeding areas characterize cavity-nesting waterfowl generally (Rohwer & Anderson 1988) and wood ducks particularly (Grice & Rogers 1965; Hepp et al. 1987b, 1989; Hepp & Kennamer 1992; Bellrose & Holm 1994). Natal philopatry also occurred at Moraine Hills. In 1991–94 we tagged 1,017 ducklings from 80 clutches on the day they hatched. Of these, 53 females and one male returned. Assuming a 1:1 sex ratio at hatching, the return rate for natal females was 0.104 versus 0.002 for natal males (Semel & Sherman, in press).

Adult nesters also were site-faithful. Among females that nested at Moraine Hills one year, 0.69 $\pm$ 0.05 (SD) returned to nest the next year (range: 0.60–0.86). Females nested for one to 7 successive years and then disappeared permanently (probably they died). On average, females nested in our population for 2.93 $\pm$ 1.62 consecutive years.

Females of some cavity-nesting waterfowl are faithful to particular nest sites (Rohwer & Anderson 1988; Gauthier 1990; Hepp & Kennamer 1992). This was certainly true of female wood ducks at Moraine Hills. Of the 64 females that nested for 2 to 7 years, 23 (0.36) used just one box and, on average, these females nested in only 2.5 $\pm$ 1.6 different boxes throughout their lifetime. This is significantly less than expected if they had switched cavities yearly.

*Occurrences of parasitism.* Of 240 nests initiated in the hidden boxes in 1989–95, 110 (0.46) increased by $\geq 2$ eggs in a 24-hour period at least once. This is an unequivocal indicator of parasitism (Drobney 1980). Most nests that were parasitized received several eggs (mean = $3.7 \pm 1.0$ parasitic eggs/nest, range = 1 to 18 eggs). This was because females typically focused egg laying on a few nests rather than scattering their eggs widely (Fig. 16.5). Returning females were observed laying in only $1.52 \pm 0.93$ different boxes per season and new recruits (most of which were one-year-olds) were observed laying in only $1.31 \pm 0.58$ different boxes. Of 56 new recruits and returning females that laid parasitically, 39 were observed to lay in only one other female's nest (0.70) and 52 (0.93) laid in only one or 2 other female's nests.

Parasitism was not due to unavailability of empty nest sites. Indeed, parasitism occurred most frequently when only 40 to 60 percent of boxes were occupied, as we (Semel & Sherman 1986; Semel et al. 1988, 1990) and others (Morse & Wight 1969; Heusmann et al. 1980) had documented previously. Moreover, there were no correlations between proportions of occupied boxes and frequencies of parasitism, either within (weekly) or among nesting seasons.

Of the 3,496 eggs that were laid during our study, 2,330 hatched (0.67). Eggs in nests in which parasitism was not detected were significantly more likely to hatch ($0.78 \pm 0.27$) than eggs in parasitized nests ($0.60 \pm 0.24$).

*Observations of parasitism.* We individually marked 173 adult females that nested in our boxes. Before first light on 158 mornings during 1992–95, observers hid in blinds erected near boxes and recorded behaviors of these birds. In total, 103 parasitic events were witnessed: 44 (0.43) by females that had nested previously in the population and 59 (0.57) by new recruits.

It quickly became apparent that parasitism was a flexible, not fixed, behavior. New recruits were especially likely to lay parasitically. They typically returned from migration after older females had already initiated clutches (Semel & Sherman, in press; also Hepp & Kennamer 1993; Bellrose & Holm 1994), and most new recruits did not lay and incubate a full clutch. Rather, they ranged widely in the woods, typically in groups, "prospecting" for cavities. Often they followed older females and parasitized opportunistically when they discovered an active nest that was temporarily unoccupied.

The longer a female's breeding tenure the more likely that sometime in her life she would have both laid in another female's nest (i.e., behaved as a parasite) and incubated a nest containing foreign eggs (behaved as a host). Of 115 females that bred for one to 4 years, 86 (0.75) had served as a host at least once and 62 (0.54) laid at least one parasitic egg. In 27 cases a female behaved as both a parasite and a host in a single season. Among the 18

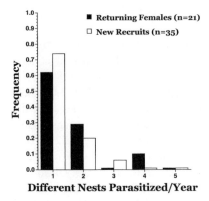

Figure 16.5. Numbers of different nest boxes in which returning females (*dark bars*) and new recruits (*open bars*) laid eggs that they did not incubate within a nesting season at Moraine Hills State Park. The data are from Semel and Sherman (in press).

females with 3- to 4-year breeding tenures, 13 (0.72) had behaved as a host and a parasite at least once during their lifetime.

*Triggers of parasitism: returning females.* Whether or not a returning female would parasitize was determined primarily by the contents of the box in which she had nested previously (Fig. 16.6). On 29 occasions that box was unoccupied when the female returned from migration, and 25 times (0.86) she nested there again (0.86). Twenty-seven of these females (0.93) laid only in their own nest, and 2 (0.07) laid parasitically before initiating their own clutch in the previous year's box (Fig. 16.6a).

On 49 occasions the previous year's box was occupied when the female returned. Ten times the occupant was a heterospecific (e.g., an owl, starling, or squirrel). All 10 females switched boxes to nest, and none was observed to lay parasitically (Fig. 16.6b).

On the remaining 39 occasions the previous year's box contained ≥ one fresh egg, laid by an earlier-arriving conspecific. Thirty times we observed the returning female to also start laying there again. In 13 of these cases (0.33) the returning resident successfully reoccupied the box. Eggs left behind by the earlier-arriving female became "parasitic" when she departed (Fig. 16.6c). Conversely, in 17 cases (0.44) it was the earlier-arriving female who stayed and the returning resident who abandoned her "parasitic" eggs.

In the remaining 9 cases (0.22), returning females nested elsewhere, and we did not see them lay in their previous year's box. However, all 9 boxes were parasitized soon after the former resident returned from migration, and every one of these females was observed nearby on the day(s) foreign eggs were laid in their previous year's box. In summary, whereas only 2 of 29 returning females (0.07) laid in a conspecific's nest if their previous year's box was unoccupied, at least 30 of 39 (0.77) and probably all 39 returning females laid in a conspecific's nest when it was in their previous year's box (Fig. 16.6c).

Among the 26 returning females that were unable to reoccupy their pre-

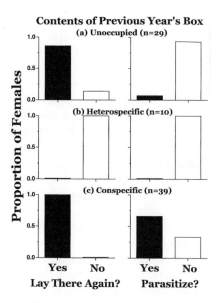

**Contents of Previous Year's Box**

Figure 16.6. Laying behaviors of females that returned to Moraine Hills State Park grouped according to the current contents of the box each female nested in the previous season: (a) the box was unoccupied, (b) the box contained a heterospecific (e.g., a starling, screech owl, or evidence of raccoon or squirrel visitation), and (c) the box contained freshly laid wood duck eggs. Note that females usually laid in their previous year's box if it was unoccupied or if a conspecific already was laying there (a, c), and that "parasitism" was associated with presence of conspecifics' eggs in the favored box. Reprinted from Semel and Sherman (in press).

vious year's box, 20 nested and 6 disappeared. The nesters did not move far: 10 (0.53) nested in the next nearest box, 18 (0.90) nested in one of the closest 5 boxes, and one nested in the favored box after the first female's eggs hatched. When a female's "second choice" box also contained fresh eggs, she nonetheless laid there and her eggs, or those of the other female, eventually became "parasitic" when one of them was evicted. All returning females that parasitized 2 or more nests in one season (Fig. 16.5) had first laid in and abandoned their previous year's box, then laid in and abandoned a nearby box, and finally incubated a clutch in an unoccupied box.

Twelve females that in year $t + 1$ were excluded from the box they had used in year $t$ returned to the population in year $t + 2$. All of them laid first in the box they had used in year $t$. Five successfully reoccupied that box and 7 were displaced, leaving behind "parasitic" eggs. All 7 then nested in one of the 2 nearest boxes. Strong preferences for previously used boxes explains why females typically nested in so few over their lifetime.

*Triggers of parasitism: returning parasites.* Every year one-third to one-half of the new recruits to the population laid a few parasitic eggs and abandoned them. Eleven of these females returned the following year. At that time, 5 of the boxes they had parasitized were unoccupied and 6 already contained eggs. Each of the former 5 nested in the unoccupied box that she had parasitized the previous year. Three of the latter 6 laid again in the occupied box: 2 of them successfully occupied it and incubated a mixed clutch of their own and the other female's "parasitic" eggs, while the third female abandoned her "parasitic" eggs and disappeared. The other 3 females switched boxes to nest, 2 in the box nearest the one they had parasitized the previous year. Although we did not observe them laying in the focal box, all 3 boxes were parasitized and the relevant female was seen nearby on the day(s) parasitic eggs were laid. Thus at least 8 of 11 (0.73) and probably all 11 returning parasitic females initially laid in the box they had parasitized previously.

*Kinship and parasitism.* We observed laying behaviors of 26 returning one-year-old females that had hatched in our population, while simultaneously monitoring the contents of their natal box. Only 2 of these females found their natal boxes empty: one nested there and the other laid 2 eggs in the closest box and disappeared. Natal boxes of the other 24 yearlings already were occupied: 7 times by the yearling's "mother" (i.e., the female that incubated her) and 17 times by a nonrelative. New recruits never laid in their natal box if their mother again was nesting there: 3 parasitized a different box and disappeared, 3 nested in the nearest unoccupied box, and one waited for her mother's clutch to hatch and then nested in her natal box.

By contrast, of the 17 yearlings whose natal box was occupied by a female other than her "mother," 4 laid ≥ one egg there and disappeared, 4 parasitized the box nearest the natal box (one in a nest initiated by her mother), and 3 parasitized a more distant box. The remaining 6 yearlings laid and incubated their own clutch (4 in one of the 2 nearest boxes to their natal box) and we did not observe them to lay parasitically. However, each of their natal boxes was parasitized and in every case the relevant yearling was seen nearby on the day(s) foreign eggs appeared in the natal box. In sum, at least 4 (0.24) and, probably, 10 of 17 (0.59) yearling females parasitized their natal box when it contained eggs laid by a female other than their mother. Thus parasitism occurred far more frequently when the box occupant was not a female's mother.

To further investigate avoidance of kin, we observed 37 pairs of ≥ 2-year-old relatives (23 "mother-daughter" pairs and 14 pairs of same-age clutch-mates), both of whom laid ≥ one egg in the same season. These females laid 28 eggs that they did not incubate: 27 (0.96) were laid in nests of nonrelatives and one female laid an egg in her mother's nest (0.04). If females had parasitized randomly they would have been expected to lay in nests of kin 14 times (based on frequencies of active nests of each female's kin and nonkin on the day she laid each egg: Semel & Sherman, in press). Thus females laid in relatives' nests significantly less often than expected by chance.

*Laying behaviors.* Adult females that arrived early in the nesting season found most boxes unoccupied. Once they had initiated clutches, females started behaving surreptitiously around their box. They arrived at or before first light, flew directly and silently to their box, and entered it on the wing (i.e., without landing at the entrance hole); they remained inside 0.5 to 1.0 hour, then departed silently. If conspecific females were nearby, nesting females waited until the others left the area before entering or exiting their box.

The behavior of new recruits was dramatically different. They appeared at boxes significantly later in the morning than adults (usually > one hour after sunrise), often in groups. They approached boxes hesitantly, first landing in nearby trees and craning their neck to inspect the box from various distances and angles. These lengthy inspections (> thirty minutes) were accompanied by noisy vocalizations, as if to determine whether the box owner was inside. When a new recruit located a temporarily unattended nest she entered, quickly laid an egg, then rapidly departed, whistling and squealing.

Adult females that had nested in the population previously behaved more like nest owners than new recruits. Returning residents appeared soon after sunrise, flew directly and silently to their previous year's box, entered it on the wing, remained inside about an hour, and departed silently. They waited for conspecific females to leave the area before entering or exiting their box. It usually was impossible to determine behaviorally which female—the later-arriving former resident or the earlier-arriving new resident—was the "host" and which the "parasite." Only when one of the females finally abandoned the box to the other could a female's eggs be designated as "parasitic."

*Female-female aggression.* On 25 mornings we witnessed what happened when one adult female was inside a box and another female attempted to enter it. On 11 occasions (0.44) the intruder was shoved backward out of the hole mouth, and 14 times (0.56) the intruder entered the box. Twice we saw the box shaking and heard sounds of fighting (e.g., wing flapping, scratching); in one case feathers were torn from the head of both the former and current residents. We did not detect aggression in the other 12 cases, but we suspect it occurred because both females left the box shortly after the second bird entered, and they departed rapidly and nearly simultaneously.

## Behavioral Flexibility

Whether or not a female wood duck laid eggs she did not incubate depended on her age and the contents of the cavity in which she had nested previously. New recruits to the population laid 0.57 of parasitic eggs. They returned from migration well after older females had initiated nests, and only about one-third of them laid and incubated a full clutch. Yearlings' energy reserves may have been insufficient to lay a full clutch (Drobney 1982; Hepp et al.

1987a; Saether 1990). We observed them "prospecting" for cavities, sometimes by following older females to active sites. As a result, it was primarily active nests (rather than unoccupied boxes) that were parasitized by new recruits. This explains the lack of relationship between proportions of occupied boxes and weekly rates of parasitism.

The behavior of returning adult females was considerably different (Fig. 16.7). They typically laid in the box they had used the previous season, unless it was occupied by a heterospecific. The presence of fresh eggs in their favored box apparently was not a deterrent to laying, as indicated by the fact that they began to lay at the same time in the season as females whose previous year's box contained no eggs (Semel & Sherman, in press). Temporary joint nesting could occur because female wood ducks are not territorial (Haramis 1990; Eadie et al. 1998) and they spend most of their time during egg laying foraging out of sight of their nest (Drobney 1980; Bellrose & Holm 1994).

When females that were laying in the same box finally met, they often fought. Bellrose and Holm (1994, 255–262) also reported altercations, serious injuries, and even deaths among females competing for preferred nest boxes. Usually (two-thirds of the time) the female that began laying first prevailed. Displaced females usually nested nearby and, the following season, attempted to return to their preferred (i.e., original) box. "Parasitism" resulted when a female departed, leaving behind eggs that were incubated by her rival; 0.43 of all "parasitic" eggs resulted from such evictions.

## Evaluating the Alternative Hypotheses

Our observations indicated that parasitism was not a unitary phenomenon, but rather a complex mixture of contingent behaviors. In wood ducks, as in all other species that parasitize conspecifics (e.g., Lank et al. 1989; Weigmann & Lamprecht 1991; Eadie & Fryxell 1992; Sorenson 1991, 1993; Lyon 1993a; McRae 1998), parasitism and nesting are flexible alternative behaviors—that is, "conditional strategies"—not a genetic polymorphism (Maynard Smith 1982) or a genetically controlled, mixed ESS (i.e., females did not parasitize with fixed probability $p$ and nest with fixed probability $q$: Parker 1984). Within and between seasons a female wood duck might both parasitize and nest, only parasitize, or only nest. During their lifetime, most females acted as both "parasites" and "hosts," and many did so in the same season, depending on the contents of favored boxes (Figs. 16.6 and 16.7).

We can now assess the applicability of the various hypothesized advantages of intraspecific parasitism (Table 16.1) to wood ducks. First, females might lay parasitically to spread predation risks, thereby "hedging their bets" against total reproductive failure (Payne 1977; Rubenstein 1982). Theoretical models (e.g., Bulmer 1984; Sorenson 1991) suggest that fitness gains from

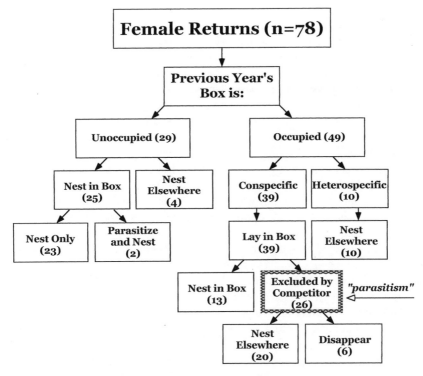

Figure 16.7. Schematic summary of the sequence of laying behaviors exhibited by forty-one returning females on seventy-eight nesting occasions (seasons) in relation to the contents of the box in which they nested previously at Moraine Hills State Park. Sample sizes are indicated in parentheses. "Parasitism" occurred primarily when one female forced another to abandon a nesting attempt, leaving behind eggs that were incubated by her victorious competitor. Reprinted from Semel and Sherman (in press).

such parasitic risk-spreading are small unless predation is severe—which it is in cavity-nesting species generally (Bellrose 1976) and wood ducks particularly (Robb & Bookhout 1995). However, to take full advantage of risk-spreading, females are expected to scatter their eggs in multiple nests (Lyon 1993a, b). Instead, new recruits and returning females laid in relatively few boxes (Fig. 16.5), implying that they were not merely attempting to hedge their bets.

Second, females might be forced to lay in active nests because all suitable cavities are fully occupied (Bellrose et al. 1964; Gowaty & Karlin 1984; Evans 1988). Under this hypothesis, females are "making the best of a bad situation" (Dawkins 1980) and "salvaging" some reproduction through parasitism (Eadie et al. 1988; Weigmann & Lamprecht 1991; Sorenson 1993, 1998). However, we found that there was no relationship between rates of parasitism and degrees of nest-site saturation. Suitable, unoccupied boxes were available throughout every nesting season, especially when parasitism was most prevalent. Thus, nest-site saturation cannot explain the parasitism we observed, in agreement with previous studies of cavity-nesting waterfowl (e.g., Eriksson & Andersson 1982; Dow & Fredga 1984; Davies & Baggott 1989; McCamant & Bolen 1979), including wood ducks (e.g., Morse & Wight 1969; Clawson et al. 1979; Heusmann et al. 1980).

Third, parasitism might be triggered by the loss of a partially completed clutch, for example, due to predation or a storm that destroyed the nest cavity. If another suitable cavity were not available or females were physiologically unprepared to lay a full replacement clutch, individuals might "salvage" some reproduction by laying their last few eggs parasitically (Haramis et al. 1983; Evans 1988; McRae 1997). However, partial clutch loss cannot explain the parasitism we observed because numbers of suitable nest boxes fluctuated very little during our study and no catastrophic losses of boxes or trees containing natural cavities occurred at Moraine Hills.

Fourth, what appeared to be "parasitism" might actually be a form of cooperative breeding (Andersson 1984; Gibbons 1986; McRae 1996). When high-quality cavities are scarce or hard to find, there may be an inclusive fitness advantage for related females to lay in the same nest, and perhaps even share incubation. Although natal philopatry and nest-site fidelity characterize wood ducks, there was no evidence of cooperative breeding or nepotism. In fact, the opposite occurred: new recruits avoided parasitizing their natal box when it was occupied by their "mother," and adult females avoided laying in nests of clutch-mates or "daughters."

Fifth, females may enhance their reproduction by laying parasitically whenever they locate an active, unattended nest. Opportunistic parasites potentially gain reproductive benefits without paying costs of incubation or parental care. The behavior of new recruits was consistent with reproductive enhancement. Interestingly, however, new recruits typically laid in only one box (Fig. 16.5) and in subsequent years they focused their reproductive activities on it. This implies that their prospecting behavior involved not only opportunism but also locating a site for future nesting (Eadie & Gauthier 1985; Bellrose & Holm 1994).

The behavior of returning nesters did not fit predictions of hypothesis 1–5. These females laid in the box they used previously regardless of whether it already contained eggs (Figs. 16.6 and 16.7). They seldom skipped laying in that site to lay elsewhere, although they must have occasionally observed conspecifics entering and leaving other boxes. If they were evicted from their favored box, returning nesters laid and incubated nearby, and the next year returned to the first box.

These behaviors support the sixth hypothesis, namely, that for returning former residents "parasitism" was a side effect of fidelity to suitable nesting cavities and competition for them. When high-quality cavities are limited in number relative to the population size, multiple females may simultaneously attempt to use the same one (Erskine 1972, 1990; Yom-Tov 1980; Andersson & Eriksson 1982). Eventually one of the competing females is evicted. She leaves behind eggs that become "parasitic" in the definitional sense (i.e., raised by a female other than their mother) when the victorious female incubates them.

The sixth hypothesis differs from the alternatives in an important way. Under the first five "adaptive" hypotheses, parasitic females benefit from the presence of the host because she will incubate their eggs. However, if parasitism is a side effect of nest-site competition neither female directly benefits from the presence of the other because each is primarily attempting to secure exclusive use of the cavity. Of course, a female that loses the competition may eventually have a few eggs successfully incubated and so recoup some reproduction. However, females did not behave as if "parasitism" were their favored option for enhancing fitness. Rather, they fought to retain the box (i.e., to be the "host"), with "parasitism" being the default that resulted from eviction.

The side effect hypothesis requires variation in the quality of cavities, resulting in competition for the best ones. Natural cavities used by wood ducks do differ considerably in physical characteristics (e.g., internal dimensions, height, entrance size: Weier 1966; Soulliere 1988, 1990; Ryan et al. 1998) and in accessibility to predators (e.g., Bellrose & Holm 1994; Robb & Bookhout 1995). In cavity-nesting species generally, a site that was used successfully, including the natal cavity, is worth returning to—and fighting for—because its location is known and its quality is proven (Dow & Fredga 1983; Nilsson 1984; Gauthier 1990; Hepp & Kennamer 1992; Pöysä 1999).

The general implication of our results is that scarcity of optimal nesting sites probably is the key ecological factor underlying the four unusual behavioral phenomena exhibited by all cavity-nesting waterfowl: natal philopatry, nest-site fidelity, nest-site competition, and, most important, intraspecific parasitism.

## Future Directions

Enough is now known about the natural history of parasitism in cavity-nesting waterfowl, especially wood ducks, mandarin ducks, and black-bellied whistling ducks, to support experimental tests of hypotheses for its occurrence. For example, I have begun conducting field manipulations with wood ducks at Montezuma National Wildlife Refuge in central New York. To test the nest-site competition hypothesis for adult females I have been removing one of the putative competitors. If the female that laid eggs in the box earlier in the season (i.e., the "usurper") were removed, the later-arriving former resident should take over and incubate the mixed clutch, and vice versa if the later-arriving returning female were removed. However, if the former resident were merely laying "parasitically" (e.g., to hedge her bets or enhance her reproduction), then she should abandon her eggs whether or not the usurper was removed, and the clutch should not be incubated.

One possible way to test the reproductive enhancement hypothesis for parasitism by one-year-old females involves associating body condition with behavior. The hypothesis predicts that yearlings in the best condition should lay a full clutch and incubate it, whereas yearlings in the poorest condition should be "present only" (i.e., they may visit boxes but not lay eggs); yearlings that parasitize but do not nest should be in intermediate condition. Condition can be estimated using residuals of regressions comparing various body measurements (e.g., wing or tarsus length versus weight or fat deposits), and also manipulated experimentally to see if the birds' subsequent laying behavior is altered in the predicted directions.

## Conclusion

My study began with the straightforward question, "Is intraspecific parasitism in wood ducks nepotistic?" It has taken sixteen years and considerable physical and emotional effort to get the answer. Clearly, it is "no."

Testing and rejecting the kin selection hypothesis has revealed that conspecific parasitism in wood ducks is actually a complicated and intriguing set of behaviors. An unanticipated side-benefit has been the management improvements that sprung from knowledge of the behavioral ecology of parasitism. Brad Semel and I have witnessed with pride the widespread adoption of our recommendation to place nest boxes at densities and in locations that mimic those of natural cavities.

Wood ducks are common "back yard" birds in many parts of the United States. Their accessibility and the extensive natural history information now available surely qualifies them as a "model system" for addressing both applied and basic questions about parasitism in cavity-nesting waterfowl. Answering these questions offers many intriguing possibilities for future research.

## Acknowledgments

Brad Semel and I have collaborated for over twenty years. I appreciate his friendship, stamina, enthusiasm, and honesty. Cynthia Kagarise Sherman and I have collaborated for over twenty-five years. I thank her for insights, encouragement, and love, especially in the dark months following the scientific "disaster" in the swamps of Missouri. Leigh H. Fredrickson facilitated my initial studies and James C. Ware helped build the catwalk. In Illinois, assistance and logistical support were provided by George V. Burger, Steven M.

Byers, John I. Laskowski, Robert A. Montgomery, and John B. Schweder. Our wood duck studies were facilitated by the Max McGraw Wildlife Foundation and Montezuma National Wildlife Refuge, and supported by the John Simon Guggenheim Foundation, Max McGraw Wildlife Foundation, Wildfowler's Association of Central New York, and the U.S. Department of Agriculture, through Hatch Grants administered by the Agricultural Experiment Station at Cornell University.

# References

Andersson M, 1984. Brood parasitism within species. In: Producers and scroungers: strategies of exploitation and parasitism (Barnard CJ, ed). London: Croom Helm; 195–228.

Andersson M, Ericksson MO, 1982. Nest parasitism in goldeneyes *Bucephala clangula*: some evolutionary aspects . Amer Nat 120:1–16.

Barber CA, Robertson RJ, Boag PT, 1996. The high frequency of extra-pair paternity in tree swallows is not an artifact of nest boxes. Behav Ecol Sociobiol 38:425–430.

Beauchamp G, 1997. Determinants of intraspecific brood amalgamation in waterfowl. Auk 114:11–21.

Bellrose FC, 1976. Ducks, geese and swans of North America. Harrisburg, PA: Stackpole Books.

Bellrose FC, 1990. The history of wood duck management. In: Proceedings of the 1988 North American wood duck symposium (Fredrickson LH, Burger GV, Havera SP, Graber DA, Kirby RE, Taylor TS, eds). St. Louis; 13–20. Privately published.

Bellrose FC, Holm DJ, 1994. Ecology and management of the wood duck. Mechanicsburg, PA: Stackpole Books.

Bellrose FC, Johnson KL, Meyers TU, 1964. Relative value of natural cavities and nesting houses for wood ducks. J Wildl Manage 28:661–676.

Brown CR, Bomberger-Brown M, 1989. Behavioral dynamics of intraspecific brood parasitism in colonial cliff swallows. Anim Behav 37:777–796.

Brown CR, Bomberger-Brown M, 1996. Coloniality in the cliff swallow. Chicago: University of Chicago Press.

Bulmer MG, 1984. Risk avoidance and nesting strategies. J Theor Biol 106:529–535.

Caro T (ed), 1998. Behavioral ecology and conservation biology. Oxford: Oxford University Press.

Clawson RL, Hartman GW, Fredrickson LH, 1979. Dump nesting in a Missouri wood duck population. J Wildl Manage 43:347–355.

Daly M, Wilson M, 1988. Homicide. Hawthorne, NY: Aldine-de Gruyter.

Davies AK, Baggott AK, 1989. Egg laying, incubation and intraspecific nest parasitism by the mandarin duck *Aix galericulata*. Bird Study 36:115–122.

Dawkins R, 1980. Good strategy or evolutionarily stable strategy? In: Sociobiology: beyond nature/nurture? (Barlow GW, Silverberg J, eds). Boulder, CO: Westview Press; 331–367.

Dow H, Fredga S, 1984. Factors affecting reproductive output of the goldeneye duck *Bucephala clangula*. J Anim Ecol 53:679–692.

Drobney RD, 1980. Reproductive bioenergetics of wood ducks. Auk 97:480–490.

Drobney RD, 1982. Body weight and composition changes and adaptations for breeding in wood ducks. Condor 84:300–305.

Eadie JM, Fryxell JM, 1992. Density dependence, frequency dependence, and alternative nesting strategies in goldeneyes. Am Nat 140:621–641.

Eadie JM, Gauthier G, 1985. Prospecting for nest sites by cavity-nesting ducks of the genus *Bucephala*. Condor 87:528–534.

Eadie JM, Kehoe FP, Nudds TD, 1988. Pre-hatch and post-hatch brood amalgamation in North American Anatidae: a review of hypotheses. Can J Zool 66:1709–1721.

Eadie JM, Sherman PW, Semel B, 1998. Conspecific brood parasitism, population dynamics and the conservation of cavity-nesting birds. In: Behavioral ecology and conservation biology (Caro TM, ed). Oxford: Oxford University Press; 306–340.

Eriksson MOG, Andersson M, 1982. Nest parasitism and hatching success in a population of goldeneyes *Bucephala clangula*. Bird Study 29:49–54.

Erskine AJ, 1972. Buffleheads. Can Wildl Serv Monogr Ser 4.

Erskine AJ, 1990. Joint laying in *Bucephala* ducks—"parasitism" or nest-site competition? Orn Scand 21:52–56.

Evans PGH, 1988. Intraspecific nest parasitism in the European starling *Sturnus vulgaris*. Anim Behav 36:1282–1294.

Fellman B, 1993. The trouble with wood ducks. Nat Wildl 31:46–51.

Gauthier G, 1990. Philopatry, nest-site fidelity, and reproductive performance in buffleheads. Auk 107:126–132.

Gibbons DW, 1986. Brood parasitism and cooperative breeding in the moorhen, *Gallinula chloropus*. Behav Ecol. Sociobiol 19:221–232.

Gowaty PA, Bridges WC, 1991. Nestbox availability affects extra-pair fertilizations and conspecific nest parasitism in eastern bluebirds, *Sialia sialis*. Anim Behav 41:661–675.

Gowaty PA, Karlin AA, 1984. Multiple maternity and paternity in single broods of apparently monogamous eastern bluebirds (*Sialia sialis*). Behav Ecol Sociobiol 15:91–95.

Gowaty PA, Wagner SJ, 1988. Breeding season aggression of female and male eastern bluebirds (*Sialia sialis*) to models of potential conspecific and interspecific egg dumpers. Ethology 78:238–250.

Greenwood PJ, 1980. Mating systems, philopatry and dispersal in birds and mammals. Anim Behav 28:1140–1162.

Grice D, Rogers JP, 1965. The wood duck in Massachusetts. Massachusetts Division of Fish and Game Final Report, Project W-19-R.

Hamilton WD, 1964. The genetical evolution of social behaviour, I and II. J Theor Biol 7:1–52.

Haramis GM, 1990. Breeding ecology of the wood duck: a review. In: Proceedings of the 1988 North American wood duck symposium (Fredrickson LH, Burger GV, Havera SP, Graber DA, Kirby RE, Taylor TS, eds). St. Louis; 45–60. Privately published.

Haramis GM, Alliston WG, Richmond ME, 1983. Dump nesting in the wood duck traced by tetracycline. Auk 100:729–730.

Haramis GM, Thompson DQ, 1985. Density-production characteristics of box-nesting wood ducks in a northern greentree impoundment. J Wildl Manage 49:429–436.

Hepp GR, Bellrose FC, 1995. Wood duck. In: Birds of North America, vol. 169

(Poole A, Gill F, eds). Philadelphia: Philadelphia Academy of Natural Sciences; 1–23.

Hepp GR, Hoppe RT, Kennamer RA, 1987b. Population parameters and philopatry of breeding female wood ducks. J Wildl Manage 51:401–404.

Hepp GR, Kennamer RA, Harvey WF, IV, 1989. Recruitment and natal philopatry of wood ducks. Ecology 70:897–903.

Hepp GR, Kennamer RA, 1992. Characteristics and consequences of nest-site fidelity in wood ducks. Auk 109:812–818.

Hepp GR, Kennamer RA, 1993. Effects of age and experience on reproductive performance of wood ducks. Ecology 74:2027–2036.

Hepp GR, Stangohr D, Baker LA, Kennamer RA, 1987a. Factors affecting variation in the egg and duckling components of wood ducks. Auk 104:435–443.

Heusmann HW, Bellville RH, Burrell RG, 1980. Further observations on dump nesting by wood ducks. J Wildl Manage 44:908–915.

Hope J, 1999. Breaking out of the box. Audubon 101:86–91.

Jones RE, Leopold AS, 1967. Nesting interference in a dense population of wood ducks. J Wildl Manage 31:221–228.

Laboratory of Ornithology, 1998. Wood duck (*Aix sponsa*) species account. In: Cornell nest box network research kit. Ithaca, NY: Cornell University; 4.1–4.3.

Lank DB, Cooch EG, Rockwell RF, Cooke F, 1989. Environmental and demographic correlates of intraspecific nest parasitism in Lesser Snow Geese *Chen caerulescens caerulescens*. J Anim Ecol 58:29–45.

Lyon BE, 1993a. Conspecific brood parasitism as a flexible female reproductive tactic in American coots. Anim Behav 46:911–928.

Lyon BE, 1993b. Tactics of parasitic American coots: host choice and the pattern of egg dispersion among host nests. Behav Ecol Sociobiol 33:87–100.

Lyon BE, Eadie JM, 1991. Mode of development and interspecific avian brood parasitism. Behav Ecol 2:309–318.

MacWhirter RB, 1989. On the rarity of intraspecific brood parasitism. Condor 91:485–492.

Maynard Smith J, 1982. Evolution and the theory of games. Cambridge: Cambridge University Press.

McCamant RE, Bolen EG, 1979. A 12-year study of nest box utilization by black-bellied whistling ducks. J Wildl Manage 43:936–943.

McRae SB, 1996. Family values: costs and benefits of communal nesting in the moorhen. Anim Behav 52:225–245.

McRae SB, 1997. A rise in nest predation enhances the frequency of intraspecific brood parasitism in a moorhen population. J An Ecol 66:143–153.

McRae SB, 1998. Relative reproductive success of female moorhens using conditional strategies of brood parasitism and parental care. Behav Ecol 9:93–100.

Morse TE, Wight HM, 1969. Dump nesting and its effects on production in wood ducks. J Wildl Manage 33:284–293.

Nesse RM, Williams GC, 1994. Why we get sick. New York: Random House.

Newton I, 1994. The role of nest sites in limiting the numbers of hole-nesting birds: a review. Biol Conserv 70:265–276.

Nilsson SG, 1984. The evolution of nest-site selection among hole-nesting birds: the importance of nest predation and competition. Ornis Scand 15:167–175.

Parker GA, 1984. Evolutionarily stable strategies. In: Behavioural ecology: an evolutionary approach, 2nd ed. (Krebs JR, Davies NB, eds). Sunderland, MA: Sinauer Associates; 30–61.

Payne RB, 1977. The ecology of brood parasitism in birds. Ann Rev Ecol Syst 8:1–28.

Pöysä H, 1999. Conspecific nest parasitism is associated with inequality in nest predation risk in the common goldeneye (*Bucephala clangula*). Behav Ecol 10:533–540.

Purcell KL, Verner J, Oring LW, 1997. A comparison of the breeding ecology of birds nesting in boxes and tree cavities. Auk 114:646–656.

Robb JR, Bookhout TA, 1995. Factors influencing wood duck use of natural cavities. J Wildl Manage 59:372–383.

Robertson RJ, Rendell WB, 1990. A comparison of the breeding ecology of a secondary cavity nesting bird, the tree swallow (*Tachycineta bicolor*), in nest boxes and natural cavities. Can J Zool 68:1046–1052.

Rohwer FC, Anderson MG, 1988. Female biased philopatry, monogamy, and the timing of pair formation in migratory waterfowl. Current Ornithology 5:187–221.

Rohwer FC, Freeman S, 1989. The distribution of conspecific nest parasitism in birds. Can J Zool 67:239–253.

Rubenstein DI, 1982. Risk, uncertainty and evolutionary strategies. In: Current problems in sociobiology (King's College Sociobiology Group, eds). Cambridge: Cambridge University Press; 91–111.

Ryan DC, Kawula RJ, Gates RJ, 1998. Breeding biology of wood ducks using natural cavities in southern Illinois. J Wildl Manage 62:112–123.

Saether B, 1990. Age-specific variation in reproductive performance of birds. Curr Ornithol 7:251–283.

Sayler RD, 1992. Ecology and evolution of brood parasitism in waterfowl. In: Ecology and management of breeding waterfowl (Batt BDJ, Afton AD, Anderson MG, Ankney CD, Johnson DH, Kadlec JA, Krapu GL, eds). Minneapolis: University of Minnesota Press; 290–322.

Semel B, Anderson DC, 1988. Vulnerability of acorn weevils (Coleoptera, Curculionidae) and attractiveness of weevils and infested *Quercus alba* acorns to *Peromyscus leucopus* and *Blarina brevicauda*. Amer Midl Nat 119:385–393.

Semel B, Sherman PW, 1986. Dynamics of nest parasitism in wood ducks. Auk 103:813–816.

Semel B, Sherman PW, 1992. Use of clutch size to infer brood parasitism in wood ducks. J Wildl Manage 56:495–499.

Semel B, Sherman PW, 1993. Answering basic questions to address management needs: case studies of wood duck nest box programs. Trans N Amer Wildl Nat Res Conf 58:537–550.

Semel B, Sherman PW, 1995. Alternative placement strategies for wood duck nest boxes. Wildl Soc Bull 23:463–471.

Semel B, Sherman PW, in press. Intraspecific nest parasitism and nest-site competition in wood ducks. Anim Behav.

Semel B, Sherman PW, Byers SM, 1988. Effects of brood parasitism and nest box placement on wood duck breeding ecology. Condor 90:920–930.

Semel B, Sherman PW, Byers SM, 1990. Nest boxes and brood parasitism in wood ducks: a management dilemma. In: Proceedings of the 1988 North American Wood

Duck Symposium (Fredrickson LH, Burger GV, Havera SP, Graber DA, Kirby RE,Taylor TS, eds). St. Louis; 163–170. Privately published.

Sherman PW, 1977. Nepotism and the evolution of alarm calls. Science 197:1246–1253.

Sherman PW, 1980. The meaning of nepotism. Amer Natur 116:604–606.

Sherman PW, 1981. Reproductive competition and infanticide in Belding's ground squirrels and other animals. In: Natural selection and social behavior: recent research and new theory (Alexander RD, Tinkle DW, eds). New York: Chiron Press; 311–331.

Sherman PW, 1985. Alarm calls of Belding's ground squirrels to aerial predators: nepotism or self-preservation? Behav Ecol Sociobiol 17:313–323.

Sherman PW, Holmes WG, 1985. Kin recognition: issues and evidence. Forts Zool 31:437–460.

Sherman PW, Morton ML, 1984. Demography of Belding's ground squirrels. Ecology 65:1617–1628.

Sherman PW, Semel B, 1989. Behavioral ecology and the management of a natural resource. NY Food Life Sci Quart 19:23–26.

Sherman PW, Semel B, 1992. Killing them with kindness. Living Bird 11:26–31.

Sorenson MD, 1991. The functional significance of parasitic egg laying and typical nesting in redhead ducks: an analysis of individual behavior. Anim Behav 42:771–796.

Sorenson MD, 1993. Parasitic egg laying in canvasbacks: frequency, success, and individual behavior. Auk 110:57–69.

Sorenson MD, 1998. Patterns of parasitic egg laying and typical nesting in redhead and canvasback ducks. In: Parasitic birds and their hosts: studies in coevolution (Rothstein SI, Robinson SK, eds). New York: Oxford University Press; 357–375.

Soulliere GJ, 1988. Density of suitable wood duck nest cavities in a northern hardwood forest. J Wildl Manage 52:86–89.

Soulliere GJ, 1990. Review of wood duck nest-cavity characteristics. In: Proceedings of the 1988 North American Wood Duck Symposium (Fredrickson LH, Burger GV, Havera SP, Graber DA, Kirby RE,Taylor TS, eds). St. Louis; 153–162. Privately published

Thornhill R, Palmer CT, 2000. A natural history of rape: biological bases of sexual coercion. Cambridge, MA: MIT Press.

Weier RW, 1966. A survey of wood duck nest sites on Mingo National Wildlife Refuge in southeast Missouri. In: Wood duck management and research: a symposium (Trefethen JB, ed). Washington, DC: Wildlife Management Institute; 91–112.

Weigmann C, Lamprecht J, 1991. Intra-specific nest parasitism in bar-headed geese, *Anser indicus*. Anim Behav 41:677–688.

Wilson SF, 1993. Use of wood duck decoys in a study of brood parasitism. J Field Ornithol 64: 337–340.

Yetter AP, Havera SP, Hine CS, 1999. Natural cavity use by nesting wood ducks in Illinois. J Wildl Manage 63:630–638.

Yom-Tov Y, 1980. Intraspecific nest parasitism in birds. Biol Rev 55:93–108.

# 17 The Mexican Jay as a Model System for the Study of Large Group Size and Its Social Correlates in a Territorial Bird

Jerram L. Brown

## How the Study Began

The Mexican jay (*Aphelocoma ultramarina*) provides an interesting system with which to study social behavior, but why would anyone give up a budding career in the new and growing field of neuro-ethology (Brown 1969a; Brown & Hunsperge 1963) to study Mexican jays, as I did in 1969 (Brown 1970)? The answer is that the allure of testing natural selection theory proved stronger to me than the appeal of physiological questions that dominated neuro-ethology, then in its infancy. "The brain" is a fine object of study, but it pales in comparison to the complexity and mystery of natural selection and evolution of social behavior. Here, at last, was a chance to make some progress in a field that had been ignored by the mechanistic approaches to animal behavior that dominated the period (for historical perspective, see Brown 1994a). It was a time when research on the fitness costs and benefits of social behavior was driving a new synthesis based on population biology (Alcock 1975; Brown 1975; Wilson 1975). Furthermore, fieldwork in pine-oak woodland of the Chiricahua Mountains offered insights and pleasures that were unobtainable in my laboratory.

I remember vividly that I became interested in the Mexican jay as a model system as a result of reading Hamilton's first paper on the evolution of altruism (Hamilton 1963) when it appeared in our university library one Saturday morning in 1963. Although Hamilton at the time was virtually unknown, because of this and later papers he became one of the most highly acclaimed evolutionary biologists since Darwin. Knowing that this species could be studied at the Southwestern Research Station in southern Arizona, which I had visited on my way to graduate school at Berkeley in 1956, I took the first opportunity that arose to study the jays there in spring during a sabbatical leave from the University of Rochester in 1969–70. I had thought naively that the terms in Hamilton's Rule could be estimated easily from field

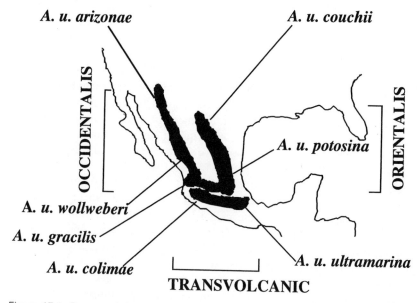

Figure 17.1. Ranges of the three groups of populations of the Mexican jay species complex, showing subspecies (after Peterson 1992). The Transvolcanic populations (*colimae, ultramarina*) are geographically disjunct. Orientalis (*couchii, potosina* or *sordida*) intergrades with Occidentalis (*arizonae, wollweberi*) in central Mexico.

observations of birds with helping behavior. I had no intention to carry the project beyond this.

The Mexican jay is a species complex of highly social populations that are found mainly in the mountains of Mexico. These populations can be divided into three groups according to the mountain ranges they inhabit (Fig. 17.1). The Orientalis populations occur in the Sierra Madre Oriental, extending north into the Chisos Mountains of Texas. The Transvolcanic populations occur in the Sierra Madre del Sur in central Mexico. The Occidentalis populations are found in the Sierra Madre Occidental, ranging north into southeastern Arizona, where our study area is located. Each of these groups includes two or more named subspecies. Orientalis and Occidentalis intergrade at the south end of their ranges.

## Helping Behavior: Is It Explained by Hamilton's Rule?

### BENEFITS OF HELPING

Hamilton's paper provided a theoretical explanation for a type of behavior in which one individual benefited another despite detriment to itself, a concept that he called altruism. Behavior that might be interpreted in this way is best known in the social insects, but also occurs in birds (Brown 1987a), mammals (Solomon & French 1996), and fish (Taborsky 1994). Altruism interested evolutionists because it could not be easily explained exclusively by conventional Darwinian selection based solely on the offspring of individual altruists. The addition of the perspective provided by kinship provided an answer, but the theory urgently needed empirical testing in natural situations. In order to determine whether a behavior should be selected according to this theory, one had to find the consequences of a social act in terms of inclusive fitness (Hamilton 1964). This required determination of the consequences of individual actions in terms of direct fitness (own offspring) and indirect fitness (offspring of other kin; Brown 1980, 1987a).

The first step with birds was to verify that aid was actually given and to determine quantitatively how much aid was given by individually identified donors, or helpers, as they were called in the literature on birds (Skutch 1935, 1961). To do this we used the technique of color-banding to enable individual recognition, a technique that I had already developed for jays as a graduate student in the 1950s (Brown 1963a, 1964). Previously we had learned that Mexican jays could be trapped and banded and that they lived all year on the same territories in large social groups composed of two to four breeding pairs plus numerous nonbreeders (Brown 1963b). We found that roughly 50 percent of the feeding of nestlings was done by individuals other than the putative parents (Brown 1970, 1972). Although accepted today and confirmed for many species, this finding aroused widespread interest and disbelief at the time. As the first study to explicitly address the application of Hamilton's theory to helping in birds, it stimulated much interest in the estimation of the terms of Hamilton's Rule.

The simplest approach was to assume that nonbreeders were helpers (feeders of nestlings not their own) and test for positive correlation between number of nonbreeders and reproductive success of the parents in situations where there were only a single breeding pair and some nonbreeders (termed *singular breeding*, Brown 1978). This was done in several species but the problem was that correlations do not prove cause/effect relationships; other hypotheses also fit the data, especially in the territorial species that were so commonly studied (Brown & Balda 1977; Gaston 1978). This possibility, which caused some observers to doubt that helping benefited the recipients at all (Zahavi 1974), could only be tested by experiment. Because they are not singular breeders, Mexican jays were unsuitable for such work. Therefore, we carried out the experiments on a different species, with positive results (Brown & Brown 1981b; Brown et al. 1982). Our work on jays then took different directions.

## Is Helping Selected?

The terms in Hamilton's Rule are costs and benefits to individuals (direct fitness) weighted by the coefficient of relatedness between participants in a dyadic interaction. It was the goal of much fieldwork on this problem by our group and others to estimate these terms in order to test whether or not the conditions for altruism, or at least for selection of helping, were satisfied. For clarification of this conceptual approach for the case of helping behavior, see Chapter 4 in Brown (1987a). It proved possible to estimate these terms unambiguously only for colonial species without all-purpose territories (Brown 1987b), such as certain kingfishers (Reyer 1980, 1984) and bee-eaters (Emlen & Wrege 1989, 1994). Still doubts remained.

Was it not possible for Hamilton's conditions to be satisfied without natural selection occurring? Perhaps helping and other social behaviors were not actually selected but were responses to changed environmental conditions, as suggested first for social insects (Eberhard 1986; West-Eberhard 1987, "flexible strategy") and later for birds (Jamieson 1989; Jamieson & Craig 1987, "unselected hypothesis"). The scenario envisioned for birds was that delayed departure of offspring placed them in a new "environment" in which begging young were found. According to this view, nonbreeders would respond to the stimuli from begging nestlings by feeding them because they already had the "neural machinery" to do so by virtue of having been selected to care for their own offspring. This proposal, which was eminently reasonable, stimulated a reaction from selectionists that led to a collection of position papers in the *American Naturalist*. It was argued that cost/benefit studies did prove that selection had occurred (Emlen et al. 1991) and that they did not (Jamieson 1991).

This controversy lay at an impasse for several years because it was nearly impossible in the case of helping to satisfy the formal conditions for demonstration that natural selection had affected the trait (Endler 1986). That objection still remains; however, an indirect approach does provide some insight. Carol Vleck and I reasoned that if natural selection had acted on the feeding of young by nonbreeders, then their physiology might also be affected. It was possible to test this hypothesis by examining levels of hormones that were already known to be associated with the feeding of young, such as prolactin. Furthermore, prolactin had already been implicated with helping in a study of Harris's hawk (*Parabuteo uncinatus*; Mays et al. 1991; Vleck et al. 1991). Therefore, we set out to test two simple predictions: (1) levels of prolactin should be higher in a species with helping than a closely related, sympatric species without helping; (2) levels of prolactin should rise *before* the appearance of begging young, in contrast to the response hypothesis, which would predict a rise *after* the appearance of begging young.

We found that levels of prolactin were much higher in the species with helpers, the Mexican jay, than in the species without helpers living in the same region, the Western scrub-jay (*Aphelocoma californica;* Brown & Vleck 1998; Vleck et al. 1996; Fig. 17.2). This was true even in the nonbreeding yearlings and in both sexes. This result could not have been predicted by the rival, "flexible strategy" or "unselected" hypothesis as delineated by Jamieson and Craig for birds.

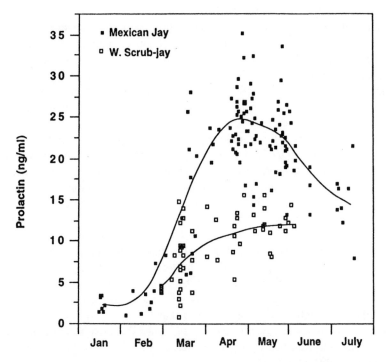

Figure 17.2. Levels of prolactin in social versus nonsocial jays. Changes in plasma level of prolactin through the breeding season in western scrub-jays (□) and Mexican jays (■) in natural populations in southeastern Arizona. Values from all birds are shown regardless of sex, age, and breeding status. Trend lines are spline-fits to the data for each species. From Brown and Vleck (1998).

Tests of the second hypothesis also supported the selectionist theory. Levels of prolactin rose in both sexes well before the first young hatched, as in another species of jay (Schoech et al. 1996). This result allows us to reject the begging-young hypothesis because the levels of prolactin rose in March and early April well before the first young appeared (May and late April). However, there remains a related hypothesis not previously considered, the begging-female hypothesis, namely, that the levels of prolactin might have risen in the helpers and fathers in response to the begging of adult females. Incubating female Mexican jays, like other jays (Brown 1964; Goodwin 1976), beg loudly to be fed on or off the nest by their mates and other flock members. Female begging reaches a peak about three weeks before young hatch, when the female switches from spending most of her time foraging for food to spending 90 percent of her time incubating. At this transition from foraging to incubating she needs to train her mate(s) and other flock members to feed her on the nest, and she is especially vociferous until she has them well trained. One must, therefore, also consider the hypothesis that such begging by adult females is the stimulus that causes prolactin to rise in mates and other feeders of the nestlings. We cannot reject this hypothesis on the basis of the temporal pattern of rise in prolactin because female begging begins at least three weeks before hatching. Nevertheless, the begging-female hypothesis fails to predict our finding that *levels* of prolactin were higher in species with helpers than in species without helpers.

## Phylogeny of Helping in Jays

A third way to study helping behavior from an evolutionary perspective (after the cost/benefit and physiological approaches) is with the use of a phylogeny. The first modern phylogeny for *Aphelocoma* was that of Peterson based on allozymes (Peterson 1992). This was a big step forward and was used to trace the evolution of behavioral traits in the genus *Aphelocoma* (Peterson & Burt 1992), but in this case allozymes gave results that were unsatisfactory in several ways. Although the addition of behavioral characters improved the phylogeny somewhat (Brown & Li 1995), doubts remained. A new phylogeny of the *A. ultramarina* species complex based on the mitochondrial control region now makes possible a clearer picture of the phylogeny of helping behavior in the species (Li et al., in press, Fig. 17.3). This phylogeny allows us to reject the hypothesis that the characteristics of the Texas population (small group size, singular breeding, rattle call, rapid maturation, speckled eggs, Fig. 17.4; Brown & Horvath 1989) are due to phylogenetic proximity to scrub-jays. The possibility of ancient or low-level hybridization with the western scrub-jay, however, remains under study (Bhagabati, in progress). Li's work indicates that the Transvolcanic populations (*colimae* in Fig. 17.3) are an outgroup that is distinct from the more northern populations of Mexican jay (Orientalis, Occidentalis) at the species level. The Occidentalis populations (*arizonae, gracilis*) are the most evolutionarily derived, with the Texas (*couchii*) population branching at an intermediate position. Thus it is quite possible for the scrub-jay-like characters of Texas Mexican jays to be derived by convergence onto the scrub-jay phenotype from an Orientalis Mexican jay stock. This result means that we may be able to study the environmental conditions that favor the secondary development of small group sizes found in Texas (*couchii*) and northeastern Mexico (*potosina* = *sordida*). A first attempt at this has shown that in northeastern Mexico group size does indeed vary with habitat quality, namely, with the richness of the oak vegetation, which supplies acorns, a principal food (Horvath & Bhagabati, unpublished data). Mexican jays in this area commonly inhabit scrub-oak communities, as do some populations of scrub-jay elsewhere.

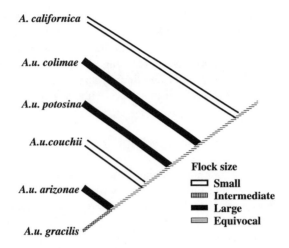

Figure 17.3. Phylogeny of group-size in the Mexican jay. Small, mostly two to four. Intermediate, mostly six to eight. Large, mostly nine to twenty-five. For details of the tree, see Li et al. Orientalis includes *potosina* and *couchii*. Occidentalis includes *arizonae* and *gracilis*. The Transvolcanic province includes *colimae*. A. *californica*, (western scrub-jay) was used as an outgroup.

## Large Group Size: Possible Influences on Social Behavior

Large group size and territorial behavior of the Mexican jay preclude the kind of analyses of components of inclusive fitness that were so nicely done on colonial, singular breeders (Emlen & Wrege 1989, 1994; Reyer 1984). Consequently, we turned our attention to questions associated with the larger group sizes found in the Mexican jay (Fig. 17.4) but not other members of the genus.

Mexican jays live in permanently resident, nonmigratory groups that are unusually large (five to twenty-five individuals) compared to most species of communally rearing birds, which live in groups of two to five. Group size did not vary much in relation to elevation, habitat, and location throughout the canyon in which our study area is located (Brown & Brown 1985), but critics doubted that these large groups were typical for the entire range of the species. We, therefore, undertook a survey of group size throughout the range of the species, which is mainly in Mexico. We found that group sizes were not significantly different from those on our study area except in two areas. Groups were significantly smaller in Texas and adjacent Mexico (*A. ultramarina couchii*) and in the Mexican state of Jalisco (*A.u. gracilis*, Fig. 17.4, Brown & Horvath 1989).

Large group size enables several behavioral changes from the pairs and small, nuclear family groups found in most species of communally rearing birds. First, it facilitates simultaneous breeding of two or more females (and

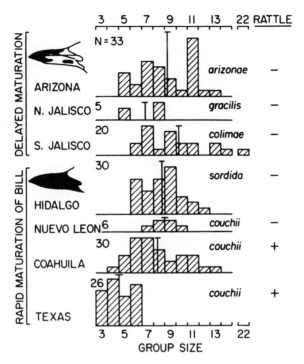

Figure 17.4. Geographic variation in group-size-related traits of Mexican jays (based on Brown & Horvath 1989). Group sizes are shown as histograms for each locality. Delayed maturation of bill color is shown by bill color. Black bills indicate rapid maturation (Orientalis). Blotched bills indicate delayed maturation (Occidentalis, Transvolcanic populations). Rattles are found only in *A.u. couchii*. *A.u. sordida* was recently renamed *A.u. potosina*.

two or more males) in the same territorial group (plural breeding). Along with plural breeding there may be changes in dominance organization and in behaviors normally associated with exclusive ownership of territories by pairs, such as certain vocalizations (Strahl & Brown 1987). When plural breeding becomes common, reciprocity between breeders becomes possible, the kin structure of groups may become looser, and extra-pair fertilization (EPF) within groups is facilitated. Dispersal also becomes more conservative, judging from comparative evidence.

## Dispersal and Kinship

A crucial component of Hamilton's theoretical explanation of social behavior is the relatedness between the participants. It was generally assumed and later confirmed for many species of birds with associated nonbreeders that the nonbreeders, especially those that helped, were usually closely related to the breeding pair (summarized in Brown 1987a) and that the association was caused by delayed departure of the offspring from the parental territory. In territorial birds, delayed departure was long ago interpreted as a behavior that facilitated acquisition of a territory when a surplus of breeding-age individuals existed (Brown 1969b). Thus delayed departure (to be distinguished from helping) was thought to be selected by direct selection, as opposed to indirect ("kin") selection (Brown 1969b; Selander 1964 and many later authors).

Therefore, we began to examine patterns of relatedness within groups of Mexican jays based on pedigrees. Perhaps not surprisingly, in view of the large group sizes on our study area (Brown & Brown 1985), we found that relatedness was often not close among randomly picked pairs of individuals (dyads) within groups (Brown & Brown 1981a). Parents had offspring in their flocks, but because there was more than one breeding female in each group (plural breeding) the breeding females were often unrelated to each other. The same was true for males. Unlike singular breeders, which tend to be in nuclear families (parents and offspring), our plural breeding jays live in alliances of extended families that include grandparents, aunts, uncles, and unrelated individuals (Fig, 17.5). Consequently, most individuals had one or more relatives, but few were related to a large number of jays within their own group.

To understand the origins of these patterns of relatedness within groups we examined the dispersal of young banded in their natal territory by tabulating the territory in which they first bred (natal dispersal; Brown & Brown 1984). In striking contrast to singular-breeding scrub-jays with non-breeding associates (Woolfenden & Fitzpatrick 1984) young Mexican jays of both sexes were most frequently found within their natal territory at the time of their first breeding attempt. If they did leave home to breed, it was almost invariably to a contiguous territory. Longer moves could have been detected but were not. We have subsequently observed some long-distance dispersal (5 to 10 kilometers); however, unlike other species of jays in the southwestern United States, long-distance dispersal is rare in the Mexican jay (Westcott 1969). Thus, in spite of the loose, multiple extended family organization, offspring remained with their parents to an even greater degree than in nuclear-family species.

## Dominance and Reproduction

In a plural breeding species that lives all year in large groups, dominance relations can be more complicated than in a singular-breeding, nuclear-family species, where typically the single breeding pair dominates (e.g., Woolfenden & Fitzpatrick 1977 and other papers reviewed in Brown 1978), thus explaining in part the nonbreeding status of the additional birds in the group.

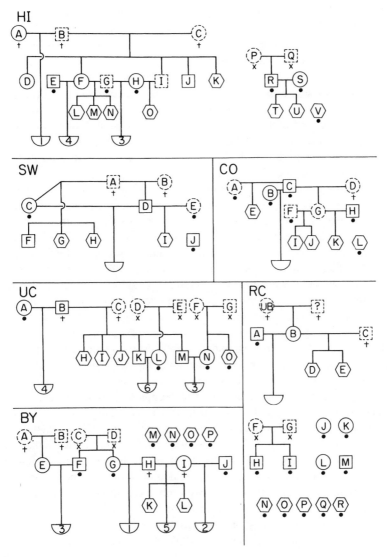

Figure 17.5. Genealogies of six social units of Mexican jays. All individuals present in May 1979 are indicated by symbols with solid borders. Selected ancestors that are no longer alive or are in another unit are indicated by dashed borders. Symbols: □ = male. ○ = female. Hexagon = unsexed. · = immigrant. † = present at start of study in 1971 so ancestry unknown. X = alive in neighboring unit. Half-circles = nests with young. From Brown and Brown (1981a).

We expected to find a similar pattern, namely, one in which parents dominate their offspring. To our surprise, however, we found much more variety. Most interesting was the observation that some individual birds of about nine to eleven months of age dominated adults at feeders (Barkan et al. 1986 and later unpublished data). Perhaps the young were more motivated (hungrier) than adults, as suggested in game theoretic models of dominance (Maynard Smith & Parker 1976), but this would not easily explain the difference between singular and plural breeding jays. Perhaps dominance has little effect on fitness so that food at feeders is not worth contesting by dominants, but we found that subordinates had to wait longer for access to food in winter when food was scarce (Craig et al. 1982).

According to general expectation, dominants should have higher reproduction than subordinates, but since plural breeding occurs regularly in the Mexican jay (Brown 1986) it is obvious that dominants do not suppress entirely the reproduction of subordinates. There was actually some doubt that subordinates suffered any loss at all on average, although anecdotal evidence suggested that some individuals did suffer (Trail et al. 1981). To test the hypothesis that dominance was related to reproduction and to estimate its effect, we examined reproductive success in relation to dominance rank measured independently in the same flock just before the breeding season (February), along with other factors, such as age, year, flock, and sex, using logistic regression (Brown et al. 1997). We examined correlates of dominance rank in two stages of nesting, early (up to and including laying) and late (after laying). The correlates of rank differed between the sexes. In the early stage, success in getting a mate and initiating a nest with eggs was correlated with rank in males but not females and with age in females and males. In the late stage, given that a bird had a nest with eggs, fledging success was significantly enhanced by rank in top-ranking birds, with no difference between the sexes. These effects of dominance at two stages of the nesting cycle, before and after egg-laying, are consistent with field observations of behavior. In the early stages of reproduction males were observed to compete aggressively and conspicuously for females. Later in the nesting season cycle, pairs competed by robbing nest lining or eggs, harassing other pairs at their nests, and even by usurping nests (Bowen & Li 1995; Brown 1963b; Trail et al. 1981).

## Extra-Pair Fertilization

The large group sizes found in the Mexican jay are associated with year-round peaceful coexistence of rival breeders of the same sex within the same groups. However, in the breeding season some rivalry over nests and females was observed right from the beginning of our work with color-banded jays (Brown 1963b). Although each nest is built by a single male joined by a single female, complications arise when a nest built by one pair is taken over by another pair or when a male from one pair takes over a female from another, hence engaging in "uncooperative breeding" (Brown & Brown 1990). We did not know the extent of departure from monogamy that this behavior caused, so we attempted to estimate the frequency of extra-pair fertilization (EPF). We were able to determine using allozymes that EPFs occurred at a fairly high rate, but we could not determine the rate exactly, and we could not identify fathers with confidence (Bowen et al. 1995). When new technology became available, we tried again using tetranucleotide DNA microsatellites developed especially for the Mexican jay (Li et al. 1997). To our surprise the rate of EPF was much higher than anyone would have anticipated. We found evidence of extra-pair fertilization in 32 of 51 complete broods (63 percent) and 55 of 139 nestlings (40 percent) for which the putative father had been identified (Brown & Li 1998; Li & Brown, in press). By identifying genetic fathers of each nestling we found that at least 96.1 percent of EPF fathers came from *inside* the territorial group.

This pattern is strikingly different from that found in other birds. In most quasi-monogamous species EPF parents come from *outside* the territory. In colonial species individuals live together in large groups and nest in pairs within breeding colonies, but there is no group territory, especially in the nonbreeding season. In polyandrous species, however, multiple paternity is expected from within the group. In cooperative polyandry two or more males may participate in a cooperative nesting effort. The Mexican jay differs from this pattern by lacking cooperation between fathers in building the same nest. In the dunnock (*Prunella modularis*) most birds nest in pairs or polyandrous trios, in which males do not aid females in nest construction. Males who have fathered some young feed nestlings in the nest of the relevant female, but nonbreeding helpers do not occur.

The revelation that social monogamy often has genetic exceptions (Westneat et al. 1990) has spurred a revolution in thinking about avian mating systems and has created a need for new hypotheses that take EPFs into account (Westneat & Sherman 1997). EPFs are surprisingly common in birds. Among seventy-two species studied genetically and reviewed by Westneat and Sherman, EPFs were documented in 74 percent, although only four species had more than 40 percent. (The Mexican jay has 40 percent.) The mean frequency for species having any EPFs was about 18 percent. Because EPFs are surprisingly common and because females appear often to seek EPCs, researchers have investigated costs and benefits of EPCs and EPFs to females. One of the most controversial issues is what benefits females gain from their choice (Andersson 1994).

The costs and benefits of EPF to female Mexican jays are not completely clear, but we found that EPF fathers were important contributors to feeding the nestlings at the nests where they had offspring (Li & Brown 1998, submitted). Regardless of the high EPF rate, putative fathers were the most consistent feeders of nestlings (Li 1997), and there was no significant difference in feeding rate between cuckolded and uncuckolded fathers. In contrast, feeding rates to nestlings by mothers with EPF young were significantly lower than by those without. On average, EPF broods had 2.31 more helpers (10.71 ± 0.54 SE helpers) than the non-EPF broods (8.4 ± 0.64 helpers), even though no significant difference between the flock sizes of EPF and non-EPF broods was observed (13.64 ± 20.96 versus 12.35 ± 20.80 jays). Although EPF broods received more total feedings from helpers, there was no significant difference in total feeding rate between EPF and non-EPF broods because the mother reduced her feeding rate.

## Dispersal and Inbreeding

We soon realized that the greatly reduced dispersal system of the Mexican jay enabled us to obtain information about many other interesting theoretical questions. Reduced dispersal should increase the chances of inbreeding. If so, then mechanisms to avoid inbreeding might also be selected for. Therefore, we first asked if inbreeding was frequent and if there was a fitness penalty for inbreeding. We found that pairs in which male and female were related on a pedigree constituted about 5 percent of all pairs, and that the most common type of inbred pair was between half-sibs (Brown & Brown 1997). More important, there were significant penalties for inbreeding mothers. Brood sizes of inbred pairs were smaller, the survival of inbred offspring was lower, and no offspring of inbred parents has yet reproduced successfully. Thus inbreeding occurred and was a danger but was not unusually common compared to other species of bird.

The occurrence of inbreeding depression in our population suggests that the high rate of EPF in Mexican jays may be related somehow. The low rate of inbreeding suggests that in spite of living usually among kin, some mechanisms are working to prevent inbreeding or reduce its effects. One obvious mechanism is plural breeding. The occurrence in the same group of several potential mates means that females can usually choose a male that is unrelated even within their own flock. If a female left her natal territory, she would not have to go as far to find a mate because Mexican jays have two to three times as many potential mates to choose from in the contiguous territories than they would in a singular breeding species since each of the neighboring territories has several potential breeders of each sex.

Since dispersal is much reduced in this species—perhaps more so than in any other continental species of passerine—the population is probably relatively viscous genetically. Thus we expect and wish to test that the probability of taking a mate that shares deleterious alleles is relatively high even if it is not related on a pedigree. If this expectation proves to be true, a further advantage of EPF would be that the probability of complete failure of a year's breeding effort due to male infertility or genetic mismatch would be reduced. Extra-pair copulation would then have the additional advantage to the female of minimizing the chance of complete failure of her brood.

## Incentives for Tolerance and Cooperation

Although we find it difficult to test, we favor the hypothesis that large group sizes in the Mexican jay are favored by natural selection to minimize the impact of the high intensity of predation experienced by this species. Selection should favor tolerance of rivals within the same territorial groups under circumstances in which long-term benefits of survival outweigh short-term benefits and costs of resource monopolization by dominant pairs. Mexican jays are a major food item in the diet of the Goshawk (*Accipiter gentilis*) and are also preyed on by Cooper's hawks (*A. cooperi*) and a variety of other hawks, mammals, snakes, and birds (Brown 1994b). These hawks not only take adult birds all year, they also take young from the nest.

Large group sizes provide protection from diurnal flying predators through "the early warning system." Large groups have more individuals to detect the approach of hawks. This is especially important in defense against *Accipiter* hawks, which fly swiftly and silently toward the foraging jays a few feet off the ground, giving them only moments to escape. The jays typically forage within hearing distance of each other, especially in the nonbreeding season. When an approaching hawk is detected, a soft, almost inaudible buzzy call is given. The usual response is to immediately and silently seek cover and "freeze."

Mexican jays also may take aggressive action against hawks and owls. The mobbing of perched owls by jays is well known. The highly social Mexican jay has been shown to mob more intensively than the nonsocial western scrub-jay, in accordance with the greater benefit expected from mobbing in a social species (Cully & Ligon 1976). Less widely appreciated is the deterrent effect that a large group of jays can have on a hawk. A hawk raiding a nest of the Mexican jay is sometimes attacked and driven off. Contrary to general expectation, even if some young are taken, others may survive to fledging from attacked nests. This is probably because of the intensity of the attack by a large group of jays. That mobbing can affect a hawk is shown by the following observation. A flock of jays was seen pursuing a young Cooper's hawk. The hawk in its haste to evade the jays collided with a screen door, got caught in the screen by its talons, and hung there upside down while the jays harassed it. The hawk escaped after a while, but might have learned a lesson about Mexican jays. Presumably the hawk had attacked the jays, but no jay was missing from the flock after the incident.

Another antipredator incentive for tolerance of sexual rivals in the same flock even in the breeding season may be the cooperative feeding of each other's fledglings by unrelated breeders. Breeding pairs rarely feed each other's nestlings (Brown 1972), but they commonly feed fledglings not their own if they are at the same stage of development (Brown & Brown 1980). A game theory model of this behavior suggests that jays that feed other fledglings that perch close to their own are protecting their own by reducing the chance that a hawk will detect the group of begging young (Caraco & Brown 1986). In the face of such an outside threat, selfishness may result in cooperation through by-product mutualism (Brown 1983b; Dugatkin 1997).

An additional benefit of group-living for Mexican jays, although probably not the principal one, is the possibility of sharing food that is discovered by other individuals in the group. Sometimes food is found that is available only briefly and which is too difficult or abundant for one individual to fully utilize. An emerging swarm of winged ants is an example (Brown 1983a). In a similar vein, young jays in the group may profit from association with older individuals who have already learned cues that may be associated with food. Mexican jays readily learn to respond to arbitrary auditory signals that are associated with food and remember them up to a year (Brown 1998).

Birds that live in winter flocks must balance the benefits of group living against the costs. For subordinates the costs may be greater than for dominants if the subordinates are victimized by dominant scroungers (Barnard 1984), as in winter flocks of sparrows. In winter flocks of juncos flock size is sensitive to energy budgets; cooler temperatures increase the tolerance of dominants for subordinates (Caraco 1979). In Mexican jays dominants have priority at food sources (Craig et al. 1982), but subordinates can wait their turn or dig up acorns stored by others, if they are careful not to be caught. For "scroungers" the ability to take food from others is a possible advantage of group-living. Mexican jays commonly retrieve food that has been stored by others.

A comparable advantage to dominants may exist with regard to breeding. Dominant males may fertilize some eggs of females mated to subordinates. Dominant females may benefit from failure of nests of subordinate females because a failed female often feeds whatever young are begging in her group.

Probably the simplest theoretical explanation of these matters is that selection favors Mexican jays that tolerate others over those that do not through the advantages, or "incentives," described above in an environment with high risk of predation. One way in which selection might act to favor such tolerance is by a reduction of the blood levels of testosterone in the group-living species. We tested this by comparing testosterone levels in a group-living species, the Mexican jay, to a pair-living species, the California scrub-jay. There was, however, no difference in the levels of testosterone among the males of the two species (Vleck & Brown 1999; Vleck et al. 1996). Thus the physiological mechanisms of tolerance remain unknown.

That avoidance of predation is important in the lives of Mexican jays is underscored by their great longevity—up to twenty years of age. Many

Mexican jays live long enough that effects of senescence are detectable, as in scrub-jays (McDonald et al. 1996) but more extreme (Shao & Brown, unpublished analysis). When the potential reproduction in any one year is low and unpredictable, as in the Mexican jay, the possibility of repeated reproductive attempts in later years becomes more important.

## Our Approach to Behavioral Biology

To follow my footsteps (or anyone's) just as taken in the 1960s and 1970s would be unwise for today's young researchers, but some elements of my strategy will always be useful. First, read the new literature avidly. When a new theory appears that can be tested empirically, consider what species might be used and where it might be studied. As far as possible considering your training, preferences, and situation, choose the species to suit the question, not vice versa. (Contrary to appearances, I have studied several other species.) A species or system that is well suited for one problem may, however, also be well suited for others. This is what kept me studying the Mexican jay. Because I could trap and band the birds, identify individuals, and stay at a permanent research station, observation of the same individuals over a long period of time was greatly facilitated. Because this was a nonmigratory species with little dispersal, observations of the same individuals over many years were possible. At the time I started in 1969, there were no long-term studies that focused on individuals. Previous long-term studies were of populations.

## Future Directions

The advantages of the Mexican jay for future work are considerable. We appear to be on the verge of understanding the significance of our discoveries that inbreeding is expensive, that dispersal is extremely conservative (more than any other continental bird species), and that extra-pair fertilization is phenomenally common. We expect to be able to test various theories of mate choice, including the indirect benefits hypothesis and the inbreeding avoidance hypothesis. We may also be able to present a convincing argument that certain aspects of dispersal in this species are adaptations to avoid inbreeding. We are investigating possible effects of climate on reproduction (Brown & Li 1996; Li & Brown 1999) and effects of global warming (Brown et al. 1999) and El Niño on our population. Because climatic phenomena may be responsible for environmental stochasticity, which in turn may affect population parameters, we are studying relationships between these factors and optimal brood size (Yoshimura et al., in preparation). Visitors to our study often ask me, "Haven't you answered all the questions about jays by now?" The truth is, however, that because of what we have learned, many new questions can be addressed using our thirty-two years of data that were unthinkable when the research began.

## Acknowledgments

I thank the National Science Foundation and the National Institute of Mental Health for major financial support, the Southwestern Research Station of the American Museum of Natural History for permission to work there, the neighboring landowners and the U.S. Forest Service for permission to work on their land, the many field assistants over the years for their expert and enthusiastic help, and the following for comments on this manuscript: S.-H. Li, N. Bhagabati, E. R. Brown.

## References

Alcock J, 1975. Animal behavior. An evolutionary approach. Sunderland, MA: Sinauer Associates.

Andersson M, 1994. Sexual selection. Princeton: Princeton University Press.

Barkan CPL, Craig JL, Strahl SD, Stewart AM, Brown JL, 1986. Social dominance in communal Mexican Jays *Aphelocoma ultramarina*. Anim Behav 34:175–187.

Barnard CJ (ed), 1984. Producers and scroungers: strategies of exploitation and parasitism. London: Croom Helm.

Bowen B, Koford RR, Brown JL, 1995. Genetic evidence for hidden alleles and unexpected parentage in the Gray-breasted Jay. Condor 97:503–511.

Brown JL, 1963a. Aggressiveness, dominance and social organization in the Steller jay. Condor 65:460–484.

Brown JL, 1963b. Social organization and behavior of the Mexican jay. Condor 65:126–153.

Brown JL, 1964. The integration of agonistic behavior in the Steller's jay *Cyanocitta stelleri* (Gmelin). Univ Calif Publ Zool 60:223–328.

Brown JL, 1969a. Neuro-ethological approaches to the study of emotional behaviour: stereotypy and variability. Ann NY Acad Sci 159:1084–1095.

Brown JL, 1969b. Territorial behavior and population regulation in birds. Wilson Bull 81:293–329.

Brown JL, 1970. Cooperative breeding and altruistic behavior in the Mexican jay, *Aphelocoma ultramarina*. Anim Behav 18:366–378.

Brown JL, 1972. Communal feeding of nestlings in the Mexican jay (*Aphelocoma ultramarina*): interflock comparisons. Anim Behav 20:395–402.

Brown JL, 1975. The evolution of behavior. New York: Norton.

Brown JL, 1978. Avian communal breeding systems. Ann Rev Ecol System 9:123–155.

Brown JL, 1980. Fitness in complex avian social systems. In: Evolution of social behavior: hypotheses and empirical tests (Markl H, ed). Weinheim: Verlag Chemie GmbH; 115–128.

Brown JL, 1983a. Communal harvesting of a transient food resource in the Mexican Jay. Wilson Bull 95:286–287.

Brown JL, 1983b. Cooperation—a biologist's dilemma. In: Advances in the study of behavior (Rosenblatt JS, Hinde RA, Beer C, Busnel M, eds). New York: Academic Press; 1–37.

Brown JL, 1986. Cooperative breeding and the regulation of numbers. Proc Internat Ornithol Congr 18:774–782.

Brown JL, 1987a. Helping and communal breeding in birds: ecology and evolution. Princeton: Princeton University Press.

Brown JL, 1987b. Testing inclusive fitness theory with social birds. In: Animal societies: theories and facts (Ito Y, Brown JL, Kikkawa J, eds). Tokyo: Japan Sci. Soc. Press; 103–114.

Brown JL, 1994a. Historical patterns in the study of avian social behavior. Condor 96:232–243.

Brown JL, 1994b. Mexican jay. In: The birds of North America (Poole A, Stettenheim P, Gill F, eds). Philadelphia; Washington, DC: Academy of Natural Sciences of Philadelphia; American Ornithologists' Union; 1–16.

Brown JL, 1998. Long term memory of an auditory stimulus for food in a natural population of the Mexican jay. Wilson Bull 109:749–752.

Brown JL, Balda RP, 1977. The relationship of habitat quality to group size in Hall's babbler (*Pomatostomus halli*). Condor 79:312–320.

Brown JL, Brown ER, 1980. Reciprocal aid-giving in a communal bird. Zeit Tier 53:313–324.

Brown JL, Brown ER, 1981a. Extended family system in a communal bird. Science 211:959–960.

Brown JL, Brown ER, 1981b. Kin selection and individual selection in babblers. In: Natural selection and social behavior: recent results and new theory (Alexander RD, Tinkle DW, eds). New York: Chiron Press; 244–256.

Brown JL, Brown ER, 1984. Parental facilitation: parent-offspring relations in communally breeding birds. Behav Ecol Sociobiol 14:203–209.

Brown JL, Brown ER, 1985. Ecological correlates of group size in a communally breeding jay. Condor 87:309–315.

Brown JL, Brown ER, 1990. Mexican jays: uncooperative breeding. In: Cooperative breeding in birds: long-term studies of ecology and behavior (Stacey PB, Koenig WD, eds). Cambridge: Cambridge University Press; 268–288.

Brown JL, Brown ER, 1997. Are inbred offspring less fit? Survival in a natural population of Mexican jays. Behav Ecol 9:60–63.

Brown JL, Brown ER, Brown SD, Dow DD, 1982. Helpers: effects of experimental removal on reproductive success. Science 215:421–422.

Brown JL, Brown ER, Sedransk J, Ritter S, 1997. Dominance, age and reproductive success in a complex society: a long-term study of the Mexican Jay. Auk 114:279–286.

Brown JL, Horvath EG, 1989. Geographic variation of group size, ontogeny, rattle calls, and body size in *Aphelocoma ultramarina*. Auk 106:124–128.

Brown JL, Hunsperger RW, 1963. Neuroethology and the motivation of agonistic behaviour. Anim Behav 11:439–448.

Brown JL, Li S-H, 1995. Phylogeny of social behavior in *Aphelocoma* jays: a role for hybridization? Auk 112:464–472.

Brown JL, Li S-H, 1996. Delayed effect of monsoon rains influences laying date of a passerine bird living in an arid environment. Condor 98:879–884.

Brown JL, Li S-H, 1998. The genetic mating system of the Mexican Jay. Ostrich 69:316.

Brown JL, Li S-H, Bhagabati N, 1999. Long term trend toward earlier breeding in an American bird: a response to global warming? Proc Natl Acad Sci USA 96:5565–5569.

Brown JL, Vleck CM, 1998. Prolactin and helping in birds: has natural selection strengthened helping behavior? Behav Ecol 9:541–545.

Caraco T, 1979. Time budgeting and group size: a test of theory. Ecology 60:618–627.

Caraco T, Brown JL, 1986. A game between communal breeders: when is food-sharing stable? J Theoret Biol 118:379–393.

Craig JL, Stewart AM, Brown JL, 1982. Subordinates must wait. Zeit Tier 60:275–280.

Cully JF, Jr, Ligon JD, 1976. Comparative mobbing behavior of scrub and Mexican jays. Auk 93:116–125.

Dugatkin LA, 1997. Cooperation among animals. An evolutionary perspective. New York: Oxford University Press.

Eberhard MJ, 1986. Alternative adaptations, speciation, and phylogeny (a review). Proc Natl Acad Sci USA 83:1388–1392.

Emlen ST, Reeve HK, Sherman PW, Wrege FL, Ratnieks FLW, Shellman-Reeve J, 1991. Adaptive versus nonadaptive explanations of behavior: the case of alloparental helping. Am Nat 138:259–270.

Emlen ST, Wrege PH, 1989. A test of alternate hypotheses for helping behavior in white-fronted bee-eaters of Kenya. Behav Ecol Sociobiol 25:303–319.

Emlen ST, Wrege PH, 1994. Gender, status and family fortunes in the white-fronted bee-eater. Nature 367:129–132.

Endler JA, 1986. Natural selection in the wild. Princeton: Princeton University Press.

Gaston AJ, 1978. The evolution of group territorial behavior and cooperative breeding. Am Nat 112:1091–1100.

Goodwin D, 1976. Crows of the world. Ithaca, NY: Cornell University Press.

Hamilton WD, 1963. The evolution of altruistic behaviour. Am Nat 97:354–356.

Hamilton WD, 1964. The genetical evolution of social behaviour. I. and II. J Theoret Biol 7:1–52.

Jamieson I, 1989. Behavioral heterochrony and the evolution of birds' helping at the nest: an unselected consequence of communal breeding? Am Nat 133:394–406.

Jamieson I, 1991. The unselected hypothesis for the evolution of helping behavior: too much or too little emphasis on natural selection? Am Nat 138:271–282.

Jamieson I, Craig JL, 1987. Critique of helping behaviour in birds: a departure from functional explanations. In: Perspectives in ethology (Bateson PPG, Klopfer PH, eds). New York: Plenum Press; 79–98.

Li S-H, 1997. The genetic analysis of paternity pattern in a natural population of Mexican Jays (*Aphelocoma ultramarina*). Ph.D. thesis, State University of New York, Albany.

Li S-H, Barrowclough G, Peterson AT, in press. The phylogeography of a sedentary species, the Mexican jay (*Aphelocoma ultramarina*). Auk.

Li S-H, Brown JL, 1998. Do extra-pair fertilizations benefit female Mexican Jays? Ostrich 69:226.

Li S-H, Brown JL, 1999. Influence of climatic variables on reproductive success in the Mexican Jay. Auk 116:924–936.

Li S-H, Brown JL, in press. High frequency of extrapair fertilization in a highly social, plural breeding species, the Mexican Jay (*Aphelocoma ultramarina*) revealed by DNA microsatellite markers. Anim Behav.

Li S-H, Brown JL, submitted. Do female Mexican Jays receive direct benefits from extra-pair copulation? Behav Ecol.

Li S-H, Huang Yi-J, Brown JL, 1997. Isolation of tetra-nucleotide repeat microsatellites from the Mexican Jay. Mol Ecol 6:499–501.

Maynard Smith J, Parker GA, 1976. The logic of asymmetric contests. Anim Behav 24:159–175.

Mays NA, Vleck CM, Dawson J, 1991. Plasma luteinizing hormone, steroid hormones, behavioral role, and nest stage in cooperatively breeding Harris' Hawks (*Parabuteo unicinctus*). Auk 108:619–637.

McDonald DB, Fitzpatrick JW, Woolfenden GE, 1996. Actuarial senescence and demographic heterogeneity in the Florida Scrub Jay. Ecology 77:2373–2381.

Peterson AT, 1992. Phylogeny and rates of molecular evolution in the *Aphelocoma* jays (Corvidae). Auk 109:133–147.

Peterson AT, Burt DB, 1992. Phylogenetic history of social evolution and habitat use in the *Aphelocoma* jays. Anim Behav 44:859–866.

Reyer H-U, 1980. Flexible helper structure as an ecological adaptation in the pied kingfisher (*Ceryle rudis rudis* L.). Behav Ecol Sociobiol 6:219–227.

Reyer H-U, 1984. Investment and relatedness: a cost/benefit analysis of breeding and helping in the Pied Kingfisher. Anim Behav 32:1163–1178.

Schoech SJ, Mumme RL, Wingfield JC, 1996. Prolactin and helping behaviour in the cooperatively breeding Florida scrub jay (*Aphelocoma coerulescens*). Anim Behav 52:445–456.

Selander RK, 1964. Speciation in wrens of the genus *Campylorhynchus*. Univ Calif Publ Zool 74:1–224.

Skutch AF, 1935. Helpers at the nest. Auk 52:257–273.

Skutch AF, 1961. Helpers among birds. Condor 63:198–226.

Solomon NG, French JA (eds), 1996. Cooperative breeding in mammals. New York: Cambridge University Press.

Strahl SD, Brown JL, 1987. Geographic variation in social structure and behavior of *Aphelocoma ultramarina*. Condor 89:422–424.

Taborsky M, 1994. Sneakers, satellites, and helpers: parasitic and cooperative behavior in fish reproduction. Adv Study Behav 23:1–100.

Trail PW, Strahl SD, Brown JL, 1981. Infanticide and selfish behavior in a communally breeding bird, the Mexican jay *(Aphelocoma ultramarina)*. Am Nat 118:72–82.

Vleck CM, Brown JL, 1999. Testosterone and social and reproductive behaviour in *Aphelocoma* jays. Anim Behav 58:943–951.

Vleck C, Brown J, Brown E, Adams J, 1996. A comparison of testosterone and prolactin in group-living Mexican jays (*Aphelocoma ultramarina*) and monogamous scrub jays (*Aphelocoma coerulescens woodhousei*). In: VI International Symposium on Avian Endocrinology (Anonymous, ed). Edmonton: University of Alberta Press; Abstract.

Vleck CM, Mays NA, Dawson JW, Goldsmith A, 1991. Hormonal correlates of parental and helping behavior in cooperatively breeding Harris' Hawks (*Parabuteo unicinctus*). Auk 108:638–648.

Westcott PW, 1969. Relationships among three species of jays wintering in southeastern Arizona. Condor 71:353–359.

West-Eberhard MJ, 1987. Flexible strategy and social evolution. In: Animal societies: theories and facts (Ito Y, Brown JL, Kikkawa J, eds). Tokyo: Japan Sci. Soc. Press; 35–51.

Westneat DF, Sherman PW, 1997. Density and extra-pair fertilizations in birds: a comparative analysis. Behav Ecol Sociobiol 41:205–215.

Westneat DF, Sherman PW, Morton ML, 1990. The ecology and evolution of extra-pair copulations in birds. In: current ornithology (Johnston RF, ed). New York: Plenum Press; 331–369.

Wilson EO, 1975. Sociobiology. Cambridge, MA: Harvard University Press.

Woolfenden GE, Fitzpatrick JW, 1977. Dominance in the Florida scrub jay. Condor 79:1–12.

Woolfenden GE, Fitzpatrick JW, 1984. The Florida scrub jay: demography of a cooperative-breeding bird. Princeton: Princeton University Press.

Zahavi A, 1974. Communal nesting by the Arabian babbler. Ibis 116:84–87.

# 18 Sexual Selection in the Barn Swallow

Anders Pape Møller

## How and Why I Became Interested in Working with Barn Swallows

There are just as many reasons for becoming a biologist as there are biologists! I have always been interested in animals and plants since early childhood so it was natural that I ended up studying biology. My interest in natural history arose from living in the countryside, where everybody had some interest in nature. My interest developed toward birds when I at fourteen got my first binoculars and at fifteen started to catch and ring birds. My supervisor provided proper guidance so my field notes from then even today are legible and contain important information. Many bird watchers get trapped in the quest for rare species, but I never suffered severely. On the contrary, I had a keen interest in the common species from the very beginning. My field notes from the summer of 1970 contain detailed information about laying dates and reproductive success in the nests of barn swallows (*Hirundo rustica*), and all nestlings were subsequently ringed. I have followed this population ever since!

I am not sure why I continued this work on this particular species. However, there are probably some aspects of sentimentality and appreciation of the elegant and exquisite flight of barn swallows; they are a sign of the arrival of spring. Furthermore, as I have indicated already, they are easy to deal with, abundant, and feasible. Thus, my master's project in 1980–82 was centered around the consequences of farming for birds and bird communities, and this continued naturally into a Ph.D. project on the causes and consequences of colonial breeding in barn swallows during 1982–85. One project followed another, and there were always many more questions than answers.

No biological system is perfect, because it is very difficult to combine all the crucial features of an ideal system: abundance, ease of capture, recapture, and manipulation, and the ability to follow individuals and their offspring under natural conditions. Fruitflies are abundant and easy to study in terms of genetics and development, but individuals cannot be studied easily under field conditions, and certainly not across generations. Barn swallows can readily be followed throughout their lives, they are easy to observe and manipulate (secondary sexual characters, reproductive effort, parasite load, health status, and immune status can be readily manipulated), and the effects on subsequent performance and phenotype, as well as the performance of the offspring, can be recorded.

# How My Studies Integrate Conceptual, Theoretical, and Empirical Work

After finishing my Ph.D. project I was quite depressed because my two applications for postdoctoral positions were bounced. This was difficult for me to understand because I already had fifteen publications in good journals, so I was wondering what else had to be done. One of my friends, Arne Lundberg, whom I had met at a graduate course on behavioral ecology, sent me information about an assistant professorship in Uppsala in Sweden, and I reluctantly applied. Luckily, I got the job, and this provided a dramatic turn in my prospects and views. Exposed to the buzzing atmosphere in Staffan Ulfstrands' department with crazy ideas about experiments on almost everything, I started wondering about what I could do. I have a clear memory as a university student while reading Darwin and later about Peter Grant's studies of Galapagos finches that I thought of selection, which is *the* mechanism of evolution, as something that could be studied in faraway places. It was only when I started looking at my own data that I realized that this could even be done in quite ordinary and mundane places like barns. You do not have to go to the Galapagos to measure selection! I had realized during my master's and Ph.D. studies that there was considerable phenotypic variation in barn swallows, and particularly males had huge variation in the size of the outermost tail feathers, but much less in other morphological traits. Since barn swallows were monogamous, I had difficulty understanding how such large variation could have evolved and be maintained, since sexual selection had to be less intense than in polygynous birds such as Malte Andersson's (1982) long-tailed widowbirds (*Euplectes progne*), not to think of lekking species such as black grouse (*Tetrao tetrix*).

Darwin had considered monogamy in his writings about sexual selection and suggested that if females varied in condition and thus timing of breeding, preferred males would experience a mating advantage simply because of their earlier timing of breeding. Offspring produced early had greater probability of survival, as clearly shown by David Lack (1966), so this could be the mechanism. Thus I contacted Malte Andersson to learn how he had made his tail manipulation experiment on widowbirds and I made a couple of trials with chicken feathers before starting on the barn swallows in 1987. The trick was to capture the males as early as they arrived. Since natural variation in tail length was only in the range from about 90 to 130 millimeters, I decided to go for a very small change of only 20 millimeters. With the help of scissors and superglue I manipulated tail feathers, and after a case where I glued my fingers together, everything went smoothly. "Smoothly" is perhaps a slight exaggeration, because it meant many evenings and nights in barns and stables trying to catch males, and many early mornings from sunrise trying to observe the birds and determine the arrival of females. My sample

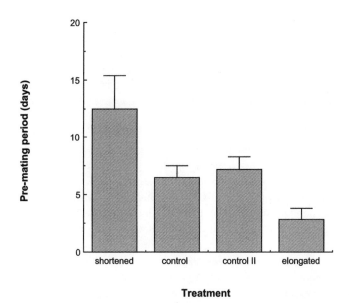

Figure 18.1. Mating latency measured as the duration of the premating period between tail length manipulation and date of mating in male barn swallows. Treatments are shortening by 20 millimeters, a sham-manipulation (control I), an untreated control (control II), and elongation by 20 millimeters. Values are mean number of days (SE). Adapted from Møller (1988).

sizes ended up relatively small (only forty-four males manipulated before mating), and it was thus out of the question to test if mating success differed among treatments. However, the mating latency worked fine and the effects on reproductive success were dramatic (Fig. 18.1; almost a doubling of reproductive success when comparing males with shortened and elongated tails). The perspectives were shocking to me because barn swallows were common, and many different experiments could be made without too much trouble.

I have a clear memory of an early morning observation in 1984 of a barn swallow male sitting in its territory after its mate had flown away. Suddenly, it flew "next door," singing while hovering over the neighboring female, and when the female attained copulation position engaged in an extra-pair copulation. The male made a couple more attempts, but without success and went back to its territory. I had read about such behavior in the literature, but not in Charles Darwin's (1871) writings. "His females" were all faithful! Having met Tim Birkhead (who was already then interested in the functional significance of multiple mating) at a meeting and having started to collect information on extra-pair copulations in birds, I wondered what was going on. Many more observations clearly showed me that it was not a random sample of females or males that participated in such copulations. Males that were successful in obtaining extra-pair copulations males had long tails, and extra-pair females generally had mates with short tails. Short-tailed male barn swallows, that often went without a mate for the season, even though they had a territory, were never successful in obtaining extra-pair copulations. There were several papers published about male birds having a mixed reproductive strategy, and I had already been exposed to Bob Trivers' (1972) paper on parental investment, which turned my world upside down. However, it was already clear to me in 1984 that it was the females who were running the show, and not the males, as suggested in the literature. I wrote and submitted a short paper based on my observations of male and female participants in extra-pair copulations and their success, suggesting that it was the females that displayed a mixed reproductive strategy (Fig. 18.2); while females responded similarly to their mates during the reproductive cycle, they differentially allowed male nonmates to engage in extra-pair copulations closer to fertilization. Hence, the mixed reproductive strategy of males appeared to be imposed by females, and I found this finding so different from "common" knowledge that I submitted the manuscript to *Nature*. The paper was returned without comments, and I could not understand why, although it struck me that probably I was just utterly wrong. Eventually, the tail manipulation experiment provided a second opportunity to test whether sperm competition (competition among the sperm of two or more males for the fertilization of the eggs of a single female) was a means of adjusting mate choice by females. This turned out to be the case according to my behavioral observations (Fig. 18.3; Møller 1988). Many attempts to collaborate on paternity studies in barn swallows eventually confirmed my observations that male tail length (as determined by experimental treatment, but also natural variation) affected paternity (Saino et al. 1997). Thus sperm competition (competition at the level of the gametes) turned out to be an important force of sexual selection. This eventually led to many other studies of sperm competition and sexual selection in birds and other organisms (e.g., Birkhead & Møller 1992, 1998).

While working on barn swallows I realized that tail length was just one of many traits that could affect sexual selection. It became clear that single signals do not exist, or at least only in unicellular organisms. I approached a couple of theoretically oriented friends to discuss these ideas, and some even said that this was not an interesting theoretical problem. Eventually, Andrew Pomiankowski and I embarked on the problem and explicitly stated the different possibilities (Møller & Pomiankowski 1993). The barn swallow has many different signals such as tail length, tail asymmetry, tail spots, red facial color, dorsal ultraviolet color, and various aspects of song. What was not clear was whether they were positively correlated. Or whether they indicated different features of phenotypic quality. Were they linearly related? Or were they multiplicative in effect? Today we know that sexual signals in barn swallows are relatively strongly positively correlated. For example,

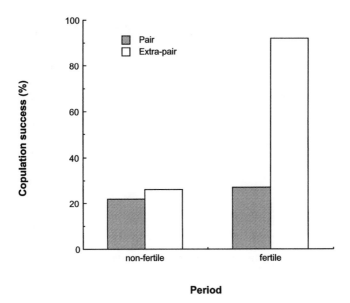

Figure 18.2. Copulation success by male mates and male nonmates with the same female barn swallows during the nonfertile (at least five days before start of laying) and the fertile period (from four days before start of laying until the penultimate date of laying). Observations made in 1983–84 on eighteen pairs of barn swallows during twenty-six reproductive cycles. The success differs significantly between periods for extra-pair copulations, but not for pair copulations. From an unpublished manuscript from 1984 by the author.

song rate interacts with tail length by mainly being important in long-tailed, but less important in short-tailed males, thus acting in an honesty enforcing manner (Møller et al. 1998a). When looking at this problem in a broader, interspecific perspective it is much less clear that all signals are honest. Many species have multiple signals that do not appear to be reliable indicators of quality. In particular, species under intense sexual selection often have many signals that are not the target of strong female mate preferences (Møller & Pomiankowski 1993). Why that is the case is far from clear.

The benefits of female mate preferences are not well known, and at least in the 1980s "good genes," which refers to the genetic effects of sires on offspring viability, were generally not considered an acceptable theoretical possibility. The case for good genes was far less clear for the empiricists. A particularly promising possibility was the suggestion by Hamilton and Zuk (1982) that secondary sexual characters might reveal important information about male resistance to parasites because of the ubiquitous nature of disease and parasitism. This hypothesis and alternatives fell out of fashion due to some comparative analyses in the 1980s, but still continued to be a real possibility to an empiricist like myself. I had been bothered by blood suck-

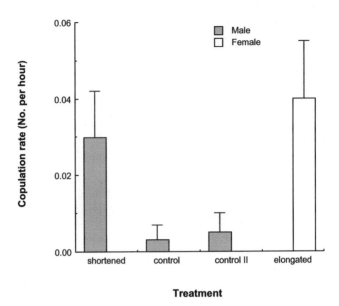

Figure 18.3. Rate of extra-pair copulations by male and female barn swallows in relation to experimental treatment of male tail length. Values are means per hour (SE). Adapted from Møller (1988).

ing mites during my barn swallow work over the years (the mites often causing sleepless nights), and it was only later that I decided to investigate the effects of the mites on swallows. These experimental attempts revealed strong effects on reproductive success, but also effects on tail length and male song (Møller 1991, 1994a). Eventually a partial cross-fostering experiment demonstrated a strong similarity in mite loads of nestlings independent of their rearing environment, and the parasite loads being negatively related to male tail length (Møller 1990a). The mechanisms causing this resemblance were not known, although analyses demonstrated host effects on blood feeding ability and hence survival and reproduction of mites (Møller 2000). Later, together with Nicola Saino I was able to demonstrate that tail length reliably reflected a measure of immunocompetence (Saino & Møller 1996), which was selectively advantageous because individuals with strong responses were more likely to make it back to the breeding grounds from their annual return trip to southern Africa (Saino et al. 1997). Changing the emphasis from parasites to general host defenses also revealed comparative evidence for a role of parasites in host sexual selection (Møller et al. 1998b, 1999), thus confirming Bill Hamilton's initial idea. Furthermore, the presence of good genes was confirmed in many different studies (Møller & Alatalo 1999), and the average importance of good genes was even of a magnitude consistent with current estimates of the heritability of fitness (M. Kirkpatrick & A. P. Møller, unpublished data).

The tail of a barn swallow is forked, with the outermost pair of feathers being elongated particularly in males. However, feathers are not always equally long, and such asymmetry is clearly visible to a human observer, but also to the swallows themselves, as demonstrated below. Asymmetry is traditionally considered a measure of developmental instability because it reflects an inability of an individual to produce a regular phenotype. Experiments and observations showed that tail asymmetry affected performance in terms of mating success, migration, and reproductive success (Møller 1992a, 1994b). If asymmetry reflected exposure to adverse environmental conditions, we could use it as a phenotypic marker of exposure to factors that perturb developmental control mechanisms. I have used this repeatedly in the barn swallow since blood sucking mites increased asymmetry in tail feathers (Møller 1992b) as did radioactivity in Chernobyl (Møller 1993b). Subsequent studies by many others have demonstrated that asymmetry often is an important determinant of mating success (Møller & Swaddle 1997), and just as important as size (Thornhill & Møller 1998). The developmental stability concept is important in many other contexts and similar approaches may be used when studying irregular phenotypes in domains other than morphology.

## Approaches to Behavioral Ecology

### OBSERVATIONS, EXPERIMENTS, AND DISTRUST OF "THE IMPOSSIBLE"

My research career has been a mixture of studies based on observations, field experiments, and comparative studies. This is a deliberate mix that I try to impose on most subjects. Observations give rise to ideas that may be testable based on available data or even better future experiments. For example, the barn swallows in 1988 had extremely asymmetric, outermost tail feathers and this posed a real problem in terms of measurements because I had previously only measured the length of the right, outermost tail feather. My colleague Rauno Alatalo had previously said that "only wildlife ecologists measure the length of both tarsi," but I still decided to proceed. The irregularity in phenotype later turned out to be negatively related to tail length, with the preferred males having the most symmetric tails (Møller 1990b). Depending on whether females assessed males based on tail length, absolute tail asymmetry, or relative tail asymmetry, this could result in reinterpretation of previous tail manipulation experiments by Andersson (1982) and myself (Møller 1988). Hence, a new experiment was required, showing that both tail length and tail asymmetry independently and in a multiplicative manner affected mate preferences (Møller 1992a). These kinds of experiments are often deemed "impossible," being considered unlikely to succeed by colleagues. It is obviously important to notice that "impossible" is a relative term; what is "impossible" is usually what has not been done so far.

This does not mean that it cannot be done in the future. The inability to do "impossible" things may either reflect a certain level of realism, or a lack of adventurous spirit that will prevent extraordinary things from happening. My own attitude is usually that nothing is impossible until it has been tried at least once!

It is important to remember that the step from theory to experimental test is not always an easy one. I have always tried to make experiments within the natural range to avoid any pathological effects. However, the step from theory to practice may be very difficult. For example, which control treatments should be included? Only an unmanipulated control or also a sham-manipulated treatment? In the treatment group, it is essential to manipulate the trait in question and not another trait correlated with the phenotypic trait. A potential example of such an unintended effect of manipulation is given below.

Sexual selection studies of barn swallows belong to the most replicated experiments in behavioral ecology (see summary in Møller 1994a). It has been my aim to repeat experiments to better understand what is going on. Tail length experiments in the barn swallow have been repeated in four different countries a total of ten different times. Only a fraction of the data has so far been published. However, repeated experiments have allowed studies of geographical variation in responses to manipulation and estimates of the consequences of manipulations in subsequent years.

Replicated tests of the same hypothesis have also been achieved by using different designs to address the same problem in a different way. For example, since the experimental manipulation of tail length and tail asymmetry affected morphology and indirectly behavior through effects on aerodynamics, I decided to manipulate perceived asymmetry by using Tipp-Ex correction fluid on the tips of the outermost tail feathers (Møller 1993c). This experiment was performed to create differences in appearance without manipulating flight. The results resembled those of the original experiment: males with apparently asymmetric tail feathers have reduced reproductive success.

A second example of a repeated test was conducted on barn swallows that rather than being unmated were already mated (de Lope & Møller 1993). The basis for this experiment was the observation that females tend to invest more in reproduction when mated to an attractive mate (Burley 1986). Such differential parental investment by females could arise from differential access by particular females to attractive males, or from differential investment. By manipulating the tail length of males that were already mated, and hoping that these manipulations would not result in divorce, it would be possible to test if females adjusted their parental investment relative to the phenotype of the mate. This was the case because females provided a relatively greater share of food to the offspring when mated to an attractive mate. Alternatively, this observation could reflect that males with elongated tails simply had greater difficulty providing parental care. However, females mated to attractive males also laid a second clutch more often, laid earlier, and had more offspring in the second clutch, suggesting that they differentially invested in reproduction (de Lope & Møller 1993).

As stated already, it is not always self-evident how to proceed from theory to implementation of an experiment. I had known for years that tail manipulations affected not only tail length, but also the size of the white spots in the tail feathers. Mati Kose, who is a young Estonian student, had observed that white tail spots were longer in adults than in juveniles and longer in males than in females, and their size was positively related to tail length (Kose & Møller 1999). Hence, the results of previous experiments could have arisen from manipulations of tail spot size rather than feather length. When Kose experimentally manipulated the size of the tail spots by painting these black, reducing their size by half, or just painting the feathers in the black part around the spots, he found that this treatment affected seasonal reproductive success. Males with larger spots had greater reproductive success (Kose & Møller 1999). Why did not all males have large white spots? Feather lice eat feathers, but prefer to eat the white, unpigmented parts since Kose found almost twice as many holes chewed by feather lice in the white spots as predicted from their area. Feather lice also preferentially settled in the white spots in a habitat selection experiment (Kose et al. 1999). Holes in feathers can be dangerous because they make feathers more prone to breakage, and many more tail feathers were broken in the white tail spots than expected by chance (Kose & Møller 1999). Studies of the abundance of feather lice in three different populations of barn swallows have shown that long-tailed males have fewer feather lice than short-tailed males (Kose & Møller 1999). Hence, long-tailed males can "afford" to have large white tail spots because they are less lousy than other males.

Scientific inquiry is about generalization: scientists want to find general descriptions of the world whereas engineers want to find solutions to specific problems. Scientific generalizations can be achieved in two different ways: (1) through a modeling approach that leaves all irrelevant detail behind while still grasping the crucial features of the problem, and (2) through a comparative approach. I have attempted to use a comparative approach to test the generality of ideas or predictions derived in a particular model system (usually the barn swallow). It is often impossible to make exactly the same kinds of tests at different levels; hence, a particular prediction cannot readily be investigated at both the intraspecific and interspecific levels. This said, information obtained at one level can often enlighten another and vice versa. For example, I have been involved in comparative studies showing that bird species with sexual dichromatism have larger immune defense organs and higher concentrations of leukocytes than closely related, sexually monochromatic species (Møller et al. 1998b); secondary sexual characters tend to be more asymmetric than homologous characters in closely related bird species without ornaments (Møller & Höglund 1991; Cuervo & Møller 1999); the presence of extravagant feather ornaments increases the rate of speciation (Møller & Cuervo 1998); sexual selection arising from sperm competition is associated with increased size of immune defense organs in birds (Møller 1997); social bird species experience stronger selection pressures from parasites than solitary species and this in turn has given rise to an evolutionary increase in immunocompetence that is acquired at a cost measured in terms of developmental time (A. P. Møller et al., unpublished data); and bird species with larger repertoire sizes have larger immune defense organs than species with small repertoires (Møller et al. 2000a). Many of these conclusions could not have reached this level of generalization without a comparative approach, and many intraspecific studies have been initiated because of the comparative results. Hence, detailed experimental studies at the intraspecific level can facilitate comparative studies and vice versa. Furthermore, the two approaches can readily be combined as shown by the comparative study of host sociality, cost of parasitism, and host immunocompetence in the swallow family (A. P. Møller et al., unpublished data).

## META-ANALYSIS AND PUBLICATION BIAS

The first empirical studies addressing a particular problem often show dramatic effects. This has to be the case almost by default because editors and referees need to be convinced, and this is not always an easy task, particularly not when it comes to very good journals. What often happens is that such ground-breaking studies tend to leave the impression that a particular relationship is what is to be expected. If another study of the same phenomenon, even on the same species, shows a weaker or no effect, such a second paper is often published immediately. For example, my own experiments on the effect of tail manipulation on mating success in the barn swallow were subsequently replicated in Canada (Smith & Montgomerie 1991). Although there was no significant effect of manipulation on offspring production during a particular year, there was a significant effect on date of reproduction. As more studies of a particular phenomenon accumulate, we can start investigating general patterns. For example, sexual selection has been extensively reviewed by Andersson (1994), who listed all quantitative studies of sexual selection on phenotypic characters (his Table 6A).

Textbook examples of sexual selection are well known, while the many examples of weak and statistically nonsignificant sexual selection, even involving extravagant secondary sexual characters, are less well known. However, there are many examples of this kind, and they are not randomly distributed across species (Møller 1993a). Lack of statistically significant selection may arise from a lack of statistical power, for example, due to a small sample size, or from the true absence of selection. A trivial, but some-times important, factor is that poor studies of poor quality also will tend to remain unpublished. Obviously, null results will often remain unpublished (Rosenthal 1991), and this may cause a bias in the literature. Reviews of the literature are often based on a narrative summary, and this can cause prob-lems because such narratives are not based on a quantitative assessment of the evidence. Meta-analysis provides such a tool for quantitative assessment of a body of literature (Arnqvist & Wooster 1995). The effect in different studies is estimated using a common statistic such as Pearson's product-moment correlation coefficient. The influence of each study is usually weighted by sample size since it is reasonable to assume that a large sample size will give rise to an effect that is closer to the real effect in the popula-tion. A meta-analysis of all studies of sexual selection on visual characters in birds gave an average correlation coefficient of 0.30 across 138 studies (Gontard-Danek & Møller 1999), while the effect size from Andersson's (1982) study of the long-tailed widowbird was 0.68 and the effect size from my own study of the barn swallow 0.78 (Møller 1988). Thus on average across all species studied the mean correlation between expression of sec-ondary sexual characters and mating success is only 0.30, which implies that secondary sexual characters explain 9 percent of the variance in mating suc-cess. This is not very much, but is still considered an effect of intermediate magnitude (Cohen 1988). The studies of Andersson and myself were experi-mental while many other studies were entirely correlational and hence may be affected by additional factors. For studies with both observations and experiments effect sizes were 0.32 and 0.42, respectively, thus accounting for 9.9 and 17.6 percent of the variance (Gontard-Danek & Møller 1999). Hence, an experimental approach almost doubles the amount of variance explained.

Publication bias is exceedingly difficult to measure because we will never know the true distribution of results obtained by scientists (Begg 1994). However, we can make some inferences based on empirical analyses. One measure of the robustness of a mean effect is the fail-safe number, which is the number of studies with null results needed to nullify the estimated effect. If this number exceeds 3 times the number of studies plus 10, the result is considered to be robust (Rosenthal 1991). We found a fail-safe number of 47,194 (Gontard-Danek & Møller 1999), which is an exceedingly large num-ber unlikely to exist as unpublished studies in the drawers of behavioral ecologists! We investigated a number of determinants of effect size such as type of character (color versus morphology), mating system, and currency (preference, reproductive success, paternity, mating success), and we found some evidence of heterogeneity related to these variables (Gontard-Danek & Møller 1999).

# Future Research Directions

If multiple studies of a particular phenomenon reveal differences, this can escalate into a conflict or a dispute. There are many such conflicts, of which the "Scandinavian flycatcher war" is a good example. Shortly summarized, this conflict arose from differences in the relative importance of male phenotype versus material resources as determinants of mating success, how important sperm competition was in determining male reproductive success, and several other features. It turned out that the two populations of pied flycatcher (*Ficedula hypoleuca*) around Oslo in Norway and Uppsala in Sweden lived in very different habitats and were influenced to different degrees by introgression from hybridization with the collared flycatcher (*Ficedula albicollis*) (Lundberg & Alatalo 1992). Hence, both parties turned out to be correct. Scientists often work with a single population, and if they have seen a particular phenomenon with their own eyes, this is "how it is." Very few scientists work in multiple populations subjected to different environmental conditions. Different populations will be subject to different selection pressures and past evolutionary histories, and hence analyses of patterns within and among populations constitute a second layer in an attempt to understand ecological and evolutionary questions.

## GEOGRAPHICAL VARIATION IN SELECTION PRESSURES

I have initiated studies of barn swallow populations in several different parts of Europe and started collaborations with local scientists. We are currently working on barn swallows in Spain, Italy, Hungary, Denmark, Estonia, and Finland. Several scientific questions are addressed in this large-scale geographical approach: (1) geographical variation in sexual selection; (2) geographical variation in immune function; (3) geographical variation in life history; and (4) geographical variation in migration.

First, barn swallows demonstrate considerable geographical variation in secondary sexual characters, with sexual size dimorphism in tail length increasing from around 5 percent on average in North Africa to more than 20 percent in Finland (Møller 1995) and a similar cline is found in North America (Brown & Brown 2000). The reason for this cline has been hypothesized to derive from the different costs of long tail feathers (Møller 1995; Barbosa & Møller 1999). Barn swallows feed on large Diptera on the wing, and such insects are difficult to catch, particularly when ambient temperature is high. Hence, large insects will be particularly difficult to catch at low latitudes with high average temperatures. Tail manipulation experiments have demonstrated stronger effects of a given level of manipulation on number and size of insects captured in Spain compared to Denmark (Møller et al. 1995), as we predicted. This preliminary study is extended to other populations to test further whether costs of sexual displays differ with latitude. We are also currently investigating whether latitudinal differences in tail length are associated with latitudinal differences in intensity of sexual selection, parasitism, and host immune responses. Finally, we are currently investigating geographical patterns of covariation between different secondary sexual characters such as tail length, tail asymmetry, size of white tail spots, red facial coloration, dorsal ultraviolet coloration, and features of the song. Such analyses in multiple populations require knowledge of the history of the populations, which will be obtained from a phylogeny of the different populations derived from molecular markers.

Second, parasites and diseases are often assumed to be more common or more virulent in warmer climates because populations of parasites and pathogens are not, or are only to a small degree, knocked down by winter conditions at low latitudes (Connell 1971; Janzen 1970). Hence, host populations closer to the tropics may be controlled in a density-dependent manner by severe selection pressures imposed by parasites and pathogens. There is evidence for this hypothesis based on larger investment in immune function among tropical as compared to temperate species of birds (Møller 1998). Studies of the barn swallow across Europe have demonstrated a significant cline in immune response (T-cell response) among nestlings, with nestlings being more immunocompetent toward the north (A. P. Møller et al., unpublished data). Preliminary studies also suggest a similar cline in immune response among adults, suggesting that larger secondary sexual characters are associated with greater immunocompetence across populations (A. P. Møller et al., unpublished data). Finally, parasite communities vary geographically, and the European barn swallow populations are only heavily infected with blood parasites in southern Spain. This provides us with an opportunity to compare morphology and physiology of birds from populations infected to different degrees with malarial parasites. A long-term objective is to investigate the geographical patterns of genetic variation in major histocompatibility (MHC) genes, and how this relates to expression of secondary sexual characters, parasite impact, and host immune responses. These patterns of geographical variation in immune function bring me naturally to the next subject of current interest: life history variation.

Third, geographical variation in life history is well described in the barn swallow, since clutch size increases toward the north while the frequency of second and third clutches decreases (Møller 1984). Thus annual productivity decreases dramatically from more than nine nestlings per pair per year in southern Spain to less than five nestlings in Finland. There is also a clear geographical decrease in adult survival from around 40 percent in southern Spain to around 30 percent in Denmark (A. P. Møller et al., unpublished data). Thus a larger proportion of nestlings are recruited locally in the north. Interestingly, this is the area where immunocompetence of nestlings is the highest. Food availability for small birds usually decreases during the season, and multibrooded species like the barn swallow often start laying before the seasonal peak of food (Crick et al. 1993). Nestling quality measured in terms of immunocompetence decreases with progressing season, and immunocompetence is consistently lower in second, as compared to first, broods of the same parents (Saino et al. 1997). We are currently testing whether it is possible to predict the frequency of second clutches from the seasonal decrease in immunocompetence of nestlings during the first clutch. Geographical patterns of immunocompetence suggest a trade-off between quality and quantity of offspring, with few and high-quality offspring being produced in the north and many and low-quality offspring being produced in the south. If this is the case, we should expect differences in costs of reproduction between northern and southern populations of barn swallows, and such differences are currently being tested using brood size manipulation experiments across Europe. A second issue in life history theory that is of much current interest is the causes of senescence, which is the deterioration of the phenotype with increasing age. Again, there are obvious links between immunology and evolutionary approaches. Preliminary studies of the barn swallow have suggested that morphology, reproduction, migration, and health status all deteriorate with increasing age (Møller & de Lope 1999). The causes for this effect remain unknown, although we are currently testing whether old individuals accumulate mutations, and whether immunocompetence decreases with increasing age. The barn swallow is a unique study organism for this purpose because "very old" individuals are only five years! Furthermore, since survival rate varies geographically, we can investigate the rates of senescence in populations with different survival rates. Studies of sexual selection provide a unique experimental model system for studying senescence because experimental manipulation of secondary sexual characters has been shown to affect the deterioration of the phenotype (Møller 1989, 1992b). We have only begun investigating the effects of phenotypic manipulations on phenotype and reproductive success in subsequent years.

Fourth, migratory birds like the barn swallow are faced with extreme conditions at least twice a year during their long-distance migration. Not only do extreme levels of exercise reduce immune responsiveness (Hoffmann-Goetz & Pedersen 1994) and increase the level of potentially dangerous and mutagenic free radicals (Leffler 1993), but long-distance migrants also face the difficulty of being "tourists" that go to distant countries without prior vaccination against local parasites and diseases. It is perhaps not surprising that migratory birds tend to invest heavily in immune function compared to resident relatives (Møller & Erritzøe 1998). However, despite intense scientific inquiry during the past hundred years, we know very little about the adaptations that allow migratory birds to achieve this feat. We do not even know whether the breeding season, the winter period, or the migration impose the most severe selection pressures on migratory birds. The barn swallow is a suitable study organism for these studies because populations range in migratory habits from residents in Egypt and Sudan, to short-distance migrants in Spain that only leap across the Sahara to West Africa, to long-distance migrants from Scandinavia that go all the way to South Africa. The ecological and evolutionary consequences of geographical variation in migration distances are high on the priority list for future studies.

## GENETICS AND BEHAVIORAL ECOLOGY

Our different research questions all require an integration of genetics, ecology, and evolution. We have already begun this attempt to unify these different branches of biology by using mini- and microsatellite markers to study evolutionary phenomena within and among populations. "Natural" variation in mutagens as in Chernobyl has been used to investigate the consequences of mutations for individuals and to investigate sex differences in mutation rates (Ellegren et al. 1997). A genetic approach will be accentuated in the coming years by focusing on genetic markers that can be used for identifying individuals from different populations, and by focusing on genes involved in parasite resistance such as the MHC genes. A different approach relating physiology and ecology is based on the role of heat shock proteins. Heat shock proteins are involved in protection of proteins under stressful conditions, and preliminary studies have demonstrated effects of parasites on heat shock protein induction (Merino et al. 1998). We are currently investigating differences in the abundance of heat shock proteins in different populations of barn swallows and different species of the swallow family to determine whether abundance of heat shock proteins is related to sexual selection, parasite impact, and distribution. Free radical scavengers are biochemicals that neutralize free radicals that can cause mutations, and some carotenoids are known to have free-radical scavenging properties (Møller et al. 2000b). Hence, a trade-off is possible between sexual coloration and free-radical scavenging, as suggested for Chernobyl barn swallows that have very pale faces (Camplani et al. 1999). Since free radicals are produced in large amounts during heavy work load, and thus should be particularly abundant in males, one can imagine a direct relationship between sexual selection and free radicals (von Schantz et al. 1999). It is perhaps interesting that particularly male barn swallows showed a reduction in coloration in Chernobyl (Camplani et al. 1999).

## UTILITARIAN APPROACHES AND CONSERVATION

Funding agencies often emphasize short-term, utilitarian approaches to research. Certainly, it is not always easy to justify the usefulness of studies of sexual selection. Utilitarian approaches to research funding may previously have been a severe impediment to progress on the use of electromagnetic research in development of electricity or many other bizarre happenings in human history during the past couple of centuries. Obviously, pure research may not necessarily be at conflict with applied aspects. For example, sexual selection has many applied aspects that range from conservation to human social problems such as substance abuse, violence, and behavior in traffic. Sexual selection tends to reduce effective population size because of nonrandom mating. However, more subtle consequences of sexual selection include greater risks of extinction among sexually selected species (Sorci et al. 1998), greater effects of demographic stochasticity in species with intense sexual selection (Legendre et al. 1999), and reductions in reproductive success in species with strong mate preferences because of absence of mates with preferred phenotypes (Møller 1999). There are potentially many other conservation consequences of sexual selection (Møller 1999), so even the most utilitarian student may find something of general interest in a study of sexual selection.

## Acknowledgments

My research has been supported by grants from the Swedish and Danish Natural Science Research Councils and an ATIPE BLANCHE from the CNRS.

## References

Andersson M, 1982. Female choice selects for extreme tail length in a widowbird. Nature 299:818–820.

Andersson M, 1994. Sexual selection. Princeton: Princeton University Press.

Arnqvist G, Wooster D, 1995. Meta-analysis: synthesizing research findings in ecology and evolution. Trends Ecol Evol 10:236–240.

Barbosa A, Møller AP, 1999. Aerodynamic costs of long tails in male barn swallows *Hirundo rustica* and the evolution of sexual size dimorphism. Behav Ecol 10:128–135.

Begg CB, 1994. Publication bias. In: The handbook of research synthesis (Cooper H, Hedges LV, eds). New York: Russell Sage Foundation; 399–409.

Birkhead TR, Møller AP, 1992. Sperm competition in birds. London: Academic Press.

Birkhead TR, Møller AP (eds), 1998. Sperm competition and sexual selection. London: Academic Press.

Brown CR, Brown MB, 2000. The barn swallow. In: The birds of North America (Poole A, Gill F, eds). Philadelphia: The Birds of North America Inc.

Burley N, 1986. Sexual selection for aesthetic traits in species with biparental care. Am Nat 127:415–445.

Camplani A, Saino N, Møller AP, 1999. Carotenoids, sexual signals and immune function in barn swallows from Chernobyl. Proc R Soc Lond B 266:1111–1116.

Cohen J, 1988. Statistical power analysis for the behavioral sciences (2nd ed.). Hillsdale: L. Erlbaum.

Connell JH, 1971. On the role of natural enemies in preventing competitive exclusion in some marine animals and in rain forest trees. In: Dynamics of numbers in populations (den Boer PJ, Gradwell GR, eds). Wageningen: Centre for Agricultural Publication and Documentation; 298–312.

Crick HQP, Wingfield Gibbons D, Magrath RD, 1993. Seasonal changes in clutch size in British birds. J Anim Ecol 62:263–273.

Cuervo JJ, Møller AP, 1999. Phenotypic variation and fluctuating asymmetry in avian feather ornaments in relation to sex and mating system. Biol J Linn Soc 68:505–529.

Darwin C, 1871. The descent of man and selection in relation to sex. London: John Murray.

de Lope F, Møller AP, 1993. Female reproductive effort depends on the degree of ornamentation of their mates. Evolution 47:1152–1160.

Ellegren H, Lindgren G, Primmer CR, Møller AP, 1997. Fitness loss and germline mutations in barn swallows breeding in Chernobyl. Nature 389:593–596.

Gontard-Danek M-C, Møller AP, 1999. The strength of sexual selection: a meta-analysis of bird studies. Behav Ecol 10:476–486.

Hamilton WD, Zuk M, 1982. Heritable true fitness and bright birds: a role for parasites? Science 218:384–387.

Hoffman-Goetz L, Pedersen BK, 1994. Exercise and the immune system: a model of the stress response? Immunol Today 15:382–387.

Janzen DH, 1970. Herbivores and the number of tree species in tropical forests. Am Nat 104:501–528.

Kose M, Mänd R, Møller AP, 1999. Sexual selection for white tail spots in the barn swallow in relation to habitat choice by feather lice. Anim Behav 58:1201–1205.

Kose M, Møller, AP, 1999. Sexual selection, feather breakage and parasites: the importance of white spots in the tail of the barn swallow. Behav Ecol Sociobiol 45:430–436.

Lack D, 1966. Population studies of birds. Oxford: Clarendon.

Legendre S, Clobert J, Møller AP, Sorci G, 1999. Demographic stochasticity and sexual selection. Am Nat 153:449–463.

Leffler JE, 1993. An introduction to free radicals. New York: Wiley.

Lundberg A, Alatalo RV, 1992. The pied flycatcher. London: Poyser.

Merino S, Martínez J, Barbosa A, Møller AP, de Lope F, Pérez J, Rodríguez-Caabeiro J, 1998. Increase in heat-shock protein from blood cells in response of nestling house martins (*Delichon urbica*) to parasitism: an experimental approach. Oecologia 116:343–347.

Møller AP, 1984. Geographical variation in breeding parameters of two hirundines. Ornis Scand 15:43–54.

Møller AP, 1988. Female choice selects for male sexual tail ornaments in the monogamous swallow. Nature 332:640–642.

Møller AP, 1989. Viability costs of male tail ornaments in a swallow. Nature 339:132–135.

Møller AP, 1990a. Effects of an haematophagous mite on the barn swallow (*Hirundo rustica*): a test of the Hamilton and Zuk hypothesis. Evolution 44:771–784.

Møller AP, 1990b. Fluctuating asymmetry in male sexual ornaments may reliably reveal male quality. Anim Behav 40:1185–1187.

Møller AP, 1991. Parasites, sexual ornaments and mate choice in the barn swallow *Hirundo rustica*. In: Ecology, behavior, and evolution of bird-parasite interactions (Loye JE, Zuk M, eds). Oxford: Oxford University Press; 328–343.

Møller AP, 1992a. Female swallow preference for symmetrical male sexual ornaments. Nature 357:238–240.

Møller AP, 1992b. Parasites differentially increase fluctuating asymmetry in secondary sexual characters. J Evol Biol 5:691–699.

Møller AP, 1993a. Patterns of fluctuating asymmetry in sexual ornaments predict female choice. J Evol Biol 6:481–491.

Møller AP, 1993b. Morphology and sexual selection in the barn swallow *Hirundo rustica* in Chernobyl, Ukraine. Proc R Soc Lond B 252:51–57.

Møller AP, 1993c. Female preference for apparently symmetrical male sexual ornaments in the barn swallow *Hirundo rustica*. Behav Ecol Sociobiol 32:371–376.

Møller AP, 1994a. Sexual selection and the barn swallow. Oxford: Oxford University Press.

Møller AP, 1994b. Sexual selection in the barn swallow (*Hirundo rustica*). IV. Patterns of fluctuating asymmetry and selection against asymmetry. Evolution 48:658–670.

Møller AP, 1995. Sexual selection in the barn swallow (*Hirundo rustica*). V. Geographic variation in ornament size. J Evol Biol 8:3–19.

Møller AP, 1997. Immune defence, extra-pair paternity, and sexual selection in birds. Proc R Soc Lond B 264:561–566.

Møller AP, 1998. Evidence of larger impact of parasites on hosts in the tropics: investment in immune function within and outside the tropics. Oikos 82:265–270.

Møller AP, 1999. Sexual selection and conservation. In: Behaviour and conservation (Gosling M, Sutherland WJ, eds). Cambridge: Cambridge University Press; 161–171.

Møller AP, 2000. Survival and reproductive rate of mites in relation to resistance of their barn swallow hosts. Oecologia 124:351–357.

Møller AP, Alatalo RV, 1999. Good genes effects in sexual selection. Proc R Soc Lond B 266:85–91.

Møller AP, Biard C, Blount JD, Houston DC, Ninni P, Saino N, Surai PF, 2000b. Carotenoid-dependent signals: indicators of foraging efficiency, immunocompetence or detoxification ability? Poultry Avian Biol Rev 11:137–159.

Møller AP, Christe P, Lux E, 1999. Parasite-mediated sexual selection: effects of parasites and host immune function. Q Rev Biol 74:3–20.

Møller AP, Cuervo JJ, 1998. Speciation and feather ornamentation in birds. Evolution 52:859–969.

Møller AP, de Lope F, 1999. Senescence in a short-lived migratory bird: age-dependent morphology, migration, reproduction and parasitism. J Anim Ecol 68:163–171.

Møller AP, de Lope F, López Caballero JM, 1995. Foraging costs of a tail ornament: experimental evidence from two populations of barn swallows *Hirundo rustica* with different degrees of sexual size dimorphism. Behav Ecol Sociobiol 37:289–295.

Møller AP, Dufva R, Erritzøe J, 1998b. Sexual selection and immune defence in birds. J Evol Biol 11:703–719.

Møller AP, Erritzøe J, 1998. Host immune defence and migration in birds. Evol Ecol 12:945–953.

Møller AP, Henry P-Y, Erritzøe J, 2000a. The evolution of song repertoires and immune defence in birds. Proc R Soc Lond B 267:165–167.

Møller AP, Höglund J, 1991. Patterns of fluctuating asymmetry in avian feather ornaments: implications for models of sexual selection. Proc R Soc Lond B 245:1–5.

Møller AP, Pomiankowski A, 1993. Why have birds got multiple sexual ornaments? Behav Ecol Sociobiol 32:167–176.

Møller AP, Saino N, Taramino G, Galeotti P, Ferrario S, 1998a. Paternity and multiple signalling: effects of a secondary sexual character and song on paternity in the barn swallow. Am Nat 151:236–242.

Møller AP, Swaddle JP, 1997. Asymmetry, developmental stability and evolution. Oxford: Oxford University Press.

Rosenthal R, 1991. Meta-analytic procedures for social research. New York: Sage.

Saino N, Møller AP, 1996. Sexual ornamentation and immunocompetence in the barn swallow. Behav Ecol 7:227–232.

Saino N, Bolzern AM, Møller AP, 1997. Immunocompetence, ornamentation and viability of male barn swallows (*Hirundo rustica*). Proc Natl Acad Sci USA 97:579–585.

Saino N, Calza S, Møller AP, 1997. Immunocompetence of nestling barn swallows in relation to brood size and parental effort. J Anim Ecol 66:827–836.

Saino N, Primmer C, Ellegren H, Møller AP, 1997. An experimental study of paternity and tail ornamentation in the barn swallow *Hirundo rustica*. Evolution 51:562–570.

Smith HG, Montgomerie RD, 1991. Sexual selection and the tail ornaments of North American barn swallows. Behav Ecol Sociobiol 28:195–201.

Sorci G, Møller AP, Clobert J, 1998. Plumage dichromatism of birds predicts introduction success to New Zealand. J Anim Ecol 67:263–269.

Thornhill R, Møller AP, 1998. The relative importance of size and asymmetry in sexual selection. Behav Ecol 9:546–551.

Trivers RL, 1972. Parental investment and sexual selection. In: Sexual selection and the descent of man, 1871–1971 (Campbell B, ed). Chicago: Aldine-Atherton; 136–179.

von Schantz T, Bensch S, Grahn M, Hasselquist D, Wittzell H, 1999. Good genes, oxidative stress and condition-dependent sexual signals. Proc R Soc Lond B 266:1–12.

# PART IV    Mammal Model Systems

# 19 Cunning Coyotes: Tireless Tricksters, Protean Predators

Marc Bekoff

## That Old Man Coyote: The Allure of Behavioral Variability

Coyote is the trickster par excellence for the largest number of American Indian cultures (Bright 1993, 4).

Coyotes (*Canis latrans*) are amazing carnivores. They are many things to many people. They wear many faces that affect different people in different ways. They are victims of their own success. Coyotes are extremely adaptable and versatile survivors who are frequently portrayed and referred to as cunning tricksters, thieves, gluttons, outlaws, spoilers, and ultimate survivors in American mythology and in Native American tales (Lopez 1977; Bright 1993, and references therein). These characterizations are based mainly on this maligned predator's uncanny ability to survive and reproduce successfully in diverse locations and in extremely harsh conditions.

Coyotes have even endured incessant and abusive onslaughts (including such brutal planned community hunts as those still held in Wyoming; *New York Times* [16 November 1998, A14] and Arizona) by humans and trapping on National Wildlife Refuges (Fox 1998). They also have openly defied expensive attempts to kill them using extremely inhumane methods (leghold traps, snares, coyote-getters, various poisons, aerial gunning) by private parties and government organizations such as the Animal Damage Control (now politically correctly called Wildlife Services, which was responsible for killing about 82,000 coyotes in 1996 alone; Wolfe & St. John 1998). Indeed, one might ask why coyotes are so numerous and "problematic" if control programs worked. Frederick Knowlton, who has been involved in coyote control for almost four decades, maintains it is necessary to kill coyotes to protect livestock even if the coyotes return (Finkel 1999). Simply put, wantonly killing coyotes has *not* and does *not* work. One reason that many control/killing programs have been ineffective is because they have not used knowledge of the behavior and behavioral ecology of these charismatic, protean predators.

Fortunately, coyotes also still attract much attention from numerous biologists who are interested in learning more about their varying lifestyles and their fundamental role in structuring ecological communities rather than in killing them. Recently it has been shown that coyotes are extremely important in regulating species at lower trophic levels, and that their decline and disappearance influence the distribution and abundance of smaller carnivores and their avian prey (Crooks & Soulé 1999; Saether 1999). Suffice it to say, although coyotes can be difficult to study and try one's patience, they are a model system in which to learn about numerous aspects of behavioral ecology.

## Getting to Where I Am Now

My journey to (and within) the field of behavioral ecology and coyotes was rather indirect. There were some fast starts and equally quick stops, and some false starts along bifurcating paths that resulted from, and resulted in, clarification of my interests and goals. I really did not care if there were false leads for I felt that a carefully built foundation would result in a sustained interest in behavioral ecology and a solid product. As soon as I began studying behavioral ecology I knew that I wanted to do it for a long time, and I also wanted an academic career combining teaching and research.

I was graduated from Washington University (St. Louis) and was unsure what to do with my newly granted bachelor's degree in anthropology. I became a record salesman at a local department store. While listening to the likes of The Beatles, Bob Dylan, Small Faces, and Janis Joplin, I began looking into graduate programs in biology and enrolled at Hofstra University because I could keep my job and go to school at the same time. While studying macromolecules and memory at Hofstra, I developed an interest in animal behavior, mainly by reading books and essays by Konrad Lorenz, Niko Tinbergen, Desmond Morris, R. F. Ewer, John Eisenberg, Robert Hinde, Iraneus Eibl-Eibesfeldt, Peter Marler and William Hamilton, and even Robert Ardrey.

After completing my master's thesis on relationships between RNA metabolism and memory, I enrolled at Cornell University Medical College and began a Ph.D. program in neurobiology and behavior. I was studying vision and learning in domestic cats (*Felis catus*) and the experimental protocol involved killing (aka "sacrificing") the cats to localize lesions made using stereoscopic methods. I hated doing this (Bekoff 1998a) and began looking for another graduate program. Quite coincidentally a newsletter from Washington University appeared in my mailbox. There was an article in it about a newcomer on campus, Michael W. Fox, who was studying the development of social behavior in various canids, including coyotes, wolves (*C. lupus*), domestic dogs (*C. familiaris*), golden-jackals (*C. aureus*), and red foxes (*Vulpes vulpes*). Much of this work is summarized in Fox (1970). Mike also had a strong background in neurobiology (Fox 1971). He had not been there when I was an undergraduate and because I really enjoyed my undergraduate days (and nights) at Washington University and wanted to join his research group, I immediately called him and a few months later I was accepted into a Ph.D. program there beginning in September 1970. Mike was an outstanding mentor. He supported me on his research grants, helped me locate animals to study, did not force me to take classes merely to fill class quotas, stressed a strongly comparative and developmental perspective, and also allowed me much-needed independence.

After gaining admission I began reading everything I could find on canids and behavior in general, and immediately was taken with the incredible adaptability of these carnivores. I also was surprised that while there was a lot of natural history and descriptive accounts of coyotes (e.g., Murie 1940; Young & Jackson 1951) there were no detailed field studies that were question-oriented and dealt mainly with behavior and ecology. Coyotes were touted as being territorial but there were no detailed data on patterns of scent-marking and territorial defense, especially between neighboring packs. I was also very interested in behavioral development and began exploring ideas about how behavioral ontogeny could be related to adult behavior, especially intraspecific variability and the ability to behave flexibly in response to changing environmental conditions (see also Fisler 1969; Cockburn 1988; Lott 1991). Previous researchers had observed coyotes living alone, in pairs, and in packs but with no detailed data. Indeed, Petr Kropotkin (1914, 41) also reported packs of coyotes in his book *Mutual Aid: A Factor of Evolution*.

After completing my doctoral research on the development of behavior in various captive canids, concentrating on social play and rank-related agonistic behavior (Bekoff 1972, 1974), I began developing interests in trying to study coyotes in the field. In 1974 I went to the University of Colorado, in Boulder, and began observing coyotes in Rocky Mountain National Park, Estes Park, Colorado. In September 1977 I began my seven-year study of coyotes in the Grand Teton National Park (summarized in Bekoff & Wells 1986). The main focus at the start was on individual differences in behavioral development and dispersal (Bekoff 1977a). I was interested not only in coyotes, but also other canids. Later my interests became more broadly comparative and I ventured into life history analyses (Bekoff et al. 1981; Bekoff et al. 1984).

I also wanted to try to bridge the gap between those "wildlifers" who wanted to control coyotes in the absence of much knowledge about their behavior and those who wanted to learn more about them from a more academic perspective that could have practical implications. Data was published that suggested that coyotes were major predators on livestock and other animals, but much information was not subjected to statistical analyses. Indeed, when I did some analyses, it turned out that coyotes were not as devastating to livestock as other causes such as other predators and disease (Bekoff 1979; interestingly, this paper was accepted and then temporarily rejected for what I learned were political reasons; some were not happy with what the statistical analyses showed).

## Wilson's Wagon: Surrounded But Not Defeated

We have little choice but to seek inspiration from gurus of the newest ideas; sometimes they turn out to be partially right. However, we should never believe them without a struggle. If an idea seems too good to be true, it is probably not true.
(Houle 1998, 1875)

In 1975 E. O. Wilson's difficult-to-tote tome, *Sociobiology: The New Synthesis*, appeared. One of the century's most important books in behavioral biology, this volume was, and remains, a gold mine of ideas. It was all-the-rage! Wilson (1975, 4) defined sociobiology as "the systematic study of the biological basis of *all* [my emphasis] social behavior." And sociobiology, along with behavioral ecology, were destined to cannibalize ethology (Wilson 1975, Fig. 1–2 and pp. 5–6; see also Rosenberg 1980; Kitcher 1985). Wilson wrote (1975, 6): "The future, it seems clear, cannot be with the ad hoc terminology, crude models, and curve fitting that characterize most of contemporary ethology and comparative psychology." This is interesting because so much of behavioral ecology is indeed distinguished by these three characteristics.

Numerous students of behavior jumped on the sociobiology bandwagon and embraced sociobiology as "the only way to go," whereas others, myself included, saw much promise to the sociobiological approach but also recognized that perhaps some of the flexibility in the social systems of various animals could be studied and understood without resorting to the vise of overbearing genetic constraints. Wilson's ideas were very challenging because they established an arena in which to exchange ideas concerning possible relationships between behavioral flexibility and genetic constraints.

## The Integration of Conceptual, Theoretical, and Empirical Work: Natural History Meets Hard Science

Most of my research has been done on coyotes near Jackson, Wyoming, and western evening grosbeaks (*Coccothraustes vespertina*; Bekoff et al. 1987, 1989; Bekoff 1995a) and steller's jays (*Cyanocitta stelleri*; Bekoff et al. 1998, 1999) in the foothills outside Boulder, Colorado. My fieldwork has centered on questions dealing with the development of behavior, various aspects of social ecology, and the evolution of different and flexible developmental and organizational patterns. In this essay I concentrate on my research on coyotes. I realize that to current readers some of what I write may seem utterly naive, but it is important to remember that many of my ideas were gestating and investigated not long after the late W. D. Hamilton's classic paper (1964) and Crook's (1970) *Social Behaviour in Birds and*

*Mammals* were published, and just prior to the publication of such books as *Behavior as an Ecological Factor* (Davis 1974), *Sociobiology* (Wilson 1975), and the first of four editions of *Behavioural Ecology* (Krebs & Davies 1978, 1984, 1991, 1997). I often ask my students to design experiments to test various hypotheses in behavioral ecology imaging themselves to be working pre- and post-Darwin. It also is an interesting exercise to develop hypotheses and to design studies in behavioral ecology pre- and post-Hamilton, Wilson, and Krebs and Davies.

My research integrates conceptual, theoretical, and empirical work in a variety of ways. First and foremost, it stresses that behavioral responses are phenotypic adaptations to various environmental conditions and that many behavior patterns are modifiable when environmental demands change. Indeed, there has been selection for plasticity rather than rigidity in many species. An understanding of behavioral flexibility is one of the most exciting and difficult challenges that faces students of behavioral ecology.

I also believe that a solid database is indispensable for driving the development of conceptual models and theories. However, there also is a strong and deep reciprocal relationship between the collecting of data and the developing of models and theories; people must always take stock of where they are going in both types of endeavors. Two examples are useful here. The first deals with urine scent-marking in coyotes. There has been a lot of theorizing (essentially, guessing) about the role of scent-marking in these canids but few people have actually done detailed observational studies. We found that while some theories about the use of scent-marking in territorial behavior are helpful for explaining some aspects of this behavior, there are other aspects of scent-marking that have been overlooked by a narrow focus on territoriality (Wells & Bekoff 1981; Gese & Ruff 1997; Allen et al. 1999). Furthermore, we have found that inferences drawn from yellow (urine-stained) snow produce spurious results and that direct observations are essential to make reliable inferences about patterns of scent-marking (Bekoff 1980; Wells & Bekoff 1981; see also Gese & Ruff 1997; Allen et al. 1999).

Another area that we have studied in considerable detail is alloparental (helping) behavior. A lot of "what animals are supposed to do" is driven by sociobiological and kin selection theory, but we found that coyotes will help not only kin but also more distantly related individuals. Indeed, at the 1979 International Ethology Congress held in Vancouver, Canada, I was asked (by a person who is now a "star") if we really knew the genealogy of the coyotes we were studying, for patterns of helping did not jibe with kin selection theory. I assured the skeptic that we did! Furthermore, concentrating on helping within the breeding season excludes many important aspects of helping behavior, especially in driving off intruders during the nonbreeding season. Thus, while theory has been useful in driving our research we have found, contrary to what might be expected, that an analysis of coyote helping takes us out of the somewhat narrow constraints of sociobiological and kin selection theory.

## Watching Coyotes: Chasing Big and Little Questions and Being Flexible

It is important to emphasize the incredible adaptability and especially the behavioral variability of coyotes; references to the way in which *the* coyote lives can be misleading. Coyotes are mobile (they can run as fast as 48 kilometers/hour), opportunistic, and adaptive carnivores who thrive in diverse habitats ranging from arid, warm deserts, to wet grasslands and plains, to colder climates at high elevations (up to about 3,000 meters), to large urban cities such as Los Angeles, California. They are now found between about 10 degrees north latitude (northern Alaska) and 70 degrees south latitude (Costa Rica) and throughout the United States and Canada. Coyotes enjoy a catholic diet including plant and animal matter and inanimate objects; diet varies greatly seasonally and geographically, as do the methods by which prey are acquired.

In this section I want to discuss different aspects of our research on coyotes in terms of big and little questions (I do not try to present an exhaustive review of available literature; for comparative data, see Bekoff 1977b, 1978, 1982; Bekoff & Wells 1980 1986; Bowen 1981, 1982; Pyrah 1984; Gese et al. 1996a, b; Gese & Ruff 1997, 1998; Gese 1998a, b; Crabtree & Sheldon 1999; and references therein). Needless to say, all of the topics we studied are interrelated as are questions of all sizes.

Much of our research was motivated using the guidelines put forth by the Nobel laureate, Niko Tinbergen. Briefly, Tinbergen (1951/1989, 1963) suggested that in studies of behavior, the following four areas need to be considered: evolution, adaptation, causation, and development. While no one study can deal in depth with questions in all of these areas, all are important for framing a long-term research program.

When we began our research we had a fairly good idea of what we would be able to observe directly, but our most optimistic predictions were surpassed. However, although we began with a keen interest in learning more about the role of development in dispersal, we soon realized that we would be able to collect novel data in many different areas of behavioral ecology, but not much in this area.

## Plasticity, Theory, and Data: Why Form Packs?

Coyotes rarely hunt in a coordinated or cooperative manner. Packs are usually found around dead ungulates such as deer, elk, and moose, and while

Figure 19.1. (a) A pack of coyotes around elk carrion on the National Elk Refuge, Jackson, Wyoming. Coyotes rarely kill large ungulates such as elk. This elk died of natural causes. (Photograph courtesy of Franz Camenzind). (b) A coyote watching out for other coyotes at elk carrion on the National Elk Refuge, Jackson, Wyoming. (Photograph courtesy of Franz Camenzind)

coyotes will occasionally kill these animals, there are few actual observations of this event. Often groups of coyotes merely congregate around and scavenge carcasses of animals that died of other causes and then individuals travel on by themselves.

Coyotes' main foods (small rodents) are easy for individuals to catch so we hypothesized that coyotes lived in packs when there was ample food for all mouths and that pack formation was an adaptation for the defense of food or land, rather than for the acquisition of food (see also Messier & Barrette 1982; Crabtree & Sheldon 1999). Indeed, whether cooperative hunting favors sociality in any or many carnivores is still being hotly debated (see Creel & Creel 1995 for detailed discussion, and references therein).

It was the variability of coyote social organization and space use and the plasticity of individual behavior that allowed us to try to tie together data with theory—*cautiously*. Clearly, generalizations about the lifestyle of *the* coyote were misleading, for normative thinking deemphasized the incredible behavioral plasticity of these canids. Why are coyotes so variable? How can they survive on their own or in packs? Is there a single factor that is causally related to their success? Based on our and others' findings it seems safe to conclude that food supply is the key. When there is ample winter food to feed adults and offspring from previous years, some, but not all, coyotes remain with their pack and do not disperse from their natal area. Pack structure is retained. When ample food is not available, young-of-the-year typically disperse and packs do not form although a mated pair may remain together. How this all might come about will be discussed below as possible relationships between social development and dispersal provide another opportunity to wed theory with data.

On our study areas, intrusions by nonpack coyotes were more numerous when food was abundant on the pack's territory and scarce in other areas around Blacktail Butte. Defense of food and territory was shared by pack members, and pack members won 75 percent of their encounters with intruders. The mean number of pack members involved in successful encounters with intruders (intruders were driven off the pack's territory) was 2.5. Significantly fewer (1.3) pack members were involved in pack losses. Sixty-seven percent of encounters with intruders by a single pack member were lost. The mean group sizes of intruders when they won and lost were 1.4 and 1.1, respectively. Thus, there was a clear advantage for more than two coyotes—more than just a mated pair—to live together for successful group defense. It is important to mention that helpers were responsible for initiating 56 percent of encounters with intruders. Clearly helpers' presence was important even when pups were not around. Bowen (1981) found that packs of two or more coyotes had better access to, and spent more time feeding at, carcasses than lone individuals. Groups also defended carcasses more successfully than single coyotes. Gese et al. (1996a, b) found that dispersing coyotes in Yellowstone National Park had little or no access to ungulate carcasses during winter.

## Social Organization and Activity Budgets

We also were able to collect data bearing on other aspects of group-living. Theory would predict that there would be differences between pack-living and less social coyotes. We found that during winter months pack members traveled less, rested more, were observed at carrion a greater proportion of time, played more, and hunted less than did nonpack members. As I just mentioned, Bowen (1981) reported that packs of two or more coyotes had better access to, and spent more time feeding at, carcasses than lone individuals and Gese et al. (1996a, b) found that dispersing coyotes in Yellowstone National Park had little or no access to ungulate carcasses during winter. When there was a lot of snow on the ground and travel was difficult, pack members traveled less than did nonpack coyotes, thus saving energy for other activities. Advantages to pack members were more pronounced during difficult living conditions. Pack reproductive females also showed energy savings when compared to reproductive females living only with their mate.

More data is needed to assess more completely possible long-term effects of variations in behavioral budgeting among individuals. There is no data that shows conclusively that economizing energy expenditure maximizes reproductive fitness. And, it might be impossible to collect this data for coyotes and most other carnivores. But theory does make some predictions that can be more or less supported by yet-to-be-conducted carefully designed studies in the field and under captive conditions.

## Social Organization and Helping Behavior

Coyotes are typically monogamous and pair bonds between a male and female can last for years. In packs, there is usually only one mated pair per season. Both males and females are able to breed during their first year of life, usually when they are about 9 to 10 months of age. Parents often receive help from other group members in rearing young, and in defending the group's territory both within and outside of the breeding season. Mortality usually is highest during the first year of life, and greatest life expectancy is about 2 to 8 years of age. Mortality depends greatly on the level of exploitation to which populations are exposed (Crabtree & Sheldon 1999) and generalizations between different populations are of questionable validity. Coyotes can live as long as 18 years in captivity, but in the wild few live longer than about 6 to 10 years.

Individuals might remain in their natal area and help if ecological conditions permit, dispersal is risky, the availability of potential mates or breeding areas is low, or initial reproductive success is low. In coyotes, dispersal by yearlings is risky (Bekoff & Wells 1986; Crabtree & Sheldon 1999), and newly reproductive individuals usually do not make a significant contribution to local populations. We still do not know much about the fate of dispersing individuals—what proportion of first-year dispersers successfully reproduce. Comparing lifetime reproductive success between young and old dispersers is at best an exercise in speculation. However, our data leads to the conclusion that both male and female coyotes who cannot breed should help if they can, but that breeding is better than helping.

One obvious advantage of group-living is that there is the potential to have helpers participate in rearing young and to partake in other activities such as the defense of territory or food. Pack formation in coyotes results from the delayed dispersal and retention of some offspring as helpers, and, as mentioned above, natal philopatry is associated with the presence of an abundant, clumped, and defendable winter food source. Coyote helpers mainly served as baby-sitters. Helpers' presence at dens could not be predicted by the number of other adults present, litter size, the percentage of time mothers spent at dens, or the amount of time fathers devoted to den sitting.

Theory predicted that helpers should make some difference in the reproductive success of young to whom they provided care, and via kin selection or reciprocity, the helpers themselves should benefit genetically. We found that there were no significant correlations between litter size and the number of adults present at dens or the number or percentage of pups surviving to five to six months of age. However, there was a positive relationship ($r_s$ = +0.39) between pup survival and the number of adults attending dens. Each helper was responsible for the survival of 0.49 additional pups. The presence of helpers was associated with high-quality winter habitat, but even these two factors working together did not significantly increase pup survival when compared to resident pairs. Perhaps a larger database would have resulted in our positive trend becoming a statistically significant relationship.

In various canids pup survival is positively correlated with the presence of helpers although the relationship is not always statistically significant and possible effects of confounding variables have not been determined (Bekoff & Wells 1982). The values of coefficients of determination in regression analyses of the relationship between the number of young surviving to a given age and the number of adults/helpers at a den are generally low, indicating that factors other than the number of attendant adults is important to consider. And, the slope of the regression lines (regression coefficients) are not statistically significantly different for the same regression analyses for black-backed jackals (*C. mesomelas*), African wild dogs, or coyotes (Bekoff & Wells 1982).

Despite the many caveats that need to be reconciled, in a number of studies the presence of adults other than parents at den sites has been associated with increased pup survival. More data is needed for all species as is detailed analyses of possible confounding factors such as the group size into which young are born and the nature of food resources. For example, in black-backed jackals, an increase in the number of young surviving may be due to help or due to the fact that females produce larger litters in groups with helpers (Montgomerie 1981). And, when food is scarce the presence of helpers might negatively influence pup survival in wolves and red foxes (Bekoff & Wells 1986).

## The Evolution of Helping: The Importance of Noisy Data

In the coyotes we studied, helping can be explained by invoking both kin selection and reciprocity. Helpers provided care to full and half-siblings and also received help from individuals to whom care had been previously provided. The treatment of full siblings could not be differentiated from treatment of half-siblings. Malcolm and Marten (1982) also noted that in African wild dogs, there was little overall relationship between an individual's contribution to rearing and its genetic relationship to pups. One male helper (Bekoff & Wells 1986; see Fig. 2, p. 260), remained in his natal group for three years, inherited a breeding territory, and finally mated with an unknown (at least to us) female who replaced his mother after his mother and father disappeared. Whether or not inbreeding avoidance was an issue is not known (see also Gese et al. 1996a, b). Whether or not territory inheritance is a cause rather than a beneficial consequence of group-living also is not known (Lindström 1986; Bekoff 1987).

The take home lesson from our data is that trying to jam the data into what animals *ought* to do provides a misleading picture of helping. Theory-driven research certainly set the stage for our study but so much interesting noise fell out of our analyses that we had to consider alternatives to views of what ought to be the case. More comparative data is certainly needed to learn about what is happening "out there" and how reliable extant theories are. With respect to helping, the complex relationships among social organization, development, ecology, space use, and dispersal need to be studied in more detail for many more species (Lindström 1986; Bekoff et al. 1981, 1984; Bekoff 1987; Baker et al. 1998).

## Development, Dominance, and Dispersal: Social Cohesion and Strife among Sibs

In the absence of how the behavior of individuals unfolds during early life, we can only guess at the supposed adaptive significance of various developmental patterns and how they may be related to the immediate situation in which a young animal finds itself and its later reproductive activities and fitness (Bekoff 1989). Therefore, studies of adult behavior conducted in the absence of developmental data are incomplete and may make unwarranted assumptions. Having made this bold and perhaps arrogant pronouncement, I am only too aware of how difficult it is to conduct developmental field studies on numerous species, including coyotes. For many carnivores there still are only scant data on behavioral development and how development is related to ecological conditions or the evolution of different life history strategies, developmental and otherwise.

In canids, there is a general correlation between species-typical trends in behavioral development and species-typical patterns of social organization. Basically, individuals of more social species fight less and play more earlier in life than individuals of less social species (Bekoff 1977a). While coyotes, red foxes, and golden jackals engage in serious fights early in life (Fig. 19.2), wolves typically do not. The delay in the appearance of rank-related aggression may be responsible for the development of strong social bonds and of a coordinated social group (for similar data on marmots, see Barash 1974). Furthermore, there seems to be a wider range of behavioral phenotypes within litters in the less social species; the more social species tend to be more behaviorally homogenous, displaying a smaller range of behavioral profiles.

Based on our and others' data, it seemed likely that the antecedents of dispersal would be interesting and important to study because aggression *at the time of dispersal* did not adequately explain the movement of individuals. Other researchers had also noted that "parental encouragement" to leave the natal site was not necessary to produce dispersal of the young. Predispersal behavioral profiles were needed. This required a concentration on *individual* animals and analyses of within-litter individual differences in behavior.

One event that is of paramount importance in the life of coyotes and many other animals is dispersal. Dispersal influences the social organization, population size, and genetics of populations. Dispersal also may reduce intraspecific competition. In coyotes and other carnivores, dispersal is very variable and it is unlikely that there is a single factor, extrinsic or intrinsic, that results in individuals leaving their natal or resident group. There are complex and important decisions that need to be made because of trade-offs between heightened mortality by dispersers, especially youngsters, and potential reproductive benefits. Furthermore, there might be advantages associated with staying in one's natal group and eventually acquiring territory, a mate, and helpers at the den (Bekoff & Wells 1986; for comparative data and discussion, see Baker et al. 1998).

In coyotes, dispersal of young-of-the-year from their natal area usually begins after they are about four to five months of age, and continues until they are about ten months of age; some individuals disperse after they are one year of age. Only rarely will an entire litter remain intact longer than a year. There do not appear to be any sex differences in the proportion of individuals who disperse or in the time period or distance individuals travel

Figure 19.2. Two ten-week old coyotes fighting near their den. (Photograph courtesy of Franz Camenzind)

before they settle down (if they do). Dispersal by young animals is highly risky; generally dispersing individuals suffer much higher mortality than do their sedentary peers who remain in their natal group in the area in which they were born (Bekoff & Wells 1986; Harrison 1992; Crabtree & Sheldon 1999). Causes of dispersal are unknown. There appears to be an association between food availability and dispersal; when there is enough food to feed more than just the mated pair, competition among individuals is reduced, social bonds are more likely to form among at least some group members, and the likelihood for dispersal is reduced.

It remains unclear whether or not aggression among young or between adults and young plays a role in dispersal, but available data does not support this notion. Some dispersing individuals continue to live as transients, whereas others join up with another individual(s) with whom they associate for varying periods of time. It is not known with any certainty whether or not coyotes who disperse are more likely to breed than are more sedentary individuals, although this possibility is still bandied around as if this *has* to be the case because of predictions of extant, but largely untested, theories.

Our initial funding was to study relationships between the development of individual behavioral phenotypes, especially dominance relationships, and dispersal (Bekoff 1977a, 1981; Fig. 19.3). This has been called the "social cohesion hypothesis" and stresses the role of affiliative behavior and social bonding in dispersal. Before I published my paper little attention had been

agonistic behavior
↓
rank-order
↓
social play
avoidance of ⟋ ⟍ avoidance by
individual(s) ⟍ ⟋ individual(s)
weak social bonds
↓
leave group

Figure 19.3. Relationships between the development of dominance, social play, and dispersal. (From: Bekoff, M. 1977a. Mammalian dispersal and the ontogeny of individual behavioral phenotypes. American Naturalist, 111 (no. 980): 715–732. © 1977 by The University of Chicago. All rights reserved.)

paid to the experiential history of individuals and how it may be related to later behavior, especially dispersal. Other workers (e.g., Armitage 1973; Svensden & Armitage 1973) had stressed the importance of learning about individual behavioral profiles, but few data existed.

## Christian Coyotes?

The late John Christian (1970) championed the view that there was a direct correlation between aggressive behavior and dispersal, especially in voles. In his social subordination hypothesis, he postulated that low-ranking subordinate individuals are forced to disperse into suboptimal habitats by being aggressively driven out by more dominant animals. Certainly, Christian's ideas have intuitive conceptual and theoretical appeal, but perhaps it was because they so readily "made sense" that they were accepted uncritically and infrequently supported by solid databases in many species (Brandt 1992; Harris & White 1992; Chitty 1996). While there is some support for Christian's theory in various rodents and other mammals (for reviews, see Gaines & McGlenaghan 1980; Brandt 1992; and other chapters in Stenseth & Lidicker 1992), his stimulating hypothesis seemed, and turned out to be, too simplistic for many animals such as coyotes that do not undergo regular population cycles. For many species, dispersal seemed to be caused, or at least to be highly correlated with, difficult-to-identify environmental and behavioral factors, some or many of which were not apparent at the time of dispersal.

## Where Are We Now?

Based on a large amount of data collected on coyotes living in various captive conditions, a number of predictions were used to ground our initial efforts. Essentially, we predicted that both dominant and subordinate individuals might disperse but for different reasons. The major "causes" for dispersal would be developmental and conditions at the time of dispersal might be of secondary importance when compared to the developmental experiences of individuals. According to the social cohesion hypothesis, dominant and subordinate individuals might not form tight social bonds with litter-mates or other pack members. Dominant animals are avoided by other animals and their attempts to engage in play are generally ignored. Also, the subordinate individuals actively avoid others and tend to be active when others are not. The end result is that weak social bonds are formed and both "types" of individuals might be predisposed to leave their natal group of their own accord. Qualitative data for black-backed jackals (Van der Merwe 1953) and red foxes (Burrows 1968) supports the notion that both dominant and subordinate animals are predisposed to disperse from their natal group.

## Social Cohesion and a Behavioral Polymorphism for Dispersal

The major points of the social cohesion hypothesis can be summarized as follows: (1) aggression does not provide the adequate stimulus for dispersal in many mammals (see also Stenseth & Lidicker 1992); (2) within-litter individual differences in behavior appear early in life and influence how different individuals interact with littermates and how siblings interact with them in turn; (3) individuals who have the most difficulty interacting with their sibs will not develop strong social bonds with sibs and other group members and will be the animals most likely to leave their natal site; (4) there are trade-offs between increased mortality for young dispersers and their attempts to breed earlier in life when compared to individuals who remain in their natal group where they cannot breed; (5) opposing selective forces could result in a behavioral polymorphism for dispersal strategies; the range of individual behavioral phenotypes within litters either will facilitate the execution of a mixed dispersal strategy or will favor the evolution of sociality.

It is simply impossible to assess this hypothesis on any broad scale with current data. Brandt (1992) notes that there is little current support for the social cohesion hypothesis in small mammals, that the lack of social bonds causes emigration in these animals. I am afraid he is right! However, there are marked differences in the dispersal patterns and mechanisms between large and small mammals (Cockburn 1992) and perhaps Brandt's findings are not surprising. The importance of social bonding, affiliative behavior, and dispersal has been demonstrated in Harris and White's (1992) study of urban red foxes for males but not females (see below; see also Meyer & Weber 1996). Gese et al. (1996a, b) concluded that their data for coyote dispersal supported Christian's social subordination hypothesis. However, they never observed dominant coyotes forcing subordinate coyotes to disperse. They also did not collect developmental data and also concluded that the social subordination and social cohesion hypotheses might both play a role at different life stages for coyotes. However, they did note that subordinate coyotes who dispersed spent little time with other pack members and that they had little chances of breeding in their natal group and little access to carcasses. Even for our, Gese's, and Crabtree and Sheldon's user-friendly populations of coyotes, not much insight could be gained concerning developmental aspects of dispersal. Furthermore, we simply do not know what role dispersal plays in outbreeding.

## A Simple and General Guide for Research: Having Fun and Developing Your Own Niche

Here is where I step onto thin ice. I have always felt that research should be fun, exciting, and challenging. We are very fortunate to be able to study behavioral ecology. Of course fieldwork is difficult, but this does not mean that it cannot be fun.

I have always suggested to my students that they develop their own niche, rather than ride my coattails. Then, in the future, others will be able to tell what was theirs and what was mine. I also tried to have them select fundable projects in which they were keenly interested, or to do "cheap research." Unfortunately, behavioral ecology is not well-funded, at least in the United States, and there is an incredible number of important and interesting projects that can be pursued for small amounts of money. "Think locally" is one maxim I have favored. In many locales there are numerous graduate projects just waiting to be done.

It also is essential to blend natural history with "soft" and "hard science." Anecdotes are extremely important in all scientific endeavors. Of course, they do not take the place of hard data, but they do serve an important role in stimulating research.

In this section I would like briefly to make a case that the ways in which we approached the study of coyotes can be useful for guiding future projects. Mistakes were made along the way, but remaining open to alternative views and novel suggestions for conducting our studies were the rules rather than the exceptions. Why be dogmatic and closed-minded when our learning curve was exponential? Discussions in which we congratulated ourselves for what we were doing and heated exchanges both were useful throughout our work.

## The Road Most Traveled: Beautiful Hypotheses and Ugly Facts

Being open to the fact that the rocky and twisting road of field research would be the road most traveled was a learning experience not only for the students working with me, but also for me. Of course there was a lot of smooth sailing, but with animals such as coyotes, surprises regularly occurred. Social plasticity was the name of the game and being unconditionally married to a specific question(s) would have truly produced a rather narrow view of the behavioral ecology of the coyotes we studied. Studying such flexible canids as coyotes would have been exasperating if we had stuck religiously to our original research protocol, and various federal and private funding agencies recognized this. Certainly, many Native Americans and ranchers could have told us that. Indeed, on numerous occasions local ranchers, some of whom were expert naturalists, thought it a joke that we were being paid to learn what they already knew! Just when thought we knew what was happening and could type coyotes in one way or another, something else would occur that would warn us against normative thinking. There really are beautiful hypotheses and ugly facts (Chitty 1996).

## Toward an Understanding and Appreciation for Natural and Individual Variation: Does *the* Coyote Exist?

One quick lesson was that talking about *the* coyote was misleading. There was so much individual variability that our view of coyotes would have been very narrow and misleading if we thought that we could characterize individuals in a normative way. Of course, there were general trends associated with different aged coyotes, males and females, and coyotes living different lifestyles. However, it was variability that continually fueled our work. We simply had to remain open-minded about competing explanations. We had to be patient and not hastily dispense of what some thought was conceptual debris that got in the way of good science. There were few "right" and "wrong" answers to some of the important questions with which we were concerned. For example, we could be right and wrong to claim that coyotes were territorial, that they lived in packs, or that there were helpers at the den. It was essential to keep our own options open just as the coyotes did.

## The Importance of Studying Development

In the 1970s there was little detailed data on the development of behavior in carnivores and how variations in development might influence later behavior. Even the indexes to the first three editions of Krebs and Davies did not contain an entry for "development." We wanted to know why coyotes displayed early rank-related aggression. Could it be explained as an adaptation to developmental social environment and preparation for the future (see also Holekamp & Smale 1998)? We also were interested in learning how individual differences in development are related to such processes as dispersal, which in turn led us to try to understand intraspecific variation in behavior and social organization.

One of the major aspects of our research has been to concentrate on individual differences in behavior and to try to understand the developmental bases for this variation. Similar approaches have been taken by other researchers studying carnivores (e.g., Drea et al. 1996; Holekamp & Smale 1998). In canids, there have been no direct field tests of relationships between individual social development and later patterns of dispersal; however, Harris and White (1992) devised an alternative means of assessing this relationship in red foxes. In these foxes, ear tags are chewed during social grooming. They studied patterns of grooming by looking at teeth marks left on ear tags and found that for males, there was a strong negative relationship between grooming (an affiliative behavior) among littermates and future dispersal. Their data supported some of the ideas put forth in my 1977 paper on individual development and dispersal (but see Meyer & Weber 1996). In another study, Woollard and Harris (1990) did not observe increasing aggression directed toward yearling red foxes at the time of dispersal.

## Bringing Together Field Research and Studies of Captive Animals

Because detailed longitudinal developmental data on individually identified coyotes (and many other carnivores) is nearly impossible to collect under field conditions (Bekoff 1989), we studied individually identified infant and juvenile coyotes in different conditions of captivity. An important lesson for all of us in the field was that while we could not observe many details of development, the patterns of behavior that we observed among infants and juveniles were very similar to those we observed in captivity. Various canids engage in serious rank-related aggression very early in life (Wandrey 1975; Bekoff et al. 1981b; Henry 1986; Bekoff 1989; Feddersson-Petersen 1991), the result of which is a stable hierarchy (Henry 1986), but it is often difficult to collect enough data under field conditions to "know" who dominates whom. However, having detailed developmental data collected under a variety of captive conditions that were consistent across studies allowed us to make reliable inferences about what happens under field conditions. Thus, although we could not be absolutely certain that a particular individual was the dominant member of its litter, based on patterns of social interaction, especially play solicitation, avoidance, and the initiation, escalation, and termination of fights (Bekoff et al. 1981b; see also Bekoff & Dugatkin 2001), we felt comfortable assigning different dominance ranks to many individuals. For example, knowing that dominant individuals were avoided in captivity and that subordinate coyotes avoided interaction in captivity allowed us to assign dominance rank to many individuals. Furthermore, knowing that dominant coyotes engaged in self-handicapping and role-reversing in play and knowing that other animals were reluctant to play with dominant coyotes if the dominant animal did not previously signal play (Bekoff 1975, 1995b) allowed us to make reliable inferences about social rank among littermates.

## The RDH, TIH, and CTSH

It is important to stress that there still is much to learn about social behavior and ecology in most carnivores. There still is disagreement about some very basic questions. For example, as I mentioned above, it still is not clear whether cooperative hunting is an adaptation for group-living in any carnivore (see Creel & Creel 1995 for discussion).

Three useful starting points for future work include rigorous testing of the resource dispersion hypothesis (RDH; Macdonald 1983; Macdonald & Carr 1989; see, for example, Baker et al. 1998), Lindström's territory inheritance hypothesis (TIH; Lindström 1986; Bekoff 1987), and von Schantz's (1984) constant territory size hypothesis (CTSH) to explain the evolution of group-living in carnivores. The RDH considers spatiotemporal distribution of food *within* a year, whereas the CTSH considers *between-year* fluctuations in food supply. According to the RDH, territory size will be determined by the dispersion of discrete patches of food, whereas group size will be determined by their richness. Also, territory size will be determined by the dispersion of food patches, whereas group size will be determined by their richness. Macdonald and Carr (1989) present a general version of the RDH which suggests that groups may form even in the absence of any functional advantage to one individual from the presence of others.

According to the TIH, territory inheritance could be an evolutionary cause rather than a beneficial consequence of group-living. Natal philopatry is implicated in the evolution of groups from pairs of animals. The TIH considers the question of "how are groups formed and maintained?" and "what factors influence group size?" Once group-living is established, it will be favored as long as adult survival is high and the population is stable.

## Where to from Here?

We know a lot about many animals, but perhaps not as much as some think. This is not a criticism of optimists or a condemnation of the field, but rather a reflection of the difficulty of studying canids and also of the difficulty and complexity of some of the questions at hand.

One area of research in which there is rapidly growing interest is cognitive ethology (see Allen & Bekoff 1997; Bekoff 1999, 2000a). We need to know what life is like for the animals we study, for the more we learn in this area the better will be our future research. The cognitive dimensions and emotional foundations of social behavior deserve considerably more attention in a wide variety of animals if we are to make progress in learning more about the evolution of cognitive capacities (Bekoff 2000a, b). Tinbergen's (1951/1989, 1963) framework is also useful for those interested in animal cognition (Jamieson & Bekoff 1993; Allen & Bekoff 1997). Burghardt (1997) has suggested adding a fifth area, *private experience*. He (p. 276) noted that "The fifth aim is nothing less than a deliberate attempt to understand the private experience, including the perceptual world and mental states, of other organisms. The term private experience is advanced as a preferred label that is most inclusive of the full range of phenomena that have been identified without prejudging any particular theoretical or methodological approach."

## Interspecific Competition among Canids

It is their flexibility and other adaptations that make carnivores potentially strong competitors. Another area that warrants more attention concerns competition between sympatric species (e.g., Johnson et al. 1996; Carbone et al. 1997; Gorman et al. 1998; Crabtree & Sheldon 1999; Palomares & Caro 1999). There is little information on how carnivore communities are structured by interspecific interactions; however, interspecific killing appears to be common in some communities (Arjo 1999; Crabtree & Sheldon 1999; Palomares & Caro 1999).

Coyotes and wolves have also been forced to interact after wolves were reintroduced to Yellowstone National Park, and studying these interactions is extremely important for learning more about how competition might structure these canid communities (Fig. 19.4; Mlot 1998; Crabtree & Sheldon 1999). A popular account of how wolves are influencing coyotes in Yellowstone (Mlot 1998) indicated that coyotes are changing their feeding habits. Furthermore, coyote packs seem to be more cohesive and they are changing how they use space to avoid wolves. Traditional coyote territories are not recolonizing after being extirpated by wolves. The core of the wolf territory once had four coyote packs and now there are none (Crabtree & Sheldon 1999).

Numerous coyotes also have been killed by the reintroduced wolves (for discussion of the ethics of reintroduction, see Bekoff 2000b). Crabtree and Sheldon (1999) report that wolf-killed coyotes during the winters of 1997 and 1998 resulted in a 50 percent reduction in coyote numbers. Pack size also has decreased from an average of six coyotes to four. Detailed data on coyote-wolf interactions still are forthcoming (but see Arjo 1999). They will be extremely useful in our learning more about this wily canid.

Figure 19.4. A wolf chasing a coyote in the Lamar Valley, Yellowstone National Park. (Photograph courtesy of Monty Dewald)

## Ethics: Doing Science and Respecting Animals

Last, but certainly not least, we also need to pay careful attention to the effects of our intrusions into animals' lives (Laurenson & Caro 1994; Bekoff 1998b, 2000b; Bekoff & Jamieson 1996). We need to do science and respect animals at the same time. Conducting behavioral studies in the field can be very challenging, and there are many ethical problems that arise in these sorts of research endeavors, even those that involve "just being there" or "just using relatively 'noninvasive' techniques" (Bekoff & Jamieson 1991, 1996).

Because of their status as a predatory pest, coyotes are routinely subjected to horrible experiments. Many serious ethical questions arise, and coyotes, more likely than not, get the short end of the stick in these deliberations. (See, for example, Knowlton et al.'s [1985] assessment of coyote vulnerability to various management techniques, a study that included, among other things, gunning down coyotes from a helicopter, a practice that "was relatively ineffective in reducing the number of coyotes under the conditions dictated by this study," p. 175.)

Those who trap animals using leghold traps only later to kill them do not often give serious thought to the ethical issues that are involved when these sorts of activities are pursued (Liss 1998). Live trapping using leghold traps can be incredibly inhumane and the experience of being caught in a trap can be very physically and psychologically painful for an animal. While this may be obvious to most people, research still continues. For example, a group of researchers at Utah State University performed studies on coyotes (Wildlife Damage Review 1993) in which individuals, some of whom were seriously injured, were kept in leghold traps for long periods of time to determine the effects of using tranquilizers to keep them calm when in pain (see also Olsen et al. 1986).

## Cunning Coyotes, Tireless Tricksters, Ultimate Survivors

As we learn more and more about coyotes, we discover what fascinating animals they are. Coyotes are wonderful animals to study for a wide variety of reasons. They offer us lessons about behavioral, morphological, and physiological adaptability and variability, as well as lessons in the ethics of research and human responsibilities to the animals we study and use. As we learn more and more about coyotes, we also learn more about ourselves. They are excellent mentors.

Coyotes have shown themselves to be great survivors despite unrelenting attempts by humans to drastically limit their numbers, and this does not appear to be a skill that is being lost. In many ways coyotes truly are victims of their own success. *By any measure they are a success story.*

## Acknowledgments

I thank all the animals whose lives we shared and also all of the people who helped with the fieldwork and the funding agencies that made this project possible (see Bekoff & Wells 1986). Special thanks to Mike Wells for his single-minded dedication to our work. Joel Berger, John Byers, Bob Crabtree, Lee Dugatkin, John Fentress, Carron Meaney, Bill Merkle, Axel Moehrenschlager, John Reed, and anonymous reviewers provided helpful comments on a draft of this chapter. Thanks also to Franz Camenzind and Monty Dewald for graciously allowing me to use their photographs.

## References

Allen C, Bekoff M, 1997. Species of mind: the philosophy and biology of cognitive ethology. Cambridge, MA: MIT Press.
Allen J, Bekoff M, Crabtree R, 1999. An observational study of coyote (*Canis la-*

*trans*) scent-marking and territoriality in Yellowstone National Park. Ethology 105:289–302.

Arjo WM, 1999. Behavioral responses of coyotes to wolf recolonization in northwestern Montana. Can J Zool 77:1919–1927.

Armitage KB, 1973. Population changes and social behavior following colonization by the yellow-bellied marmot. J Mamm 54:842–854.

Baker PJ, Robertson CPJ, Funk SM, Harris S, 1998. Potential fitness benefits of group living in the red fox, *Vulpes vulpes*. Anim Behav 56:1411–1424.

Barash DP, 1974. The evolution of marmot societies: a general theory. Science 185:415–420.

Bekoff M, 1972. The development of social interaction, play, and metacommunication in mammals: an ethological perspective. Quart Rev Biol 47:412–434.

Bekoff M, 1974. Social play and play-soliciting by infant canids. Am Zool 14:323–340.

Bekoff M, 1975. The communication of play intention: are play signals functional? Semiotica 15:231–239.

Bekoff M, 1977a. Mammalian dispersal and the ontogeny of individual behavioral phenotypes. Am Nat 111:715–732.

Bekoff M, 1977b. *Canis latrans*. Mammalian Species 79:1–9.

Bekoff M (ed), 1978. Coyotes: biology, behavior, and management. New York: Academic Press.

Bekoff M, 1979. Coyote damage assessment in the west: review of a report. BioScience 29:754.

Bekoff M, 1980. Accuracy of scent-mark identification for free-dogs. J Mammal 61:150.

Bekoff M, 1981. Mammalian sibling interactions: genes, facilitative environments, and the coefficient of familiarity. In: Parental care in mammals (Gubernick D, Klopfer PH, eds). New York: Plenum Press; 307–346.

Bekoff, M, 1982. Coyote, *Canis latrans*. In: Wild mammals of North America: biology, management, and economics (Chapman J, Feldhamer G, eds). Baltimore: Johns Hopkins University Press; 447–459.

Bekoff M, 1987. Group living, natal philopatry, and Lindström's lottery: it's all in the family. Trends Ecol Evol 2:115–116.

Bekoff M, 1989. Social development of terrestrial carnivores. In: Carnivore behavior, ecology, and evolution (Gittleman JL, ed). Ithaca, NY: Cornell University Press; 89–124.

Bekoff M, 1995a. Vigilance, flock size, and flock geometry: information gathering by Western Evening Grosbeaks (Aves, fringillidae). Ethology 99:150–161.

Bekoff M, 1995b. Play signals as punctuation: the structure of social play in canids. Behaviour 132:419–429.

Bekoff M, 1998a. Minding animals. In: Responsible conduct of research in animal behavior (Hart L, ed). New York: Oxford University Press; 96–116.

Bekoff M (ed), 1998b. Encyclopedia of animal rights and animal welfare. Westport, CT: Greenwood Publishing Group, Inc.

Bekoff M, 1999. Social cognition: exchanging and sharing information on the run. Evol Cog 5:128–136.

Bekoff M (ed), 2000a. The smile of a dolphin: remarkable accounts of animal emotions. New York: Random House/Discovery Books.

Bekoff M, 2000b. Ethical enrichment, compassion, and human-carnivore interactions: adopting proactive strategies. In: Conservation in carnivores (Gittleman JL, Funk SM, Macdonald DW, Wayne RK, eds). Cambridge: Cambridge University Press.

Bekoff M, Allen C, Grant MC, 1999. Feeding decisions by Steller's jays (*Cyanocitta stelleri*): the utility of a logistic regression model for analyses of complex choices. Ethology 105:393–406.

Bekoff M, Allen C, Wolfe A. 1998. Feeding behavior in Steller's jays (*Cyanocitta stelleri*): effects of food type and social context. Bird Behavior 12:79–84.

Bekoff M, Daniels TJ, Gittleman JL, 1984. Life history patterns and the comparative social ecology of carnivores. Ann Rev Ecol System 15:191–232.

Bekoff M, Diamond J, Mitton JB, 1981a. Life-history patterns and sociality in canids: body size, reproduction, and behavior. Oecologia 50:386–390.

Bekoff M, Dugatkin LA, 2001. Winner and loser effects and the development of dominance in young coyotes: an integration of data and theory. Evol Ecol Res 2:871–883.

Bekoff, M, Jamieson D, 1991. Reflective ethology, applied philosophy, and the moral status of animals. Persp Ethology 9:1–47.

Bekoff M, Jamieson D, 1996. Ethics and the study of carnivores: doing science while respecting animals. In: Carnivore behavior, ecology, and evolution, vol. 2 (Gittleman JL, ed). Ithaca, NY: Cornell University Press; 15–45.

Bekoff M, Scott AC, Conner DA, 1987. Nonrandom nest site selection in evening grosbeaks. The Condor 89:819–829.

Bekoff M, Scott AC, Conner DA, 1989. Ecological analyses of nesting success in Evening Grosbeaks. Oecologia 81:67–74.

Bekoff, M, Tyrrell M, Lipetz VE, Jamieson RA, 1981b. Fighting patterns in young coyotes: initiation, escalation, and assessment. Agg Behav 7:225–244.

Bekoff M, Wells MC, 1980. The social ecology of coyotes. Sci Am 242:130–148.

Bekoff M, Wells MC, 1982. Behavioral ecology of coyotes: social organization, rearing patterns, space use, and resource defense. Z Tierpsychol 60:281–305.

Bekoff M, Wells MC, 1986. Social ecology and behavior of coyotes. Adv Study Behav 16:251–338.

Bowen WD, 1981. Variation in coyote social organization. Can J Zool 59:639–652.

Bowen WD, 1982. Home range and spatial organization of coyotes in Jasper National Park. Can J Wildl Manage 46:201–216.

Brandt CA, 1992. Social factors in immigration and emigration. In: Animal dispersal: small mammals as a model (Stenseth NC, Lidicker WZ, eds). New York: Chapman & Hall; 86–141.

Bright W, 1993. A coyote reader. Berkeley: University of California Press.

Burghardt GM, 1997. Amending Tinbergen: a fifth aim for ethology. In: Anthropomorphism, anecdote, and animals (Mitchell RW, Thompson NL, Miles L, eds). Albany: SUNY Press; 254–276.

Burrows R, 1968. Wild fox. New York: Tapplinger.

Carbone C, Du Toit JT, Gordon IJ, 1997. Feeding success in African wild dogs: does kleptoparasitism by spotted hyaenas influence hunting group size? J Anim Ecol 66:318–326.

Chitty D, 1996. Do lemmings commit suicide? Beautiful hypotheses and ugly facts. New York: Oxford University Press.

Christian JJ, 1970. Social subordination, population density, and mammalian evolution. Science 168:44–90.

Cockburn A, 1988. Social behaviour in fluctuating populations. New York: Croom Helm.

Cockburn A, 1992. Habitat heterogeneity and dispersal: Environmental and genetic patchiness. In: Animal dispersal: small mammals as a model (Stenseth NC, Lidicker WZ, eds). New York: Chapman & Hall; 65–95.

Crabtree, RL, Sheldon JW, 1999. Coyotes and canid coexistence in Yellowstone. In: Carnivores in ecosystems (Clark TW, Curlee AP, Minta SC, Kareiva PM, eds). New Haven, CT: Yale University Press; 127–163.

Creel S, Creel M, 1995. Communal hunting and pack size in African wild dogs, *Lycaon pictus*. Anim Behav 50:1325–1339.

Crook JH (ed), 1970. Social behaviour in birds and mammals. New York: Academic Press.

Crooks KR, Soulé ME, 1999. Mesopredator release and avifaunal extinctions in a fragmented system. Nature 400:563–566.

Davis DE (ed), 1974. Behavior as an ecological factor. Stoudsburg, PA: Dowden, Hutchinson & Ross, Inc.

Drea CM, Hawk JE, Glickman SE, 1996. Aggression decreases as play emerges in infant spotted hyaenas: preparation for joining the clan. Anim Behav 51:1323–1336.

Fedderson-Petersen D, 1991. The ontogeny of social play and agonistic behaviour in selected canids species. Bonn Zool Beitr 42:97–114.

Finkel M, 1999. The ultimate survivor. Audubon 101:52–59.

Fisler GF, 1969. Mammalian organizational systems. Contrib Sci Los Angeles County Museum, No. 167, 32.

Fox CH, 1998. Trapping on national wildlife refuges. Sacramento, CA: Animal Protection Institute.

Fox MW, 1970. Integrative development of brain and behavior in the dog. Chicago: University of Chicago Press.

Fox MW, 1971. The behaviour of wolves, dogs, and related canids. New York: Harper. Gaines MS, McClenaghan LR, 1980. Dispersal in small mammals. Ann Rev Ecol System 11:163–196.

Gese EM, 1998a. Response of neighboring coyotes (*Canis latrans*) to social disruption in an adjacent pack. Can J Zool 76:1960–1963.

Gese EM, 1998b. Threat of predation: do ungulates behave aggressively towards different members of a coyote pack? Can J Zool 77:499–503.

Gese EM, Ruff RL, 1997. Scent-marking by coyotes, *Canis latrans*: the influence of social and ecological factors. Anim Behav 54:1155–1166.

Gese EM, Ruff RL, 1998. Howling by coyotes (*Canis latrans*): variation among social classses, seasons, and pack sizes. Can J Zool 76:1037–1043.

Gese EM, Ruff RL, Crabtree RL, 1996a. Social and nutritional factors influencing the dispersal of resident coyotes. Anim Behav 52:1025–1043.

Gese EM, Ruff RL, Crabtree RL, 1996b. Foraging ecology of coyotes (*Canis latrans*): the influence of extrinsic factors and a dominance hierarchy. Can J Zool 74:769–783.

Gorman, ML, Mills, MG, Raath, JP, Speakman JR. 1998. High hunting costs make African wild dogs vulnerable to kleptoparasitism by hyaenas. Nature 391:479–481.

Hamilton WD, 1964. The genetical evolution of social behaviour, I, II. J Theoret Biol 7:1–52.

Harris, S, White PCL, 1992. Is reduced affiliative rather than increased agonistic behaviour associated with dispersal in red foxes? Anim Behav 44:1085–1089.

Harrison DJ, 1992. Dispersal characteristics of juvenile coyotes in Maine. J. Wildl Manage 56:128–138.

Henry JD, 1986. Red fox: the catlike canine. Washington, DC: Smithsonian Institution Press.

Holekamp KE, Smale L, 1998. Behavioral development in the spotted hyena. BioScience 48:997–1005.

Houle D, 1998. High enthusiasm and low r-squared. Evolution 52:1872–1876.

Jamieson D, Bekoff M, 1993. On aims and methods of cognitive ethology. Phil Sci Assoc 2:110–124.

Johnson WE, Fuller TK, Franklin WL, 1996. Sympatry in canids: a review and assessment. In Carnivore behavior, ecology, and evolution (Gittleman JL, ed). Ithaca, NY: Cornell University Press; 189–218.

Kitcher P, 1985. Vaulting ambition: sociobiology and the quest for human nature. Cambridge, MA: MIT Press.

Knowlton FF, Windberg LA, Wahlgren CE, 1985. Coyote vulnerability to several management techniques. Proc 7th Great Plains Animal Damage Control Workshop; 165–176.

Krebs JR, Davies NB, 1978, 1984, 1991, 1997. Behavioural ecology: an evolutionary approach. New York: Blackwell Science.

Kropotkin P, 1914. Mutual aid: a factor of evolution. Boston: Expanding Horizons Press.

Laundré JW, Keller BL, 1984. Home range of coyotes: a critical review. J. Wildl Manage 48:127–139.

Laurenson MK, Caro TM, 1994. Monitoring the effects of non-trivial handling in free-living cheetahs. Anim Behav 47:547–557.

Lindström E, 1986. Territory inheritance in the evolution of group-living in carnivores. Anim Behav 34:1540–1549.

Liss K, 1998. Trapping. In: Encyclopedia of animal rights and animal welfare (Bekoff M, ed). Westport, CT: Greenwood Publishing Group, Inc; 338–340.

Lopez BH, 1977. Giving birth to thunder, sleeping with his daughter: coyote builds North America. New York: Avon.

Lott DF, 1990. Intraspecific variation in the social systems of wild vertebrates. New York: Cambridge University Press.

Macdonald DW, 1983. The ecology of carnivore social behaviour. Nature 301:379–384.

Macdonald DW, Carr GM, 1989. Food security and the rewards of tolerance. In: Comparative socioecology: the behavioural ecology of humans and other mammals (Standen V, Foley RA, eds). London: Blackwell Scientific Publications; 75–99.

Malcolm J, Marten K, 1982. Natural selection and the communal rearing of pups in African wild dogs (Lycaon pictus). Behav Ecol Sociobiol 10:1–13.

Meyer S, Weber J-M, 1996. Ontogeny of dominance in free-living red foxes. Ethology 102:1008–1019.

Messier F, Barrette C, 1982. The social system of the coyote (Canis latrans) in a forested habitat. Can J Zool 60:1743–1753.

Mlot C, 1998. The coyotes of lamar valley. Sci News 153:76–78. Montgomerie RD, 1981. Why do jackals help their parents? Nature 289:824–825.

Murie A, 1940. Ecology of the coyote in the Yellowstone. Fauna Nat Park US, Bull No 4.

Myers G, 1990. Writing biology: texts in the social construction of scientific knowledge. Madison: University of Wisconsin Press.

Olsen GH, Linhart SB, Holmes RA, Dash GJ, Male CB, 1986. Injuries to coyotes caught in padded and unpadded steel foothold traps. Wildl Soc Bull 14:219–223.

Palomares F, Caro TM, 1999. Interspecific killing among mammalian carnivores. Am Nat 153:492–508.

Pyrah D, 1984. Social distribution and population estimates of coyotes in north-central Minnesota. J Wildl Manage 48:679–690.

Rosenberg A, 1980. Sociobiology and the preemption of social science. Baltimore: Johns Hopkins University Press.

von Schantz T, 1984. 'Non-breeders' in the red fox, *Vulpes vulpes*: a case of resource surplus. Oikos 42:59–65.

Saether B-E, 1999. Top dogs maintain diversity. Nature 400:510–511.

Stenseth NC, Lidicker WZ (eds), 1992. Animal dispersal: small mammals as a model. New York: Chapman & Hall.

Svensden GE, Armitage, KB. 1973, Mirror-image stimulation applied to field behavioral studies. Ecology 54:623–627.

Tinbergen N, 1951/1989. The study of instinct. New York: Oxford University Press.

Tinbergen N, 1963. On aims and methods of ethology. Z Tierpsychol 20:410–433.

van der Merwe NJ, 1953. The jackal. Fauna and Flora 4:4–77.

Wandrey R, 1975. Contributions to the study of social behaviour of captive golden jackals (*Canis aureus* L.). Z Tierpsychol 39:365–402.

Wells MC, Bekoff M, 1981. An observational study of scent-marking in coyotes, *Canis latrans*. Anim Behav 29:332–350.

Wildlife Damage Review, 1993. ADC animal research exposed. Spring:3.

Wilson EO, 1975. Sociobiology: the new synthesis. Cambridge, MA: Harvard University Press.

Wolfe P, St. John J, 1998. Waste, fraud & abuse in the U.S. Animal Damage Control Program. Wildlife Damage Review, *http://www.azstarnet.com/~wdr/wfa.html*.

Woollard T, Harris S, 1990. A behavioural comparison of dispersing and non-dispersing foxes (*Vulpes vulpes*) and an evaluation of some dispersal hypotheses. J Anim Ecol 59:709–722.

Young SP, Jackson HHT, 1951. The clever coyote. Washington, DC: The Wildlife Management Institute.

# 20 Bottlenose Dolphins: Social Relationships in a Big-Brained Aquatic Mammal

Richard C. Connor

Every dolphin researcher can recall being schooled by well-intentioned mentors and advisors in all of the reasons not to choose an aquatic mammal as a subject for research, and certainly not for a thesis topic. It is not easy for a terrestrial mammal to observe an aquatic one and moreover, dolphins are large, long-lived, and capable of traveling great distances—not a promising choice for a young behavioral ecologist striking out to test the prominent hypotheses of the day. Better to choose something small, abundant, with short, tidy generations that is easy to observe and manipulate.

There are, however, some very good reasons to include dolphins in our comparative framework. Most generally, to what degree can our understanding of ecological influences on social evolution in mammals, learned mostly from studies of terrestrial species, be extended to the aquatic sphere? Will a three-dimensional world nearly devoid of refuge from predators and without the weight of gravity reveal novel solutions to social living? What commonalties might explain convergence in the social systems of terrestrial and aquatic mammals? And then there are those big brains—why do dolphins have them?

Many cetologists began their studies not seeking to test a hypothesis from behavioral ecology but with the taxon in mind (Gans 1978), drawn to their subjects by a stubborn interest in broad issues such as cetacean conservation, behavior, ecology, and cognition (Connor et al. 2000). Against the risk of exploring a relatively unknown and difficult to observe taxa, there were potential rewards of a new discovery—a discovery which could then become the subject of "question-focused" science. Indeed, cetacean field studies initiated in the 1970s and1980s have yielded some remarkable discoveries, including convergence between sperm whale and elephant social organization, natal philopatry by both sexes in killer whales, and evidence of vocal learning in several species (Connor et al. 1998a).

It's a fair bet that more students want to study dolphins than all of the other animals mentioned in this volume combined. Those few cetologists with teaching positions are constantly reminded of this by a continual barrage of letters and e-mails from interested students. Only a very small minority of those students end up conducting research on whales or dolphins.

Those that succeed probably will do so with a combination of luck and perseverance or "stick-to-itiveness," as one of my advisors used to say.

## Uncle Charlie's Summer Camp

Coming from an industrial town in the frozen Midwest, the University of California at Santa Cruz seemed like paradise—2,000 acres of redwood forest overlooking beautiful Monterey Bay. UCSC was rigorous, laid back, and irreverent with the banana slug as its unofficial mascot. When I arrived on campus, I already knew that I wanted to study the behavior of wild dolphins and that Kenneth S. Norris was the man to see. An extraordinary naturalist, biologist, and conservationist, Norris was an eminent dolphin researcher who delighted in the company of undergraduates and encouraged their dreams with an infectious enthusiasm for the natural world.

### DISCOVERING "SOCIOBIOLOGY"

What little I learned of animal behavior during freshman biology at Santa Cruz in 1977—just a taste of classical ethology—struck me, I have to admit, as boring (my interests have broadened considerably since then). Then I heard of a new professor who was teaching a controversial subject called "sociobiology." That professor was Robert L. Trivers, one of the primary architects, along with William D. Hamilton and George Williams, of the paradigm shift in thinking about ultimate causation of behavior that was originally called sociobiology before it became blended with Tinbergen's other levels of analysis into modern behavioral ecology. Trivers's course, "social evolution," was a revelation. I can't describe the excitement I got from learning the theories; they were so logical, so powerful, and Trivers was a brilliant lecturer—still the best (and funniest) I have ever heard.

To Trivers I owe my interest in the evolution of altruism and cooperation, which I have maintained as a "sideline" over the years (Connor 1986, 1992, 1995a, b, c). It would make a nice story to say that my interest in the evolution of altruism and cooperation stemmed from my observations of wild dolphins in Shark Bay, but that is simply not the case. I was initially impressed with the potential of reciprocal altruism to explain much of human social exchange and even cognitive abilities (see Trivers 1971). My own initial exuberance in applying Trivers's theory of reciprocal altruism to previously published accounts of dolphin behavior (Connor & Norris 1982) evolved into skepticism at claims for tit-for-tat in animals and a search for alternative models (Connor 1985, 1992, 1995 a, b, c, 1996; Connor & Curry 1995). As it turned out, we did discover fascinating cooperative behavior among dolphins, and my work on models of cooperation has certainly aided my thinking about dolphin behavior.

## VISIONS OF A DOLPHIN GOMBE

In Ken Norris I found a mentor who seemed as enthusiastic about my goals as I was, but I still had to find somewhere to study wild dolphins. A small picture in a 1979 *National Geographic Magazine* article about dolphins caught my eye and imagination. The picture showed a seaside camp in remote Shark Bay, Western Australia, where a group of bottlenose dolphins (*Tursiops aduncus*, see LeDuc et al. 1999) interacted with tourists in the shallows. A few years later, Elizabeth Gawain, a regular visitor to that camp, named Monkey Mia, visited Santa Cruz to tell Ken Norris and his students about the wonderful untapped research potential of the Monkey Mia dolphins. She described a dolphin Gombe, the famous camp where Jane Goodall first habituated wild chimpanzees, leading to a remarkable series of studies on chimpanzee behavior and ecology. I knew immediately that I had to go to Shark Bay.

Unsuccessful at raising funds to go to Australia that first summer, I postponed graduate school, remaining in Santa Cruz to try for Shark Bay again the following year. Had I rushed straight into graduate school, I'm sure I would be doing something else today.

## YOU CAN'T STUDY DOLPHINS AT THE UNIVERSITY OF MICHIGAN!

My application to the University of Michigan was a source of great amusement to many in the biology department. Not surprisingly, dolphin research was not a major item at Michigan. I was attracted to Michigan because of its strengths in evolutionary biology and primatology, and I figured that I could handle the dolphin part. During my first campus visit, one normally placid mammalogist pounded on his desk shouting, "You cannot study dolphins at the University of Michigan; Michigan expects large sample sizes!" Fortunately, his opinion was not universally shared. Richard Wrangham, a prominent primatologist and behavioral ecologist, was clearly excited about a place where habituated wild dolphins could be observed and was fascinated by the potential comparisons with his first love, chimpanzees and other primates. To my surprise and delight, Richard D. Alexander, a biologist who had written one of the foundation papers of sociobiology and who sponsored students working on a range of taxa, was a strong advocate of the position that there are important questions for every species (Gans 1978). He was more interested in students who took risks in order to do something important than those who played it safe just to ensure a "large sample size."

## Beyond the Beach: The True Value of Shark Bay

Shark Bay has emerged as a premier dolphin research site with biologists from several countries working on a number of projects. There are a number of reasons for this. The initial attraction of Shark Bay was the half-dozen dolphins that visited Monkey Mia. During our first visit to Shark Bay in1982, Rachel Smolker and I were able to observe some new behaviors among these "beach dolphins" (Connor & Smolker 1985). However, it was two days in a borrowed dinghy that revealed the true potential of Shark Bay. Protected from the ocean swell, when the wind died the sea surface turned to glass. Looking down into the water was like peering into an aquarium. The dolphins themselves were abundant and not at all shy—even then most of the dolphins we approached did not seem disturbed by the presence of our small dinghy. The dolphins often rolled belly-up when riding our bow, allowing us to view their genitals and determine their sex. We identify individual dolphins by individual characteristics of their dorsal fins—and even here we had an advantage. The original paper describing photographic identification of bottlenose dolphins focused on the nicks and scars along the trailing edge of fins that were all basically cut from the same pointy, sickle-shaped mold (Wursig & Wursig 1977). Shark Bay dolphins offered an incredible range of fin shapes, which when combined with fin scars, greatly aided our task of identifying individuals. In later years we would come to recognize an additional factor that makes Shark Bay an exceptional research site: the sheer number of dolphins appears to be higher than in other locales. I will return later to the possible significance a high density of dolphins in Shark Bay has for understanding the remarkably complex pattern of male alliance formation we find there.

I hasten to add that while we were amazed at what we could see offshore, a biologist spoiled on tame terrestrial animals might find Shark Bay's glass half-empty. After all, the dolphins are constantly moving up and down, in and out of sight, and those glassy days are rare as Shark Bay is a windy place where one can be landlocked for days at a time. The solution to this latter problem was easy to swallow: one simply had to stay in Shark Bay for long periods.

When we began our offshore studies in Shark Bay, systematic observations on wild dolphins were largely limited to the "survey" method, in which one recorded the composition and activity of dolphins encountered, then moved on in search of other groups. Over time, it is possible to accumulate a substantial database on individual association patterns that can provide considerable information about social structure (Whitehead 1997). This is especially true for species such as the bottlenose dolphin that live in fission-fusion societies in which the typical group size is manageably small (less than ten individuals, Smolker et al. 1992; Wells et al. 1987). Missing from such an approach is information on *social relationships*, which is built on repeated observations of social interactions between individuals: not simply who is with whom but who does what with whom; the pattern, content, and quality of interactions over time (Hinde 1976). Methods for learning about individual social relationships had been pioneered by primatologists (Altmann 1974). In particular, primatologists found an ethological gold mine in focal follows—systematically recording specific behaviors of recognized individuals for either a whole day or specified part of a day (as brief as ten minutes).

My initial research was aimed as much at trying a new method, focal follows, as it was shaped by specific questions about dolphin social relationships. It was an exciting but daunting prospect. Dolphins spend much of their time underwater; at best following them would be like watching monkeys high up in trees, moving in and out of sight. Even in a place as conducive to watching wild dolphins as Shark Bay, would I be able to stay with wild dolphins long enough to systematically record behavior and social interactions? The answer turned out to be both yes and no. Obviously, observations were restricted to behaviors that occurred at or near the surface. Given that, I still found that strict focal behavior sampling is usually not possible on a consistent basis. We rely on a good view of dolphin fins from the side to identify individuals, and in many contexts, we lose that view. While underwater, dolphins often move ahead, fall behind, or change direction. In any of those circumstances, it may not be possible to tell which individual in a group, the focal dolphin or another, performed a given behavior. Deteriorating sea-state further exacerbates the problem. When dolphins are leaping around chasing fish or on choppy days, just keeping track of a focal dolphin can be a challenge—and in such circumstances we often lose them. In spite of these difficulties, it is possible to obtain systematic records of particular behaviors of interest that occur at the surface. Male dolphins often surface side by side synchronously, a reflection of their social bond (Connor, unpublished data), and it is likely that the sample we obtain traveling in a position alongside the group is unbiased. Other behaviors, such as petting and stroking, may be observed just under the surface, where it is often difficult to tell who is involved. In such situations, we are often limited to brief glimpses of behaviors that in reality may have a substantial duration. Because of these difficulties, it is more realistic to record the predominant activity (Hutt & Hutt 1970; Mann 2000) of the group a focal dolphin is in, rather than the focal's activity.

On the other hand, I was able to remain with focal dolphins for an average of three to four hours and up to ten hours at a time. In spite of the incomplete behavioral record obtained, these follows turned out to be enormously rewarding. Only by following dolphins for periods of hours (rather than minutes), can you record the dynamics of grouping in their fission-fusion society—and it is often in the context of groups joining together and splitting up that many important aspects of social relationships are revealed. The first several months of my research were spent following roughly equal numbers of males and females, ostensibly to compare their activity patterns. At least that is the story I sold to my thesis committee. In reality, I was hoping to find something more exciting to focus on. That something turned out to be male alliances.

## Male Alliances in Birds and Mammals

Two or more individuals that join forces against conspecifics constitute an alliance or coalition (Harcourt & de Waal 1992). Harcourt and deWaal (1992) suggested that the term *alliance* should be restricted to coalitions that recur on a regular basis, but this dichotomy is not usefully applied to male dolphins in Shark Bay (see below).

In general, female reproductive success is limited by access to resources and male reproductive success by the number of successful fertilizations (Trivers 1972; Emlen & Oring 1977; Bradbury & Vehrencamp 1977). Resources are often divisible, so members of successful female alliances can often expect to obtain a share of the spoils. Alliance formation is expected to be less common among males because fertilizations are not divisible (van Hoof & vanSchaik 1992, 1994). In mammals, male alliances are common among primates and scattered about other taxa (reviewed in Caro 1994; Harcourt & de Waal 1992; van Hoof & van Schaik 1994).

### Discovering Dolphin Alliances in Shark Bay

Rather than simply summarizing our knowledge of alliance formation among male bottlenose dolphins in Shark Bay, I present a rough chronological account of what I consider to be our three major discoveries about male alliances: (1) herding of females by male alliances, (2) two levels of alliance formation, and (3) two patterns of alliance formation. I hope this narrative conveys some of the sense of discovery that I have found to be the most exciting and rewarding aspect of my research career.

When I began my thesis research in Shark Bay in 1986, it was already apparent from surveys that males associated strongly with one or two other males. Males in these pairs and trios were nearly always together. Further, we knew that each pair or trio was found often with another pair or trio, but we had no idea what these male associations were based on. Toward the end of that first field season of focal follows on males and females, the two provisioned or "beach" males began visiting the shallows at Monkey Mia accompanied by single unhabituated dolphins. Mostly strangers of unknown sex, the visitors would swim back and forth slowly just beyond the males

Figure 20.1. An alliance of three male bottlenose dolphins in Shark Bay, Western Australia.

who were seeking fish handouts from people. I wasn't paying much attention to the "beach" dolphins at that point, as I was intent on obtaining a good sample of offshore focal follows and was in the boat every possible minute. Then one morning the males came in with a stranger, and the male Snubby began behaving very oddly. People were offering his favorite fish, bony herring, but he would not come in close enough to get any. Instead, he swam in halfway between the fish buckets and the stranger and stayed there, whirling back and forth, orienting to the fish and the stranger and back again in a highly excited or agitated manner. It certainly appeared that Snubby was concerned that the stranger would leave if he swam in close to get a fish. What was going on? Unfortunately, that field season was over and I was left with a nagging feeling that I had failed to adequately document something important. Then, back at Michigan, I read Jane Goodall's (1986, 465) description of an interaction between the provisioned male Goliath and a female he was leading to the camp where the chimps were fed bananas. As they approached the camp, the unhabituated female saw Goodall and fled, only to be chased and attacked by Goliath. When his efforts failed to convince the female to follow him to the waiting bananas, Goliath spent five minutes rushing back and forth between the camp and the female. Jane Goodall was describing exactly what I had seen with Snubby and the stranger! Chimpanzee males aggressively herd females who are in estrous, but the female with Goliath was just too frightened by people to follow him to the bananas.

Fortunately, I did not have to wait very long for confirmation of herding by male dolphins when we returned the following year. One morning the males came in with a female and newborn infant in tow. The female was clearly agitated and had been in the shallows with people standing in the water for only a minute or two before she turned and bolted offshore with her infant—and the males right behind. The males caught up to her, considerable splashing followed, and they all came back. Again the female bolted, again the males caught her and escorted her back to the beach. Over the next few years we would document literally hundreds of cases of herding by the provisioned males (Connor et al. 1996), whose relationships with each other were a fascinating (and highly amusing) study in conflict and cooperation (Connor et al. 1992b; Connor & Smolker 1995).

Herding was clearly a complex behavior with a specific set of signals. Males produced a "popping" vocalization, backed up by aggression, that induced the female to turn toward them (Connor & Smolker 1996). Research on cetacean vocalizations is notoriously hard because of the difficulty of localizing sound underwater. However, because the beach males spent considerable time at the surface in such shallow water (less than a meter) they often produced the popping vocalization in the air, where it was clearly audible so we could tell which male produced the sound and observe the reaction of the female.

The complexity of herding suggested that it was unlikely a new invention of provisioned males faced with the novel conflict of getting fish and keeping unhabituated females nearby, but it remained to be seen whether herding was an important male strategy offshore. Here the importance of focal follows came into play. Associations between male alliances and individual females were readily observable offshore, but it was only by following males for hours at a time that we occasionally observed evidence that such associations were coerced. Evidence of coercion involved the males capturing a female, the female bolting from the males, aggression in the form of biting or hitting, or males making the popping vocalization—all events of relatively short duration that are unlikely to be observed during surveys (Connor et al. 1996). I didn't have a hydrophone at the time—how could we hear the males popping? Again, another bit of luck. Shark Bay dolphins often rest at the surface, backs exposed, a behavior that is apparently unusual in many other populations (Shane et al. 1986). It turned out that the males would often pause at the surface and produce pop-trains just before changing direction. Thus even offshore we were often able to tell which dolphin made the sound. It turns out that consortships are often maintained by aggressive herding, but we cannot say that all of them are (Connor et al. 1996).

The pattern of consortships over time was puzzling. Females may be consorted for days to over a month and for varying periods over many months during the year they conceive (Connor et al. 1996). The aggressive nature of herding suggests a cost to females. Why not just cycle once, conceive, and be done with it? It seemed especially odd that females that conceive during

Figure 20.2. When traveling with a herded female, males typically swim on either side and a bit behind the female.

the peak breeding season start being herded by males several months earlier, during the Austral winter when very few births, and—given a one-year gestation—conceptions, occur (Connor et al. 1996). Herding is clearly a strategy on the part of males to monopolize a female and it is possible that multiple cycling allows females that are unable to escape from undesirable males more choice in who fertilizes their egg. Another explanation is suggested by similar systems in some primates; multiple cycling may allow females to mate with many males, thereby confusing paternity and reducing the risk of infanticide (Connor et al. 1996). We have not observed aggression toward infants by males in Shark Bay since making this prediction, but evidence of infanticide has been found in at least two other populations (Dunn et al. 1998; Patterson et al. 1998). It is worth noting that if infanticide risk explains not only multiple cycling but out-of-season cycling (some of which may be anovulatory, see Connor et al. 1996) then the probability of fertilization required to inhibit male infanticidal behavior must be very small.

Herding could explain what males were doing in pairs and trios, but we still didn't know why different alliances were sometimes found traveling together. If males are competing for females, then relationships between alliances should be exclusively hostile. Clearly, they were not. During follows I found that pairs of alliances spent considerable time together in all activities: foraging, resting, traveling, and socializing. For the most part, relations between males of different but associating alliances were amicable and it was not unusual to see males of different alliances stroking each other. Again, it was a combination of observations at the beach and during follows that

Figure 20.3. Together, these two alliances, a pair and a trio, make up a "second-order" alliance. Second-order alliances attack other alliances to take females and defend against such attacks.

revealed the raison d'etre of affiliative bonds between alliances. To our great surprise, an unhabituated trio of males showed up at the beach one morning and approached but did not interfere with the beach males who were herding a female. One of the visitors was the next male on my target focal list, so after they left we followed them. They traveled offshore and joined up with a pair of males, whereupon both alliances traveled back to the beach where they attacked the beach males and took the female. That day, 19 August 1987, remains the most exciting of my research career because it marked our discovery of two levels of alliance formation in bottlenose dolphins (Connor et al. 1992a). In that particular engagement, two alliances attacked one and the skirmish was over quickly; on another occasion the alliance being assaulted was aided by a second alliance and the chasing and fighting ranged over several miles and lasted over seventy minutes (Connor et al. 1992b).

After several years of observation, we thought we had a pretty good handle on male alliance formation in Shark Bay. Most adult males appeared to fit the mold; they formed alliances of two to three males that were generally stable within and across years (fourteen years in one case). Stable alliances formed teams of two alliances to attack or defend against other alliances. One group of ten males didn't seem to fit this pattern, but we suspected that they were just immature or maturing and would eventually break up into smaller groups.

In 1994 we set out to examine vocal exchanges between alliances using two boats and towing hydrophones. This required increasing the number of alliances we would sample and many of the males we had worked with in the 1980s had disappeared. So we expanded our study range to encompass the ranges of a larger number of male dolphins. Many individuals we "scouted" for the new study were already known to us from occasional survey records over the years. These efforts to find new alliances produced a wonderful case of serendipity—a discovery that had nothing to do with our planned study on vocalizations.

During those first two days offshore in 1982, we had photographed a male we called "Wow," based on our initial reaction to the impressive shark-bite scar draped over both sides of his back behind his dorsal fin. We thought Wow and some of his more or less well known male associates would be good candidates for the new study, so we went looking for them north of our usual range. We found Wow, remarkably, in a group with thirteen other males! This was very surprising, finding fourteen adult males together without hostilities breaking out. The fourteen males were clearly associating in several trios and pairs, one of which was consorting a female, but they would also come together into one large group. Over the next few hours we carefully recorded the composition of the different alliances in Wow's group. These males would be worth another look. Did these alliances associate together often?

The next time we found them we got our second surprise: the composition of some of the trios in the group had changed. Switching alliance partners? This was different! What was going on? We decided to spend a lot more time watching Wow's group.

Over the next 3 years (1995–97), counting only cases in which an alliance was consorting a female, Mike Heithaus, Lynne Barre, and I documented 39 different alliances among the 14 males, including 35 trios and 4 pairs (Connor et al. 1999). In one 4-month season alone (1996) each male was observed in 3 to 7 different alliances and had 4 to 8 of the 14 males as alliance partners. However, males were not forming alliances randomly within the superalliance, as each male had certain males he preferred and others he avoided (Connor et al. 2001).

The 14 males spent a lot of time together—in 25 percent of our survey records at least 10 of the 14 males were found together. During a typical follow, the entire group might be together for a time resting, traveling, or socializing, but they would also spread out into their separate alliances. Some days we might spend hours with only a few members of the group and never see the rest. Evidence that the 14 males formed a very large second-order alliance, or "superalliance," came from a few observations of conflicts between 6 to 14 members of the group and other second-order alliances (teams of pairs or trios). The "superalliance" was clearly victorious each time, chasing away the other males and responding to conflicts involving group members from up to 3 kilometers (Connor et al. 1999).

Over the years we have had some individual males and alliances suddenly disappear from the study area. Although we cannot rule out emigration, we have clear evidence of philopatry in some males and suspect that males that disappear are usually dead. Observations such as the conflict described above make us wonder if some deaths might result from a male or an alliance finding themselves in the wrong place at the wrong time, outnumbered substantially by rival males.

## The Basis for Cooperation in Male Dolphins

We do not know what role, if any, that kin selection might play in alliance formation in Shark Bay. In Sarasota, Florida, observations and genetic testing indicate that males in pairs have different mothers and fathers (Wells, unpublished data; Connor et al. 2000c). In Shark Bay, we have recorded two cases of membership changes in stable alliances. In 1989 the male "BOH" joined the pair REA and HII after his previous two partners disappeared. By 1994 the male POI left his partner LUC to form a pair with BOH. LUC has survived without an alliance partner since 1994, traveling mostly in groups of females and juveniles. In the superalliance, males formed first-order alliances with five to eleven of the fourteen males. It is virtually impossible that all of the males in alliances in Shark Bay are close relatives. Unlike lions and cheetahs that can produce an alliance in a single litter or synchronous litters of closely related females, dolphins have one offspring at a time. The most successful females in Shark Bay have weaned only three offspring over a ten-year period (Mann et al. 2000a). If we allow ten years to be an upper range for the age spread within alliances, then a female that managed to wean three males consecutively might produce a trio of maternal brothers. Such males would likely have different fathers, given the multiple-mating habits of females. Given female philopatry, there might be more opportunities for sons of related females, perhaps half sisters or mother-daughter pairs, to form alliances. Such males would be maternally related by only one-eighth or less.

Further, an unequal knowledge of maternal versus paternal relatedness could disfavor kin-based alliances. If interacting males are sometimes paternally related but do not know it (and it is hard to see how they could), then a male choosing alliance partners based on maternal relatedness might find himself unknowingly opposing paternal relatives. I suggest that such kin recognition errors (see Keller 1997), if they occurred often enough, might weaken maternal relatedness as a basis for alliance formation in favor of nonkin factors such as rank. This may explain the surprising recent discovery that alliance formation in male chimpanzees is generally not kin-based (Goldberg & Wrangham 1997; Mitani et al. 2000).

Kin selection is even less likely to be of major importance for second-order alliance relationships. A male's second-order alliance relationships include the number of males in other alliances outside his own that he cooperates with. Over several years, males in stable alliances may enjoy second-order alliance relationships with up to ten different males from four different alliances. The fourteen-member superalliance is a very large second-order alliance.

The most likely mechanism sustaining cooperation among males in first-order alliances is by-product mutualism (sensu Connor 1995c). Interactions *between* alliances are not limited to cooperation for mutual benefit as they include acts of altruism. An alliance that has a female consort will sometimes help another alliance steal a female (Connor et al. 1992b). Altruism between alliances might be based on reciprocity (which would require that the assisted alliance return the favor) or pseudo-reciprocity (which would require that the assisting alliance be dominant and have mating access to females herded by the alliance they assist).

## THREE IS COMPANY, FOUR IS A CROWD

The number 3 appears to be an upper limit for male dolphins sharing a female. This number appears in some other systems—3 is the maximum number of male cheetahs or lions in an alliance that contains a nonrelative (Caro 1994; Packer et al. 1988, 1991). Packer et al. (1988) suggests that 3 is limiting in lions because the variance in male reproductive success increases with alliance size. This he attributes to (1) the rarity of mixed paternity in lions and the fact that (2) the number of females in a pride does not increase with alliance size. Because females breed synchronously and males guard females in estrous, each male may monopolize a female in a small alliance. As alliance size, but not the number of females, increases, an increasingly large number of males will not have any reproductive success. Thus it pays males to be in larger alliances only if they are related to all of the other males, so that if they do not reproduce at least their efforts at pride defense produce inclusive fitness benefits.

Such an explanation does not help us understand male dolphins, which consort single females that have a "litter" of only one infant. Dolphins refocus our attention on the number 3 because they cooperate in larger groups to defend females but only consort them in groups of 2 or 3. This issue is brought home most forcefully by the superalliance of 14 males. The average male group of superalliance males was larger than in the stable alliances (6.1 versus 3.6) and alliances were labile as males often switched alliance partners (Connor et al. 1999). However, 3 was still the largest group consorting a female and partner switches occurred only between consortships. Of 39 different alliances recorded, 35 were trios and 4 were pairs. Ninety-five of 100 consortships were by male trios and only 5 cases involved male pairs. It was common for members of the group to have one to 4 females, leaving some members without a female. It is not clear why do we not find 4 individuals consorting a female (one possible case lasted about one hour before one of the males swam off by himself).

The interesting question is whether there is an explanation for 3 being the maximum size for dolphin alliances that is, like Packer's explanation for lions, specific to the dolphins, or if there is a more general explanation that might apply to dolphins, lions, cheetahs, and other species. A dolphin-specific hypothesis might focus on the task of coercing females; perhaps the herding efforts of a third (but not a fourth) male makes sharing worthwhile to the other alliance members. A more general hypothesis might focus on male relationships or variance in reproductive success with increasing alliance size. Perhaps the critical difference between 3 and 4 might be the greater ease with which one individual can be excluded by the rest or the fact that an alliance of 4 can divide into 2 alliances but an alliance of 3 cannot. Finding out about relatedness and reproductive skew in alliances will be critical to understanding whether the number 3 is of great or minor significance. Our observations indicate that males were not equal players in the superalliance. Stability of alliances formed among the 14 males varied and is correlated with the percentage of time males spent in consortships (Connor et al. 2001). These observations suggest dominance relationships and some degree of skew.

## Alliance Formation in Bottlenose Dolphins: Cognition and Ecology

### THE ECOLOGICAL PERSPECTIVE: ALLIANCE FORMATION IN THREE DIMENSIONS

Maneuverability may be an essential ingredient for success when two or more individuals with low resource holding power form a coalition to challenge a higher-ranking individual. Wrangham and Firos (manuscript) suggest that the paucity of intragroup male alliances in arboreal primates might reflect a constraint on movement during fights that take place in trees, a "discontinuous three dimensional habitat." Male baboons may successfully use coalitions to take females from higher-ranking males on the ground during the daytime, but often lose them at night in trees or on cliff ledges (e.g., see Smuts 1985). A fight on a tree branch or narrow cliff ledge is essentially a fight in one dimension. Cetaceans live and fight in a three-dimensional habitat that should be highly conducive to alliance formation. Is alliance formation commonplace in odontocetes and other cetaceans? At this early stage of cetacean field studies we cannot say. However, scattered observations of coalitionary behavior or of high levels of association among males are suggestive (Connor et al. 2000b). For a student bound and determined to study cetaceans, such observations might provide enough of an incentive or an excuse to find out.

Or a student might explore the fascinating variation in alliance formation among populations of bottlenose dolphins. In Sarasota Bay, Florida, male bottlenose dolphins also form stable pairs that consort with individual females (Wells 1991; Connor et al. 2000c). However, no adult trios are known and some males remain single. It is not established whether consortships in Sarasota are maintained by coercion or whether males form second-order alliances, which are ubiquitous and obvious to observers in Shark Bay. In the Moray Firth, Scotland, there are no high-level associations among any adults that are in the same range as those found between alliance partners in Sarasota and Shark Bay (Wilson 1995; Connor et al. 2000c). Such variation between sites will greatly aid our efforts to understand the primary ecological influences on alliance formation in bottlenose dolphins. What are some of the relevant parameters?

Males may form first-order alliances in competition with rival males or in order to coerce females. A higher rate of interaction among males in competition over receptive females should favor alliance formation (Connor et al. 2000b; Connor et al. 2000c). The rate of interaction may, in turn, be affected by a number of factors such as population density, habitat "openness," day range, operational sex ratio, and differences in patterns of females' receptivity. The density of dolphins in Shark Bay appears to be higher than at other sites (Connor et al. 2000c) likely because of their smaller size (Box 20.1) and possibly a greater abundance of resources in Shark Bay, which has the largest seagrass beds in the world. Shark Bay is also a more open habitat than Sarasota, which might allow rival males to detect females and each other at greater distances (Smolker et al. 1992; Connor et al. 2000c). There is some evidence that interbirth intervals may be longer in Shark Bay, which might reduce the relative number of females receptive to adult males, favoring alliance formation (Connor et al. 2000c). Differences in female receptivity and male or female ranging patterns could also affect the rate at which males come into conflict over females. Finally, even if all of the factors listed above are equal, predation risk could tip the balance in favor of alliance formation. Shark attack risk appears high in Shark Bay, but minimal or absent in the Moray Firth.

Differences in body size and sexual size dimorphism should significantly impact male-female interactions across sites. Moray Firth adults grow to nearly twice as long as adults in Shark Bay. Sarasota dolphins are intermediate in length but still much larger than adults in Shark Bay, which are the size of two- to three-year-old calves in Sarasota. A Shark Bay male that is scarcely larger than a female might need assistance to herd her. On the other hand, relatively larger males in Sarasota and the Moray Firth might be less maneuverable than females, making herding a more difficult option. Thus it

Box 20.1
The Bottlenose Dolphin

Familiar to the public as the most common dolphin on display in captivity, the bottlenose dolphin is found worldwide in temperate and tropical waters and is notably abundant in shallow coastal and estuariane waters heavily used by people. The systematics of the various populations of bottlenose dolphins remains uncertain and the genus might not be monophyletic (LeDuc et al. 1999). Body size varies dramatically among populations, ranging from just over 2 m in Shark Bay to 4 m in the northeast Atlantic. Sexual size dimorphism may also vary among populations, being more pronounced in populations with larger individuals. Longevity is unknown in Shark Bay but one female that died from a stingray spine that penetrated her heart was about 35. Maximum ages of $>$ 50 for females and $>$ 40 for males are reported from Sarasota Bay, Florida (Connor et al. 2000). Females are polyestrous; in Shark Bay multiple cycles are suggested by patterns of female attractiveness to males (Connor et al. 1996). Females give birth to single offspring, beginning at age 12 or older in Shark Bay. Gestation lasts for one year. While births can occur during any month, a clear breeding season (births and conceptions) lasts from September to February (Mann et al. 2000). Intervals between births for females with surviving calves is usually 3 to 6 years; 4-year intervals are most common (16/33, 47%). Young remain with their mothers from 3 to 6 years typically, the maximum age of offspring dependency was 7. Mortality of infants by age 3 is 44 percent (n = 110). Over a 10-year period, the most successful females in Shark Bay managed to raise 3 offspring to dependency (Mann et al. 2000).

Bottlenose dolphins exhibit a classic fission-fusion grouping pattern, similar to primates such as chimpanzees or spider monkeys. In Shark Bay, groups average 4 to 5 individuals and change in composition on a daily or hourly basis. Over 400 individuals have been identified on the basis of dorsal fin shape and scars but there is no evidence of a closed community of dolphins in Shark Bay. We therefore refer to a *social network* of individuals with a mosaic of overlapping home ranges in Shark Bay, rather than a "community" or "group."

The fission-fusion grouping pattern of bottlenose dolphins is likely related to their diet of patchily distributed schooling and solitary fish, squid, and occasional crustaceans. Prey are pursued and captured throughout the water column as well as into the air above (for skipping and jumping fish) and under the bottom sand and even onto the beach. Individual foraging specializations appear to be common in Shark Bay, and include unusual behaviors such as "sponge-carrying" (Smolker et al. 1997) or "kerplunking," in which individuals hunting in shallow water use fluke-slaps to startle fish hiding in seagrass (Connor et al. 2000). Shark-bite scars are abundant on bottlenose dolphins and while infants are obviously much more vulnerable to predation, fresh scars are sometimes seen on adults in Shark Bay. Dolphin reactions to the presence of large sharks vary from flight to no apparent reaction at all (Connor & Heithaus 1996).

becomes critical not only to find out whether males are forming alliances in each habitat, but the nature of the male-female mating relationship.

## Social Cognition and Alliance Formation

Shark Bay bottlenose dolphins might exhibit such complex alliance formation not only because particular ecological conditions favor it, but because they *can*. Several authors have suggested that the relatively restricted distribution of within-group alliances among mammals might reflect cognitive constraints.

Bottlenose dolphins and several other delphinids have large brains—larger than any other nonhuman mammal when body size is taken into account (Ridgway 1986; Ridgway & Brownson 1984; Connor et al. 1992a). Jerison (1973, 1983) argued that large brains provide their owners with a model of the external world. A larger brain might allow an animal to model a complex but predictable distribution of resources in space and time (e.g., Milton 1988) or social relationships (e.g., Byrne & Whiten 1988), which is arguably a more challenging task given that the subjects being modeled are trying to outwit you but resources are not. Recognizing that the brain is not a general-purpose computer, most reject overall brain size as the appropriate comparative measure, focusing instead on structures considered most relevant to the type of information processing under question (e.g., the size of the hippocampus and foraging habits in birds, Krebs 1990). However, if the question is which selective factor has favored large brain evolution, then the obvious brain structure to use in comparisons is the neocortex (which comprises 41 to 76 percent of the primate brain, see Barton & Dunbar 1997). Using group size as an index of social complexity, several authors have found a relationship between neocortex size and social group size in primates (Dunbar 1992; Sawaguchi 1992; Barton 1996; Barton & Dunbar 1997) and carnivores (Barton & Dunbar 1997). The scarcity of information on social group size in dolphins precludes a similar analysis (Connor et al. 1998b).

The formation of coalitions and alliances within groups is common in primates compared to other taxa, leading to the suggestion that complex alliances require some minimum level of information-processing capacity. What specifically is complex about primate alliances? The fact that they often occur within groups—a rarity in other mammals—allows for what Kummer (1967) called triadic interactions: two individuals can ally against one. Additionally, alliances may be based on affiliative bonds and individuals can recruit and compete for alliance partners. Knowledge of third-party relationships, involving individuals engaged in a contest (e.g., their relative dominance ranks) or absent individuals (does the subordinate individual have a powerful friend or relative nearby?), may require much greater information-processing capacity than interactions between groups (Harcourt

1992). Intergroup interactions—the basis for most alliances in animals—are basically dyadic (= group against group) interactions that are exclusively hostile or, at best, tolerant.

We do not know if dolphins have knowledge of third-party relationships; such data would be very difficult to obtain from free-ranging dolphins. What is most impressive about dolphin alliances is the existence of both two levels and two patterns of alliance formation within one social network. The importance of levels of alliance formation within a social group has been largely neglected by primatologists, perhaps because it is unknown among nonhuman primate males and surfaces only occasionally with overthrows of matrilines in females (see Connor et al. 1992a). The "within group" nature of interactions between dolphin alliances is revealed by our observations of alliances that cooperate in one context but become rivals in another (Connor et al. 1992a). This kind of switching is characteristic of interactions between individuals within primate groups but not between groups. Multiple-level male alliances *within a social group* are obviously of extreme importance in our own species, and dolphins provide the first known example in nonhumans. It is easy to imagine how additional levels of alliance formation might complicate the social landscape enormously and select for greater information-processing capacity. Individuals must take into consideration how their behavior toward partners in one level of alliance will impact their relations with males at another level.

The two patterns of alliance we have discovered will, I suspect, turn out to be extremes of a continuum. Nonetheless, the different alliance types coexist in the same habitat, eliminating ready explanations based on habitat differences in predator or prey abundance (Connor et al. 1999). We do not know why alliance formation is so labile in the superalliance compared to the stable alliances. Perhaps the partner switching is a mechanism by which the fourteen males in the superalliance maintain their affiliative bonds. Whatever the reason for differences in first-order alliance stability and the size of second-order alliances, the variation adds an additional axis of complexity to male alliance formation in Shark Bay.

## The Future: Relatedness, Ecology, and Variation in Alliance Formation

I am wedded to dolphin research but not to Shark Bay. In my mind there is an inviting romance to starting a new study site—learning about an unknown species or a previously studied species in a new habitat—that is very compelling. When I complete a project in Shark Bay, I make a decision about whether the next project in Shark Bay looks more tantalizing than this or that great new study site I have learned about. But the lure of Shark Bay only grows stronger. As our knowledge of individual histories and social relationships grows, so does the value of each new observation.

Our discovery of the superalliance completely throws open the door on alliance formation in Shark Bay. I am as excited about going back to take a fresh look at Shark Bay alliances as I was going there for the first time.

By far the most frequently asked question is if the males in alliances are related. As I have argued above, given that all males are in alliances and the demographics, it is not possible that all of the males are closely related to their alliance partners. It is very possible, however, that some are related and this will affect the behavior of males in alliances. Michael Kreutzen and Bill Sherwin, out of the University of New South Wales, are leading a team collecting tissue from individual dolphins for microsatellite analysis of relatedness and paternity. It took us three frustrating field seasons to develop a working dart for our petite bottlenose dolphins, but through 1999 over three hundred samples have been collected. Tissue samples have already been collected from the superalliance and a number of stable alliances for comparison of relatedness.

My next set of tasks will be to find out if other "superalliances" or perhaps, "not-so-super-alliances" exist (we already have a few candidates), if there is a correlation between first-order alliance stability and the size of second-order alliances, and if the correlation between alliance stability and consortship rate that we found in the superalliance is a general phenomenon that predicts differences in reproductive success. Another, and much more difficult task, will be to find out why such variation in alliance size exists. The extensive overlap in home range of stable alliances and the superalliance negates inviting hypotheses that invoke habitat differences such as predator or prey abundance. One possible explanation derives from observations of baboons, where alliance formation is a conditional strategy used mostly by mid-ranking males against high-ranking males. It is possible that males in the superalliance have to travel in a larger group to remain competitive with smaller groups of stable alliances. Thus far, the superalliance has clearly won the encounters we have observed. Members of the superalliance may spread out over many kilometers and it may be possible to observe conflicts between smaller subsets of the superalliance when the rest of the group is not near enough to offer assistance.

Another hypothesis derives from the extensive foraging specializations we observe in Shark Bay. In addition to the striking behavioral specializations such as "sponge-carrying" (Smolker et al. 1997) and "kerplunking" (Connor et al. 2000d), we find differences between alliances in the kind of habitat (shallows versus deep water) they prefer to forage in (Connor, unpublished data). While the deep-water foraging superalliance shares their habitat with stable alliances, it remains possible that deep-water stable and superalliance males differ in the kind of prey they seek, reducing the cost of grouping for males in the superalliance groups.

Finally, it is possible that there are minimal differences in grouping costs without foraging differences. Low travel costs in dolphins might translate into grouping costs that rise relatively slowly with increasing group size (Connor 2000; Williams et al. 1992). If so, then benefits such as predator avoidance or greater access to females may be able to keep pace with rising costs across a wider range of group sizes.

How might our future investigations on dolphin alliances impact behavioral ecology? I anticipate important comparative data on the cooperative basis for male alliance formation (kinship, reciprocity, reproductive skew, etc.). Ecological influences on alliance formation will be revealed in comparisons of Shark Bay with other populations where alliances are less abundant (Connor et al. 2000c). Variation within populations, such as the two patterns of alliance formation in Shark Bay or the presence of single males and alliances in Sarasota Bay, Florida, might reveal the basis for conditional alliance strategies (Connor et al. 2000c). Comparisons with terrestrial species and smaller-brained cetaceans will shed light on the role cognition plays in alliance formation and may help us answer the important question of why we find nested levels of male alliances in human and dolphin social networks but not elsewhere. Finally, observations of strong male-male associations and group conflicts suggest that male alliance formation might be found in a wide range of cetaceans, greatly expanding the comparative possibilities (Connor et al. 2000c).

## Dolphins, Whales, and Behavioral Ecology

I would be remiss to speculate about the future of my dolphin alliance studies without broadening the discussion, if only briefly, to include the rest of the order cetacea. As we enter the new millennium, we can confidently describe the social structure of only a few cetaceans—most notably the four species featured in the recent volume *Cetacean Societies: Field Studies of Dolphins and Whales* (Mann et al. 2000b). These four species—sperm whale (*Physeter catadon*), killer whale (*Orcinus orca*), humpback whale (*Megaptera novaenglia*), and bottlenose dolphin (*Tursiops* spp.)—can be used as a yardstick to estimate the future; studying cetaceans isn't easy, so is it worthwhile?

In addition to complex alliances in bottlenose dolphins, we have discovered possible tool use, remarkable foraging specializations, and an unusual combination of dependency and precociality in infants (see Connor et al. 2000c). In "resident" killer whales, neither males nor females disperse from their natal group. The genetically distinct but sympatric "transient" killer whales hunt seals cooperatively (Baird 2000). Humpback whales lek and use bubble nets to catch fish. Sperm whale social organization and life history have converged to a remarkable degree with that of elephants (Weilgart et al.

1996). Vocal learning, a rarity in terrestrial mammals, has been documented in bottlenose dolphins and humpback whales (Janik & Slater 1997).

Discoveries of novel social adaptations in cetaceans will require novel explanations, thereby expanding the scope of theory in behavioral ecology. Hal Whitehead (1998) suggests that culture may account for low mitochondrial diversity in matrilineal whales. Another example concerns the question of philopatry: primatologists assume that one sex or the other must emigrate to avoid the costs of inbreeding. Female options are considered first. Where female fitness does not depend on kin support there appears an option for female emigration and a relaxation of the need for male emigration. If male philoptary is favored, females will be forced to emigrate (van Hoof & van Schaik 1994). Primatologists can apply this formula because it works for them, but it obviously fails for resident killer whales. How can adult male and female resident killer whales afford to remain in their mother's group? Compared to terrestrial mammals, killer whales experience a low cost of locomotion and have enormous day and home ranges. Connor et al. (1998a; Connor 2000) suggested that these characteristics in combination with a lack of dependence on a breeding site allow both sexes of killer whales to enjoy the benefits of philopatry while avoiding the cost of inbreeding.

It is unlikely that any behavioral ecologist would have predicted multiple-level male alliances in bottlenose dolphins or natal philopatry by male and female resident killer whales. If the past is any predictor of the future, we can expect many exciting new discoveries in cetacean behavioral ecology.

## Acknowledgments

Several members of the dolphin research group at Shark Bay spent time working on the male alliance project. Thanks to Lynne Barre, Mike Heithaus, and Rache Smolker. Many others, too numerous to name here, provided essential help as assistants. I gratefully acknowledge assistance from the Monkey Mia Resort, the Department of Human Biology at the University of Western Australia (notably Richard Host and Ron Swan), and C.A.L.M. Happy halloween! This work has been supported by the National Geographic Society, NSF, NIH, and the University of Michigan.

## References

Altmann J, 1974. Observational study of behavior: sampling methods. Behaviour 49:227–267.

Baird RW, 2000. The killer whale—foraging specializations and group hunting In: Cetacean societies: field studies of dolphins and whales (Mann J, Connor RC, Tyack PL, Whitehead H, eds). Chicago: University of Chicago Press.

Barton RA, 1996. Neocortex size and behavioural ecology in primates. Proc R Soc Lond B 263:173–177.

Barton RA, Dunbar RIM, 1997. Evolution of the social brain. In: Machiavellian intelligence II: extensions and evaluations (Whiten A, Byrne RW, eds). Cambridge: Cambridge University Press; 240–263.

Bradbury JW, Vehrencamp S, 1977. Social organization and foraging in emballonurid bats. III: mating systems. Behav Ecol Sociobiol 2:1–17.

Byrne R, Whiten A, 1988. Machiavellian intelligence (Byrne R, Whiten A, eds). Oxford: Oxford University Press.

Caro TM, 1994. Cheetahs of the Serengeti Plains. Chicago: University of Chicago Press.

Connor RC, 1986. Pseudoreciprocity: investing in mutualism. Anim Behav 34:1562–1566.

Connor RC, 1992. Egg-trading in simultaneous hermaphrodites: analternative to tit-for-tat. J Evol Biol 5:523–528.

Connor RC, 1995a. Impalla allogrooming: tit-for-tat or parcelling? Anim Behav 49:528–530.

Connor RC, 1995b. Altruism among non-relatives: alternatives to the Prisoner's Dilemma. Trends Ecol Evol 10:84–86.

Connor RC, 1995c. The benefits of mutualism: a conceptual framework. Biol Rev 70:427–457.

Connor RC, 1996. Partner preferences in by-product mutualisms and the case of predator inspection in fish. Anim Behav 51:451–454.

Connor RC, 2000. Group living in whales and dolphins. In: Cetacean societies: field studies of whales and dolphins (Mann J, Connor RC, Tyack P, Whitehead H, eds). Chicago: University of Chicago Press.

Connor RC, Curry R, 1995. Helping non-relatives: a role for deceit? Anim Behav 49:389–393.

Connor RC, Heithaus MR, 1996. Approach by great white shark elicits flight response in bottlenose dolphins. Mar Mamm Sci 12:602–606.

Connor RC, Heithaus MR, Barre LM, 1999. Superalliance of bottlenose dolphins. Nature 397:571–572.

Connor RC, Heithaus MR, Barre LM, 2001. Complex social structure, alliance stability and mating access in a bottlenose dolphin "super-alliance." Proc R Soc Lond B 268:263–267.

Connor RC, Heithaus MR, Berggren P, Miksis J, 2000. Surface flukeslaps during shallow water bottom foraging by Indian Ocean bottlenose dolphins. Mar Mamm Sci 16:646–653.

Connor RC, Mann J, Tyack PL, Whitehead H, 1998a. Social evolution in toothed whales. Trends Ecol Evol 13:228–232.

Connor RC, Mann J, Tyack PL, Whitehead H, 1998b. Quantifying brain-behavior relationships in odontocetes: a reply. Trends Ecol Evol 13:408.

Connor RC, Mann J, Tyack PL, Whitehead H, 2000a. Introduction: the social lives of whales and dolphins. In: Cetacean societies: field studies of whales and dolphins (Mann J, Connor RC, Tyack P, Whitehead H, eds). Chicago: University of Chicago Press.

Connor RC, Norris KS, 1982. Are dolphins reciprocal altruists? Am Nat 119:358–374.

Connor RC, Read A, Wrangham RW, 2000b. Male reproductive strategies and social bonds. In: Cetacean societies: field studies of whales and dolphins (Mann J, Connor RC, Tyack P, Whitehead H, eds). Chicago: University of Chicago Press.

Connor RC, Richards AF, Smolker RA, Mann J, 1996. Patterns of female attractiveness in Indian Ocean bottlenose dolphins. Behaviour 133:37–69.

Connor RC, Smolker RA, 1985. Habituated dolphins (*Tursiops* sp.) in Western Australia. J Mamm 36:304–305.

Connor RC, Smolker RA, 1995. Seasonal changes in the stability of male-male bonds in Indian Ocean Bottlenose dolphins (*Tursiops* sp.). Aquatic Mamm 21:213–216.

Connor RC, Smolker RA, 1996. "Pop" goes the dolphin: a vocalization male bottlenose dolphins produce during consortships. Behaviour 133:643–662.

Connor RC, Smolker RA, Richards AF, 1992a. Two levels of alliance formation among male bottlenose dolphins (*Tursiops* sp.). Proc Nat Acad Sci 89:987–990.

Connor RC, Smolker RA, Richards AF, 1992b. Dolphin alliances and coalitions. In: Coalitions and alliances in humans and other animals (Harcourt AH, de Waal FBM, eds). Oxford: Oxford University Press.

Connor RC, Wells R, Mann J, Read A, 2000c. The bottlenose dolphin: social relationships in a fission-fusion society. In: Cetacean societies: field studies of whales and dolphins (Mann J, Connor RC, Tyack P, Whitehead H, eds). Chicago: University of Chicago Press.

Dunbar RIM, 1992. Neocortex size as a constraint on group size in primates. J Human Evol 20:469–493.

Dunn DG, Barco S, McLellan WA, Pabst DA, 1998. Virginia Atlantic bottlenose dolphin (*Tursiops truncatus*) stranding: gross pathological finding in ten traumatic deaths. Abstract: Atlantic coast dolphin conference, Sarasota, FL.

Emlen ST, Oring LW, 1977. Ecology, sexual selection, and the evolution of mating systems. Science 197:215–223.

Gans C, 1978. All animals are interesting! Am Zool 18:3–9.

Goldberg T, Wrangham R, 1997. Genetic correlates of social behavior in wild chimpanzees: evidence from mitochondrial DNA. Anim Behav 54:559–570.

Goodall J, 1986.The chimpanzees of Gombe. Patterns of behavior. Cambridge, MA: Harvard University Press.

Harcourt AH, 1992. Coalitions and alliances: are primates more complex than nonprimates? In: Coalitions and alliances in humans and other animals (Harcourt AH, de Waal FBM, eds). Oxford: Oxford University Press.

Harcourt AH, de Waal FBM, 1992. Coalitions and alliances in humans and other animals. Oxford: Oxford University Press.

Hinde RA, 1976. Interactions, relationships, and social structure. Man 11:1–17.

Hutt SJ, Hutt C, 1970. Direct observation and measurement of behavior. Springfield, IL: Charles C. Thomas.

Janik VM, Slater PJB, 1997. Vocal learning in mammals. Adv Study Behv. 26:59–99.

Jerison HJ, 1973. Evolution of the brain and intelligence. New York: Academic Press.

Jerison HJ, 1983. The evolution of the mammalian brain as an information-processing system. Adv Stud Mamm Behav.

Keller L, 1997. Indiscriminate altruism: unduly nice parents and siblings. Trends Ecol Evol 12:99–103.

Klettenheimer BS, 1997. Father and son sugar gliders: more than a genetic coalition? J Zool Lond 242:741–750.

Krebs JR, 1990. Food-storing birds: adaptive specialization in brain and behaviour? Phil Trans Royal Soc, Series B 329:153–160.

Kummer H, 1967. Tripartite relations in hamadryas baboons. In: Social communication among primates (Altman SA, ed). Chicago: University of Chicago Press.

LeDuc RG, Perrin WF, Dizon AE, 1999. Phylogenetic relationships among the delphinid cetaceans based on full cytochrome b sequences. Mar Mamm Sci 15:619–646.

Mann J, 2000. Unraveling the dynamics of social life: long-term studies and observational methods. In: Cetacean societies: field studies of whales and dolphins (Mann J, Connor RC, Tyack P, Whitehead H, eds). Chicago: University of Chicago Press.

Mann J, Connor RC, Barre LM, Heithaus MR, 2000a. Female reproductive success in bottlenose dolphins (*Tursiops* sp.): life history, habitat and group size effects. Behav Ecol 11:210–219.

Mann J, Connor RC, Tyack P, Whitehead H, 2000b. Cetacean societies: field studies of dolphins and whales. Chicago: University of Chicago Press.

Milton K, 1988. Foraging behaviour and the evolution of primate intelligence In: Machiavellian intelligence (Byrne R, Whiten A, eds). Oxford: Oxford University Press; 285–305.

Mitani J, Merriwether DA, Zhang C, 2000. Male affiliation, cooperation, and kinship in wild chimpanzees. Anim Behav 59:885–893.

Noe R, 1994. A model of coalition formation among male baboons with fighting ability as the crucial parameter. Anim Behav 47:211–213.

Packer CL, Herbst AE, Pusey JD, Bygott JP, Hanby SJ Cairns, Borgerhoff Mulder M, 1988. Reproductive success in lions. In: Reproductive success (Clutton-Brock TH, ed). Chicago: University of Chicago Press; 363–383.

Packer C, Gilbert DA, Pusey AE, O'Brien SJ, 1991. A molecular genetic analysis of kinship and cooperation in African lions. Nature 351:562–565.

Patterson IAP, Reid RJ, Wilson B, Grellier K, Ross HM, Thompson PM, 1998. Evidence for infanticide in bottlenose dolphins: an explanation for violent interactions with harbour porpoises? Proc Roy Soc Lond B265:1–4.

Ridgway SH, 1986. Physiological observations on dolphin brains. In: Dolphin cognition and behavior: a comparative approach (Schusterman RJ, Thomas JA, Wood FG, eds). Hillsdale, NJ: Lawrence Erlbaum Associates; 31–59.

Ridgway SH, Brownson RH, 1984. Relative brain sizes and cortical surface areas in odontocetes. Acta Zool Fennica 1972:149–152.

Sawaguchi T, 1992. The size of the neocortex in relation to ecology and social structure in monkeys and apes. Folia Primatol 58:131–145.

Shane SH, Wells RS, Wursig B, 1986. Ecology, behavior and social organization of the bottlenose dolphin: a review. Marine Mammal Science 2:34–63.

Smolker RA, Richards AF, Connor RC, Mann J, 1997. Sponge carrying by dolphins (Delphinidae, *Tursiops* sp.): a foraging specialization involving tool use? Ethology 103:454–465.

Smolker RA, Richards AF, Connor RC, Pepper J, 1992. Association patterns among bottlenose dolphins in Shark Bay, Western Australia Behaviour 123:38–69.

Smuts, BB, 1985. Sex and friendship in baboons. Hawthorne, NY: Aldine.

Trivers RL, 1971. The evolution of reciprocal altruism. Q Rev Biol 46:35–57.

Trivers RL, 1972. Parental investment and sexual selection. In: Sexual selection and the descent of man (Campbell B, ed). Chicago: Adline; 1–31.

van Hoof JARAM, van Schaik CP, 1992. Cooperation in competition: the ecology of primate bonds. In: Coalitions and alliances in humans and other animals (Harcourt AH, de Waal FBM, eds). Oxford: Oxford University Press; 357–389.

van Hoof JARAM, van Schaik CP, 1994. Male bonds: affiliative relationships among nonhuman primate males. Behaviour 130:309–337.

Weilgart LS, Whitehead H, Payne K, 1996. A colossal convergence. Am S. 84:278–287.

Wells RS, 1991a. The role of long-term study in understanding the social structure of a bottlenose dolphin community. In: Dolphin societies (Pryor K, Norris KS, eds). Berkeley: University of California Press.

Wells RS, Scott MD, Irvine AB, 1987. The social structure of free-ranging bottlenose dolphins. In Current mammalogy, vol. 1 (Genoways HH, ed). New York: Plenum Press.

Whitehead H, 1997. Analyzing animal social structure. Anim Behav 53:1053–1067.

Whitehead H, 1998. Cultural selection and genetic diversity in matrilineal whales. Science 282:1708–1711.

Williams TM, Friedl WA, Fong ML, Yamada RM, Dedivy P, Haun JE, 1992. Travel at low energetic cost by swimming and wave-riding bottlenose dolphins. Nature 355:821–823.

Wilson DRB, 1995.The ecology of bottlenose dolphins in the Moray Firth, Scotland: a population at the northern extreme of the species' range. Ph.D. diss., University of Aberdeen, Scotland.

Wrangham RW, Firos S. Distribution of within-group alliances in primates: the arboreal constraint hypothesis. Submitted.

Wursig B, Wursig M, 1977. The photographic determination of group size, composition, and stability of coastal porpoises (*Tursiops truncatus*). Science 198:755–756.

# 21  Bonnet Macaques: Evolutionary Perspectives on Females' Lives

Joan B. Silk

In Chapter 2, Tom Seeley explains what comprises a model system in behavioral ecology: "for a species to be a model system you must be able to get and stay close to its members living in the wild. It is also essential that, while making your observations, you can recognize particular individuals (through natural or artificial markings) and do not disturb them by your presence. Also, a model system species must be amenable to experimental work, that is, to delicate alterations of the animal itself, the animal's environment, or both. Unless you can perform manipulations of specific properties of the animal or its environment, you will be unable to conduct controlled experiments. And without experiments, your analysis of the causes and the effects of the animal's behavior will be crippled. Finally, for a species to be a model system for behavioral-ecological studies, you must be able to measure the lifetime reproductive success of individuals in nature, or at least reasonable proxies of reproductive success, so that you can see what behavioral factors contribute to variation in reproductive success."

By this standard, no primate species qualifies as a model system for behavioral-ecological analyses. It's possible to recognize individuals in most species from natural markings and idiosyncratic features, but primates don't meet any of Seeley's other criteria. Our presence certainly influences the animals that we watch—it often takes months or years for animals to get used to us being near them. For both practical and ethical reasons, we are quite limited in the kinds of experiments that we can conduct on primates, and most behavioral-ecological research on primates is based on naturalistic observations, not experiments. Information about the lifetime reproductive success of individuals is available from very few field sites because primates live for decades, much longer than the span of the typical field project.

So, primates can't be considered model systems by Seeley's criteria. A quick scan of the web pages of biology departments across the country would suggest that behavioral ecologists might be best advised to avoid primates altogether. Although primatology is a thriving academic specialization, there are relatively few primatologists employed in the biological sciences.

At this point you are probably thinking that you should skip this chapter altogether. But there are at least two reasons that you might want to read this chapter (or study primates) anyway. The first reason is that certain kinds of questions can only be asked in certain kinds of species. If you want to understand the dynamics of echolocation, you have to study bats or dolphins, no matter how difficult it may be to work with these creatures. Primates *are* model systems for studying the dynamics of complex behavior in long-lived, highly social creatures. In primates, that live in stable social groups, we can examine how kin selection and reciprocal altruism have shaped the evolution of social interactions. Researchers can examine the evolution of life history strategies in animals with long life spans, overlapping generations, and extensive parental investment. We can study the ecological factors that influence the evolution of social organization and the dynamics of social relationships within social groups. Since primates have large brains and rely heavily on learning, we can ask how evolution has shaped cognitive abilities to meet environmental needs and social challenges. These are important and interesting questions, particularly if we want to understand the evolution of similar phenomena in our own species.

The second reason that you should read this chapter is because it will give you some idea of the impact that behavioral ecology has had in disciplines outside the biological sciences. Behavioral ecology has become the dominant paradigm in biological anthropology and is a growing intellectual force in psychology. In biological anthropology, the academic discipline that I know best, behavioral ecology has profoundly altered our understanding of primate behavior, the evolutionary history of our hominid ancestors, and the behavior of modern humans. Biological anthropologists now rely on a rich body of theory and behavioral data to guide the construction of models of human evolution, to interpret the behavior of closely related species, and to examine the causal processes that shape the morphology and behavior of contemporary humans.

## Starting Out

I began graduate school at an exciting time. It was 1976. E. O. Wilson's massive tome on evolutionary biology, *Sociobiology: The New Synthesis*, had just been published. Wilson popularized the theoretical work of W. D. Hamilton, Robert Trivers, John Maynard Smith, and many others, and compiled a vast amount of empirical information about the form and function of behavior. This body of work generated considerable intellectual excitement within biology and the behavioral sciences, provoking intense debates about the units of selection, the nature of adaptation, and the effects of selective forces on the evolution of social behavior. A flood of empirical and theoretical work followed. These ideas became the foundation of my graduate training.

Like many students, I was anxious to apply my academic knowledge to my own research project. My choice of taxa was limited to primates for several reasons. As an undergraduate, I had spent nearly a year studying chimpanzees at the Gombe Stream Reserve under the supervision of Jane Goodall. That experience led me to the graduate program in anthropology at the University of California at Davis. One branch of anthropology, biological anthropology, is concerned with human evolution and adaptation. Biological anthropologists study primates because we share a common evolutionary history and we are similar to other primates in our physiology, morphology, life history, and behavior. Academic custom dictates that anthropologists study primates, but not other kinds of animals.

So, I was limited to primates. I was lucky that U.C. Davis is the site of the California Regional Primate Research Center (CRPRC), which houses a large population of Old World monkeys. My graduate advisor, Peter Rodman, suggested that I join him and another student, Amy Samuels, in studying a group of monkeys there. Anxious to begin doing research, I accepted his offer. I passed the obligatory TB test, was issued a white lab coat, and started bicycling out to the primate center several days a week.

The monkeys were members of an obscure species, the bonnet macaque (*Macaca radiata*). Bonnet macaques (Fig. 21.1) are indigenous to southern India, where they live in groups of about thirty. Many bonnets were imported to the United States from India for biomedical research in the 1960s. Some of these animals found their way into behavioral research projects, including the founding members of a group of monkeys that I observed at the California Regional Primate Research Center at the University of California, Davis. In the Davis group, bonnets reared offspring, developed relationships, conducted courtships, and grew old under the watchful eyes of observers.

Over the years, I spent hundreds of hours standing outside the bonnets' enclosure, sweating under the hot Central Valley sun and shivering in the damp Valley fog. Over the years, I got to know dozens of monkeys. I learned to recognize their faces, admire their punk hair-do's, and appreciate their distinctive personalities.

I originally decided to study female mate choice in the bonnet group. I wanted to know how females' preferences affected male reproductive success and why females preferred some males over the others (or vice versa). This project was grounded in evolutionary theory, but it turned out to be intractable empirically. Three fundamental problems emerged. First, female bonnet macaques provide no visible external signs of their ovulatory status. This meant I couldn't determine when females were sexually receptive or likely to conceive. Second, I couldn't assess whether females' preferences were effective in determining who fathered their offspring because newly

Figure 21.1. A female bonnet macaque holds her newborn infant.

developed immunological techniques could not be used to assess paternity in this species. Third, it became clear that females' choices were sometimes constrained by male-male competition, making it difficult to discern their true preferences. By the end of the first mating season, I knew the project would not work.*

But in those months, I'd learned a lot about the bonnets. Life in a macaque group is like a soap opera in which all the leading roles are played by females: relationships are formed, conflicts erupt, alliances are forged, and compromises are negotiated. Like despots everywhere, the dominant females in the group enjoyed the advantages of their exalted positions. They regularly supplanted lower-ranking females from drinking fountains, were first to inspect newly delivered monkey chow, and disrupted peaceful grooming parties. High-ranking females appropriated the shadiest resting spots in the summer and installed themselves in the warmest and driest refuges in the winter.

*Today, the situation is very different because newly developed genetic techniques allow questions about paternity to be resolved. One of my graduate students has successfully examined female mate choice in Japanese macaques (*Macaca fuscata*; Soltis et al. 1997a, 1997b) and resolved nearly all of the questions that plagued my research on bonnets.

We noticed that high-ranking females had larger families than low-ranking females. The relationship between females' rank and the size of their families interested us because variation in reproductive success provides the raw material on which natural selection acts. Thus, any factors that create systematic differences in reproductive success among individuals should be subject to strong selective pressure. However, there had been very few attempts to measure the source and significance of variation in reproductive success among mammalian females. This was partly due to the fact that the reproductive success of mammalian males potentially varies more than that of females, who are constrained by the demands of gestation and lactation (Bateman 1948). Since variance in female reproductive success was relatively low, it was assumed to be of little adaptive importance.

But the conventional wisdom didn't seem to fit what we saw among the bonnets—reproductive success among females did vary. The highest-ranking family was the largest one in the group, while low-ranking families tended to be quite small. Thus, we set out to identify the sources of variation in female reproductive success and to examine the processes that contributed to this variation.

## Likely Sources of Variation in Female Reproductive Success

It is important to realize that not all the factors that affect female reproductive success are likely to produce systematic variation in reproductive success among females. This is because some factors, such as age, have roughly the same effect on all females. Among primates, young females routinely reproduce less successfully than mature females do (Fairbanks & McGuire 1995; Small & Rodman 1981; Wilson et al. 1983), while females' fertility becomes progressively more variable as they reach old age (Caro et al. 1995). All other things being equal, age-related changes in fertility do not produce variation in reproductive success among females because all females grow old.

Some factors generate variation in reproductive success, but are not subject to natural selection because they are not based on heritable traits. Thus, environmental events (like floods, droughts, predation, and disease) may generate variation in individual reproductive success. But if vulnerability to environmental catastrophes is not linked to heritable traits, then this variation will not be subject to selection.

However, there are some factors that have consistent effects on different individuals, and may contribute to systematic variance in reproductive success among females. Thus, the ability to acquire valuable resources, the choice of a good mate, or the ability to protect offspring from environmental hazards may cause female reproductive success to vary systematically. If the traits that enhance females' reproductive success are transmitted genetically to offspring, they will be subject to natural selection.

For macaque females, there is good reason to think that dominance rank might contribute to differential reproductive success. The reproductive success of female mammals is limited by their access to food (Wrangham 1980). Many primates, including macaques, rely on foods that occur in clumps that can be monopolized, making competition over access to food sites profitable (Isbell 1991; van Schaik 1989; van Hooff & van Schaik 1992). In many of these species, high-ranking animals have a distinct advantage because they are able to monopolize scarce and valuable resources (e.g., Cheney et al. 1981; Dittus 1979, 1986; Whitten 1983; Wrangham 1981). Thus, a female's rank may influence her access to food, which in turn may affect her nutritional condition and ability to reproduce.

Although this logic seems compelling now, there was little evidence that it was correct when we began our work. In one provisioned population established on an island in the Caribbean, Drickamer (1974) found that high- and middle-ranking females' daughters produced their first infants at slightly earlier ages than low-ranking females' daughters did. Moreover, he found that high-ranking females were more likely to produce infants each year than lower-ranking females, and the infants of high-ranking females were more likely to survive to six months than the infants of low-ranking females were. However, these trends emerged only among females who had been born in the wild, not females who had been born in the colony. Drickamer (1974) suggested that this discrepancy was due to the fact that the wild-born females were all older than the captive-born females, but he was unable to assess the effects of birthplace, age, and rank independently. Shortly afterwards, Sade et al. (1976) reported that high-ranking lineages grew at faster rates than low-ranking ones, providing further evidence that dominance rank influenced females' reproductive performance. We were encouraged by these results because they showed that there was substantial variation in female reproductive success and suggested that female dominance rank contributed to that variation.

## Sources of Variation in Female Reproductive Success

All other things being equal, a reproductively successful female will be one who has a long reproductive career, produces infants at regular intervals, and rears her infants successfully. Variation in any of these parameters can create variation in females' reproductive success.

Females' reproductive careers began when they conceived their first infant at about 3 1/2 years of age (Silk 1988a; Silk et al. 1981a). Most females gave birth each year, although the length of the interval between births (interbirth intervals) depended on their age and the outcome of their pregnancies. Interbirth intervals following surviving infants lasted, on average, 15 months, while intervals following nonsurviving infants lasted 13 months on average (Silk 1988a; Silk et al. 1981a). This difference in the length of IBI following surviving and nonsurviving infants reflects the energetic costs of rearing infants. These costs appear to be most pronounced for young females (Wilson et al. 1983). Females whose firstborn infant survived had IBIs that lasted on average 21 months, while females whose firstborn infant did not survive had IBIs that lasted 14 months on average. Fertility was generally highest among middle-aged females, but as females aged their fertility declined (Caro et al. 1995). Our analysis revealed little systematic variation in age at first birth, interbirth intervals, or age-specific fertility rates (Silk 1988a; Silk et al. 1981a). We had to look elsewhere to explain the observed variation in lineage size.

We discovered that females varied considerably in their success in raising offspring. About one-half of all infants born in the group survived to the age of 6 months (Silk 1988a; Silk et al. 1981a). Mothers' rank was directly linked to the likelihood that their infants would survive. Sixty-two percent of the offspring of high-ranking females survived their first 6 months of life, while only 38 percent of the offspring of low-ranking females survived to this age (Silk 1988a). Differences in infant survivorship to 6 months were also reflected in differences in survival to reproductive age. Forty-six percent of the offspring of high-ranking females survived to the age of 4 years, while only 27 percent of the offspring of low-ranking females survived to this age (Silk 1988a). Moreover, rank-related variation in reproductive success was perpetuated in the next generation. The daughters of high-ranking females were more likely to reproduce successfully than the surviving daughters of low-ranking females (Silk 1988a).

## Adaptive Significance of the Relationship between Female Dominance Rank and Reproductive Success

We have subsequently discovered that dominance rank is related to female reproductive success in many primate species (reviewed by Harcourt 1987; Silk 1987, 1993). This does not mean that rank is positively correlated with all components of reproductive success in every group (e.g., Altmann et al. 1988; Gouzoules et al. 1982), or that statistically significant correlations between dominance rank and reproductive success are established in every study (e.g., Cheney et al. 1988; Packer et al. 1995). However, there is no evidence that *low* rank ever confers unambiguous advantages on females.

For macaques, the adaptive significance of the relationship between dominance rank and reproductive success is compounded by the relationship between female dominance rank and matrilineal kinship. Juvenile macaques typically acquire ranks just below their mothers, and females maintain these ranks when they mature (reviewed by Chapais 1992). The inheritance of maternal rank creates female dominance hierarchies in which members of the same matriline occupy adjacent ranks and all members of a given matriline rank above or below all the members of other matrilines. Matrilineal dominance hierarchies have now been documented in at least seven species of macaques (Chapais 1992), including the bonnet macaques in Davis (Silk et al. 1981c). Matrilineal dominance hierarchies are also observed among savannah baboons and vervet monkeys (Chapais 1992).

The acquisition of maternal dominance rank has important reproductive consequences for females because matrilineal dominance hierarchies tend to be very stable over time (Hausfater et al. 1982; Isbell & Pruetz 1998; Kawai 1958; Lee 1983; Sade 1967). In the bonnet group, the dominance hierarchy remained relatively stable for over a decade. This means that a female who is born into low-ranking lineage is likely to become low-ranking and reproduce unsuccessfully for most of her life, while a female born into a high-ranking lineage is apt to become high-ranking and raise many healthy offspring.

The formation of matrilineal dominance relationships is generally thought to be the product of kin selection (Hamilton 1964), which favors altruistic behavior toward kin. Altruistic acts, which are defined as acts that increase the genetic fitness of the recipient at some cost to the actor, are not favored by natural selection because they reduce the fitness of the actor. However, altruism can evolve via kin selection. The general logic of kin selection is based on the fact that individuals who behave altruistically to their relatives have some chance of conferring benefits on individuals who carry copies of their own genes which they have acquired through descent from a common ancestor. Hamilton showed that altruism can evolve when the benefits to the recipient (b) devalued by the degree of relatedness between the actor and the recipient (c) exceed the costs to the actor (c), or $br > c$. This has come to be known as Hamilton's Rule.

Macaques behave as if they had studied Hamilton's Rule themselves. For example, they spend much of their time in close proximity to their relatives, devote more time to grooming kin than nonkin, and are more tolerant of kin than nonkin when they are feeding (reviewed by Dugatkin 1997; Gouzoules & Gouzoules 1987). Females also support their offspring and other close kin when they are involved in agonistic encounters. Both naturalistic and experimental studies have demonstrated that support from kin is largely responsible for the acquisition of maternal rank and the formation of matrilineal dominance hierarchies (reviewed by Chapais 1992).

## Local Resource Competition and Harassment of Females

In nature, the growth of primate populations is generally limited by the availability of resources in the local area, or local resource competition

(LRC). In matrilineal species, including macaques, baboons, and vervets, males are the dispersing sex, while females are philopatric (Pusey & Packer 1987). All other things being equal, LRC is likely to have a bigger impact on the fitness of philopatric females than dispersing males. Thus, natural selection may favor behaviors that enable females to minimize the amount of competition that they and their daughters will encounter in the future (Silk 1983).

One way for females to limit the extent of local resource competition is to reduce the number of unrelated infants born and raised in their groups. There is good evidence that bonnet females do just that. Infants and juveniles were sometimes threatened, chased, and attacked by adult females. Not all infants were equally vulnerable to harassment. The offspring of low-ranking females were harassed at higher rates than the offspring of high-ranking females, and rates of aggression toward immature females were generally higher than rates of aggression toward immature males (Silk et al. 1981b; Fig. 21.2). This aggression was sometimes quite severe, and a number of infants suffered serious injuries. The pattern of injuries reflected the pattern of harassment: offspring of low-ranking females were more likely to be injured than the offspring of high-ranking females, and immature females were more likely to be injured than immature males (Silk 1991; Silk et al. 1981b).

Infant survivorship was a function of both maternal rank and infant sex. While maternal rank influenced the survival of both male and female offspring, the effects of maternal rank were much more pronounced for daughters than for sons (Fig. 21.3). Differences in survivorship to one month became even more exaggerated by the time that individual reached two years of age. At that point, daughters of low-ranking females were particularly disadvantaged. Adult females were at least partly responsible for the high rates of mortality among the daughters of low-ranking females.

The selective harassment of immature females that we observed in the bonnet group may be a form of reproductive competition that results from local resource competition. This idea is supported by the fact that data from other studies also indicate that juvenile females are harassed at higher rates than males (Dittus 1977, 1979, 1980; Maestripieri 1994; Pereira 1988; Simpson & Simpson 1985). Comparative analyses indicate that in female-bonded species, like macaques, baboons, and vervets, juvenile females suffer higher mortality than males do when food is limited (van Schaik & de Visser 1990). This suggests that local resource competition favors behaviors which reduce the viability of young females, and limits the recruitment of potential competitors.

## What's a Mother to Do?

In the bonnet group, daughters of low-ranking females faced daunting odds: they were most likely to be harassed, most likely to be injured, and least

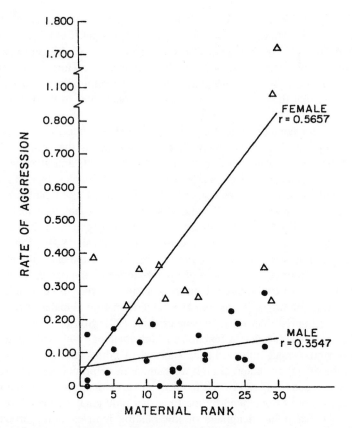

Figure 21.2. Juvenile females received more aggression from unrelated females than did juvenile males, when maternal rank was held constant. These two slopes are both significantly different from zero and from each other.

likely to survive to adulthood. If they did survive to reproductive age, they were apt to inherit their mother's low rank and reproduce unsuccessfully themselves.

The dismal prospects for the daughters of low-ranking females may be linked to skews in the secondary sex ratio. Over the years, more males than females were born in the bonnet group (1.29 males per female, n = 245; Silk 1988a). Low-ranking females were largely responsible for this imbalance. While high-ranking females produced approximately equal numbers of male and female offspring (0.98 males per female), low-ranking females produced significantly more males than females (1.67 males per female).

It seemed plausible that the relationship between offspring sex and maternal rank was linked to the fact that the benefits derived from producing male and female offspring differ for high- and low-ranking females. High-ranking females' daughters are likely to become high-ranking and reproductively

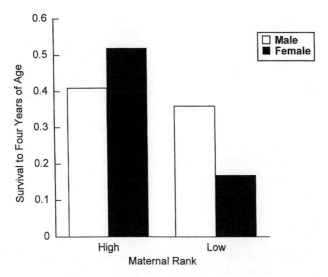

Figure 21.3. Survivorship to one month and two years is a function of maternal rank and sex. The daughters of low-ranking females were less likely to survive than the sons of low-ranking females or the offspring of high-ranking females.

successful, while their sons disperse and meet uncertain fates in non-natal groups. Low-ranking females' daughters are unlikely to survive to adulthood, and if they do, they are likely to become low-ranking and reproduce unsuccessfully. Their sons will disperse and may have a greater chance of reproducing successfully than their daughters. Thus, low-ranking females who invest more heavily in sons than daughters might achieve higher fitness than low-ranking females who produce equal numbers of sons and daughters (Altmann et al. 1988; Gomendio et al. 1990; Silk 1983; van Schaik & Hrdy 1991).

Although this hypothesis made sense from what we know about matrilineal primate species, it did not fit the conventional understanding of how natural selection shapes sex ratios. R. A. Fisher (1930) showed that sex ratios will be balanced at equilibrium because every infant has two parents. When males are less common than females, the average fitness of males in the population will be higher than the average fitness of females, and vice versa. Natural selection will always favor parents who bias investment in offspring of the less common sex (because they will on average have higher fitness than members of the more common sex), leading to balanced sex ratios at equilibrium. Thus, there was a general consensus that sex ratios would hover around unity.

However, as we published our first analyses of the relationship between maternal rank and infant sex, the consensus was being eroded by new theoretical developments. Trivers and Willard (1973) imagined a situation in which offspring of one sex benefited more from maternal investment than the other sex, and females varied in their ability to invest in their offspring. In this situation, they argued, females in good condition would invest more heavily in the sex that benefited most from that investment, while females in poor condition would invest more heavily in the other sex. Thus, natural selection would favor the ability to adjust offspring sex ratios in relation to maternal condition. Trivers and Willard originally imagined that males would be the sex that benefited most from additional investment, but in matrilineal primate societies the opposite pattern might hold because high-ranking females can have a greater impact on the fitness of their daughters than of their sons. Trivers and Willard's model is faithful to Fisher's basic logic because sex ratios remain balanced within the population.

Skewed sex ratios may evolve if the population is subdivided into discrete groups (Hamilton 1967; Silk 1984; Wilson & Colwell 1981), a factor that Fisher did not consider. In these models, there is a balance between natural selection operating within groups and natural selection operating between groups. Forces operating within groups may favor biased sex ratios. Thus, in some haplodiploid insects, females lay all of their eggs in one spot, and offspring compete for mates with their siblings. When there is local mate competition (LMC), females benefit by producing just enough male off-spring to fertilize each of their female offspring (Hamilton 1967). In this situation, selection within local groups favors balanced sex ratios because males have higher fitness than females. However, selection between groups favors biased sex ratios because females who produce few sons have many daughters who carry the sex-ratio-biasing allele and will contribute to the global population.

Using the same logic, Clark (1978) argued that when there is local re-source competition, females might benefit from producing a surfeit of sons who will disperse to form new groups and limiting production of daughters who will remain in the natal area and compete for local resources. We can take this argument one step further to consider what happens when the competitive abilities of females vary, and some females are able to protect their daughters better than others (Silk 1983, 1984). For females whose daughters are most vulnerable to harassment, it will be very costly to rear female off-spring successfully. These females are likely to bias investment most heavily in favor of males. Under these conditions, natural selection will favor adaptations that enable females to adjust the sex ratio of their progeny in relation to their own dominance rank (Silk 1984).

I suggested that the interaction between maternal rank and infant sex that we observed in the bonnet group reflected the effects of local resource competition (Silk 1983). Similar patterns among free-ranging baboons in Amboseli, Kenya (Altmann 1980; Altmann et al. 1988) and captive rhesus macaques (Simpson & Simpson 1982; Gomendio et al. 1990) seemed consistent with this explanation. However, as more and more information about the relationship between maternal rank and infant sex became available, the picture became more complicated (reviewed by Clutton-Brock & Iason 1986; van Schaik & Hrdy 1991). Some of the new data fit the local resource competition hypothesis. Thus, in chimpanzees and spider monkeys, where males are the philopatric sex rather than females, high-ranking females invested more heavily in sons than daughters (Boesch 1997; Symington 1987). However, other data did not fit the local resource competition hypothesis. In some populations, high-ranking females produced more sons than daughters, the opposite of the pattern that we had seen among the bonnets (e.g., Meikle et al. 1984; Paul & Kuester 1987, 1988; van Schaik et al. 1989). In other populations, there was no consistent relationship between maternal rank and infant sex (e.g., Berman 1988; Small & Hrdy 1986).

This contradictory body of data produced considerable controversy among primatologists, and generated considerable skepticism about the adaptive manipulation of sex ratios among primates. This dispute generated more heat than light, until van Schaik and Hrdy (1991) found one way to make sense of the observed variation. They discovered that the extent of local resource competition, measured in terms of the population growth rate, was associated with sex ratio biases. When local resource competition is intense, and reproductive opportunities for females are limited, high-ranking females invest more heavily in daughters than low-ranking females do. When local resource competition is relaxed, high-ranking females shift investment more toward sons. This explanation fits the data reported in the literature, but it is not clear that it will hold up as more data on sex ratios are published and examined in this framework.

## The Environment of Evolutionary Adaptedness for Bonnet Macaques

I have argued that harassment of immature offspring of unrelated, lower-ranking females is a form of competition over reproductive opportunities. One way that females respond to these pressures is to adjust investment in male and female offspring. However, it may seem odd that captive monkeys, who have abundant food, are motivated to compete for reproductive opportunities. Indeed, the intensity of competition and the variation in reproductive success among females is more pronounced than we see in many free-ranging groups (Harcourt 1987). This raises important questions about how we can study evolutionary processes in captive environments.

We know that evolution provides animals with adaptations that enable them to survive in particular environments. When we remove animals from the environments in which they evolved, we are likely to disrupt the link between morphological or behavioral traits and their adaptive function. If we are interested in understanding the adaptive basis of behavior, this presents both a dilemma and an opportunity.

The dilemma arises because it is difficult, if not impossible, to understand the functional basis of behaviors without understanding the context in which these adaptations evolved. To understand why low-ranking female bonnet macaques skew their sex ratios in favor of males, we rely on our knowledge of their lives in the wild. Knowledge of the patterns of dispersal in the wild and the nature of competition within and between groups played a critical role in explaining these behavioral patterns. Thus, captive studies are most useful when they are grounded in good natural history.

If knowledge of natural history is needed to understand the behavior that we observe in captivity, then you might wonder why we bother to study animals in captivity at all. Why not concentrate our efforts on fieldwork? There are several reasonable answers to this question.

First, much of the behavior that we see in captive animals is very similar to what we see in the wild. Most of the behavioral patterns that we documented among the bonnets in Davis were consistent with what researchers working on other species in captivity and the wild had observed. Certain features of the behavior of female macaques, such as formation of linear, matrilineal dominance hierarchies, and nepotistic patterns of social interactions, are apparently unaffected by environmental details in the short term. This suggests that the behavior that we observe in captivity reflects an evolved predisposition to respond to certain conditions in certain ways. Sensitivity to changes in demographic parameters, such as group size or population density, may have been favored by natural selection if these variables provided an accurate measure of the intensity of present and/or future reproductive competition. In captive situations, the same evolved mechanism may shape behavior, even though food is abundant.

Second, when we study animals in captivity, we are able to control certain conditions and test functional hypotheses explicitly. Some behavioral ecologists take animals into the laboratory so that they can conduct carefully controlled experiments and test adaptive hypotheses. Several of the chapters in this book provide outstanding examples of this approach. Even though primatologists are rarely able to design experiments with the same precision and rigor as behavioral ecologists studying other taxa, researchers can study some phenomena experimentally. Thus, Bernard Chapais has been able to examine in detail how the presence of kin influences the acquisition and maintenance of dominance rank among macaque females (Chapais 1992). In the bonnet group, stochastic variation in the size and composition of the group from year to year enabled me to examine the influence of certain demographic variables on female reproductive success. I discovered that female fertility was highest when there relatively few adult females present, and infant survivorship was highest when the size of the annual birth cohorts was smallest (Silk 1988b; Fig. 21.4).

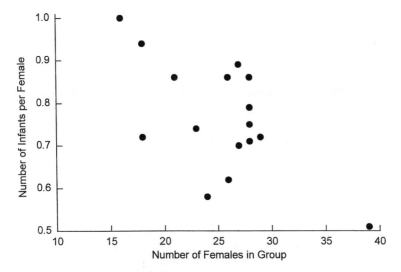

Figure 21.4. During years in which there were many adult females in the group, the proportion of females who produced infants declined. In general, female fertility was correlated with the number of adult females in the group, but not with the overall size of the group. (Adapted from Table III in Silk 1988b)

The third reason to study captive animals is that free-ranging animals are not always available or accessible. The bonnets, for example, are residents of southern India, a region that was largely off limits to American primatologists for many years. A number of researchers have been forced to abandon field projects when their safety was threatened by guerilla attacks or gun battles. Some primate species, like the bonobo, live in politically volatile countries, making fieldwork quite dangerous. It is also important to remember that many species are endangered in the wild. Primates are particularly vulnerable because most species live in tropical forests that are being destroyed at an alarming rate (Harcourt 1998). Many primate species now number less than a thousand individuals in the wild. For some species, captive studies may provide our only source of knowledge about behavior. (For a fuller discussion of how behavioral ecology research is relevant to conservation biology, see Caro 1998.)

## Looking into the Future

A few years ago, the bonnet group in Davis was disbanded, a casualty of changing funding priorities and new management policies. Perhaps Seeley should add another criteria for model systems in behavioral ecology: systems that can be maintained at minimal cost or reconstituted from scratch when funding is interrupted.

When the bonnet project ended, I shifted my empirical focus to another species on another continent—baboons in Africa. This is, of course, another system that falls far short of Seeley's criteria for a model system in behavioral ecology. However, it is a system that I know well, because there are many similarities in the social organization and behavior of baboons and macaques. Thus, many of the questions that we asked among bonnets can be transferred directly to baboons. Part of the great satisfaction of watching animals with complex social lives is that new questions arise more quickly than we can design studies to answer them. But if we watch carefully, the animals will lead us to the answers.

## Acknowledgments

My graduate advisor, Peter Rodman, invited me to join ongoing research on the bonnet group at Davis, and I am grateful to him for providing theoretical and empirical training that became the foundation of my research. Amy Samuels spent hundreds of hours observing the bonnets with me. Her superb observational skills, extraordinary patience, and deep affection for the bonnets made long hours outside the bonnet cage both productive and entertaining. Many people have contributed to my understanding of bonnet females' lives—too many to acknowledge properly. But special thanks should go to Jeanne Altmann, Dorothy Cheney, Lynn Fairbanks, Sarah Blaffer Hrdy, and Robert Seyfarth for their mentorship, feedback, and support. The bonnet group was supported by the California Regional Primate Research Center base grant NIH RR00169. I am grateful to the administration and animal care staff of the CRPCR for their assistance and support. My work was supported by grants from the Harry Frank Guggenheim Foundation, Wenner Gren Foundation, Sigma Xi, UCLA Academic Senate, and the National Science Foundation.

## References

Altmann J, 1980. Baboon mothers and infants. Cambridge, MA: Harvard University Press.

Altmann J, Hausfater G, Altmann SA, 1988. Determinants of reproductive success in savannah baboons, *Papio cynocephalus*. In: Reproductive success (Clutton-Brock TH, ed). Chicago: University of Chicago Press; 403–418.

Bateman AJ, 1948. Intra-sexual selection in *Drosophila*. Heredity 2:349–368.

Berman CM, 1988. Maternal condition and offspring sex ratios in a group of free-ranging rhesus monkeys: an eleven-year study. Am Nat 131:307–328.

Boesch C, 1997. Evidence for dominant wild female chimpanzees investing more in sons. Anim Behav 54:811–815.

Caro TM (ed), 1998. Behavioral ecology *and* conservation biology. Oxford: Oxford University Press.

Caro TM, Sellen DW, Parish A, Frank R, Brown DM, Voland E, Borgerhoff Mulder M, 1995. Termination of reproduction in nonhuman and human female primates. Int J Primatol 16:205–220.

Chapais B, 1992. The role of alliances in social inheritance of rank among female primates. In: Coalitions and alliances in humans and other animals (Harcourt SA, de Waal FBM, eds). Oxford: Oxford Science Publications; 29–59.

Cheney DL, Seyfarth RM, Andelman SJ, Lee PC, 1988. Reproductive success in vervet monkeys. In: Reproductive success (Clutton-Brock TH, ed). Chicago: University of Chicago Press; 384–402.

Cheney DL, Seyfarth RM, Lee PC, 1981. Behavioral correlates of nonrandom mortality among free-ranging female vervet monkeys. Behav Ecol Sociobiol 9:153–161.

Clark AB, 1978. Sex ratio and local resource competition in a prosimian primate. Science 201:163–165.

Clutton-Brock TH, Iason GR, 1986. Sex ratio variation in mammals. Q Rev Biol 61:339–374.

Dittus WPJ, 1977. The social regulation of population density and age-sex distribution in the toque monkey. Behaviour 63:281–322.

Dittus WPJ, 1979. The evolution of behaviors regulating density and age-specific sex ratios in a primate population. Behaviour 69:281–322.

Dittus WPJ, 1986. Sex differences in fitness following a group take-over among toque macaques: testing models of social evolution. Behav Ecol Sociobiol 19:257–266.

Drickamer LC, 1974. A ten-year summary of reproductive data for free-ranging *Macaca mulatta*. Folia Primatol 2:61–80.

Dugatkin LA, 1997. Cooperation among animals. Oxford: Oxford University Press.

Fairbanks LA, McGuire MT, 1995. Maternal condition and the quality of maternal care in vervet monkeys. Behaviour 132:733–754.

Fisher RA, 1930. The genetical theory of natural selection. London: Dover.

Gomendio M, Clutton-Brock TH, Albon SD, Guinness FE, Simpson MJ, 1990. Mammalian sex ratios and variation in costs of rearing sons and daughters. Nature 343:261–263.

Gouzoules S, Gouzoules H, 1987. Kinship. In: Primate societies (Smuts BB, Cheney DL, Seyfarth RM, Wrangham RW, Struhsake, TT, eds). Chicago: University of Chicago Press; 299–305.

Gouzoules H, Gouzoules S, Fedigan L, 1982. Behavioural dominance and reproductive success in female Japanese monkeys (*Macaca fuscata*). Anim Behav 30:1138–1151.

Hamilton WD, 1964. The genetical evolution of social behavior. J Theor Biol 7:1–51.

Hamilton WD, 1967. Extraordinary sex ratios. Science 156:477–488.

Harcourt AH, 1987. Dominance and fertility among female primates. J Zool London 213:471–487.

Harcourt AH, 1998. Ecological indicators of risk for primates, as judged by species' susceptibility to logging. In: Behavioral ecology *and* conservation biology. Oxford: Oxford University Press; 56–79.

Hausfater G, Altmann J, Altmann SA, 1982. Long-term consistency of dominance relations among female baboons (*Papio cynocephalus*). Science 217:752–755.

Isbell L,1991. Contest and scramble competition: patterns of female aggression and ranging behavior among primates. Behav Ecol 2:143–55.

Isbell LA, Pruetz JD, 1998. Differences between vervets (*Cercopithecus aethiops*) and patas monkeys (*Erythrocebus patas*) in agonistic interactions between adult females. Int J Primatol 19:837–856.

Kawai M, 1958. On the system of social ranks in a natural troop of Japanese monkeys. I. Basic rank and dependent rank. Primates 1:111–130.

Lee PC, 1983. Context-specific unpredictability in dominance interactions. In: Primate social relationships: an integrated approach (Hinde RA, ed). Sunderland, MA: Sinauer Associates; 35–44.

Maestripieri D, 1994. Influence of infants on female social relationships in monkeys. Folia Primatol 63:192–202.

Meikle DB, Tilford BL, Vessey SH 1984. Dominance rank, secondary sex ratio, and reproduction of offspring in polygynous primates. Am Nat 124:173–188.

Packer C, Collins DA, Sindimwo A, Goodall J, 1995. Reproductive constraints on aggressive competition in female baboons. Nature 373:60–63.

Paul A, Kuester J, 1987. Dominance, kinship, and reproductive value in female Barbary macaques (*Macaca sylvanus*) at Affenberg, Salem. Behav Ecol Sociobiol 21:323–331.

Paul A, Kuester J, 1988. Life history patterns of barbary macaques (*Macaca sylvanus*) at Affenberg Salem. In: Ecology and behavior of food-enhanced primate groups (Fa JE, ed). New York: Alan R. Liss; 199–228.

Pereira ME, 1988. Agonistic interactions of juvenile savanna baboons. I. Fundamental features. Ethology 79:195–217.

Pusey AE, Packer C, 1987. Dispersal and philopatry. In: Primate societies (Smuts BB, Cheney DL, Seyfarth RM, Wrangham RW, Struhsaker TT, eds). Chicago: University of Chicago Press; 250–266.

Sade DS, 1967. Determinants of dominance in a group of free-ranging rhesus monkeys. In: Social communication in primates (Altmann SA, ed). Chicago: University of Chicago Press; 99–114.

Sade DS, Cushing K, Cushing G, Dunaif J, Figueroa A, Kaplan JR, Lauer C, Rhodes D, Schneider J, 1976. Population dynamics in relation to social structure on Cayo Santiago. Yrbk Phys Anthropol 20:253–262.

Silk JB, 1983. Local resource competition and facultative adjustment of sex ratios in relation to competitive ability. Am Nat 121:56–66.

Silk JB, 1984. Local resource competition and the evolution of male-biased sex ratios. J Theoret Biol 108:203–213.

Silk JB, 1987. Social behavior in evolutionary perspective. In: Primate societies (Smuts BB, Cheney DL, Seyfarth RM, Wrangham RW, Struhsaker TT, eds). Chicago: University of Chicago Press; 318–329.

Silk JB, 1988a. Maternal investment in captive bonnet macaques (*Macaca radiata*). Amer Nat 132:1–19.

Silk JB, 1988b. Social mechanisms of population regulation in a captive group of bonnet macaques (*Macaca radiata*). Am J Primatol 14:111–124.

Silk JB, 1991. Mother-infant relationships in bonnet macaques: sources of variation in proximity. Int J Primatol 12:21–38.

Silk JB, 1993. The evolution of social conflict among primate females. In: Primate social conflict (Mason WA, Mendoza S, eds). Albany: SUNY Press; 49–83.

Silk JB, Clark-Wheatley CB, Rodman PS, Samuels A, 1981a. Differential reproductive success and facultative adjustment of sex ratios among captive female bonnet macaques (*Macaca radiata*). Anim Behav 29:1106–1120.

Silk JB, Samuels A, Rodman PS, 1981b. The influence of kinship, rank, and sex upon affiliation and aggression among adult females and immature bonnet macaques (*Macaca radiata*). Behaviour 78:112–137.

Silk JB, Samuels A, Rodman PS, 1981c. Hierarchical organization of female *Macaca radiata*. Primates 22:84–95.

Simpson MJA, Simpson AE, 1982. Birth sex ratios and social rank in rhesus monkey mothers. Nature 300:440–441.

Simpson AE, Simpson MJA, 1985. Short-term consequences of different breeding histories for captive rhesus macaque mothers and young. Behav Ecol Sociobiol 18:83–89.

Small MF, Hrdy SB, 1986. Secondary sex ratios by maternal rank, parity, and age in captive rhesus macaques (*Macaca mulatta*). Am J Primatol 11:359–365.

Small MF, Rodman PS, 1981. Primigravidity and infant loss in bonnet macaques. J Med Primatol 10:164–169.

Soltis J, Mitsunaga F, Shimuzu K, Yanagihara Y, Nozaki M, 1997a. Sexual selection in Japanese macaques I: female mate choice or male sexual coercion? Anim Behav 54:725–736.

Soltis J, Mitsunaga F, Shimuzu K, Nozaki M, Yanagihara Y, Domingo-Roura X, Takenaka O, 1997b. Sexual selection in Japanese macaques II: female mate choice and male-male competition. Anim Behav 54:737–746.

Symington MM, 1987. Sex ratio and maternal rank in wild spider monkeys: when daughters disperse. Behav Ecol Sociobiol 20:421–425.

Trivers RL, Willard DE, 1973. Natural selection of parental ability to vary the sex ratio of offspring. Science 179:90–92.

van Hooff JARAM, van Schaik CP, 1992. Cooperation in competition: the ecology of primate bonds. In: Coalitions and alliances in humans and other animals (Harcourt AH, de Waal FBM, eds). Oxford: Oxford University Press; 357–390.

van Schaik, CP, 1989. The ecology of social relationships amongst female primates. In: Comparative socioecology, the behavioural ecology of humans and other mammals (Standen V, Foley RA, eds). Oxford: Blackwell; 195–218.

van Schaik CP, de Visser JAGM, 1990. Fragile sons or harassed daughters? Sex differences in mortality among juvenile primates. Folia Primatol 55:10–23.

van Schaik CP, Hrdy SB, 1991. Intensity of local resource competition shapes the relationship between maternal rank and sex ratios at birth in cercopithecine primates. Am Nat 138:1555–1562.

van Schaik CP, Netto WJ, van Amerongen AJJ, Westland H, 1989. Social rank and sex ratio of captive long-tailed macaque females (*Macaca fasicularis*). Am J Primatol 19:147–161.

Whitten P, 1983. Diet and dominance among female vervet monkeys (*Cercopithecus aethiops*). Am J Primatol 5:139–159.

Wilson DS, Colwell RK, 1981. Evolution of sex ratio in structured demes. Evolution 35:882–897.

Wilson EO, 1975. Sociobiology: the new synthesis. Cambridge, MA: Harvard University Press.

Wilson ME, Walker ML, Gordon TP, 1983. Consequences of first pregnancy in rhesus monkeys. Amer J Phys Anthropol 61:103–111.

Wilson ME, Walker, ML, Pope NS, Gordon TP, 1988. Prolonged lactational infertility in adolescent rhesus monkeys. Biol Reprod 38:163–174.

Wrangham RW, 1980. An ecological model of female-bonded primate groups. Behaviour 75:262–300.

Wrangham RW, 1981. Drinking competition in vervet monkeys. Anim Behav 29:904–910.

# 22 Chimpanzee Hunters: Chaos or Cooperation in the Forest?

Christophe Boesch

## Field Notes

October 11, 1990: Late in the afternoon, a large party of twenty chimpanzees arrives under a large group of red colobus monkeys. The two brothers, Kendo and Fitz, have climbed toward the monkeys and push them quickly toward the east. Fitz keeps driving them in a long and quick move in the trees over 150 meters, while all the others follow the prey on the ground. Some of the males accelerate on the ground and Macho as well as Ulysse anticipate the escape movement of the prey by climbing in two different trees way ahead of Fitz. Their judgment was correct, as one of the adult monkeys arrives in a tree at the same time as the three chimpanzee males coming from different directions and he is quickly captured by Fitz. Kendo, the alpha male, and Rousseau and Darwin, watching this from the ground where they were following the progression of the hunt, make loud screams. The females farther away understand immediately that this means "capture" and with excited calls they join the males.

The simultaneous presence of six adult males and eight adult females at the kill site is accompanied by a tremendous pandemonium around the dead colobus, that meanwhile has been brought to the ground. Fitz, not yet fully adult, has lost the prey to Macho and Ulysse, and his older brother Kendo tries hard to steal it from them. Rousseau allies with the two brothers to displace Ulysse and Macho from the prey, but the higher-ranking females, as so often, support the best hunters and make sure the hunters gain access to the meat despite the regular interruption they have to make to respond to the constant charges of Kendo and Rousseau. Ulysse at the time by far the best hunter, judging from his unfailing participation in all hunts, but only middle-ranking, is granted free access to the prey that is kept under perfect control by Ondine, the highest-ranking female. Macho, the beta male, has to remain at some distance from the meat but still manages to gather some pieces. He is actively replying to all attacks made by Kendo, who jumps several times right over the scrum without ever gaining any piece of meat. It is not until the situation quiets down some thirty minutes later, as the prey is divided in two by Ulysse and the sharing clusters are formed, that Kendo obtains his first pieces of meat.

During this wild scrum I wondered whether I was about to gain some true insight into another animal species. What some time ago had seemed like sheer chaos, seemed to be indeed well organized and to follow some precise pattern. I just had to "see" it. While the night fell in and as I still tried to identify the black shadows eating meat in the dark, I remembered Konrad Lorenz's *Conversations with Birds, Fishes and Mammals*. I read this book as a teenager and I recall very well that I said to myself: "Now, that is exactly what I would like to do with my life!" And so my life in the rainforest.

When we first saw a chimpanzee hunt for monkeys in the early days of the study, we were impressed by the prevailing confusion. First, a sense of excitement dominated the whole scene as it was so obvious that all the hunters had their attention totally focused toward capturing one of the elusive and quickly moving monkeys in the trees above. Second, we also felt there was a sort of chaos, with the hunters moving in all directions, as well as the prey, and it was difficult to see any organization. This feeling was even stronger when they succeeded in capturing a monkey, as the forest then filled with loud and high-pitched, totally uninhibited screams, the chimpanzees more excited than before, running around, attacking one another and embracing at the same time, some of them with blood from the prey on their faces. How could one conceivably make sense of such a wild scrum? Other chimpanzee observers had the same feelings and data on the amount of meat eaten by group members was badly missing (Teleki 1973; Busse 1978; Nishida et al. 1983).

During the ten years prior to such observations, we had been following this chimpanzee population constantly and progressively gained the feeling that their life was in fact very organized and planned. How then could such an important event as meat eating be chaotic? We set out to address this question and to examine whether we were simply unable to see the organization in hunts.

After having read Konrad Lorenz's book, I decided to study biology and was lucky enough to be able to realize a second dream that I had after reading George Schaller's book *One Year with the Mountain Gorillas*. This account of the mountain gorillas impressed me so much that I wanted to observe them. In 1973, I was able to undertake my master thesis on the mountain gorillas in Rwanda under the supervision of Diane Fossey. This first field experience was very important to me because it gave me a precise idea of what fieldwork means. On one hand it can be hard to work and live under primitive conditions in remote areas, but on the other hand the fact of being able to be in a real pristine forest with such wonderful great apes is simply an unbelievable privilege. The first of these is often underestimated but it represents a dramatic change one has to adapt to. The balance between the two was positive for me and I decided with my wife, Hedwige, to find a new study site to start a project on wild chimpanzees in a habitat where they have not been studied before. Why chimpanzees? Because many observations were available at the time from two different populations reporting how variable the chimpanzee behavior was and how important it is to increase our knowledge about this species (Goodall 1968, 1970; Nishida 1979, 1990).

Being human, we are all selfishly interested in understanding more about ourselves and what really differentiates us from other animal species. We were animal lovers, and I strongly felt that the scientific knowledge and the image in our society of animals was still a very limited and sketchy representation of the complexity and sophistication of the animal world. This was especially apparent in the prevailing scenarios of human evolution presented in the early 1980s, where such behaviors like the ability to use and make tools, to cooperate in large groups, to make war, to share food actively, to possess the competence to use elaborate mental maps, and to have a concept of self or to plan actions were considered unique human abilities. Chimpanzees, as our closest living relatives, could possess some of the most elaborate abilities that animals develop in order to survive and reproduce in the wild, and we wanted to study them. One of the most popular theories of human evolution gives a central role to the ability of hunting (Dart 1925; Leakey 1961). According to the "Man the Hunter" scenario, our ancestors were forced to hunt to survive in the new dry environment of East Africa some 3 to 4 million years ago and by doing so acquired some of the most typical human abilities (Washburn & Lancaster 1968). In all subsequent scenarios of human evolution, hunting remains very important mainly because of all primates, humans are the only ones in which all known populations hunt regularly. The hunting abilities of chimpanzees, known since the early 1960s (Goodall 1963), presented a real challenge and one that could potentially provide some answers to questions about the complexity and the extent of their hunting abilities.

## The Question of Cooperation

From our observations such as the one on 11 October 1990, chimpanzees in the Taï forest can hunt in groups and collaborate to capture a prey in the trees. Once the capture is done, some meat-sharing rules seem to be enforced so that some of the male hunters obtain meat. The modern theory of evolution is based on the premise that individuals are egoists and perform strategies so as to increase their own lifetime reproductive success. But we saw that the Taï chimpanzees were cooperating. How can we explain such an observation? Are individual chimpanzees not egoists but rather altruists?

The evolution of cooperation and altruism has been for more than three decades a major research area, because if it is true that individuals are not acting in a selfish way then this contradicts one of the basic premises of the evolution theory. Some in the human sciences suspect that cooperation in humans is understandable in our own species as, the argument goes, we have succeeded in freeing ourselves from the influence of natural selection and are able to acquire behaviors that have no obvious benefit for the individual. Biologists, on the other side, wanted to study to what extent the evolution of such behaviors can be explained at an individual level. To understand the evolution of cooperation and altruism, a number of mathematical models have flourished and some major advances in modeling animal behavior have been made. Game theory models have been introduced in the study of animal behavior (Axelrod & Hamilton 1981; Maynard Smith 1982) and had a major and positive effect on our present approach to animal behavior.

The first models showed that cooperation by selfish individuals was possible under some limited conditions. Since these first models, many more have been presented and it would be much too tedious to review all of them (see review in Dugatkin 1997; Harcourt & Waal 1992). The main lessons gained from them are that cooperation may evolve under three conditions: kin selection, reciprocity, and mutualism. Under the kin selection argument, related individuals would accept that kin receive a share from their own part, since by doing so they are helping carriers of the same genes. Thus, the more closely related one is to another one, the more willing one should be to share or cooperate.

Under the reciprocity argument, unrelated individuals should share with each other on the condition that at a future occasion the situation is reversed and the givers will receive a share from the individuals with whom they shared before (Trivers 1971; Axelrod & Hamilton 1981). This situation has attracted special attention, because individuals are apparently acting altruistically when taking all instances separately and it is only through the repetition of interactions that they all benefit. Two problems are associated with this special case. Reciprocity is dependent both on individuals being able to recognize their partners and on a certain likelihood that two individuals will meet again. These proved to be major hurdles since the more potential cooperators, the more difficult it is to keep track of the interactions they have with each other, and the more difficult it is to judge if an additional encounter is likely to occur.

Under the mutualism argument, individuals directly profit from the act of cooperating, because they achieve a higher benefit than they would if they were acting alone (Brown 1983). This makes the evolution of cooperation easy to understand, and has therefore attracted much less interest from theoreticians than the reciprocity argument. To understand the place of cooperation in nature, it should, however, not be neglected.

Under both the reciprocity and mutualism arguments, the mere presence of cheaters threatens cooperation. This proved to be the major problem for the stability of cooperation and much theoretical work has been devoted to understanding how cooperation could persist in face of individual cheaters. Basically, the more individuals that are cooperating, the more they are at risk from cheaters, and the less likely it is that cooperation will persist. This has proved so problematic that it has been proposed that for cooperation to evolve interactions have to start between close kin before cooperators could outcompete a population of cheaters and be able to resist the appearance of new cheaters (Nowak & Sigmund 1992; Dugatkin 1997).

The results of these models are worrisome since cooperation seems very unstable, and should therefore be unlikely to be present in nature. But cooperation has been observed in many animal species. The common characteristic of game theoretical models is that they seek to explain the evolution of populations containing only cooperating individuals, when each individual in the model follows the same strategy all the time, and where the benefit of each strategy in the game is fixed and stable all the time. Some biologists have then started to wonder if the emphasis in this field has been properly based, as mutualism might be more appropriate to explain cooperation in animals, and the importance attached to reciprocity might not reflect its role in nature (Clements & Stephens 1995). In addition, observations about hunting in social carnivores and chimpanzees show that cooperation can involve many individuals, a condition under which cooperation was unlikely to evolve in most reciprocity models.

The evolution of cooperation is not only interesting to understand on its own, but cooperation has also been proposed to be one of the factors that led to the evolution of sociality in the first place, at least for hunting animal species (Caraco & Wolf 1975; Pulliam & Caraco 1984). The evolution of sociality has long been the subject of discussion because it is easy to observe the costs animals face when living together, for example, higher competition for food and other resources and higher risks of transmission of diseases. But the benefit of living in groups has been much more difficult to demonstrate and is still a source of debate between biologists. Initially, the analysis of the hunting behavior of different social carnivores concluded that they benefit from group hunting (Schaller 1972; Kruuk 1972) and, thus, cooperation as a factor favoring sociality was a viable explanation. However, more recent analyses have raised serious concerns regarding this issue, arguing that with more detailed observations we find that group hunting does not directly benefit the individual hunters. As such we should look at cooperation as being a by-product of individuals living in groups for other reasons (Packer & Ruttan 1988; Busse 1978; but see Creel 1997).

The situation with the Taï chimpanzees could be quite instructive in light of these theoretical questions, since on average four to five chimpanzees hunt together for their prey (the colobus monkey) and 42 percent of the individuals eating meat do not participate in the hunt. Thus, in Taï chimpanzees, cooperation between large groups and the constant presence of cheaters are the rule when hunting (Boesch & Boesch, in press; Boesch 1994). Under such a situation, theoretical models would find it difficult to predict a persistence of cooperation.

## Cooperation When Hunting among Taï Chimpanzees

Taï chimpanzees, like other chimpanzee populations, live in small parties of about eight individuals that constantly join with others and split again, producing a very fluid social structure called "fission-fusion." In such a community, many adult males live with an equal or larger number of females. Community size fluctuates from thirty to eighty individuals and individuals live in a territory of about 20 km². Females have an infant every five years on average and roam widely through the territory with the males of the community to look for fruits and leaves that constitute the majority of their diet. On such forays, they regularly encounter arboreal monkeys that are very abundant in the Taï forest (Boesch & Boesch-Achermann 2000).

Meat is clearly a part of Taï chimpanzees' diet, providing 186 grams of meat per day on average for the males in this community (Boesch & Boesch-Achermann 2000). While meat can be an important source of food, it needs to be captured. All the monkey species hunted by the chimpanzees are arboreal, moving in the highest trees, and since they are about one-quarter the weight of an adult male chimpanzee, they can move and jump on much thinner branches than the chimpanzees. In addition, the Taï forest is a dense and intact forest with two continuous layers of trees, allowing the colobus to move in all directions without much hindrance. To solve this challenge, the Taï chimpanzees have specialized in group hunting: 84 percent of the 326 hunts we followed in detail between 1987 and 1995 were group hunts with two or more hunters acting at the same time against the same group of prey. In 77 percent of them the hunters were collaborating by adopting different complementary roles to capture the prey (Boesch & Boesch 1989, in press).

We discuss two questions: Is group hunting in Taï chimpanzees a cooperative action? Is cooperation stable, by assuring hunters more meat than cheaters?

To answer these questions we needed to observe the hunt closely enough to differentiate between individuals just looking at the hunt but eating some meat (the cheaters) from those actively taking part in it (the hunters). We also needed to be able to quantify the energy invested in hunting by measuring the time spent hunting by each participant, as well as evaluate the quantity of meat eaten by each individual present at the meat-eating site (Boesch 1994). This was possible only after habituation of the chimpanzees allowed us to move within the group of the hunters without affecting their behavior. This took seven years, during which we observed many hunts and had the opportunity to train ourselves in following group hunts taking place high in trees and over long distances. The Taï forest has the decisive advantage of being a primary forest in which the undergrowth is relatively open and where we could run whenever needed to follow the hunt. Such perfect observational conditions are rare for chimpanzee habitats and explains why similar data from other populations are still sparse.

## Benefits of Cooperation

Taï chimpanzee hunting success is related to the number of hunters (Fig. 22.1: $r_s = 0.78$, $p = 0.05$), and the increase in the hunting success of larger groups is more than additive (Chi-square of fit for 1 to 5 hunters, $X^2 = 9.95$, $p < 0.05$). This indicates a synergism between the individual hunters, which probably results from the better organization of the individuals when hunting in larger groups (frequency of collaborative hunts and hunting group size, $r_s = 0.99$, $N = 7$, $p < 0.02$). In addition, the meat return per hunt shows a positive correlation with increasing group size, with two maxima at 4 and at more than 6 hunters (Fig. 22.1: $r_s = 0.82$, $p < 0.05$). All measures of hunting success in Taï chimpanzees show an increase with larger hunting group sizes, favoring the evolution of cooperation. But as mentioned, this does not mean that hunters profit from cooperation, since the cheaters may obtain enough meat for the benefit of cooperation to disappear.

## Stability of Cooperation

Once cooperation has evolved, cheaters might invade the Taï chimpanzee community if they have the same access to meat as hunters. Does a social mechanism in Taï chimpanzees limit the success of cheaters? If it does, is it strong enough to make cooperation stable?

To test this, we compared the amount of meat eaten by hunters versus the type of cheaters present in Taï chimpanzees. We define two types of cheaters: a "bystander" as an individual present in the party during the hunt, but who does not take an active part in it, and a "latecomer" as an individual absent from the party that made the hunt, but that joins it after a capture has been achieved. If we were to consider only the amount of meat eaten by each individual, we would underestimate the true success of the individual, for part of the secured meat is shared with other group members. On average, hunters share 45 percent of their meat with other group members, while bystanders share only 15 percent. Thus, meat sharing is very important and not including it in this analysis penalizes hunters (if we would analyze the data by only counting the amount of meat eaten by an individual, the results would go in the same direction but no significant difference would be observed; see Boesch 1994). We compare the amount of meat secured, which includes the meat eaten by the owner, and the meat shared by individuals performing the 3 strategies observed at Taï. Being a hunter is a better strategy than bystander for groups of 3 and 4 hunters (Wilcoxon signed rank test

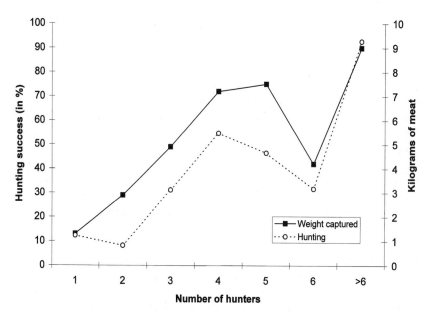

Figure 22.1. Hunting success and amount of meat secured by Taï chimpanzees hunting colobus monkeys.

for group size 3; T+ = 32, N = 8, p < 0.05, group size 4; T+ = 31, N = 8, p < 0.05) and better than latecomer for most groups larger than 2 hunters (Wilcoxon signed rank test for group size over 2: T+> = 30, N = 8, p < 0.05, except for 6 hunters where T+ = 29, p = 0.07) (see Boesch 1994 for details about energy estimations). Thus, a social mechanism regulating meat access to the individuals does exist, but it is a complex one, for it is the group members that allow an individual access to meat depending on its contribution during the hunt, and it is partly hidden by the fact that meat owners share large portions of their meat. This social mechanism makes cooperation stable in Taï adult male chimpanzees.

The puzzle in this system is that so many cheaters are present that do not invest any energy in the hunting and they, nevertheless, receive in most hunts a share of meat. Why do so many individuals cheat, and why are they so readily accepted by hunters? The answer is easy: they are accepted by the hunters because on other occasions they are hunters as well. All male chimpanzees use a mixed strategy, as they are sometimes hunters and at other times bystanders or latecomers (Fig. 22.3). In all cases, they have access to less meat when they do not hunt (Wilcoxon signed ranks test: Hunter versus Bystander; T+ = 34, N = 8, p < 0.02, Hunter versus Latecomer; T+ = 36, N = 8, p < 0.01). Thus, participation in the hunt is the prime factor affecting meat access for males. However, for a given strategy meat

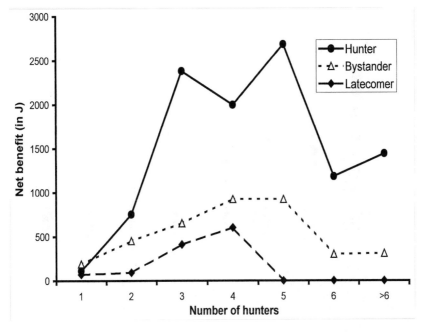

Figure 22.2. Net benefit of the different strategies for Taï chimpanzees when hunting colobus monkeys.

access is quite different for each male (Fig. 22.3), and at least one other factor must explain this. I consider three of them: hunting time, age, and dominance. The amount of meat eaten by each hunter shows a significant correlation with age and dominance. However, there is no correlation with hunting time. As the correlation between age and dominance is not significant, both age and dominance contribute to the meat access of hunters, but the data indicate that dominance is more important than age (see Boesch 1994): old dominant hunters gain more meat than young dominant hunters, which have more meat than old subdominant hunters. For the two other strategies, bystander and latecomer, neither age nor dominance correlates with the amount of meat eaten by the males.

In conclusion, group hunting in Taï chimpanzees represents a clear case of cooperation between hunters that increases their own benefit and where the stability of cooperation is guaranteed thanks to a social mechanism that limits the amount of meat that is eaten by cheaters. Mutualism is at work. We can exclude a kin selection argument at Taï, as an analysis of the genetic relationship among the males revealed that they are not more related one with another than the females, which transfer between communities, are among themselves (Gagneux et al. 1999).

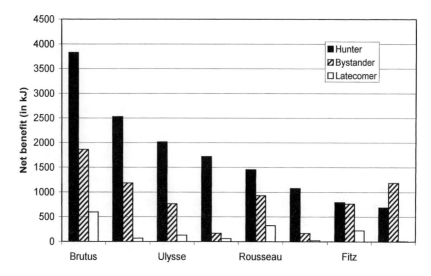

Figure 22.3. Benefit during the hunt of the three strategies at Taï for each individual male chimpanzee.

## Lessons from the Field Observations

These observations on the chimpanzees of the Taï forest have some direct implications for the theoretical debate about cooperation and its importance in the evolution of sociality in animals. The major implication is that cooperation can remain stable within a mixed population of cheaters and cooperators. Chimpanzee cooperation suggests that the goal of much of the theoretical work on cooperation, producing models that result in populations of pure cooperators, may be misplaced. The solution adopted by the Taï chimpanzees, which is based on mutualism and a different meat reward according to the contribution made by each individual during the hunt, shows a simple and efficient mechanism to stabilize cooperation (see Boesch & Boesch-Achermann 2000). This cooperation works even under the most difficult conditions—in large groups of unrelated individuals with cheaters always present and mostly successful. Social life for the chimpanzees in Taï is more complex than models assume.

The fact that cheaters and cooperators coexist in a social group shows that calculations of payoffs have to be flexible. In Taï chimpanzees, hunter success is a function of the number of other hunters acting in common, as well as of the number of cheaters obtaining meat. The payoff matrices used in all game theory models have fixed and rigid values that do not vary according to the number of participants, contrary to what we have seen in the Taï chimpanzees. This would make modeling more difficult but at the same time shows how theories may model more realistically what happens in an animal species. Varying payoffs depending on the conditions within the game have rarely been considered and would pave the way to modeling more natural situations and open new tracks to broader conditions for the analysis of evolution and stability of cooperation. At the same time, this implies that animals involved in such a cooperative task have to be able to make such flexible evaluations.

Many cheaters are females in Taï chimpanzees, and they hardly ever hunt. This seems possible for different reasons. Females actively help to enforce the meat-sharing rules during the wild scrum period following the capture of a large prey (as seen so clearly in the above 11 October 1990 observation), they are sexual partners for most males, and, most important, the dominant females have long-term relations with the males playing an active social role in the males' interactions (Boesch & Boesch, in press). At Taï, a type of cheater is tolerated, the adult females, despite the fact that some of them are not sexually active and do not present a direct benefit to the hunter. This stresses the fact that animals can exchange benefits of different natures and this complicates the discussion about the conditions for the evolution of cooperation.

## Evolution of Cooperation

Observations on the Taï chimpanzees and comparison with other chimpanzee populations where cooperation is rarer opens the way for understanding the factors that favor the evolution of cooperation among hunters. In a recent review of the evidence for the evolution of cooperation, including the observations of social carnivores, we showed that the harder the conditions under which the hunt is performed (less abundant prey species, easier escape routes for the prey), the more hunters tend to cooperate (Boesch & Boesch-Achermann 2000). For example, the Taï forest, with its continuous and dense forest cover where monkeys can easily escape, is a more difficult habitat for hunting than the Mahale or Gombe woodlands where chimpanzees can corner monkeys or hunt at closer range. Taï chimpanzees have been observed to hunt more in groups and more cooperatively than their East African counterparts. Similarly, lions in the Serengeti do not need to cooperate to be successful, while lions in Etosha, Botswana, cooperate regularly to hunt for less common and more difficult-to-capture prey. The ecology of the population under investigation clearly affects the importance attributed to cooperation in shaping the evolution of sociality. If group-living evolved first in a low-prey-density environment, than cooperation could have been the driving force behind it.

These remarks point to future priority that should be given to research in this field: How precise are the mechanisms controlling cheating? If a mechanism to control for obvious cheaters exists, some animals may be smart enough to pretend to hunt without really doing so, and by this try to circumvent some of the meat-sharing rules prevailing in their group. Such an approach leads to another question: What are the cognitive capacities required to implement such sophisticated tactics during hunting and meat sharing? Ideally, to answer these questions different populations, hunting prey under different conditions, should be studied, as we expect hunting strategies to vary and individuals to react flexibly to them. With the support of Jane Goodall, I was able to observe the hunting behavior of the chimpanzees at the Gombe Stream National Park, and I will do the same in the near future in collaboration with Toshisada Nishida with the chimpanzees living in the Mahale National Park in Tanzania. I hope to identify the factors affecting the hunting strategies and the evolution of cooperation in this species. Ideally, field observers should study different study sites and thus have the possibility to compare behavior patterns directly; hunting is so complex that comparison from published data might be sometimes misleading and thus of limited value.

Lastly, I like to mention here that one should try to clarify the different contribution of hunting and tool use to the evolution of intelligence, since both are so typically important in chimpanzee and human populations and might have played an important role in our evolution (see Boesch & Boesch-Achermann 2000). Intriguingly, hunting in chimpanzees and many human societies is mainly a male activity, in contrast to tool use, which is mainly a female activity. What this means for the evolution of cooperation and intelligence remains to be explored.

# References

Axelrod R, Hamilton WD, 1981. The evolution of cooperation. Science 211:1390–1396.

Boesch C, 1994. Cooperative hunting in wild chimpanzees. Animal Behaviour 48:653–667.

Boesch C, Boesch H, 1989. Hunting behaviour of wild chimpanzees in the Taï National Park. American Journal of Physical Anthropology 78:547–573.

Boesch C, Boesch-Achermann H, 2000. The chimpanzees of the Taï forest: behavioural ecology and evolution. Oxford: Oxford University Press.

Brown JL, 1983. Cooperation—a biologist's dilemma. Advances in the Studies in Behavior 13:1–37.

Busse C, 1978. Do chimpanzees hunt cooperatively? Am Nat 112:767–770.

Caraco T, Wolf LL, 1975. Ecological determinants of group sizes of foraging lions. American Naturalist 109:343–352.

Clements K, Stephens D, 1995. Testing models of non-kin cooperation: mutualism and the Prisoner's Dilemma. Animal Behaviour 50:527–535.

Creel S, 1997. Cooperative hunting and group size: assumptions and currencies. Animal Behaviour 54:1319–1324.

Dart R, 1925. *Australopithecus africanus*, the man-ape of South Africa. Nature 115: 195–199.

Dugatkin L, 1997. Cooperation among animals: an evolutionary perspective. Oxford: Oxford University Press.

Gagneux P, Boesch C, Woodruff D, 1999. Female reproductive strategies, paternity, and community structure in wild West African chimpanzees. Animal Behaviour 57:19–32.

Goodall J, 1963. Feeding behaviour of wild chimpanzees: a preliminary report. Symposium of the Zoological Society, London 10:39–48.

Goodall J, 1968. Behaviour of free-living chimpanzees of the Gombe Stream area. Animal Behaviour Monograph 1:163–311.

Goodall J, 1970. In the shadow of man. London: Collins.

Harcourt AH, de Waal FBM, 1992. Coalitions and alliances in humans and other animals. New York: Oxford University Press.

Kruuk H, 1972. The spotted hyena. Chicago: University of Chicago Press.

Leaky LSB, 1961. The progress and evolution of man in Africa. London: Oxford University Press.

Maynard Smith J, 1982. Evolution and the theory of games. Cambridge: Cambridge University Press.

Nishida T, 1979. The social structure of chimpanzees of the Mahale Mountains. In: The great apes (Hamburg DA, McCown E, eds). Menlo Park: Benjamin/Cummings; 73–122.

Nishida T, 1990. The chimpanzees of the Mahale Mountains: sexual and life history strategies. Tokyo: University of Tokyo Press.

Nishida T, Uehara S, Nyondo R, 1983. Predatory behaviour among wild chimpanzees of the Mahale Mountains. Primates 20:1–20.

Nowak M, Sigmund K, 1992. Tit for rat in heterogeneous populations. Nature 355:50–252.

Packer C, Ruttan L, 1988. The evolution of cooperative hunting. American Naturalist 132:159–198.

Pulliam HR, Caraco T, 1984. Living in groups: is there an optimal group size? In: Behavioural ecology: an evolutionary approach (Krebs JR, Davis NB, eds). London: Blackwell Science; 122–147.

Maynard Smith J, 1982. Evolution and the theory of games. Cambridge: Cambridge University Press.

Schaller GB, 1972. The Serengeti lion. Chicago: University of Chicago Press.

Teleki G, 1973. The predatory behavior of chimpanzees. Lewisburg: Bucknell University Press.

Trivers RL, 1971. The evolution of reciprocal altruism. Quaterly Review of Biology 46:35–57.

Washburn SL, Lancaster C, 1968. The evolution of hunting. In: Man the hunter (Lee RB, DeVore I, eds). Chicago: Aldine; 293–303.

# 23 Cooperative Hunting and Sociality in African Wild Dogs, *Lycaon pictus*

Scott Creel

I began studying African wild dogs for two reasons. First, wild dogs are obligately social. A wild dog is rarely alone, and if it is, chances are it will be looking for its packmates. The groups of other social species are generally far less cohesive, so wild dogs present a raft of interesting questions about the ecological forces that drive them to live in groups, and the consequences of sociality for their behavior, physiology, and demography. Second, wild dogs are endangered, with perhaps five thousand surviving in the wild. Although conservation is not the focus of this chapter, I aim for a research program that simultaneously answers questions posed by scientific curiosity and questions driven by conservation concerns. More and more, I feel that I'm fiddling while Rome burns if my research does not include components that are relevant to conservation. At the same time, social behavior is an engaging puzzle to me, just an interesting thing to examine from a new angle. Wild dogs satisfied both of these interests.

I first encountered wild dogs while studying dwarf mongooses in Serengeti National Park. A pack would occasionally appear, engage in a wild display of social behavior or an electrifying hunt, then vanish as abruptly as it arrived. At a personal level, I simply found wild dogs fascinating. It's difficult to put a finger on this, just as it is hard to explain why you have a preference for certain colors or foods. For reasons I can't explain, wild dogs are magnetic for me. As a behavioral ecologist, I saw fundamental questions about cooperative hunting and social evolution that could be addressed by research on wild dogs. Finally, as a conservation biologist, I saw an opportunity to examine why wild dogs are so rare when compared to sympatric carnivores with similar ecological requirements. Wild dogs attain very low population densities under the best of conditions, and consequently live in small populations with a high risk of local extinction. The reasons for this pattern were not clear in the 1980s, when I first began thinking about wild dogs, although subsequent work in several ecosystems has come a long way in providing answers (McNutt 1995; Mills & Gorman 1997; Creel & Creel 1998).

Because wild dogs are endangered, it is difficult to find a population large enough for quantitative research in behavioral ecology. The well-known population in Serengeti National Park (where I was living when I began to consider studying the dogs) held less than 50 adults in 18 of 20 years, and typically held only 20 to 30 individuals. While looking for a study site, Markus Borner of the Frankfurt Zoological Society told me that wild dogs were still seen regularly in southern Tanzania, particularly in the Selous Game Reserve. The population was thought to be one of the largest remaining, although there were no estimates of its size. The Selous is the largest protected area in Africa, but its remoteness and logistical headaches have created a research vacuum, and the idea of working in a huge unstudied wilderness was exciting.

The Selous Ecosystem covers 70,000 km², with 43,600 km² inside the game reserve itself. Most of the reserve is covered by deciduous woodland known as *miombo*. We began fieldwork in 1991, focusing on an area of 2,600 km² in the northern part of the reserve, where woodland and grassland intermingle and visibility is sufficient for behavioral observations. We did not collect any data in our first five months, simply because we had no good opportunities to attach radiocollars. The home ranges of wild dogs in Selous average 380 km², and daily travel distances average 12.3 km (up to 45 km), so opportunistic encounters can't be used to find and follow wild dogs. As frustrating weeks rolled by, I began to envision our first annual report to funding agencies: "Everyone says there are lots of dogs here, but we never saw any (P < 0.01)." After several months, we finally had a safe opportunity to dart. We double-checked everything and Nancy eased the vehicle forward. As the pack stood, I aimed for the nearest dog's rump, squeezed the trigger—and with a gentle *foop* the dart slid to the end of the barrel and hung there like a little flag that says *bang*. The words that ran through my head will probably cause me to come back as a sea slug. Eventually we fixed the rifle, darted our first dog, and began gathering data. Other packs fell into place, and our study ultimately included 366 dogs in 13 packs. Fitting radiocollars did not affect stress hormone levels over the long term, nor did it affect mortality rates (Creel et al. 1997).

## Conceptual, Theoretical, and Empirical Issues

### COOPERATIVE HUNTING AND GROUP SIZE

About 85 percent of the species in the order Carnivora are solitary. This poses an interesting question about the remaining 15 percent—what ecological and demographic forces favor group-living? Prior to quantitative research, it was uncritically accepted that group living in large carnivores arose through the benefits of cooperative hunting. Grouping can confer several advantages to hunters, including the ability to tackle larger prey, a higher probability of catching prey once a hunt begins, shorter chase distances, and an increased likelihood of catching several prey in one hunt. On the other hand, larger groups must obtain more food to hold per capita food intake constant. The first quantitative field studies (Schaller 1972; Kruuk 1972) tended to support the hypothesis that cooperative hunting promotes sociality. In Serengeti lions, the hunting success of groups was double that of singletons (groups: 82 kills in 273 hunts = 30 percent; solitaries: 37 kills in 249 hunts = 15 percent). These data have been widely interpreted as showing that group hunts were twice as likely to end in a kill, but Schaller (1972, 254) states that the difference arose because groups were more likely to kill several prey when they did succeed. Schaller (1972, 445) also presented direct data on the outcome of hunts by singletons (14 percent successful) and groups (30 percent successful). These data suggest that groups are twice as likely to make a kill, on a per hunt basis. I cannot rectify the two data sets to give a single estimate of the relationship between kills and hunting attempts, but both analyses show an advantage to grouping. Differences in prey mass also favored group hunting, because singletons hunted small prey more often than groups did (e.g., 74 percent versus 65 percent of hunts were Thomson's gazelle).

Kruuk (1972, 1975) found similar advantages to group hunting for spotted hyenas in Serengeti. When alone, hyenas usually hunted small prey such as Thomson's gazelles, but formed groups when hunting large prey such as wildebeest or zebra. For hyenas hunting calves in wildebeest herds, the success rate of groups was five times that of singletons (74 percent versus 15 percent hunting success). This benefit arose mainly because groups of hyenas never failed to kill a calf when its mother chose to stop and fight, while solitary hyenas never succeeded in this situation (Kruuk 1972).

These studies suggested that cooperative hunting could favor the evolution of sociality, or could provide selection to maintain sociality once it arose for other reasons. Thus, the previously uncritical acceptance of cooperative hunting as a force in social evolution was backed by some quantitative data. Little new data on the issue appeared for several years, although Schaller's lion data was reanalyzed with models that highlighted some unresolved issues. First, while Schaller's data showed that groups were more successful than singletons in some aspects of hunting, there was little benefit to forming groups larger than two. Thus, for Serengeti lions, most hunting parties were substantially larger than predicted, regardless of the method of analysis (Caraco & Wolf 1975; Packer 1986; Mangel & Clark 1988). The second problem was more fundamental; Schaller's data did not include estimates of all variables needed to measure the net benefits of hunting as a function of group size (Packer 1986).

Even if large groups benefit from higher hunting success, larger prey, or a higher probability of killing multiple prey, there still might be a decline in per capita food intake because the prey is divided among a large number of hunters (Kruuk 1972, 1975). This is fundamentally an empirical issue—many of the variables affecting food intake change with group size, and they must be quantified simultaneously to determine how costs and benefits vary with group size. In the 1990s, several studies measured daily per capita food intake as a function of group size, and tested whether individuals typically adopted the group size that maximized intake (Table 23.1). This approach was taken in studies of African wild dogs (Fanshawe & Fitzgibbon 1993), cheetahs (Caro 1994), and two lion populations (Packer et al. 1990; Stander 1992). Grouping improved foraging success in several tests (Table 23.1), but only lions in Etosha National Park showed grouping patterns that closely matched predictions based on the benefits of communal hunting (Stander 1992).

Consequently, recent reviews have argued that communal hunting has little power to explain grouping patterns in carnivores, particularly in felids (Packer et al. 1990; Caro 1994). These reviews (and others, e.g., Cooper 1991) emphasized that many variables are affected by group size, including protection of offspring, defense of carcasses from scavengers, and the maintenance of high-quality territories. While these variables are clearly important in explaining carnivore grouping patterns (Rood 1990; Packer et al. 1990; Cooper 1991; Caro 1994), I don't agree that current data supports a "growing consensus that cooperative hunting cannot account for patterns of group living in carnivores" (Packer & Caro 1997). Theoretical work on optimal foraging has emphasized that grouping can affect both the benefits and the costs of foraging (Stephens & Krebs 1986; Packer & Ruttan 1988), but carnivore studies have focused almost exclusively on the benefits, by using per capita food intake as their measure of fitness. Per capita food intake is a measure of gross benefit—it does not account for variation in the effort required to maintain a given rate of food intake (Fig. 23.1). For example, suppose individuals in small and large groups obtain equal per capita food intake, but individuals in small groups must make a greater hunting effort. Clearly, selection would favor large groups, but an analysis based on per capita food intake would not detect the difference. If all groups hunt at a rate sufficient to meet their needs (which seems plausible), then per capita food intake would show little variation across group sizes. Consequently, conclusions based on per capita food intake rates do not constitute a strong test of the role of cooperative hunting in the evolution of sociality. Models of optimal foraging focus on the net rate of energy intake as a measure of fitness, and a comparable measure is needed in field studies (Creel 1997; Packer & Caro 1997).

In my view, changes in the net rate of energy intake provide a strong test of the role of communal hunting in social evolution (although the risk of injury may also be important). Caro (1994) and others place more emphasis on evidence that hunting behavior is finely coordinated among group members. Based on behavioral data, Caro concluded that cooperative hunting was relatively unimportant for the grouping patterns of male cheetahs, even though trios obtained significantly more food than singletons or pairs. In my view, behavioral cooperation can be difficult to evaluate, particularly for a

Table 23.1

A Meta-analysis to Test the Relationship between Foraging Success and Group Size in Cooperatively Hunting Species

| Species | Test | P | ln P | Measure of Intake | Notes & References |
|---|---|---|---|---|---|
| Chimpanzee | Spearman | 0.05 | −2.9957 | Net | Taï forest, low singleton hunting success, Boesch 1994 |
| Chimpanzee | Spearman | 0.55 | −0.5978 | Net | Gombe, high singleton hunting success, Boesch 1994 |
| Killer Whale | Kruskal-Wallis | 0.001 | −6.9078 | Gross | Transient killer whales, Baird & Dill 1996 |
| Harris's Hawk | Kendall's Tau | 0.05 | −2.9957 | Gross | Data from Bednarz 1988 |
| African Wild Dog | Mann-Whitney | 0.02 | −3.912 | Gross | Comparing packs of 1–2 vs 3+, hunts of wildebeest, the most important prey |
| | | | | | Data from Fanshawe & Fitzgibbon 1993 |
| African Wild Dog | Mann-Whitney | 0.9 | −0.1054 | Gross | Comparing packs of 1–2 vs 3+, hunts of Thomson's gazelle, second most important prey |
| | | | | | Data from Fanshawe & Fitzgibbon 1993 |
| African Wild Dog | OLS Regression | 0.007 | −4.9618 | Net | Creel & Creel 1995; Creel 1997 |
| Female Lion | Mann-Whitney | 0.98 | −0.0202 | Gross | Serengeti, prey scarce, comparing prides of 1 vs 2–4, Packer et al. 1990 |
| Female Lion | Mann-Whitney | 0.002 | −6.2146 | Gross | Serengeti, prey scarce, comparing prides of 2–4 vs 5+, Packer et al. 1990 |
| Female Lion | Kruskal-Wallis | 0.48 | −0.734 | Gross | Serengeti, prey abundant, P value derived from chi-squared, Packer et al. 1990 approximation for reported H statistic. |
| Female Lion | Spearman | 0.05 | −2.9957 | Gross | Etosha, Stander 1992—lower P value in Stander & Albon 1993 |
| Cheetah | Mann-Whitney | 0.7 | −0.3567 | Gross | Adolescent females only, Caro 1994 |
| Cheetah | Mann-Whitney | 0.08 | −2.5257 | Gross | Adult males only, Caro 1994 |
| Wolf | OLS regression | 0.98 | −0.0202 | Gross | Schmidt & Mech 1997; combines data from several ecosystems |

Fisher's statistic = 70.7, df = 28; P < 0.001

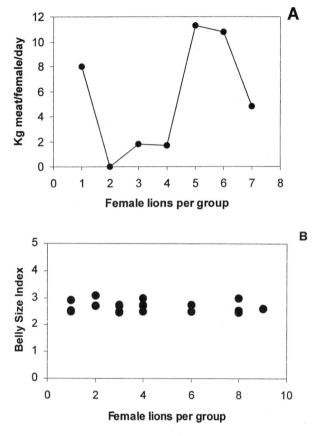

Figure 23.1. Measures for quantifying the role of communal hunting in the evolution of carnivore sociality, with an example from female lions. (A) Kilograms of meat obtained per individual per day. (B) An index of belly distension. Although lionesses in groups of two to four apparently have low hunting success, they are as well-fed as females in other group sizes. Variations in scavenging or in hunting effort both provide a possible explanation for the difference. Data from Packer 1986 and Packer et al. 1990.

complex and rapidly changing task such as isolating and killing one wildebeest from a herd of hundreds (Fig. 23.2). For example, in an analysis of cooperation during simulated territorial intrusions, Heinsohn and Packer (1995) suggested that a high rate of glances between a pair of lions indicates "distrust." While frequent glances could serve to detect cheating by a group mate, they might also serve to coordinate movements with a group mate that is cooperating fully. Similarly, Grinnell et al. (1995) argued that a tendency for male lions to walk side by side when approaching a simulated territorial intrusion was driven by the need to monitor pride mates that might otherwise slink away. While this is logical, it also seems plausible that a coordinated

Figure 23.2. Behavioral cooperation in hunting can be difficult to evaluate. (a) Two male cheetahs killing a yearling wildebeest, redrawn from Caro (1994, 277), who notes "the lack of coordination. The male on the left of the picture is using his weight to twist the victim's neck and so make it fall to its left, while the male on the right of the picture has bitten the wildebeest's right hindleg and is pulling backward in order to make it topple to the right." (b) Two wild dogs killing a yearling wildebeest, using positions similar to the cheetahs'. In my view, the victim's options are more constrained by the presence of two hunters than they would be if either hunter was absent, for both the cheetahs and the wild dogs.

side-by-side approach could deter intruders more effectively. If one accepts that these interpretations of intention are debatable, then many behavioral measures of cooperation are difficult to evaluate, even in experiments.

The hunts of large carnivores play out quickly, with many individuals moving simultaneously, sometimes over large areas. To assess behavioral cooperation, each individual's movements must be evaluated relative to the movements of other hunters and prey. Because many players are in motion at once, the simple presence of an additional predator, even one that is apparently doing little, might limit the options for prey. When a hunt begins, it is hard to say what might occur if an additional hunter was present, or if a specific hunter behaved differently (Fig. 23.2). For these reasons, I view the *outcome* of hunts (food obtained, effort expended, injuries sustained) to be the most reliable measure of the selection pressures that shape communal hunting.

## Quantifying the Costs and Benefits of Hunting

African wild dogs are well-suited subjects for measuring the costs and benefits of hunting. They are cursorial, with fast chases ranging from 50 meters to 4.6 kilometers, so the length of a chase is a good measure of the energetic cost of hunting. For predators that rely on stealth, passive or opportunistic hunts make it difficult to determine the costs of a hunt, or even to evaluate whether an individual is hunting or not (Packer et al. 1990; Scheel & Packer 1991). In stalking felids, a single hunter is capable of killing most types of prey, by clinging to the victim while suffocating it (although single lions cannot kill buffalo, as groups do: Schaller 1972; Caro 1994). Unlike the stalkers, most coursers do not employ a specific killing bite, and therefore kill by disemboweling, which may provide a greater benefit to cooperation. For example, a single wild dog cannot simultaneously restrain and kill a wildebeest that outweighs it by a factor of 10. One or more dogs restrain the prey, usually holding the nose or ears, while others disembowel it (Fig. 23.3). Disemboweling requires considerable force, and this is achieved by several dogs simultaneously pulling in opposite directions. Cooperation in making first contact with the prey may also be more beneficial in long, open pursuits (typical of canids and hyenids), than it is in short stalks or ambushes (typical of felids). In short, selection pressures on group size may prove to be very different in stalkers and coursers. Thinking about these patterns suggested to me that cooperative hunting was likely to be beneficial for wild dogs if it is beneficial to any species.

Obtaining reasonable sample sizes can be a serious problem for field studies of hunting behavior, but wild dogs normally hunt twice a day, and this allowed us to observe many chases and kills (Fig. 23.4). In Selous, wild dogs often traveled but rarely hunted in the night, reducing our worries about

Figure 23.3. For predators that lack a killing bite or stranglehold, prey are usually killed by disemboweling. For prey that are large relative to the predator, one hunter cannot simultaneously restrain the prey and kill it.

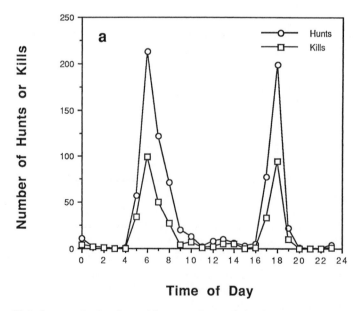

Figure 23.4. Crepuscular hunting, with two active periods of about three hours each, allows a large number of hunts to be observed for wild dogs.

missing chases or kills during round-the-clock follows (although some packs did hunt under moonlight, particularly when denning.) Group size varied widely (from three to twenty adults in our study), letting us test for differences in foraging success over a broad range of group sizes.

Offsetting these benefits, wild dogs are hard to find and follow. We spent many fruitless days motoring slowly through home ranges with the headset on, trying to detect a faint radiotransmitter beat behind deafening static. Listening for a sound that doesn't come eventually becomes torture. Low population density (compared to other large carnivores) made it difficult to monitor more than six to ten packs at once. The dogs traveled up to 45 kilometers in a day, and our observations were occasionally cut short when it was impossible to follow them across a river or rocky hillside. For much of the year, muddy floodplains made for daily digging, winching, and hauling to get the vehicle unstuck. I recall lying under a Suzuki jeep up to my eyebrows in mud, well into the second day of trying to work loose, watching an elephant sink 3 feet into the muck as it passed by and thinking "well, in the worst case we can chip it out next dry season." Splintering ebony stumps made punctured tires a constant annoyance, with a record of thirteen flats in one *very* long day.

Despite these impediments, the advantages of studying wild dogs greatly outweighed the difficulties, and it was possible to gather a large data set. Over a period of 2.5 years, we observed 404 kills in 905 complete hunts on 310 days. For most analyses of cost and benefit, we used daily means as data points, restricting the data to 266 days for which we observed all hunts. Means for successive days were not autocorrelated.

## PACK SIZE AND PREY MASS

In the Selous, wild dogs hunted seventeen species and killed ten species, but more than 90 percent of the diet was composed of wildebeest, impala, and warthog (Fig. 23.5a, b). Packs smaller than the median size (ten adults) preyed more heavily on impala than on wildebeest, while larger packs preyed most heavily on wildebeest (Fig. 23.5c, d). Mean prey mass increased significantly as pack size increased (Fig. 23.6), owing mainly to a shift in the proportions of wildebeest and impala killed. As pack size increased, there was a decline in the proportion of prey killed that is small and vulnerable to a single dog, and a parallel increase in the proportion of prey that is large and requires more than one dog to kill (Fig. 23.7). Small prey can be killed by a single dog because the dog can pick the prey up and shake it. Larger prey can stay on their feet when grabbed by a single dog (or even several dogs), and consequently can break free when a dog shifts its grip to attempt disemboweling. Even with several hunters cooperating, it is not unusual to see large prey escape after having been pulled to a halt.

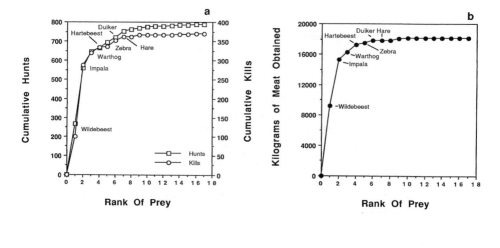

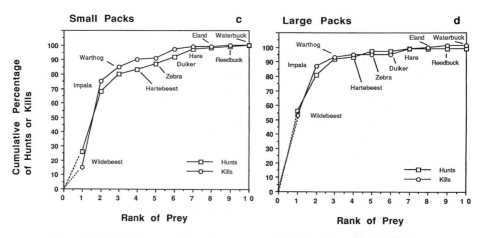

Figure 23.5. Wild dogs in Selous hunted seventeen species (part a) and killed ten of these (part b). Order of species along abscissa: wildebeest, impala, warthog, Lichtenstein's hartebeest, zebra, common duiker, African hare, eland, common reedbuck, African buffalo, greater kudu, bushbuck, sable antelope, bushpig, waterbuck, banded mongoose, and yellow baboon. Packs smaller than the median size preyed most heavily on impala (part c), while larger packs preyed most heavily on wildebeest (part d).

Interestingly, there is no detectable relationship between pack size and the frequency of killing *very* difficult prey (Fig. 23.7). This is probably because the hardest prey types are hunted opportunistically, only when an vulnerable individual is detected or the circumstances make a kill unusually likely. If so, then the hardest prey class in Figure 23.7 can be thought of as windfall. Regardless, these prey constitute a small part of the diet, and there is a strong positive relationship between prey mass and pack size (Fig. 23.6).

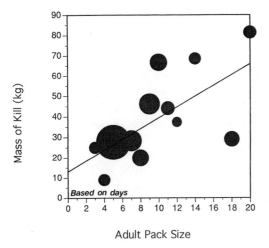

Figure 23.6. The relationship of prey mass to pack size. Points are means, weighted by days of observation for each pack size, with point size proportional to sample size. Line shows OLS regression ($b = 2.79 \pm 0.14$ *SE*, $r^2 = 0.45$, $t = 6.96$, $P < 0.001$).

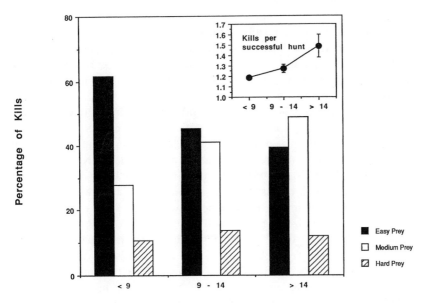

Figure 23.7. Patterns of prey selection for small, medium, and large wild dog packs. As pack size increases, easy prey form less of the diet, and prey of medium size form more of the diet. The hardest class of prey is rarely killed by packs of any size. In addition to larger prey, multiple kills are common for large packs (*inset*). *Easy* (small) prey includes impala fawns and yearlings, wildebeest calves, warthog piglets, African hares, duikers, bushbuck, and juveniles or yearlings of all other species. *Medium* prey includes adult female impala, yearling wildebeest, and yearling warthogs. *Hard* (large) prey includes adult male impala, and adult wildebeest, warthog, zebra, hartebeest, reedbuck ,and waterbuck. Reedbucks are classified as hard (despite being small) because wild dogs cannot see or run well in the very long grass that reedbucks favor. Warthog are classified as hard (despite moderate size) because their defensive tusk slashing is very effective compared to other prey.

## Pack Size, Hunting Success, and Multiple Kills

If measured as number of kills divided by number of hunts (following Schaller 1972), hunting success increases as pack size increases. Packs smaller than the median ("small packs" hereafter) made 248 kills in 658 hunts (kills/hunts = 0.38) while large packs made 127 kills in 247 hunts (kills/hunts = 0.51). If restricted to common prey species, hunting success shifts to 227 kills in 556 hunts for small packs (0.41), compared to 120 kills in 218 hunts for large packs (0.55). These proportions differ for large and small packs due to differences in the probability of making multiple kills in a single hunt (as with Serengeti lions: Schaller 1972).

Multiple kills were common. Of 404 animals killed, 89 (22%) were captured with at least one other victim. Maximum numbers killed in a single hunt were 7 impala, 6 wildebeest, and 3 warthogs. The most common type of multiple kill was 2 juveniles, but for impala it was not unusual for adult females and fawns (presumably mothers and their offspring) to be killed simultaneously. If one defines hunting success as the proportion of hunts resulting in one or more kills (ignoring the number of prey killed) small packs (232 of 658 hunts = 0.35) and large packs (83 of 247 hunts = 0.34) had identical hunting success.

Digressing a bit, it is interesting to consider the "confusion effect" in wild dog hunts. For some predator-prey interactions, prey are less vulnerable in large groups, because it is difficult for a predator to single out and track an individual target when the prey scatter (Neill & Cullen 1974; Kenward 1978; Fitzgibbon 1988). Crisler (1956, in Caro & Fitzgibbon 1993) suggested that confusion might work against the prey in some cases, as in wolves hunting caribou. In hunts of impala or wildebeest, wild dogs were more likely to make a kill as herd size increased (wildebeest: $\chi^2 = 5.71$, P = 0.039, impala: $\chi^2 = 7.37$, P = 0.007, logistic regressions fit by maximum likelihood: Fig. 23.8). One simple explanation is that large herds are likely to hold vulnerable individuals, but in addition, large herds were sometimes apparently confused when an attack began. Quite often, prey in large herds impede one another at the start of their flight, particularly in thick habitats, where the vegetation constrains the paths of escape. Fast flight through woodland requires many decisions about how to pass obstacles. When many individuals are making these decisions simultaneously, coordination becomes difficult. A good analogy is that small airfields don't require air traffic controllers, but imagine Heathrow or O'Hare with hundreds of pilots making decisions independently. Confusion is more apparent for impala than for wildebeest, because impala bolt in many directions, while wildebeest are likely to run parallel to one another if possible. When impala herds scattered in woodland fawns were sometimes captured after stopping at a barrier, even though all of the wild dogs were behind them. It is interesting that confusion appears to favor the prey in fish and birds (Neill & Cullen 1974; Kenward

1978), but may favor the predator among mammals. Among birds and fish, predators pursue prey in three dimensions, while mammals operate in two dimensions. All else being equal, the loss of one dimension would increase the odds of prey impeding one another. Confusion effects in different contexts could be an interesting subject for further study.

## PACK SIZE AND HUNTING EFFORT

Larger packs are more likely to make a kill, and they take larger prey, but these observations do not exclude the possibility that larger packs simply work harder. Large packs engage in more chases per day than small packs (Creel & Creel 1995; and see below), but also made significantly shorter chases than small groups (Fig. 23.9). Detecting that larger packs make shorter chases is important, because it identifies another aspect of hunting in which large packs are more effective. Detecting that large packs hunt more often is also important, because it restores the focus on the basic question—are the benefits of cooperative hunting large enough to offset the increased food requirements of larger groups?

## PACK SIZE, GROSS BENEFIT, AND NET BENEFIT

Pack size was correlated with prey mass, hunting success (due to changes in the probability of multiple kills), chase distance, and hunts per day. Ideally, the goal is to combine all of these effects into a single measure of foraging success. As discussed above, the most widely used measure of foraging success in empirical studies of cooperative hunting is kilograms of meat obtained per individual per day (gross benefit) (Packer et al. 1990; Stander 1992; Caro 1994), but this does not account for variation in the frequency of hunts or in effort per hunt. To account for variation in effort, one might measure kilograms of meat obtained per individual per kilometer traveled. This measure of foraging success removes the implicit assumption that the costs of hunting do not vary with group size, but creates three new difficulties.

*Difficulty One:* It is not obvious what measure of distance traveled is most appropriate to measure hunting effort. Total distance traveled per day is an appropriate measure of hunting costs if all movements are related to hunting. If some movements are not related to hunting (e.g., territorial patrolling), then it might be more accurate to focus on distance traveled while chasing prey. Distance traveled per day has the advantage of being more inclusive, while distance traveled in chases has the advantage of focusing on costs that unequivocally pertain to hunting.

Ideally, one would start with daily travel distance, remove the travel unrelated to hunting, then convert distances traveled to their energetic costs. In

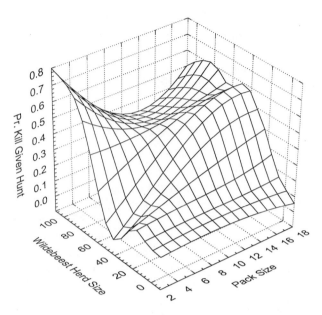

Figure 23.8. The probability of a kill as a function of wildebeest herd size and wild dog pack size.

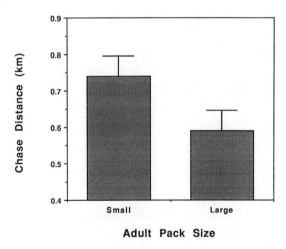

Adult Pack Size

Figure 23.9. Mean chase distance for wild dog in packs smaller and larger than (or equal to) the median pack size (10). Data from 775 pursuits measured by odometer.

practice, the distinction between searching for prey and other activities is not obvious, and in reality there may be no distinction. Suppose a predator moves 3 kilometers directly from a rest site to a water hole, without encountering prey. It would be difficult to conclude that the intention of the movement was to get a drink and not to hunt. The predator might have struck water by chance, or perhaps it moved intentionally to water, but would have hunted if it had encountered prey. Total distance traveled and distance spent in direct pursuit of prey are two objective measures of hunting effort, but they side-step these ambiguities rather than resolving them.

*Difficulty Two:* Raw data on the costs and benefits of hunting are recorded in different currencies. Benefit is measured by the mass of meat obtained, while cost is measured in distance covered or time spent hunting. If costs and benefits can't be converted to common units (see "Difficulty Three"), then the currency of foraging success will probably be kilograms of food obtained per kilometer traveled or chased, which is a ratio of benefits ($B$) to costs ($C$). In the optimal foraging literature, the ratio $B/C$ is called efficiency, and it has several shortcomings (Stephens & Krebs 1986). First, efficiency does not distinguish between small gains at small cost and large gains at large cost. Consequently, the group size that maximizes efficiency will not necessarily maximize the net rate of energy intake ($[B-C]$/time). Second, efficiency is affected by scaling problems if $B$ and $C$ differ greatly: two strategies that have very different efficiencies could yield similar net benefits. While both of these problems are potentially important for analyses of cooperative hunting (Packer & Caro 1997), efficiency is a reasonable currency of foraging success under certain conditions (Creel 1997). First, if $B$ and $C$ are similar in value, then efficiency and net rate of intake will give similar results, and it is likely that the costs and benefits of hunting are of similar magnitude over the long run. My second point relates to a thought problem raised by Packer and Caro (1997), as follows. Suppose hunters in one group expend 1,000 calories to obtain 5,000 calories (netting 4,000), while individuals in another group expend 5 calories to obtain 500 calories (netting 495). The strategy with the greater efficiency yields a lower net benefit, so Packer and Caro concluded that the less efficient group is optimal, if daily requirements are greater than 495 calories. However, Packer and Caro are implicitly assuming that hunters in the more efficient group are constrained to expend only 5 calories hunting. If they exerted an effort equal to that of the less efficient group, they would obtain a much greater net benefit. If group size does not constrain hunting effort in social carnivores, this scenario does not present a problem for the use of efficiency. While hunting effort ($C$) might *vary* across group sizes enough to be evolutionarily important, I see no evidence that group size *constrains* hunting effort.

*Difficulty Three:* Efficiency is an adequate measure of foraging success under certain circumstances, but the net rate of energy intake is better (Stephens & Krebs 1986). Measuring net benefit requires that costs and benefits both be converted to units of energy. For large carnivores, these conversions undoubtedly introduce some undesirable process error. That said, converting kilograms of food obtained into kilojoules of energy is conceptually straightforward, using estimates of the composition of carcasses (Sachs 1967; Blumenschine & Caro 1986), the proportions of each part of a carcass that are edible, and the energy content of various tissues (USDA 1990). Using this information, I estimated that wild dogs gained 6.9 *MJ* per *kg* of impala or wildebeest killed (Creel 1997). Of this, about 5.8 *MJ* per *kg* should be available after digestion (Gorman et al. 1998). In Selous, packs killed 4.0 ± 0.35 *kg*/dog/day (Creel & Creel 1995), which equates to an energetic benefit of 23 *MJ*/day.

For comparison, Gorman et al. (1998) used doubly labeled water to estimate that wild dogs in Kruger National Park burned 15.3 *MJ* per day, when active for 3.45 hours per day. From Gorman et al.'s estimates of energy expenditure at rest and while hunting, the 23 *MJ*/day acquired by wild dogs in Selous would support 6.0 hours of activity per day, which is a reasonable match to measured daily activity in Selous (Fig. 23.4). This match supports my method of converting from kilograms of meat to megajoules of energy to estimate of the benefits of hunting. Having converted benefits into units of energy, it is still necessary to convert costs.

There are two approaches for converting the costs of hunting into units of energy. The first approach is to use doubly labeled water to measure expenditure directly, as Gorman et al. (1998) did. Doubly labeled water has the advantage of giving a direct measure of field metabolic rate. Its drawbacks include the difficulty of obtaining large sample sizes, and the fact that expenditure is usually measured over an interval during which individuals are both active and inactive. Consequently, Gorman et al. (1998) did not directly measure expenditure while hunting, but inferred it from three quantities: daily energy expenditure, an estimate of basal metabolic rate (derived from domestic dogs), and a time budget with the proportions of time active and time inactive. This approach yielded an estimate of 3.14 *MJ* expended per hour of hunting and travel (Gorman et al. 1998).

A second approach to measuring the costs of hunting relies on measurements of oxygen consumption. Assuming that anaerobic contributions to energy production are negligible, measurements of oxygen consumption on a treadmill can be used to estimate energy expenditure as a function of speed and body mass (Taylor et al. 1982; Calder 1984). Amazingly enough, oxygen consumption has been measured for an African wild dog running on a treadmill (Taylor et al. 1971), but the data came from a half grown puppy running at low speeds. Rather than using this data, I used an allometric equation based on treadmill tests with several terrestrial carnivores (Taylor et al. 1982). For a carnivore of the wild dog's mass trotting at 5 *m/s*, this equation estimates the cost of travel at 3.04 *MJ* per hour of travel (Creel 1997). This method allows distance *traveled* per day to be converted into units of energy, because traveling dogs move at speeds that can be examined in a treadmill test. Treadmill oxygen-consumption tests cannot be used to convert the distance of direct, high-speed *chases* into units of energy, for three reasons: (1) treadmills do not reach the speeds at which wild dogs chase prey (up 17 *m/s*); (2) oxygen consumption is not a linear function of running speed; and (3) anaerobic pathways of energy production are probably important during chases. Despite these limitations, the two methods of converting travel costs into units of energy gave similar results, 3.04 *MJ* and 3.14 *MJ* per hour of travel (Creel 1997; Gorman et al. 1998).

## THE TAKE HOME MESSAGE

After converting costs and benefits into a common currency, one can examine the per capita net rate of energy intake as a function of pack size (Fig. 23.10). The regression of net rate of energy gained on pack size is significant (OLS regression, $P = 0.01$, $r^2 = 0.53$), supporting the hypothesis that cooperative hunting is a force in the evolutionary maintenance of group-living in wild dogs.

I could compare the observed distribution of pack sizes to that predicted by Figure 23.10. Most analyses of cooperative hunting (including mine—Creel 1997) have done this, but in retrospect I think it is not particularly useful. Foraging success is one variable that affects fitness, but it is virtually certain that other factors also affect the fitness payoffs to living in different pack sizes. For example, larger packs are better able to defend their kills against kleptoparasitism by spotted hyenas (Fanshawe & Fitzgibbon 1993). Conceptually, it is clear that the frequency distribution of pack sizes will be shaped by selection acting simultaneously on many components of fitness that are affected by pack size. Comparing the pack size distribution to the predictions of a single variable is a weak test of that variable's importance (even if there is a nice match), unless the predictions for other variables are known. Implicitly, the null hypothesis for this comparison includes an assumption that no other components of fitness vary with group size. Because of this problem, a better test is to identify a variable that is likely to affect fitness (such as foraging success or predation risk), quantify the variable, test whether it varies significantly across group sizes, and stop there. This approach makes no assumptions about the relationship of other variables to pack size. An alternate (and extraordinarily hard) approach is to identify *all* of the variables likely to affect fitness, quantify each as a function of group size, convert them to a common currency (or establish weighting factors for the impact of each variable on fitness), and make a joint prediction to which the frequency distribution of group sizes could be compared. The latter approach would be a life's work for many species.

My data suggest that cooperative hunting favors living in packs as large as possible (up to twenty adults), with other factors setting an upper limit on pack size. Reproductive suppression of social subordinates is one factor that is likely to limit pack size. In a wild dog pack, only the socially dominant individuals maintain normal reproductive hormone levels (Creel et al. 1997), and subordinates are unlikely to reproduce (Girman et al. 1997). As pack size increases, the breeding queue gets longer, as do the odds of surviving long enough to become a breeder. Pack size will eventually cross a threshold at which subordinates would do better to disperse in search of breeding opportunities. Consistent with this argument, young adults are the most common dispersers, and dispersal is most likely from large packs that have just recruited a large cohort of same-sexed young adults (McNutt 1996; Creel, unpublished data).

Tentatively, the data for wild dogs suggests that ecological interactions favor large packs, with opposing selection due to reproductive competition within packs. In addition to hunting more effectively, large packs are better able to defend kills against competitors (Fanshawe & Fitzgibbon 1993). Interference competition at carcasses is intense in some wild dog populations, and the density of wild dogs varies inversely with the density of spotted hyenas, their main kleptoparasite (Creel & Creel 1996). Consequently, variation in the ability to defend food could create strong selection in favor of large groups. Owing to benefits such as these, reproductive success is positively correlated with pack size (Fig. 23.11: OLS regression, $r^2 = 0.23$, $F_{1,28} = 8.53$, $P = 0.007$). If the benefits of group-living were shared equally among pack members, it is likely that wild dog packs would be larger than they are. Because reproductive success is highly skewed within a pack, large packs yield low immediate fitness benefits to social subordinates, despite being favorable with respect to obtaining and defending food.

## A Meta-analysis of the Relationship between Group Size and Foraging Success in Communal Hunters

In the most recent review of cooperative hunting, Caro (1994, 342), concluded that

> few studies report per capita foraging returns, but in the majority of those that do, per capita foraging success did not increase with group size. In populations in which it did, grouping patterns did not reflect optimal foraging group sizes. Though limited, current evidence therefore suggests that cooperative hunting is not responsible for group living in any carnivore.

For seven species that hunt communally, foraging success has now been measured as a function of group size (Table 23.1). These studies provide fourteen independent tests of the relationship between group size and foraging success, because some species have been studied in several locations (lions, wild dogs, chimpanzees), or the data were partitioned by season or group composition (lions: Packer et al. 1990; cheetahs: Caro 1994). With these fourteen tests, I conducted a quantitative meta-analysis to test the relationship between foraging success and group size, using Fisher's method for combining probabilities from independent tests (Fisher 1954; Sokal & Rohlf

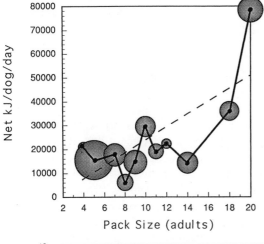

Figure 23.10. Net rate of energy intake as a function of pack size.

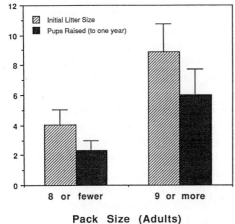

Figure 23.11. Reproductive success (yearlings raised per year) as a function of pack size.

1995). This meta-analysis is based primarily on data on gross daily intake, because few studies have estimated net rate of intake, and this is a serious limitation. Like all meta-analyses, there is room for concern that negative results can be difficult to publish, biasing the set of primary studies available for inclusion. With these caveats, there is a positive and strongly significant relationship between group size and foraging success (Table 23.1: Fisher's statistic $= 70.7$, $df = 28$, $P < 0.001$). Although small, this set of species includes stalkers and coursers that hunt on land, in water, in the trees, and in the air, pursuing their prey by swimming, climbing, running, and flying. A positive result across such dissimilar species suggests that cooperative hunting is beneficial under a broad range of ecological conditions. There are too few studies to make strong arguments about differences among species, but these studies are broadly consistent with published hypotheses that the benefits of cooperation are more consistent for hunters that do not rely on stealth, and under ecological conditions where solitary individuals have low foraging success (Kruuk 1975; Packer & Ruttan 1988; Boesch 1994).

## Future Avenues for Research

The evolution of sociality has been a dominant issue in the study of animal behavior, and cooperative hunting is one of the most widely distributed forms of cooperation among animals (Packer & Ruttan 1988). Because cooperative hunting is employed by a phylogenetically broad set of species over a wide range of ecological conditions, it is natural to ask whether hunting plays a role in the origin or maintenance of sociality. Perhaps part of our fascination with this question arises from our own origin as social hunters—the selection pressures that now act on wild dogs and chimpanzees once acted on us. Field studies have made considerable progress in measuring the costs and benefits of hunting, and current data generally supports the hypothesis that cooperative hunting favors group-living (Table 23.1).

Nonetheless, there are serious shortcomings in our current understanding of cooperative hunting. Many studies have examined the relationship between hunting success (kills per hunt) and group size, but few studies have produced direct measures of foraging success (intake/individual/day). Fewer still have produced estimates of the net benefit of hunting as a function of group size. I am aware of only two such studies to date (Table 23.1). In my opinion, the most profitable avenue of research on this topic is to test the relationship between the net benefit of hunting and group size for a wider range of species. Some communal hunters such as the dhole (*Cuon alpinus*) have received little study, and we still have no empirical tests of this issue for some well-studied carnivores such as the spotted hyena. A second useful approach would be to reexamine species already in the data set under a new set of ecological conditions. For the few species that have been studied in more than one ecosystem, data suggests that the relationship between foraging success and group size varies with ecological conditions. For example, Packer et al. (1990) found little benefit to grouping in Serengeti lions, while Stander (1992; Stander & Albon 1993) found a strong benefit. Some variables other than group size are logically expected to affect the effectiveness of cooperation (e.g., the relative abundance of large and small prey: Caro 1994), but we have not accumulated enough data to estimate the importance of these variables. Of more excitement, empirical studies might reveal unpredicted relationships between ecological variables and the effectiveness of cooperation.

Hopefully, studies of new species in new habitats will eliminate some of the practical problems with measuring the energetic costs and benefits of hunting. In this vein it may be useful to study species other than large carnivores. Logistical constraints often limit studies of large carnivores to small sample sizes and essentially paper-and-pencil methods. While I believe that purely observational research provides a critical test of the adaptive significance of natural behaviors in the real world, species that allow for more detailed measurements or manipulations might provide an additional way forward. For example, I can imagine conducting research with cooperatively hunting birds that included daily measurements of body weight using perches attached to balances (as has been done for birds and mammals in other contexts: Creel & Creel 1991). Animals that are readily captured without harm might allow for studies employing doubly-labeled water or portable body composition scanners.

Even with better estimates of the energetic costs and benefits of hunting, future studies will be improved by incorporating estimates of the physical risks of hunting. The risk of injury is probably important in decisions about hunting, but this is difficult to demonstrate. For example, I have applied standard energetic models of prey selection (Stephens & Krebs 1986) to data from wild dogs. In these analyses, two species are attacked substantially more often than the models predict: African hares and common duikers. Both are small and pose no threat of injury to attackers, compared to other prey. It is difficult to estimate the role that injury plays in shaping hunting decisions, because serious, overt injuries are relatively rare. Once a small data set on injury rates is partitioned by group size, it is likely to have little statistical power. But injuries may have large effects on fitness, despite being infrequent. The difference between a 2 percent chance and a 6 percent chance of getting your head kicked in by a wildebeest is probably of evolutionary significance, despite being hard to quantify.

Finally, future studies might develop better methods for simultaneously testing how several variables that affect fitness vary across group sizes. As discussed above, comparing the distribution of group sizes to the predictions from a single variable is of limited value, because it is unlikely that a single variable will determine the relationship between fitness and group size. This remains a difficult problem for strong, multifaceted approaches to social evolution.

Given the title of this book, I'll close by suggesting three general points that students might draw from wild dogs as a model for research on social evolution and cooperative hunting. First, this work illustrates the August Krogh principle: for every question, there are certain species that are well suited to providing answers, and it is worth the effort to find a good match. Second, I'd emphasize that my approach to a question almost always evolves during the research. It is obviously important to identify your hypothesis before going to the field, but it is also important to recognize that your initial hypothesis might not be the best available. In this case, the realization that daily food intake is not a good currency for analyses of cooperative hunting did not strike me while reading prior studies, but while watching wild dogs hunt. When one pack eats a wildebeest after a single ten-minute chase, and a second pack eats a wildebeest after a long day of running, it's not much of a reach to see the difference. Finally, I'd emphasize the value of looking for research opportunities that allow for interesting approaches to fundamental questions, while simultaneously addressing questions of immediate importance for conservation or management. For me, one of the biggest satisfactions of studying wild dogs was to answer the simple question "How many live here?" and find out the answer was "more than we thought."

# References

Baird RW, Dill LM, 1996. Ecological and social determinants of group size in transient killer whales. Behav Ecol 7:408–416.

Bednarz JC, 1988. Cooperative hunting in Harris' hawks (*Parabuteo unicinctus*). Science 239:1525–1527.

Boesch, C, 1994. Cooperative hunting in wild chimpanzees. Anim Behav 48:653–667.

Blumenschine RJ, Caro TM, 1986. Unit flesh weights of some East African bovids. Afr J Ecol 24:273–286.

Calder WA, 1984. Size, function, and life-history. Cambridge, MA: Harvard University Press.

Caraco T, Wolf LL, 1975. Ecological determinants of group sizes of foraging lions. Amer Nat 109:343–352.

Caro TM, Fitzgibbon CD, 1993. Large carnivores and their prey: the quick and the dead. In: Natural enemies: the population biology of predators, parasites and diseases (Crowley MJ, ed). Oxford: Blackwell Scientific; 117–142.

Caro TM, 1994. Cheetahs of the Serengeti Plains. Chicago: University of Chicago Press.

Cooper SM, 1991. Optimal hunting group size: the need for lions to defend their kills against loss to spotted hyaenas. Afr J Ecol 29:130–136.

Creel S, 1997. Cooperative hunting and group size: assumptions and currencies. Anim Behav 54:1319–1324.

Creel SR, Creel NM, 1991. Energetics, reproductive suppression and obligate communal breeding in carnivores. Behav Ecol Sociobiol 28:263–271.

Creel SR, Creel NM, 1995. Communal hunting and pack size in African wild dogs (*Lycaon pictus*). Anim Behav 50:1325–1339.

Creel SR, Creel NM, 1996. Limitation of endangered African wild dogs by competition with larger carnivores. Cons Biol 10:526–538.

Creel SR, Creel NM, 1998. Six ecological factors that may limit African wild dogs, *Lycaon pictus*. Anim Cons 1:1–9.

Creel S, Creel NM, Mills MGL, Monfort SL, 1997. Rank and reproduction in cooperatively breeding African wild dogs: behavioral and endocrine correlates. Behav Ecol 8:298–306.

Creel S, Creel NM, Monfort SL, 1997. Radiocollaring and stress hormones in African wild dogs. Cons Biol 11:544–548.

Fanshawe JH, Fitzgibbon CD, 1993. Factors influencing the hunting success of an African wild dog pack. Anim Behav 45:479–490.

Fisher RA, 1954. Statistical methods for research workers (12th ed.). Edinburgh: Oliver & Boyd.

Fitzgibbon CD, 1988. The antipredator behaviour of Thomson's gazelles. Ph.D. diss., University of Cambridge.

Girman DJ, Mills MGL, Geffen E, Wayne RK, 1997. A molecular genetic analysis of social structure, dispersal, and interpack relationships of the African wild dog (*Lycaon pictus*). Behav Ecol Sociobiol 40:187–198.

Gorman ML, Mills MGL, Raath JP, Speakman JR, 1998. High hunting costs make African wild dogs vulnerable to kleptoparasitism by hyaenas. Nature 391:479–481.

Grinnell J, Packer C, Pusey AE, 1995. Cooperation in male lions: kinship, reciprocity or mutualism? Anim Behav 49:95–105.

Heinsohn R, Packer C, 1995. Complex cooperative strategies in group-territorial African lions. Science 269:1260–1262.

Kenward RE, 1978. Hawks and doves: factors affecting success and selection in goshawk attacks on woodpigeons. J Anim Ecol 47:449–460.

Kruuk H, 1972. The spotted hyena. Chicago: University of Chicago Press.

Kruuk H, 1975. Functional aspects of social hunting in carnivores. In: Function and evolution in behavior (Baerends G, Beer C, Manning A, eds). Oxford: Oxford University Press; 119–141.

Mangel M, Clark CW, 1988. Dynamic modeling in behavioral ecology. Princeton: Princeton University Press.

McNutt JW, 1995. Sociality and dispersal in African wild dogs, *Lycaon pictus*. Ph.D. diss., University of California, Davis.

McNutt JW, 1996. Sex-biased dispersal in African wild dogs, *Lycaon pictus*. Anim Behav 52:1067–1077.

Mills MGL, Gorman ML, 1997. Factors affecting the density and distribution of wild dogs in the Kruger National Park. Cons Biol 6:1397–1406.

Neill SR, Cullen JM, 1974. Experiments on whether schooling by their prey affects the behaviour of cephalopod and fish predators. J Zool Lond 172:549–569.

Packer C, 1986. The ecology of sociality in felids. In: Ecological aspects of social evolution (Rubenstein DI, Wrangham RW, eds). Princeton: Princeton University Press; 429–451.

Packer C, Caro TM, 1997. Foraging costs in social carnivores. Anim Behav 54:1317–1318.

Packer C, Ruttan L, 1988. The evolution of cooperative hunting. Amer Nat 132159–132198.

Packer C, Scheel D, Pusey AE, 1990. Why lions form groups: food is not enough. Am Nat 136:1–19.

Rood, JP 1990. Group size, survival, reproduction, and routes to breeding in dwarf mongooses. Anim Behav 39:566–572.

Sachs R, 1967. Liveweights and body measurements of Serengeti game animals. E Afr Wildl J 5:24–36.

Schaller GB, 1972. The Serengeti lion. Chicago: University of Chicago Press.

Schmidt PA, Mech LD, 1997. Wolf pack size and food acquisition. Am Nat 150:513–517.

Scheel D, Packer C, 1991. Group hunting behaviour of lions: a search for cooperation. Anim Behav 41:697–709.

Sokal RR, Rohlf FJ, 1995. Biometry (3rd ed.). New York: W. H. Freeman.

Stander PE, 1992. Foraging dynamics of lions in a semi-arid environment. Can J Zool 70:8–21.

Stander PE, Albon SD, 1993. Hunting success of lions in a semi-arid environment. Symp Zool Soc Lond 65:127–143.

Stephens DW, Krebs JR, 1986. Foraging theory. Princeton: Princeton University Press.

Taylor CR, Schmidt-Neilsen K, Dmi'el R, Fedak M, 1971. Effect of hyperthermia on heat balance during running in the African wild dog. Am J Physiol 220:823–827.

Taylor CR, Heglund NC, Maloiy GMO, 1982. Energetics and mechanics of terrestrial locomotion: I. Metabolic energy consumption as a function of speed and body size in birds and mammals. J Exp Biol 97:1–21.

USDA, 1990. Composition of food. Beef products: raw, processed, prepared. United States Department of Agriculture, Agriculture Handbook Series, Number 8–13.

# 24 Gorilla Socioecology: Conflict and Compromise between the Sexes

A. H. Harcourt

## Origins

I began my studies of gorilla socioecology in the early 1970s. Vertebrate socioecologists then were using two main levels of explanation in their attempts to understand the social systems of individual species, as well as differences between species in their social system. One level of explanation was functional. The other was equivalent to the Tinbergian proximate level of explanation (Tinbergen 1963). In primatology especially, both were to quite a large extent concentrated on understanding grouping. Grouping was interesting to primatologists because primates were seen as unusual in the proportion of species that lived in more or less stable groups of individuals of both sexes outside the breeding season.

Functional explanations for grouping concerned the costs and benefits of grouping for individuals. Given that grouping necessarily involves close, long-term proximity to competitors, what are its benefits, and what is the balance of benefits and costs (Alexander 1974; Crook 1970)? In addition, how do we explain differences between species in their grouping? Crook's (1965) analysis of weaver bird socioecology was particularly crucial. While David Lack (1968) had led the way in functional explanations of intertaxonomic differences in breeding ecology of birds, John Crook very cleanly and explicitly provided an ecological explanation of differences between weaver bird species in their social system. For instance, small, insectivorous birds, living in woodland, face dispersed, small patches of food, and thus forage alone at low density. Breeding tends to be monogamous, the sexes are morphologically similar, and males, who cannot defend the range of more than one female, help with provisioning of the offspring. By contrast, larger, graminivorous birds, living in more open habitat, forage in large patches of ephemeral food, and use each other to find the patches. Birds forage and roost in groups; females are densely distributed; males have the time and energy to attract a number of females; males are more colorful than females; and breeding is polygynous (and far later shown to be polyandrous too), with males doing little or no provisioning of young. Importantly for primate socioecology, Crook then moved to primatology. He and others explained variation in primate social systems in terms of the quality and distribution of food (Crook 1970; Crook & Goss-Custard 1972; Eisenberg et al. 1972). Body size was also a crucial parameter, because it influenced antipredator behavior. In the early 1970s, the integration of antipredator and foraging strategies to explain variation in social systems in relation to body size could perhaps be seen as culminating in Jarman's (1974) analysis for African ungulates. Small animals require rich foods, which are thinly distributed; the social system is thus solitary, and the animals hide from predators. Conversely, large animals can feed on poor foods, which occur in large patches; the animals group, and use both their size and grouping to deter predators.

The Tinbergian proximate level of explanation of grouping addressed the nature of the relationships between individual group members. Robert Hinde, and through him his students (of which I was one), were main proponents of this level of analysis (Hinde 1972, 1976, 1983). A classic distinction illustrates this approach. Hamadryas baboons (*Papio hamadryas*) and gelada baboons (*Theropithecus gelada*) both live in open habitat as one-male breeding units of a male and several females and their offspring within bands of several such one-male units. On the surface, the species look as though they have a similar social system. However, the two species differ fundamentally in the nature of the relationships among individuals within the one-male units. Hamadryas females interact most with the male, and little with one another; gelada females, by contrast, interact most with one another, and only a minority of them have strong relationships with the male (Dunbar 1984; Kummer 1968). This difference indicated crucial differences between the species in payoffs of association to individuals of the two sexes. The significance of the proximate level of analysis to behavioral ecology's functional approach is, I suggest, the proximate level's focus on the individual. Socioecology's focus in the early 1970s was, by contrast, the species, and differences between species, even if individuals were the underlying unit of selection (Crook et al. 1976; Eisenberg et al. 1972; Gartlan 1968). An important advance of behavioral ecology was the quantitative demonstration that individual animals behave differently (make different "decisions"), depending on the situation and on the history that they bring to the situation (McCleery 1978). A focus on the individual, and on the nature of social relationships among individuals, was easily and immediately convertible into an understanding of social systems in terms of the adaptive strategies of their constituent members (Clutton-Brock & Harvey 1976).

Behavioral ecology is concerned with the decision rules of individuals for survival, mating, and rearing, in situations where the payoffs depend on the nature of the environment, and on the number and nature of potential competitors, mates, and partners in rearing. To some extent, a successful behavioral ecology can ignore phylogeny, because the most general rules will be largely independent of the species studied. What, then, were the reasons for me to study primates in general, the apes and the gorilla in particular?

First, a crucial stimulus and source of funding for many studies of primate socioecology is the supposition that by studying our closest relatives, we can learn about humans. As Tinbergen (1963) so importantly argued, phylogeny is one level of explanation for behavior, and for differences in behavior. Organisms are not infinitely, or instantaneously, malleable. If they were, behavioral ecology would not have to pay as much attention to phylogenetic analysis as it now does (Harvey et al. 1996; Harvey & Pagel 1991; Martins 1996).

Second, at the start of the 1970s, we knew extraordinarily little about gorillas. A very good reason to study anything, but especially to study our closest relative in the animal kingdom, is ignorance. Studying unknown species is especially important in a comparative discipline, such as socioecology, in which variety is informative.

Third, the little knowledge that was available indicated that the great apes, Pongidae, might be a particularly useful taxon for increasing socioecological understanding, because they apparently differed so greatly in their social systems, despite the fact that they were closely related (Purvis 1995; Ruvolo et al. 1994). They varied from the largely solitary orangutan, through the fission-fusion parties of the chimpanzee, to the stable family groups of the gorilla (DeVore 1965) (the bonobo was effectively unknown in the early 1970s). In addition, the main monograph on the gorilla (Schaller 1963) indicated that the gorilla was an exception to one current generalization about Old World primate social organization. In several Old World primates, strong attachments (cooperative relationships) among females were obvious, and females were described as the core of social groups (Bramblett 1970; Imanishi 1963; Kawai 1965; Kawamura 1965; Missakian 1972; Sade 1967; Vandenburgh 1967). By contrast, in the gorilla, it seemed that male-female relationships were more important than female-female relationships. And exceptions are useful for proving rules. As William Harvey, the elucidator of the circulatory system, apparently said, treasure your exceptions, because you will learn more from them than from the common run of evidence.

Finally, there was a sense in the 1970s that primates were more socially complex than were nonprimates. Subsequent comparative studies have substantiated that sense, even if at the same time they have demonstrated the high social complexity of nonprimates too (Cheney & Seyfarth 1990; Harcourt & de Waal 1992; Marler 1996). For instance, primates apparently make more complex use of allies in competition than do nonprimates (Harcourt 1992). And such social complexity is intrinsically interesting to someone concerned with explaining group-living either proximately, in terms of the nature of social relationships among individuals, or functionally, in terms of the nature of competitive and cooperative tactics of individuals.

# Integrating Theory and Observation

## A SOCIOECOLOGICAL FRAMEWORK

I will try to demonstrate the integration of theory and observation in primate socioecology with a brief account of current understanding of gorilla socioecology in terms of the interaction of males' and females' survival, mating, and rearing strategies. The framework that I will use interprets females as dispersed in relation to the distribution of food, and males as dispersed in relation to the distribution of females. In a taxon in which the female invests a lot of time and energy in production of young, females can probably make most difference to their reproductive success by seeking resources to support their reproductive capabilities; males, on the other hand, can most advance their reproductive success by seeking to mate with many females (Bradbury & Vehrencamp 1977; Clutton-Brock 1989b; Clutton-Brock & Parker 1992; Davies 1991, 1992; Dunbar 1984; Emlen & Oring 1977; Wrangham 1980; Wrangham & Rubenstein 1986).

This account thus starts with a presentation of our understanding of females' competitive and cooperative strategies as they relate to exploitation of environmental resources, particularly food. I then discuss current understanding of males' strategies as they relate to competition for access to mates. The section closes with an analysis of the interaction of male and female strategies, in particular with the influence of males' mating strategies on females. Most of the description of gorilla socioecology refers to the Virunga Volcano population of Rwanda, Uganda, and Zaire, because it is still the longest and most intensively studied gorilla population.

## FEMALE COMPETITION FOR FOOD

I started my study of gorillas in part because, as I said, gorillas appeared to differ from other primates. In Old World monkeys, bonds between related females explained, at a proximate, sociological level, the stability of groups. In gorillas, by contrast, Schaller's qualitative evidence indicated that bonds between the dominant male and females explained stability. My initial studies quantitatively confirmed the distinction. Female gorillas merely tolerated one another (Harcourt 1979b). Furthermore, they were also usually unrelated, because most females left the group in which they were born (Harcourt 1978; Harcourt et al. 1976). Until then, mammals, including primates, were characterized as female-resident/male-emigrant (females remained in the group or area in which they were born, and males dispersed). The explanation for the sex difference was that males could risk the obvious costs of emigration more than could females, because the potential benefits to males were greater given a greater variance in male reproductive success (Clutton-

Brock & Harvey 1976). The same explanation should have applied to gorillas. The reason it did not emerged in the 1970s, and was closely tied to the distribution of food and the consequent competition among females for it.

In brief, if food is rich and concentrated, then it is worth defending and defensible. If food occurs in patches large enough for more than one animal, then cooperation in defense is advantageous, especially if a relatively large number of animals are competing for it. And if cooperation is advantageous, then it is especially advantageous to cooperate with kin. Thus Old World frugivorous or omnivorous monkeys live in relatively stable associations of female kin, who defend rich resources against other kin-based groups. By contrast, if food is of relatively poor quality and widely distributed, it is not worth defending, and is indefensible in any case; thus cooperation in its defense is not beneficial (Clutton-Brock 1974, 1989b; Davies 1991; Davies & Houston 1984; Isbell 1991; van Hooff & van Schaik 1992; Wrangham 1980; Wrangham & Rubenstein 1986). The gorilla was a prime example of a folivorous primate that fell into the latter category, and for which we therefore had to search for another benefit to grouping of females than cooperative defense of food (Harcourt 1979b; Harcourt & Stewart 1987; Watts 1990, 1996, 1997, 1998; Wrangham 1979).

Observations from the early 1970s indicated that almost all gorilla females emigrated from their natal group (Harcourt 1978; Harcourt et al. 1976; Sicotte 1993; Watts 1990). However, by the early 1980s, enough females had either emigrated together, or remained in the group of their birth, for us to be able to contrast the competitive and cooperative interactions of kin with those of nonkin in gorilla groups. Furthermore, we had more detailed data on ecology and interindividual competition in gorillas. It turned out that except for emigration by females, the difference between gorillas and Old World monkeys was more one of degree than kind. Female gorillas, like female Old World monkeys, compete over food sufficiently that dominance hierarchies can be recognized; frequency of competition is proportional to number of competitors; and kin cooperate more frequently and intensely than do nonkin (Harcourt & Stewart 1989; Stewart & Harcourt 1987; Watts 1985, 1988, 1991b, 1994, 1996, 1997). Instead of an absolute distinction between female emigrant and female resident species, we now had to think of a behavioral ecology of competition and cooperation in which a balance of payoffs of staying is compared to those of emigrating. Residence is, in a sense, the default for females. If an animal has reached adulthood, its natal site is presumably relatively benign. By contrast, the inevitable consequence of emigration is loss of potential help from kin, and exposure to unfamiliar competitors and habitat (Clutton-Brock 1989b; Clutton-Brock & Harvey 1976; Isbell 1991; Isbell & Van Vuren 1996; Sterck et al. 1997; van Hooff & van Schaik 1992; Wrangham 1980; Wrangham & Rubenstein 1986). At the same time, even in female-resident species, females sometimes emigrate (Moore & Ali 1984). Clearly, there are costs to residence and benefits to emigration.

The studies on the gorilla showed that despite benefits of being with kin, nevertheless in a situation in which those benefits were not so great (because of the relative quality and distribution of food) females could afford to respond to other influences on decisions to stay or leave.

A major influence on female distribution is, potentially, male behavior and distribution. In early socioecological theory, males influenced the distribution of females in several ways. I return to them, and the gorilla male's influence on female distributions, after a consideration of male strategies.

### MALE COMPETITION FOR FEMALES

All evidence indicates the existence in gorillas of contest competition for long-term, sole access to females. We see males fight intensely (Harcourt 1978); their energetic threat displays are classic male displays of size, strength, and health (Schaller 1963); males are far larger and louder than are females, with obvious secondary sexual coloration (Schaller 1963); and group compositions of several females per male indicate a polygynous mating system (Harcourt et al. 1981a; Schaller 1963). Some early arguments, and statistical tests, concerning the correlations between degree of sexual dimorphism and adult sex ratio did not take enough account of allometric relationships, or of phylogenetic dependence. Phylogenetic effects on the relationship are strong, but when phylogeny is controlled for, allometric effects are weak, indeed effectively nonexistent in some primate taxa (Abouheif & Fairbairn 1997; Ford 1994; Kappeler 1991). Mitani et al. (1996) not only took account of allometry as well as phylogeny in their analysis of the relationship between dimorphism and mating system, but also incorporated a more precise measure of intensity of competition than socionomic sex ratio. Instead of using the usual measure of the number of adult males per fertile female in a social group, they calculated the operational sex ratio as the number of males per total fertile days of available females per year. The results confirmed the original analyses: mating system (intensity of male mating competition) correlates with degree of sexual dimorphism (Mitani et al. 1996).

Further evidence for long-term competition for sole access to females in gorillas comes from measures of testes mass in relation to body mass. Single-male taxa (whether monogamous, or uni-male polygynous) have smaller testes than do multimale taxa; in addition, the males of single-male taxa usually inseminate fewer sperm per ejaculate, produce the sperm more slowly, and exhaust their supplies more quickly (Dixson 1998; Gomendio et al. 1998; Harcourt et al. 1981b; Møller 1988a, 1989; Short 1979). The explanation is that in single-male systems, in which usually only one male mates with a fertile female per ovulatory period, males need to inseminate only enough sperm to ensure that sufficient sperm reach the egg to fertilize it. By contrast, in multimale species, in which several males mate with an estrous female, scramble competition among sperm to reach the ovum means that the male who inseminates the most sperm is the one probabilistically most likely to fertilize the female. Data on the mating behavior of gorillas and of other primates in the wild was crucial to development of sperm competition theory in primatology, which did not develop until the late 1970s (Short 1979). This now well-established theory, applicable in taxa as diverse as primates, passerines, and pierid butterflies (Birkhead & Møller 1998; Møller 1988b; Svärd & Wiklund 1989), has in turn increased our understanding of the long-term evolutionary processes behind the gorilla's social system.

Intense competition among gorilla males for sole access to females, and hence a highly polygynous mating system, should correlate with emigration of males in search of females, and aggressive takeover of groups of females (Clutton-Brock & Harvey 1976; Dobson 1982; Greenwood 1980). Nevertheless, about one-third of gorilla groups in the Virunga Volcano region and elsewhere contain more than one male (Harcourt et al. 1981a). It seems that the multimale groups exist, not because males independently join groups of females, as in other species, but because some males do not emigrate (Harcourt 1978; Harcourt et al. 1976). Thus the males of multimale gorilla groups appear to be kin that have remained in the group of their birth (Harcourt 1978; Harcourt et al. 1976; Robbins 1995; Stewart & Harcourt 1987; Yamagiwa 1987). Both a proximate analysis of the nature of social relationships between males, and between males and females, as well as a functional analysis of payoffs, can inform understanding of males' decisions to remain or leave. A high number and reproductive value of nonkin females in the natal group, and a low number or competitive ability of the other resident male(s) (old father, younger siblings) might encourage residence, while the opposite might encourage emigration (Harcourt 1979a; Robbins 1995; Sicotte 1994; Stewart & Harcourt 1987; Yamagiwa 1987). Both strategies, residence on the one hand, and emigration and establishment of a new group on the other, are potentially successful (Robbins 1995; Stewart & Harcourt 1987). As Dunbar (1984) nicely showed for gelada baboons, the relative contributions to success probably differ between the strategies. While residents might have to share mating, they might at the same time be more guaranteed access to females than a male who emigrates and has to establish a group on his own, and thus the residents might start to breed earlier. In addition, the residents can potentially gain inclusive fitness benefits by protecting related infants. And if the residents breed earlier, they are more likely to produce a son early enough to be useful as an ally in defense of the group later on, and therefore they might maintain tenure of the group for longer (Robbins 1995; Yamagiwa 1987). Furthermore, if the father dies, a son is unlikely to use infanticide as a mating strategy. By contrast, a male that establishes his own group and dies before a son is old enough to take over loses his youngest infants to the males that the females subsequently join (Fossey 1984; Watts 1989).

## INTERACTION OF FEMALE AND MALE STRATEGIES

In the basic socioecological framework of food influencing the distribution of females, which then influences the distribution of males, there is little room for males to influence the distribution of females. This absence might be the reason why the gorilla does not tidily fit into the Clutton-Brock scheme of mammalian mating systems (Clutton-Brock 1989b; Davies 1991). The gorilla female's range is not defensible (Fossey & Harcourt 1977; Mitani & Rodman 1979), and female groups are unstable (we know this from instances when a male dies, and the females disperse). In such a case, four mating systems are possible (Clutton-Brock 1989b). However, the gorilla, as represented by the Virunga population at least, does not fit any of them: it does not have mating territories, leks, or temporary harems, and male gorillas do not rove in search of females. In this section, I discuss the possible influence of male strategies on both female emigration and male-female associations.

Why do females emigrate? For a long time, avoidance of inbreeding was a main hypothesis for why one sex emigrates (Clutton-Brock & Harvey 1976; Packer 1979; Pusey & Packer 1987). Usually the male emigrates, because the potential benefits outweigh the costs more than they do for the female, especially if she benefits from residence with kin in a familiar range (Clutton-Brock & Harvey 1976). Ever since female emigration was first described in gorillas, the explanation for it has been avoidance of inbreeding (Harcourt 1978). It is the female, not the male, who moves, because males retain breeding tenure longer than the time it takes a female to mature (Harcourt 1978). The explanation seems to be general. Of nine species in which males had relatively long breeding tenure (including the gorilla), eight were female-emigrant, compared to none of sixteen species in which tenure was shorter than the time to maturity (Clutton-Brock 1989a). In the case of the gorilla, evidence for the inbreeding argument comes also from observation of mating behavior, and from timing of emigration in relation to group composition. Fathers and daughters are manifestly uninterested in mating with one another (Harcourt 1979c). Thus, if the only adult male in the group when a female matured was the father, all females emigrated. Females remained to give birth in their natal group only if there was more than one male. And after the first offspring, half those females then emigrated (Watts 1991a).

The argument that male tenure influences emigration by females begs the question of why species differ in the duration of tenure. In the case of the gorilla, the argument now returns to females' effects on male distribution. First, the usual small number of females per male coupled with infrequent estrus per female means that the rewards of a takeover are not high (Harcourt 1978). However, if the rewards are not high enough for an attempted takeover, why are they high enough for males to stay with females? Why do males not travel in search of more females than they can defend, the roving strategy of Clutton-Brock's system (1989b)? It is possible that a diet that involves half the day foraging leaves too little time for the roving strategy (Clutton-Brock 1989b; Fossey & Harcourt 1977; Watts 1988). Second, in a situation in which the distribution of food is such that continued residence in a known range with cooperative kin is not very advantageous to females, females have the option of abandoning an incompatible new male. If a contender cannot be guaranteed to retain all the expensively gained resource that he has just won, perhaps males cannot afford to fight furiously enough to oust the incumbent male (Watts 1989).

Avoidance of inbreeding explains why females leave the group in which they were born in those species in which males retain a long breeding tenure. It does not explain emigration by breeding females. The observation that breeding females emigrate, along with the observation that once a gorilla female has emigrated, she immediately moves to another male, brings us to the question of why female gorillas travel with males. In plenty of mammals, females travel on their own, as they do in the close taxonomic relatives of the gorilla, the orangutan and chimpanzee (Nishida & Hiraiwa-Hasegawa 1987; Rodman & Mitani 1987). That female gorillas travel with males is particularly puzzling in light of the fact that the male is twice the size of a female, and a main competitor for food: female gorillas receive six times as much competition for food from a male as they do from the average female group member (Stewart & Harcourt, 1987). Until the mid-1970s, primate socioecologists argued that predator defense was one explicit benefit that males could provide females (Crook 1970; Crook & Gartlan 1966). By the late 1970s, the role of predation was being questioned. For instance Wrangham (1979, 1980) explicitly excluded predation as an explanation of grouping, because it did not explain variation in the nature of groups. Nevertheless, predation remains a potential explanation for females' association with one another (Sterck et al. 1997; van Schaik 1983; van Schaik & Hörstermann 1994). Several lines of evidence indicate that female gorillas use males for protection against predation (Harcourt 1978; Stewart & Harcourt 1987). For instance, detection of danger sends females clustering around the male; and male gorillas have a very specific, roaring, charging antipredator display given only in the presence of potential predators.

Conversely, several reviews suggest that a main benefit to female primates of traveling with males is protection against infanticide (Palombit 1999; Sterck et al. 1997; van Schaik & Dunbar 1990; van Schaik & Kappeler 1997; Wrangham 1979). Such is argued to be especially the case for species where the male apparently helps little in raising of the young, and could cover the ranges of more females than the number with which he associates (van Schaik & Dunbar 1990). Why else, other than remaining to protect offspring from infanticidal males, should the males remain with just a few females? Baboons and gorillas are prime examples for this argument concerning the influence of grouping on male-female association (Palombit 1999). Furthermore, observation of male-female associations within species show that they vary with risk of infanticide, as would be expected were risk of infanticide affecting the association (Palombit et al. 1997). Nevertheless, while lions (*Panther leo*) are the classic infanticidal species (Bertram 1975), and infanticide has been argued to affect grouping in the species (Packer 1990; Packer & Puset 1983), infanticide is not yet argued to be a prevalent cause of male-female associations in mammals in general (Clutton-Brock 1989b).

Whether or not infanticide is more advantageous to primates than nonprimates, a possible problem with the infanticide hypothesis for male-female association is that if a female joins a single male, and breeds with him, then infanticide becomes a potential reproductive tactic for the other males in the population. If instead, the female did not associate with one male, but mated with all the local males, the female could inhibit infanticide (Hrdy 1979). In the case of the Virunga gorilla population, the female's association with one male indeed correlates with a high rate of infanticide. Infant mortality rates are 35 to 40 percent, of which infanticide accounts for about 35 percent. In other words, about 13 percent of all infants in the Virunga region are killed by nonfather males (Fossey 1984; Watts 1989). Why, then, do gorilla females not avoid infanticide (and competition for food) by traveling on their own, and mating with most or all of the local males, who would then act as fathers (Hrd 1979)? After all, chimpanzee females apparently follow this strategy (Goodall 1986; Tutin 1980; Wallis 1997). The answer appears to be that gorilla females cannot. A simulation of this lone ranging strategy indicates that because of the female's short estrus period, and short daily distance traveled, she cannot mate with enough males in the local population to prevent her meeting—before her infant is weaned—a male with whom she has not previously mated (Harcourt & Greenberg, in press). Altering either duration of estrus, or daily travel distance, to the chimpanzee's values reduces infanticide to nearly zero. If the gorilla female cannot change either trait, she is forced to associate with a protective male. In other words, the anti-infanticide hypothesis for female-male association in gorillas is correct. Females associate with males, in part, for the anti-infanticidal protection the males afford; males associate with females, in this hypothesis, because if they did not, their offspring would all die (Sterck et al. 1997; Watts 1996; Wrangham 1979). Gorillas are thus a species in which male help in raising offspring is required by the females. Groups exist, instead of male-female pairs, because several females independently choose the same powerful male to protect them from infanticidal males, as well as from predators (Orians 1969; Stewart & Harcourt 1987; Watts 1990).

In summary, then, both proximate analyses of competitive and cooperative behavior among individuals, and comparative functional analysis of payoffs of association, suggest that much of the gorilla's social system fits the framework of food influencing the distribution of females, who then influence the distribution of males. However, in this female-emigrant species, male distribution and mating strategies also influence female distribution and strategies.

## The Approach

We are all encouraged to make our science hypothetico-deductive. If we want a grant, we need specific hypotheses from which to deduce specific, testable predictions. And the larger the sample size, the more likely we are to get the grant. As most readers are used to the hypothetico-deductive approach described, and it is so obviously a powerful method of procedure in science, I will suggest here some advantages of a more open approach. A goal of my study was to describe a relatively unknown, unusual, primate species. Because, as said above, socioecology is a comparative enterprise, good theory in socioecology needs a lot of good description, far more than one person can supply. Hypothetic-deductivism works well in a well-known field. Socioecology in the early 1970s was not a well enough known field, I suggest. Thus, it was not until 1977 that there appeared the first good, quantified comparative analyses in vertebrate socioecology. And they did not appear until then, because they had to wait for good, quantitative studies from a large variety of species (Clutton-Brock 1977). Study of variety for its own sake was a useful approach at the time. If describing and understanding variety is the goal, exceptions are especially useful because they are, by definition almost, extremes in the range of variety. More than that, exceptions are explicitly good tests of current hypotheses. Science is the production of testable hypotheses. How better to test the generality of a hypothesis than to investigate how an exception fits into the hypothesis? If you want to find out why the strongest cooperative relationships are seen among females, go to a species where the strongest cooperative relationships are not among females.

Another reason to support open, descriptive studies in a new field is to ensure collection of a sufficiently broad array of data that new connections can be seen, new deductions can be tested. Sixty years ago, Adolph Schultz was doing descriptive comparative anatomy, not hypothetico-deductive science. Among many other findings, he pointed out that the testicular weight of primates relative to their body weight differed enormously across species (Schultz 1938) . He had no idea why. None of us studying the behavior of wild primates in the field in the 1970s even knew of Schultz's study, let alone thought about his question. If Schultz, and Ph.D. students in the 1970s had been constrained to only hypothetico-deductive science on species that could produce large sample sizes quickly, Schultz would not have collected data on testicular weight of wild primates, the students would not have spent years watching wild primates, and we would still not be able to answer

Schultz's question. As it is, we now know the answer to why relative testicular size differs so much among species. It is sperm competition.

Geoff Parker (1970) was the originator of sperm competition theory. However, among primatologists, Roger Short's (1979) was the seminal contribution, with its demonstration that in apes and humans, testicular size correlated with mating system. A later analysis showing the same correlation for thirty species of primates strongly confirmed the theory (Harcourt et al. 1981b). This latter demonstration was possible because the large number of descriptive, primatological field studies available by the end of the 1970s finally allowed the answer to Schultz's question. Those field studies provided convincing, quantitative proof that in taxa with large testes relative to their body size, several males mated with potentially fertile females, whereas in taxa with small testes, usually only one male mated with a potentially fertile female. This demonstration across the primates indicated the probable ubiquity of sperm competition, and hence its worth as a subject of study. And empirical confirmation has continued unabated (Birkhead & Møller 1998). Not one of the primatological field studies, including Schultz's, had been done to test deductions from hypotheses related to sperm competition theory. Rather, two completely independent sets of data, collected in ignorance of one another fifty years apart, had done precisely what studies of variety for its own sake are meant to do. They had produced a fundamental shift in understanding that might not have occurred if we had all stuck to only hypothetico-deductive science, especially if we had stuck to deduction from hypotheses using only species on which we could obtain large sample sizes. If we study only those species that produce large sample sizes, we will provide a description of the world based on *Drosophila* and mice. That is not good enough, especially in a comparative discipline such as socioecology.

Finally, for most biologists, primates are not good subjects for behavioral-ecological study. Sample size is inevitably small, and the lifetime reproductive success of alternative decisions is in effect unobtainable. I gave in the "Origins" section a scientific explanation for my study of a primate. Personal history also surely played a part. I was an undergraduate in Cambridge, England, in the early1970s, and Cambridge University then was full of innovative primatology graduates, thanks largely to Robert Hinde.

## The Future

Vertebrate socioecology is still a wide open field of inquiry, as I have indicated throughout this chapter. We do not know why, for instance, species differ in the time that males can retain breeding tenure. Even for one species, the gorilla, much remains unknown. A major problem confronts vertebrate socioecology, and especially the socioecology of long-lived species such as primates. Socioecology requires field studies. But as pointed out by Dobson and Lyles (1989), most field studies are done by Ph.D. students; and therefore most field studies last just 1.5 years, which is a tiny fraction of a primate's life span. Furthermore, most studies are of just one or two social groups, which might have at the most a few tens of individuals. Thus for gorillas, it will surely be another two or three decades before we begin to get a large enough sample size to test theories concerning the relative success of different emigration strategies of males and females. We have not even yet obtained good data on which male in multimale gorilla groups is doing most of the fertile mating. Intraspecific variation might be a particularly useful avenue of socioecological study, because there should be fewer confounding variables. As I mentioned, most of our knowledge of gorilla socioecology comes from one population. It is at the extreme eastern and extreme altitudinal range of the gorilla's distribution (Schaller 1963). Gorilla socioecology needs to increase its intraspecific analysis (Yamagiwa 1998). For instance, it can take advantage of the fact that the west African gorilla's diet is more frugivorous than that of the eastern populations (Doran & McNeilage 1998; Harcourt et al. 1981a; Remis 1997; Tutin 1996; Tutin & Fernandez 1985; Tutin et al. 1992; Williamson et al. 1990). That difference, and our current understanding of primate socioecology, allows precise predictions of contrasts that we should observe in individuals' competitive and cooperative regimes. In addition, we can predict in the western gorilla populations more contest competition over food, and more cooperation among kin in that competition (Isbell 1991; van Hooff & van Schaik 1992; Wrangham 1980; Wrangham & Rubenstein 1986).

Many areas of primate (and mammalian) socioecology need further development. We still know far too little about conflict between male and female strategies. As just one example, immense variation exists within and between primate species in the degree to which overt competitive ability correlates with either mating success or reproductive success (e.g., Harcourt 1987; Smuts 1987). In the case of male success, it seems that preference by females for unfamiliar males often reduces the potential reproductive success of older males within a group (Smuts 1987). Are females avoiding inbreeding by choosing unfamiliar males, or are they somehow identifying beneficial alternative traits of mates (Smuts 1987)? How do they avoid coercion by dominant males (Clutton-Brock & Parker 1995; Smuts & Smuts 1993)? And how much does the conflict, or its resolution, explain the fact that prosimians are essentially monomorphic whatever their apparent mating system (Kappeler 1993)?

Two particular aspects of conflict between the sexes' strategies that the study of gorilla socioecology raised are dispersal of individuals from their natal group and infanticide. Male strategies force dispersal by females, it is suggested, whereas the beneficial default is to remain in the well-known region of birth, with cooperative kin. And, of course, the males' strategy of infanticide is a severe constraint on female strategies. Not only does it diminish females' reproductive output, but it apparently forces females to accept association with large, competitive males. Sex bias in dispersal of individuals from their natal group, and infanticide, are crucial components of not just the gorilla's unusual social system, but of primate socioecology as a whole. Improving our understanding of dispersal would improve our understanding of many aspects of whole-animal biology besides socioecology, including demography, epidemiology, and conservation biology (Dobson & May 1986; Hanski 1998; Hanski & Gilpin 1997; Mollison 1987). Yet, because dispersal almost by definition takes organisms out of the study area, good data on dispersal is sorely lacking, let alone a good behavioral ecology of dispersal. Thus the recorded distance of dispersal of organisms seems to be determined far more by our ability to detect the distance traveled than by the distance actually traveled (Koenig et al. 1996); and we still lack a cohesive theory of sex biasing in dispersal (Greenwood 1980; Koenig et al. 1996). Gorilla studies are normal in the poverty of information that we have on dispersing individuals. With regard to infanticide, its behavioral ecology is well understood, as is, for at least some species, its proximate causation (Parmigiani & vom Saal 1994). However, we still sorely need a socioecology that explains why species differ so greatly in the use by males of infanticide as a reproductive strategy. Why do the Virunga gorillas have such a high rate of infanticide, whereas infanticide is essentially unknown in orangutans, despite the fact the females appear to mate with a minority of males, and do not travel with males for protection (Rodman & Mitani 1987)? If primate socioecology needs a single, practicable goal, an explanation of the variation in interspecific incidence of infanticide might be an illuminating one.

While primate socioecology is a well-substantiated body of theory (Janson 2000; Lee 1999; Smuts et al. 1987), even so, it is largely missing a crucial component of scientific procedure: experimentation. Experimentation in mammalian socioecology has been successfully done (Clutton-Brock 1989c, 1999; Janson 1996; Noë & Bshary 1997; Palombit et al. 1997). However, for the most part, manipulation on the scale necessary is not easy. And for vulnerable species, such as the gorilla, it might even be unethical. An alternative to experimentally manipulating the animals or the environment itself is to manipulate it on the computer. While primate socioecology has such modeling and simulations (e.g., Dunbar 1996; Janson & Goldsmith 1995; Wrangham et al. 1993), they are nevertheless perhaps unusually rare (Harcourt 1998). In the case of the gorilla, I have mentioned the simulation of a dispersed gorilla society to test the hypothesis that protection from infanticide is the main benefit to females of association with males (Harcourt & Greenberg, in press). Such an "experiment" is, of course, possible only with a simulation. Not only are modeling and simulations thus valuable additional tools for the primate socioecologist, it might be that they will become a main tool. If tropical countries' human populations continue to increase, and the increase is associated with resultant political and economic instability (Homer-Dixon et al. 1993), fieldwork on tropical taxa might become a thing of the past.

## Acknowledgments

I thank Kelly Stewart and referees for criticism that considerably improved the manuscript.

## References

Abouheif E, Fairbairn DJ, 1997. A comparative analysis of allometry for sexual size dimorphism: assessing Rensch's rule. Am Nat 149:540–562.

Alexander RD, 1974. The evolution of social behavior. Ann Rev Eco Sys 5:325–383.

Bertram BCR, 1975. Social factors influencing reproduction in wild lions. J Zool Lond 177:463–482.

Birkhead TR, Møller AP (eds), 1998. Sperm competition and sexual selection. London: Academic Press.

Bradbury JW, Vehrencamp SL, 1977. Social organization and foraging in emballonurid bats, III: mating systems. Behav Ecol Sociobiol 2:1–17.

Bramblett CA, 1970. Coalitions among gelada baboons. Primates 11:327–333.

Cheney DL, Seyfarth RM, 1990. How monkeys see the world. Inside the mind of another species. Chicago: University of Chicago Press.

Clutton-Brock TH, 1974. Primate social organisation and ecology. Nature 250:539–542.

Clutton-Brock TH (ed), 1977. Primate ecology. London: Academic Press.

Clutton-Brock TH, 1989a. Female transfer and inbreeding avoidance in social mammals. Nature 337:70–72.

Clutton-Brock TH, 1989b. Mammalian mating systems. Proc Roy Soc Lond B 236:339–372.

Clutton-Brock TH, 1989c. Mate choice on fallow deer leks. Nature 340:463–465.

Clutton-Brock TH, 1999. Selfish sentinels in cooperative mammals. Science 284:1640–1644.

Clutton-Brock TH, Harvey PH, 1976. Evolutionary rules and primate societies. In: Growing points in ethology (Bateson PPG, Hinde RA, eds). Cambridge: Cambridge University Press; 195–237.

Clutton-Brock TH, Parker GA, 1992. Potential reproductive rates and the operation of sexual selection. Q Rev Biol 67:437–456.

Clutton-Brock TH, Parker GA, 1995. Sexual coercion in animal societies. Anim Behav 49:1345–1365.

Crook JH, 1965. The adaptive significance of avian social organisations. In: Sexual selection and the descent of man (Campbell B, ed). London: Heinneman; 231–281.

Crook JH, 1970. The socio-ecology of primates. In: Social behaviour in birds and mammals (Crook JH, ed). London: Academic Press.

Crook JH, Ellis JE, Goss-Custard JD, 1976. Mammalian social systems: structure and function. Anim Behav 24:261–274.

Crook JH, Gartlan JS, 1966. Evolution of primate societies. Nature 210:1200–1203.

Crook JH, Goss-Custard JD, 1972. Social ethology. Annual Review of Psychology 23:277–312.

Davies NB, 1991. Mating systems. In: Behavioural ecology. An evolutionary approach (Krebs JR, Davies NB, eds). Oxford: Blackwell Scientific Publications; 263–294.

Davies NB, 1992. Dunnock behaviour and social evolution. Oxford: Oxford University Press.

Davies NB, Houston AI, 1984. Territory economics. In: Behavioural ecology. An evolutionary approach (2nd ed.) (Krebs JR, Davies NB, eds). Oxford: Blackwell Scientific Publications; 148–169.

DeVore I (ed), 1965. Primate behavior. Field studies of monkeys and apes. New York: Holt, Rinehart and Winston.

Dixson AF, 1998. Primate sexuality. Oxford: Oxford University Press.

Dobson AP, Lyles AM, 1989. The population dynamics and conservation of primate populations. Conserv Biol 3:362–380.

Dobson AP, May RM, 1986. Disease and conservation. In: Conservation biology. The science of scarcity and diversity (Soulé ME, ed). Sunderland, MA: Sinauer Associates, Inc.; 345–365.

Dobson FS, 1982. Competition for mates and predominant juvenile male dispersal in mammals. Anim Behav 30:1183–1192.

Doran DM, McNeilage A, 1998. Gorilla ecology and behavior. Evolutionary Anthropology 6:120–131.

Dunbar RIM, 1984. Reproductive decisions. An economic analysis of gelada baboon social strategies. Princeton: Princeton University Press.

Dunbar RIM, 1996. Determinants of group size in primates: a general model. In: Evolution of social behaviour patterns in primates and man (Runciman WG, Maynard Smith J, Dunbar RIM, eds). Oxford: Oxford University Press; 33–57.

Eisenberg JF, Muckenhirn NA, Rudran R, 1972. The relation between ecology and social structure in primates. Science 176:863–874.

Emlen ST, Oring LW, 1977. Ecology, sexual selection, and the evolution of mating systems. Science 197:215–223.

Ford SM, 1994. Evolution of sexual dimorphism in body weight in Platyrrhines. A J Primatol 34:221–244.

Fossey D, 1984. Infanticide in mountain gorillas (*Gorilla gorilla beringei*) with comparative notes on chimpanzees. In: Infanticide: comparative and evolutionary perspectives (Hausfater G, Hrdy SB, eds). New York: Aldine Publishing Co.; 217–236.

Fossey D, Harcourt AH, 1977. Feeding ecology of free ranging mountain gorilla. In: Primate ecology (Clutton-Brock TH, ed). London: Academic Press; 415–447.

Gartlan JS, 1968. Structure and function in primate society. Folia Primatol 8:89–120.

Gomendio M, Harcourt AH, Roldan E, 1998. Sperm competition in mammals. In: Sperm competition (Birkhead TM, Møller AP, eds). London: Academic Press; 667–755.

Goodall J, 1986. The chimpanzees of Gombe. Cambridge: Belknap Press.

Greenwood PJ, 1980. Mating systems, philopatry and dispersal in birds and mammals. Anim Behav 28:1140–1162.

Hanski I, 1998. Metapopulation dynamics. Nature 396:41–49.

Hanski IA, Gilpin ME (eds), 1997. Metapopulation biology. Ecology, genetics, and evolution. San Diego: Academic Press.

Harcourt AH, 1978. Strategies of emigration and transfer by primates with particular reference to gorillas. Zeitschrift für Tierpsychologie 48:401–420.

Harcourt AH, 1979a. Contrasts between male relationships in wild gorilla groups. Behav Ecol Sociobio 5:39–49.

Harcourt AH, 1979b. Social relationships among adult female mountain gorillas. Anim Behav 27:251–264.

Harcourt AH, 1979c. Social relationships between adult male and female mountain gorillas. Anim Behav.

Harcourt AH, 1987. Dominance and fertility in female primates. J Zool Lond 213:471–487.

Harcourt AH, 1992. Coalitions and alliances: are primates more complex than non-primates? In: Coalitions and alliances in humans and other animals (Harcourt AH, de Waal FBM, eds). Oxford: Oxford University Press; 445–472.

Harcourt AH, 1998. Does primate socio-ecology need non-primate socio-ecology? Evol Anthrop 7:3–7.

Harcourt AH, de Waal FBM (eds), 1992. Coalitions and alliances in humans and other animals. Oxford: Oxford University Press.

Harcourt AH, Fossey D, Sabater Pi J, 1981a. Demography of *Gorilla gorilla*. J Zool Lond 195:215–233.

Harcourt AH, Greenberg J, in press. A model of mate searching strategies, infanticide, and grouping for gorillas. Anim Behav.

Harcourt AH, Harvey PH, Larson SG, Short RV, 1981b. Testis weight, body weight and breeding system in primates. Nature 293:55–57.

Harcourt AH, Stewart KJ, 1987. The influence of help in contests on dominance rank in primates: hints from gorillas. Anim Behav 35:182–190.

Harcourt AH, Stewart KJ, 1989. Functions of alliances in contests within wild gorilla groups. Behaviour 109:176–190.

Harcourt AH, Stewart KJ, Fossey D, 1976. Male emigration and female transfer in wild mountain gorilla. Nature 263:226–227.

Harvey PH, Leigh Brown AJ, Maynard Smith J, Nee S (eds), 1996. New uses for new phylogenies. Oxford: Oxford University Press.

Harvey PH, Pagel MD, 1991. The comparative method in evolutionary biology. Oxford: Oxford University Press.

Hinde RA, 1972. Social behavior and its development in subhuman primates. Eugene: Condon Lectures, Oregon State System of Higher Education.

Hinde RA, 1976. Interactions, relationships, and social structure. Man 11:1–17.

Hinde RA (ed), 1983. Primate social relationships. An integrated approach. Oxford: Blackwell Scientific Publishers.

Homer-Dixon TF, Boutwell JH, Rathjens GW, 1993. Environmental change and violent conflict. Sci Amer February:38–45.

Hrdy SB, 1979. Infanticide among animals: a review, classification, and examination of the implications for the reproductive strategies of females. Ethol Sociobiol 1:13–40.

Imanishi K, 1963. Social behavior in Japanese macaques, *Macaca fuscata*. In: Primate social behavior (Southwick CH, ed). New York: Van Nostrand Reinhold Co.; 68–81.

Isbell LA, 1991. Contest and scramble competition: patterns of female aggression and ranging behavior among primates. Behav Ecol 2:143–155.

Isbell LA, Van Vuren D, 1996. Differential costs of locational and social dispersal and the consequences for female group-living primates. Behaviour 133.

Janson CH, 1996. Toward an experimental socioecology of primates. In: Adaptive radiations of neotropical primates (Norconk MA, Rosenberger AL, Garber PA, eds). New York: Plenum Press; 309–325.

Janson CH, 2000. Primate socio-ecology: the end of a golden age. Evol Anthrop 9:83–86.

Janson CH, Goldsmith ML, 1995. Predicting group size in primates: foraging costs and predation risks. Behav Ecol 6:326–336.

Jarman PJ, 1974. The social organisation of antelope in relation to their ecology. Behaviour 48:215–267.

Kappeler PM, 1991. Patterns of sexual dimorphism in body weight among prosimian primates. Folia Primatol 57:132–146.

Kappeler PM, 1993. Sexual selection and lemur social systems. In: Lemur social systems and their ecological basis (Kappeler PM, Ganzhorn JU, eds). New York: Plenum Press; 225–242.

Kawai M, 1965. On the system of social ranks in a natural troop of Japanese monkeys. In: Japanese monkeys. A collection of translations (Altmann SA, ed). Atlanta: Yerkes Regional Primate Center.

Kawamura S, 1965. Matriarchal social ranks in the Minoo B group: a study of the rank system of Japanese monkeys. In: Japanese monkeys. A collection of translations (Altmann SA, ed). Atlanta: Yerkes Regional Primate Center.

Koenig WD, Van Vuren D, Hooge PN, 1996. Detectability, philopatry, and the distribution of dispersal distances in vertebrates. Trends Ecol Evol 11:514–517.

Kummer H, 1968. Social organization of Hamadryas baboons. Chicago: University of Chicago Press.

Lack D, 1968. Ecological adaptations for breeding in birds. London: Methuen.

Lee PC (ed), 1999. Comparative primate socioecology. Cambridge: Cambridge University Press.

Marler P, 1996. Social cognition: are primates smarter than birds? In: Current Ornithology 13 (Nolan V, Jr., Ketterson ED, eds). New York: Plenum Press; 1–32.

Martins EP (ed), 1996. Phylogenies and the comparative method in animal behavior. New York: Oxford University Press.

McCleery RH, 1978. Optimal behaviour sequences and decision making. In: Behavioural ecology. An evolutionary approach (Krebs JR, Davies NB, eds). Oxford: Blackwell Scientific Publications; 377–410.

Missakian EA, 1972. Genealogical and cross-genealogical dominance relations in a group of free-ranging rhesus monkeys (*Macaca mulatta*) on Cayo Santiago. Primates 13:169–180.

Mitani JC, Gros-Louis J, Richards AF, 1996. Sexual dimorphism, the operational sex ratio, and the intensity of male competition in polygynous primates. Am Nat 147:966–980.

Mitani JC, Rodman PS, 1979. Territoriality: the relation of ranging pattern and home range size to defendability, with an analysis of territoriality among primate species. Behav Ecol Sociobiol 5:241–251.

Møller AP, 1988a. Ejaculate quality, testes size and sperm competition in primates. J Hum Evol 17:479–488.

Møller AP, 1988b. Testes size, ejaculate quality and sperm competition in birds. Biol J Linn Soc 33:273–283.

Møller AP, 1989. Ejaculate quality, testes size and sperm production in mammals. Func Ecol 3:91–96.

Mollison D, 1987. Population dynamics of mammalian diseases. Symposium of the Zoological Society of London 58:329–342.

Moore J, Ali R, 1984. Are dispersal and inbreeding avoidance related? Anim Behav 32:94–112.

Nishida T, Hiraiwa-Hasegawa M, 1987. Chimpanzees and bonobos: cooperative relationships among males. In: Primate societies (Smuts BB, Cheney DL, Seyfarth RM, Wrangham RW, Struhsaker TT, eds). Chicago: University of Chicago Press; 165–177.

Noë R, Bshary R, 1997. The formation of red colobus-diana monkey associations under predation pressure from chimpanzees. Proc Roy Soc Lond B 264:253–259.

Orians GH, 1969. On the evolution of mating systems in birds and mammals. Am Nat 103:589–603.

Packer C, 1979. Inter-troop transfer and inbreeding avoidance in *Papio anubis*. Anim Behav 27:1–36.

Packer C, 1990. Why lions form groups: food is not enough. Am Nat 136:1–19.

Packer C, Puset AE, 1983. Adaptations of female lions to infanticide by incoming males. Am Nat 121:716–728.

Palombit RA, 1999. Infanticide and the evolution of pair bonds in nonhuman primates. Evol Anthrop 7:117–129.

Palombit RA, Seyfarth RM, Cheney DL, 1997. The adaptive value of 'friendships' to female baboons: experimental observational evidence. Anim Behav 54:599–614.

Parker GA, 1970. Sperm competition and its evolutionary consequences in the insects. Biological Review 45:525–567.

Parmigiani S, vom Saal FS (eds), 1994. Infanticide & parental care. Chur, Switzerland: Harwood Academic Publishers.

Purvis A, 1995. A composite estimate of primate phylogeny. Philosophical Transactions of the Royal Society of London B 348:405–421.

Pusey AE, Packer C, 1987. Dispersal and philopatry. In: Primate societies (Smuts BB, Cheney DL, Seyfarth RM, Wrangham RW, Struhsaker TT, eds). Chicago: University of Chicago Press; 250–266.

Remis MJ, 1997. Ranging and grouping patterns of a western lowland gorilla group at Bai Hokou, Central African Republic. Am J Primatol 43:111–133.

Robbins MM, 1995. A demographic analysis of male life history and social structure of mountain gorillas. Behaviour 132:21–47.

Rodman PS, Mitani JC, 1987. Orangutans: sexual dimorphism in a solitary species. In: Primate societies (Smuts BB, Cheney DL, Seyfarth RM, Wrangham RW, Struhsaker TT, eds). Chicago: University of Chicago Press; 146–154.

Ruvolo M, Pan D, Zehr S, Goldberg T, Disotell TR, Dornum von M, 1994. Gene trees and hominoid phylogeny. Proc Natl Acad Sci 91:8900–8904.

Sade DS, 1967. Determinants of dominance in a group of free-ranging rhesus monkeys. In: Social communication among primates (Altmann SA, ed). Chicago: University of Chicago Press; 99–114.

Schaller GB, 1963. The mountain gorilla. Ecology and behavior. Chicago: University of Chicago Press.

Schultz AH, 1938. The relative weight of the testes in primates. The Anatomical Record 72:387–394.

Short RV, 1979. Sexual selection and its component parts, somatic and genital selection, as illustrated by man and the great apes. Adv St Behav 9:131–158.

Sicotte P, 1993. Inter-group encounters and female transfer in mountain gorillas: influence of group composition on male behavior. Am J Primatol 30:21–36.

Sicotte P, 1994. Effect of male competition on male-female relationships in bi-male groups of mountain gorillas. Ethology 97:47–64.

Smuts BB, 1987. Sexual competition and mate choice. In: Primate societies (Smuts BB, Cheney DL, Seyfarth RM, Wrangham RW, Struhsaker TT, eds). Chicago: University of Chicago Press; 385–399.

Smuts BB, Cheney DL, Seyfarth RM, Wrangham RW, Struhsaker TT (eds), 1987. Primate societies. Chicago: University of Chicago Press.

Smuts BB, Smuts RW, 1993. Male aggression and sexual coercion of females in nonhuman primates and other mammals: evidence and theoretical implications. Adv St Behav 22:1–63.

Sterck EHM, Watts DP, van Schaik CP, 1997. The evolution of female social relationships in nonhuman primates. Behav Ecol Sociobiol 41:291–309.

Stewart KJ, Harcourt AH, 1987. Gorillas: variation in female relationships. In: Primate societies (Smuts BB, Cheney DL, Seyfarth RM, Wrangham RW, Struhsaker TT, eds). Chicago: University of Chicago Press; 155–164.

Svärd L, Wiklund C, 1989. Mass and production rate of ejaculates in relation to monandry/polyandry in butterflies. Behav Ecol Sociobiol 24:395–402.

Tinbergen N, 1963. On aims and methods of ethology. Z Tierpsychol 20:410–433.

Tutin CEG, 1980. Reproductive behaviour of wild chimpanzees in the Gombe National Park, Tanzania. J Reprod Fertil, Suppl 28:43–57.

Tutin CEG, 1996. Ranging and social structure of lowland gorillas in the Lopé Reserve, Gabon. In: The great apes (McGrew WC, Marchant LF, Nishida T, eds). Cambridge: Cambridge University Press; 58–70.

Tutin CEG, Fernandez M, 1985. Foods consumed by sympatric populations of *Gorilla g. gorilla* and *Pan t. troglodytes* in Gabon. Int J Primatol 6:27–43.

Tutin CEG, Fernandez M, Rogers ME, Williamson EA, 1992. A preliminary analysis of the social structure of lowland gorillas in the Lopé Reserve, Gabon. In: Topics in primatology. 2. Behavior, ecology, and conservation (Itoigawa N, Sugiyama Y, Sackett GP, Thompson RKR, eds). Tokyo: University of Tokyo Press; 245–253.

van Hooff JARAM, van Schaik CP, 1992. Cooperation in competition: the ecology of primate bonds. In: Coalitions and alliances in humans and other animals (Harcourt AH, de Waal F, eds). Oxford: Oxford University Press; 357–389.

van Schaik CP, 1983. Why are diurnal primates living in groups? Behaviour 87:120–144.

van Schaik CP, Dunbar RIM, 1990. The evolution of monogamy in large primates: a new hypothesis and some crucial tests. Behaviour 115:30–62.

van Schaik CP, Hörstermann M, 1994. Predation risk and the number of adult males in a primate group. Behav Ecol Sociobiol 35:261–272.

van Schaik CP, Kappeler PM, 1997. Infanticide risk and the evolution of male-female association in primates. Proc Roy Soc Lond B 1997:1681–1694.

Vandenburgh JG, 1967. The development of social structure in free-ranging rhesus monkeys. Behaviour 29:179–194.

Wallis J, 1997. A survey of reproductive parameters in the free-ranging chimpanzees of Gombe National Park. J Reprod Fertil 109:297–307.

Watts DP, 1985. Relations between group size and composition and feeding competition in mountain gorilla groups. Anim Behav 33:72–85.

Watts DP, 1988. Environmental influences on mountain gorilla time budgets. Am J Primatol 15:195–211.

Watts DP, 1989. Infanticide in mountain gorillas: new cases and a reconsideration of the evidence. Ethology 81:1–18.

Watts DP, 1990. Ecology of gorillas and its relation to female transfer in mountain gorillas. Int J Primatol 11:21–44.

Watts DP, 1991a. Mountain gorilla reproduction and sexual behavior. Am J Primatol 24:211–225.

Watts DP, 1991b. Strategies of habitat use by mountain gorillas. Folia Primatol 56:1–16.

Watts DP, 1994. Social relationships of immigrant and resident female mountain gorillas, II. Relatedness, residence, and relationships between females. Am J Primatol 32:13–30.

Watts DP, 1996. Comparative socio-ecology of gorillas. In: Great ape societies (McGrew WC, Marchant LF, Nishida T, eds). Cambridge: Cambridge University Press; 16–28.

Watts DP, 1997. Agonistic intervention in wild mountain gorilla groups. Behaviour 134:23–57.

Watts DP, 1998. Long-term habitat use by mountain gorillas (*Gorilla gorilla beringei*). 1. Consistency, variation, and home range size and stability. Int J Primatol 19:651–680.

Williamson EA, Tutin CEG, Rogers ME, Fernandez M, 1990. Composition of the diet of lowland gorillas at Lopé in Gabon. Am J Primatol 21:265–277.

Wrangham RW, 1979. On the evolution of ape social systems. Soc Sci Inform 18:334–368.

Wrangham RW, 1980. An ecological model of female-bonded primate groups. Behaviour 75:262–300.

Wrangham RW, Gittleman JL, Chapman CA, 1993. Constraints on group size in primates and carnivores: population density and day range as assays of exploitation competition. Behav Ecol Sociobiol 32:199–209.

Wrangham RW, Rubenstein DI, 1986. Social evolution in birds and mammals. In: Ecological aspects of social evolution (Rubenstein DI, Wrangham RW, eds). Princeton: Princeton University Press; 452–470.

Yamagiwa J, 1987. Male life history and the social structure of wild mountain gorillas (*Gorilla gorilla beringei*). In: Evolution and coadaptation in biotic communities (Kawanao S, Connell JH, Hidaka T, eds). Tokyo: University of Tokyo Press; 31–51.

Yamagiwa J, 1998. Socioecological factors influencing population structure of gorillas and chimpanzees. Primates 40:87–104.

# 25 Cheetahs and Their Mating System

T. M. Caro and M. J. Kelly

## The Cheetah as a Study Animal

The cheetah (*Acinonyx jubatus*) is a blatantly charismatic species that is highly endangered. What student would not jump at the opportunity to work on a species that should easily attract research funding and might even be prevented from declining further as a result of their research? After each spending ten years of our lives working on this species, we are in a strong position to identify the strengths and shortcomings of using the cheetah as a model system.

How did we get involved in cheetah behavioral ecology and later in cheetah conservation biology? In 1979, TMC was finishing his Ph.D. on the behavior of domestic cats (*Felis domesticus*) in Cambridge and wanted to place his understanding of behavioral development in cats in an ecological framework. He had visited eastern Africa before starting as an undergraduate and had been an assistant to researchers at the Serengeti Research Institute, where he had met Brian Bertram. After working with large mammals in Africa, TMC was keen to find a way to go back and Brian, now writing up his lion (*Panthera leo*) work at Cambridge, mentioned that there was no one continuing the cheetah study in Serengeti. It seemed a natural fit: here was an habituated population of a species related to domestic cats, a familiar ecosystem where observations were easy, and there was long-term demographic data on individually recognized animals. For MJK, a senior undergraduate in 1990, the chance to conduct an internship working on a high-profile species was too good to miss. She soon realized that she was looking at a twenty-five-year virtually unexplored demographic data set (Kelly et al. 1998). Early on in her graduate career she nearly switched to a different Ph.D. project related to jaguars (*Panthera onca*) in Central America. But after a rather unsuccessful first field season, she realized that she would probably never get the chance again to analyze twenty-five years of data on a large carnivore with known individual life histories. Despite not studying cats in the wild, the opportunity was too good to pass up, especially for a student interested in conservation topics such as population viability analyses (PVAs) and effective population size ($N_e$). So from undergraduate to research assistant to master's to Ph.D. student, MJK stayed with the cheetah demography. She has yet to see a cheetah in the wild, whereas TMC has seen more than enough.

What, then, are the conceptual issues that have held the attention of Serengeti cheetah workers over the past twenty years? Initially, TMC was interested in the social and ecological factors that shape the course of behavioral development, particularly of predation (Caro 1994) and play (Caro 1995). However, as he learned more about his subjects in the field, especially about its strange social system compared to other mammals, he realized there were many other interesting questions that could be asked about cheetahs. These centered on the benefits and costs of sociality (Caro 1994) for families, adolescents, and especially males (see below), all of which live in small groups and in the end took up more of his attention than developmental questions. Later, in the second half of the 1980s, Clare FitzGibbon investigated the hunting behavior of cheetahs and the factors that promote hunting failure and success, focusing on many aspects of antipredator behavior in Thomson's gazelles (*Gazella thomsoni*), cheetahs' main prey on the Serengeti Plains (FitzGibbon 1989, 1990; FitzGibbon & Fanshawe 1988). Next, Karen Laurenson discovered that juvenile mortality in cheetahs is extremely high, principally because of lion predation (Laurenson 1994, 1995, 1996). Sarah Durant has extended these analyses to show that cheetahs avoid areas of high lion density (Durant 1998). In the 1990s, we switched attention to conservation issues, challenging prevailing dogma that cheetah populations are vulnerable as a result of genetic monomorphism (Caro & Laurenson 1994; Caro 2000). Now we are interested in the long-term viability of the Serengeti cheetah population (Kelly & Durant, in press) and the influence of demographic variables on its effective population size (Kelly, in press).

Throughout the years, cheetah researchers used a simple methodology (Caro 1994): driving to rises and hilltops in our 2,500 $km^2$ study area in the central Serengeti Plains; scanning for cheetahs using $10 \times 50$ binoculars; driving slowly to our subjects so as not to disturb them; and identifying them immediately from their pattern of spots or their black-and-white bands on the tail (Caro & Durant 1991), or else taking photographs for subsequent identification.

In essence we have used this study animal in the Serengeti to understand what seemed to us to be breaking issues in behavioral ecology and conservation biology at the time. Below we summarize one of the foci of our research, the cheetah's mating system, before reflecting on the successes and mistakes of working with this study animal.

## The Mating System

The general conceptual problem in the behavioral ecology that we will address in this chapter is the complexity of mammalian mating systems (Clutton-Brock 1989; Davies 1991). While most mammals are polygynous, others are monogamous, promiscuous, polyandrous, or even lek breeders. The best way of starting to understand mammalian mating systems is to remember that a male's reproductive success (RS) is limited by the number of females that he can inseminate because his potential rate of reproduction is faster than that of a female, whereas a female's RS is limited by her access to resources (Trivers 1972; Clutton-Brock & Parker 1992). Within this framework, the economics of female monopolization by males are influenced by four key factors: the extent to which female RS can be improved by male assistance, female group size, female range size, and seasonality of breeding (Clutton-Brock 1989). In most mammals males contribute little to parental care for reasons that are poorly understood (Clutton-Brock 1991). In Serengeti cheetahs, females live alone except when they have dependent cubs (Frame 1984), have enormous home ranges of over 800 km$^2$ that follow the annual movements of Thomson's gazelles (Durant et al. 1988; Caro 1994), and breed throughout the year (Laurenson et al. 1992). In the first respect, cheetah females are similar to antelopes such as Coke's hartebeest (*Alcelaphus buselaphus*) or Grant's gazelles (*Gazella granti*), where females live in small groups on ranges that are too large to be defended by males (Gosling 1986). In these ungulates, single males defend small mating territories that are visited by females in search of resources.

In cheetahs, however, some males live in small permanent coalitions of two or three animals whereas others live alone. Most of these coalitions are composed entirely of littermates, but approximately 30 percent included an unrelated male (Caro 1994). For cheetah males, there are some parallels with waterbuck (*Kobus defassa*) (Wirtz 1982), white rhinoceros (*Ceratotherium simum*) (Owen-Smith 1972, 1975; Rachlow et al. 1998), and oribi (*Ourebia ourebi*) (Arcese 1999; Arcese et al. 1995), species in which territorial males living on small territories tolerate satellite males that contribute to territorial defense. Cheetah male coalitions also resemble species in which groups of (usually related) males defend groups of females against other males as in chimpanzees (*Pan trogolodytes*) (de Waal 1992) or horses (*Equus caballus*) (Berger 1986; Feh 1999) and perhaps bottlenose dolphins (*Tursiops* sp.) (Conner et al. 1999). The most closely related of these species is the lion, where females live in social groups (prides) and permanent groups of males, often composed of relatives, jointly defend the pride against other coalitions (Packer et al. 1988). Answering the question of whether the cheetah's mating system resembles the mating territory system of certain ungulates, or the multimale spatial defense system of lions, or neither, will help to broaden our understanding of how common ecological factors produce similar social systems in different species, and whether social organization is constrained by phylogeny. Thus we are using the strange mating system of our study animal as the starting point to answer a theoretical question. This is the way that many field-oriented behavioral ecologists work: they use the organism's behavior to generate new questions that they would never have thought of sitting in a library or in front of a computer screen.

## Male Reproductive Tactics

We found that adult male cheetahs exhibited two distinct behavioral tactics. Resident males held and urine-marked small territories whereas nonresident (floater) males roamed over large parts of the Serengeti study area and rarely urine-marked. Nonresidents were less relaxed than residents in that they sat up and lay alert more often; they exhibited signs of physiological stress, specifically elevated cortisol levels; and they were in poor condition as determined from higher white blood cell counts, higher eosinophil levels, lower muscle mass, and more sarcoptic mange (Caro et al. 1989). Resident male territories were 37 km$^2$ on average whereas nonresident ranges were huge, 777 km$^2$ on average. Territories were not occupied continuously by males throughout the study. All males started out as floaters. Whereas some remained nonresidents all their lives, others became territorial; yet others first encountered as a resident subsequently floated. Some of these findings started out as anecdotal observations, which led TMC to record quantitative information on these behaviors, supplement them with additional measures, and finally analyze the numbers statistically. Quantifying what appears to be biologically significant, be they morphological, behavioral, or physiological features, allows scientists to test their intuition objectively.

During the first five years of TMC's study, coalitions of males were more likely to obtain a territory than were singletons (9 percent of 35 singletons versus 60 percent of 25 coalitions). The most plausible explanation was coalitions' numerical advantage in fights. Fights over territories were an important source of mortality (Fig. 25.1) as males were more likely to die inside or on the immediate borders of territories than outside them and many males died on territories at the time they were occupied (called active territories) (Caro 1994). Coalitions were more likely to displace residents from a territory than singletons, the latter of which usually acquired territories obtained by taking over a vacancy (Table 25.1a).

Nevertheless, there was no statistical effect of coalition size on the length of time that residents held territories. This was surprising since single residents were more likely to be displaced by other males than were coalitions (Table 25.1b). In contrast, resident coalitions were rarely displaced, implying they vacated their territories voluntarily. Lack of association between tenure length and coalition size was probably due to reduced competition over territories in the second five years of the field study (see below). In addition, larger groups of males did not hold larger territories than smaller groups (Caro 1994). Thus the key benefit of being a coalition member was that it gave a male a greater chance of acquiring a territory. Data on fights was particularly difficult to obtain because fights were seen so rarely; instead TMC used territory takeovers and location of dead males to piece together the dynamics of intrasexual contests.

Other analyses showed that per capita foraging returns were greater for

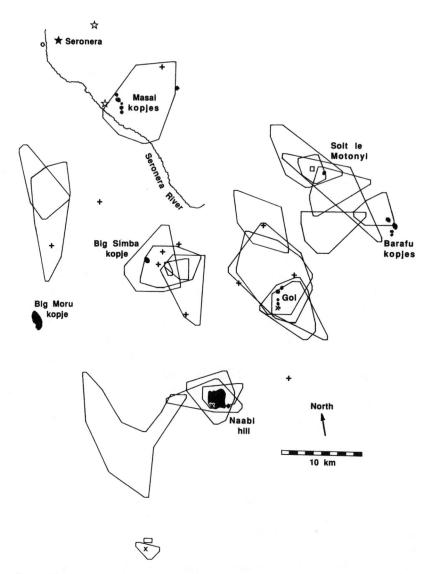

Figure 25.1. Location and extent of twenty-six territories on the Serengeti Plains, derived by the minimum polygon method for all males that were residents from March 1980 to July 1990. (The most southwesterly territory was not derived in this way.) Note that territories overlap as a result of sequential not simultaneous residence. Also shown are locations of dead cheetahs: radiocollared males (+, N = 11); males that were not radiocollared (*solid diamonds*, N = 2); females (*open circles*, N = 1); cheetahs of unknown sex located by chance (*crosses*, N = 3); and male cheetahs that died after being very ill (*open stars*, N = 2) (from Caro 1994).

Table 25.1

(a). Methods by Which Male Cheetahs Took Up Residence on a Territory. (b) Number of Instances of Termination of Residence by Males Under Different Circumstances

| Male group size: | 1 | 2 | 3 |
|---|---|---|---|
| **(a)** | | | |
| Displaced residents | 0 | 4 | 3 |
| Took up a vacant territory | 5 | 2 | 2 |
| Retained even though companion(s) disappeared | 2 | 1 | 0 |
| Unknown | 3 | 7 | 2 |
| **(b)** | | | |
| Ousted from territory by other males | 5 or 4 | 1 | 1 |
| Territory left vacant because male or his coalition partner died | 2 | 1 | 2 |
| Territory left vacant because male or his coalition partner disappeared | 0 | 2 | 0 |
| Territory left vacant but reasons unknown | 2 or 3 | 7 | 0 |
| No longer held a territory but circumstances completely unknown | 1 | 2 | 1 |

*Note*: Number of entries in (b) is four fewer than in (a) because three instances of coalition partners retaining the territory after loss of companions are omitted, and one territory was still occupied at the end of the study period. a: it was unknown whether one singleton vacated his territory or was displaced by another singleton male (from Caro 1994).

coalition members as a result of choosing to hunt larger prey items (rather than increased hunting success) but it was the competitive arena that had the key influence on survival, not food intake. Recent analyses have shown that male survival was strongly affected by the interaction of two variables: number and size of coalitions present in the study area (Durant et al., under review). Coalitions had higher survivorship than singletons when, and only when, there were many other coalitions on the Plains, pointing to the import of enhanced competitive abilities in promoting survivorship.

## BENEFITS OF TERRITORIALITY

What were the advantages of territoriality? Since theory suggests that male distribution will depend on that of females and since TMC observed males rushing toward females and investigating their reproductive status at almost every opportunity, we examined the distribution of females in and outside territories, taking the area that TMC spent searching for cheetahs into account. Results showed that there were significantly more females sighted on active territories than were seen outside territories or on those that were not occupied. Moreover, TMC was more likely to see females pass by when he was conducting behavioral observations of territorial than of nonterritorial males. In addition, sightings of males guarding females were 4 times more common on active territories than elsewhere, devalued for areas searched

(Caro 1994). Thus territories held by males were female "hotspots." In this set of analyses, we were merely trying to determine whether empirical data on male and female ranging patterns conformed to theoretical expectations (e.g., from Trivers 1972) as to where males should distribute themselves in relation to females.

## Do Females Choose Mating Partners?

Did females take note of male presence on territories? For seven territories for which there were data, females were sighted significantly more often on territories when males were in residence than when they were not, but on three there was no difference, and on one females were encountered by TMC more often when residents were absent (Caro 1994). If females had visited territories to choose mates, they should have left vacant territories and searched elsewhere. Since they did not, we can be confident that cheetah territories were not leks despite their small dimensions (Hoglund & Alatalo 1995). Second, there was no relationship between the number of females on active territories and resident male coalition size. This shows that males were not settling on territories in a ideal free fashion (Fretwell 1972). If they had been, 3 times as many females should have been found on territories occupied by trios as those occupied by singletons. Here, we were again trying to match empirical data to theoretical expectations in order to determine what class of mating system we were dealing with.

## Reproductive Payoffs for Coalitions

It was not possible to estimate the reproductive payoffs of coalition formation because matings were never observed by TMC in the wild despite collecting over five thousand hours of behavioral data. Matings that have been witnessed since are very rapid affairs lasting less than a minute. Moreover, in captivity, coalitions often mate sequentially, which would hinder estimates of per capita breeding success. DNA analyses would have been the best method (Packer et al. 1991) but in practice it was difficult to locate wide-ranging floaters over such a large study area, and many residents and non-residents were too shy to approach to within the 15 meters necessary to immobilize them using a blowpipe or dart rifle. Using a gun to obtain biopsy samples might have splintered a bone or injured a muscle. These difficulties were a major impediment to the study but abandoning the idea of large-scale blood or tissue sampling is the appropriate stance in relation to conservation ethics. Instead, we carried out rough calculations as follows.

The ratio of females inside and outside active territories was 0.0128 and 0.0078, respectively, when observer's searching was taken into account. The average number of sightings of floaters devalued by searching was 0.0026 inside and 0.0014 outside active territories, that is, a 0.65:0.35 ratio. Thus floaters had potential access to $(0.0128 \times 0.65) + (0.0078 \times 0.35) = 0.0111$ devalued females whereas resident males had access to $(0.0128 \times 1) = 0.0128$ devalued females per unit time. Resident males were estimated to live 1.78 times as long as floaters as determined from times of disappearance from the study area individually adjusted to incorporate each male's intersighting intervals (Caro 1994; Kelly et al. 1998). Lifetime encounter rates with devalued females had to be adjusted accordingly (residents $0.0128 \times 1.78 = 0.0228$, non-residents $0.0111 \times 1 = 0.0111$). These calculations assume that floaters inside active territories were as likely to encounter females as were residents, which seems reasonable as there was no significant difference in the percentage of sighting that the two sorts of males were seen with females once they were inside active territories.

To estimate fitness payoffs of living in groups of differing size, the proportion of singletons, pairs, and trios that were residents and nonresidents must be known. These were 8.8 percent and 91.2 percent, 70.6 percent and 29.4 percent, and 37.5 percent and 67.5 percent, respectively, during the period when data on female distribution was taken. Multiplying lifetime encounter rates for the two strategies for each male group size ($\times$ 100) gives payoffs of $0.021 + 1.101 = 1.213$ for singleton males, $(1.610 + 0.326)/2 = 0.968$ for each member of a pair, and $(0.855 + 0.694)/3 = 0.516$ for each member of a trio. These calculations assume that reproductive payoffs were divided equally among coalition members and we have no direct evidence for this. In the absence of witnessing matings, the best we can say is that there was no obvious behavioral dominance in relation to initiating social activity or starting hunts, in sharing food, or in obtaining proximity to females (Caro 1993). Thus per capita estimated lifetime reproductive payoffs for males in each group size was 1.101:0.968:0.516 or 45.0 percent for singletons, 35.9 percent for males in pairs, and 19.1 percent for males in trios. The distribution of payoffs corresponds closely to the proportion of males in different sized groups. Of 110 males, 40.9 percent were singletons, 40.0 percent lived in pairs, and 19.1 percent lived in trios (Caro & Collins 1986). In short, these crude calculations suggest that males were behaving in an ideal free way by distributing themselves according to group size and territorial status in such a way that each encountered equivalent numbers of females. In essence, for single males, reduced reproductive benefits of floating were balanced by not having to share matings with coalition partners.

The problem with this data is that we do not believe it! First, ideal free models assume individuals are free to go to their area of choice but we know from direct observations and locations of dead males that intruding males were prevented from occupying a territory by the residents. Second, floaters were in poor condition and physiologically stressed compared to residents, suggesting they were disadvantaged (Caro et al. 1989). Third, it is likely that floating coalitions took up residence outside the study area because they quickly passed through it; it was the floating singletons that remained in the study area. If coalitions settled outside the study area, it means our calculations of their reproductive returns are an underestimate of unknown magnitude. Fourth, the calculations are extremely crude, simply a product of mean values that can produce great error. Currently, then, the reproductive payoffs of coalition formation are not known with accuracy. This rather censorious self-analysis reflects our belief that it is important to be critical of one's own results even to the point of refuting them in later publications. Scientists are not judged by sticking doggedly to a point, be it right or wrong, but whether their observations stand up to scrutiny.

### COMPARATIVE DATA ON MATING SYSTEMS

Given data on territoriality and group size, how do cheetahs fit in with other mammals? Parallels with ungulates are few. First, there was no behavioral dominance in cheetahs but this was characteristic of relations between waterbuck males and oribi males. Second, male associations lasted many years in cheetahs but for much shorter periods than in ungulates, for example, less than 2 years in oribi (Arcese 1999). Third, territory acquisition was different: in only 3 of 18 (17 percent) instances did a cheetah coalition partner inherit his territory after his partner disappeared whereas this occurred in 42 percent of instances in waterbuck. Fourth, only 2 out of 17 (12 percent) new occupations of adjacent territories were by cheetah males that had previously held a territory in the study area. This contrasts with satellite male waterbuck, white rhinoceros, and oribi acquiring adjacent territories in 17 percent, 27 percent, and 43 percent of cases, respectively. Thus the benefits experienced by supernumerary male ungulates did not apply well to cheetahs.

In regard to primates, male coalitions appear far less egalitarian than in cheetahs. In species in which males form coalitions that repel extra-group males such as gelada baboons (*Theropithecus gelada*) (Dunbar 1984) or expel a breeding male from a single-male group as in gray langurs (*Presbytis entellus*) (Hrdy 1977), only one of the two males eventually obtains access to females. In savannah baboons (*Papio anubis*), where males cooperate directly for females, it is unclear whether reproductive benefits are shared evenly between coalition partners (Bercovitch 1988; Noe 1990). In addition, alliances between primates are commonly short-lived, terminating when a rank reversal or takeover has occurred. Turning to lions, larger coalitions are better able to obtain a territory and hence access to females. Moreover, behavioral dominance between males is absent in this species. The main difference between lions and cheetahs, however, is that larger coalitions of male lions enjoy greater per capita RS on average whereas larger coalitions of cheetahs do not encounter greater numbers of females on territories. Thus the cheetah seems to emerge as a species with no direct parallel mating system among mammals, being different from ungulates in the way males obtain reproductive benefits, from primates in regards to male relationships, and from lions in the way females are distributed.

## MALE SOCIALITY

If localized or high densities of females are responsible for group-living in both cheetahs and lions, then females should be widely dispersed or live at low densities in all the other felids where males are solitary. When MJK collated ranging data on the felids, we were able to separate species into those in which densities were higher than the median, and species in which female ranges overlapped each other. We found that males usually lived in groups in these species where both these factors pertained but that they lived alone in species in which these two factors were not congruent (Table 25.2). In conclusion, high female densities and extensive home range overlap together apparently drive male sociality in felids.

## FEMALE ASOCIALITY

The reasons that female cheetahs and all other felids except lionesses live alone is poorly understood. Packer (1986) proposed that in those species that usually capture large prey and that live in open habitats, and where female density is high, females will be social. This is because large carcasses will last for some time and would be seen and stolen by conspecifics especially if they were numerous. It therefore benefits a female to live with relatives and

Table 25.2
Species of Felids Separated According to Whether Densities Were Higher Than the Median and Female Ranges Overlapped (from Caro 1994)

|  | Densities higher than median and female ranges overlap | Densities lower than median and/or female ranges are exclusive |
|---|---|---|
| Males may live in groups | Cheetah<br>Lion |  |
| Males live alone | Serval | Bobcat<br>Cougar<br>European lynx<br>Leopard<br>North American lynx<br>Ocelot<br>Snow leopard<br>Tiger |

share food with them rather than inevitably relinquish it to nonrelatives. Packer argued that these conditions pertain only to lions. Cooperative defense against infanticidal males in lions is not sufficient reason for female sociality because other male felids commit infanticide (Caro 1994). An alternative hypothesis is that females of most species cannot afford to share prey because prey items 1 to 2 times the weight of an adult female are unavailable in most ecosystems (Caro 1989). This argument pertains to seventeen out of twenty-one field sites where felids have been studied. Only for lions are large prey sufficiently numerous to support groups of females living together. While the hypotheses differ, and the reasons for felids being asocial are not yet resolved, both hypotheses stress foraging costs as preventing the formation of groups. A subsequent, more formal model of female felid sociality (Macdonald et al., in press) suggests that relative prey size, felid population density, day range, prey capture rate, maximum prey consumption rate per day, and rate of searching for prey all need to be incorporated in order to predict when daily per capita food intake requirements can be met by females living in groups.

Always it is important to compare one's own observations to those of related species since they provide an additional test of whether conclusions are robust. If a behavioral ecologist's findings stand in marked contrast to those of others, they need to be reexamined. An additional benefit of the comparative approach at a small scale (e.g., within felids) or at a large scale (e.g., across mammals) is that it may throw out generalizations that the researcher would not have considered otherwise.

## Strengths of the Study

Our cheetah study has three main strengths. First, the detailed behavioral work on male territoriality spanned eleven years, which allowed us to examine changes in patterns of residency over time. For example, between 1980 and 1985, very few singleton males became resident on the Plains, but from 1986 on single males began to acquire territories at an increasing rate (Fig. 25.2). The most convincing hypothesis for this change was reduced competition over territories: sightings of floating coalitions declined from the first to the second half of the study from 23 percent. This may have been related to increasing lion and spotted hyena (*Crocuta crocuta*) numbers, the main cause of mortality for cheetah cubs (Laurenson 1994). Relaxed competition would allow more singletons to hold territories.

Second, inheriting a long-term demographic data set allowed us to determine demographic parameters such as lifetime reproductive success (LRS) and annual rates and relate these to changing environmental variables. For example, we could show that average litter size at independence was 2.5 between 1969 and 1979 when lion abundance was low on the Plains but

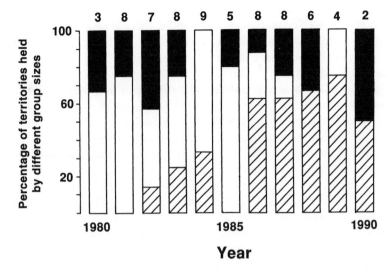

Figure 25.2. Percentage of territories held by resident singletons (*hatched*), pairs (*open*), and trios (*solid*) each year of the study. Number of territories held each year is shown above the bars (from Caro 1994).

averaged 2.0 between 1980 and 1994 when lion abundance increased by 60 percent (Kelly et al. 1998). In contrast, other environmental factors such as changing rainfall or prey abundance did not seem to be responsible for these changes.

Third, the fact that we are working on an endangered species has allowed us to address questions of conservation significance (Caro 1998). For example, we can use observations of behavior in the wild to inform captive breeding plans in zoos. As an illustration, Laurenson (1993) documented the type of lairs that cheetah mothers favor for giving birth in Serengeti and the rapidity with which litters are moved between them, allowing zoo breeders to mimic these situations in captivity. Also, we used demographic data to construct a PVA. (Kelly and Durant first built a deterministic model of the population and found that, in spite of increasing lion density, on average, the cheetah population is nearly self-replicating, i.e., $\lambda = 0.997$.) Nevertheless, we also know that lion density and presumably lion predation has increased over a twenty-year period. Although we would expect cheetahs to be able to withstand predation as they have evolved with sympatric larger predators, and their litter sizes and rapid reproduction are thought to be adaptations to intense predation (Caro 1994), it is surprising that $\lambda \approx 1.0$ given the marked recent increase in lion numbers (Hanby et al. 1995). Either lion densities are returning to "normal " as their main prey, wildebeest (*Connochaetes taurinus*), recover from rinderpest (Sinclair 1995), or cheetahs may move between areas of differing lion densities. For example, the Maswa Game Reserve to the southwest of Serengeti National Park, where lions are hunted (Caro et al. 1998), could provide a refuge for cheetahs. Alternatively, if lion densities are higher than "normal " because they are being forced into protected areas, as some have argued, $\lambda$ values may be more optimistic now than in the future.

Populations with a $\lambda = 1.0$ are still subject to extinction due to stochasticity (Shaffer 1990; Burgman et al. 1993; McCarthy et al. 1995). Hence Kelly and Durant constructed a stochastic model of the cheetah population and used our long-term records to compare actual cheetah population size to the model's predictions under demographic and environmental stochasticity. They then conducted a sensitivity analysis of extinction risk. They found extinction risk to be sensitive to adult survival, but juvenile survival, especially of 0 to one-year-olds and not one- to two-year-olds, also had a strong effect on extinction risk. Since adult cheetahs are well protected within Serengeti National Park, it is unlikely that adult survival could be enhanced. Juvenile survival, on the other hand, is likely to fluctuate with amount of predation. In fact, by combining the cheetah and lion long-term data sets, we determined the influence of different lion numbers on cheetah recruitment through a generalized linear model (Durant et al., under review). Then Kelly and Durant (in press) simulated different levels of lion abundance and found that maximum lion abundance (120 lionesses) and average lion abundance (98 lionesses) resulted in the extinction of nearly all cheetah populations in 50 years, but that cheetah populations remain extant when lioness numbers were low (72 lionesses) (Fig. 25.3). Parameterizing vital rates in areas of different lion abundance and quantifying immigration to and from such areas would add greatly to population modeling effort by including the spatial heterogeneity that likely contributes to the coexistence of these predators (Durant 1998).

## Weaknesses of the Study

Over the course of the project, we have made a number of mistakes and it is important for researchers to make these explicit so that they can be rectified or avoided in future. One of these was failing to recognize cheetahs in the field. At the time, there were good reasons to do this: pictures took up to four months to be incorporated into the photographic reference index because of processing time, sifting through photographs could take an hour or more, and time was limited to only four hours of searching in the mornings before cheetahs went to sleep under bushes. Nonetheless we were left with a backlog of ten thousand pictures at the end of the project! We solved this problem by using a matching program (Hiby & Lovell 1990). This involved capturing black-and-white photographs of cheetahs using video stills fed into a desktop computer. Digital images were then processed to extract a sample

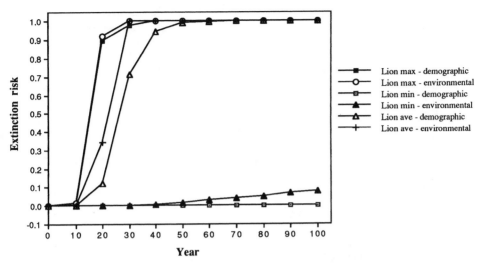

Figure 25.3. Projected extinction risk out of a hundred simulations under both demographic and environmental stochasticity for cheetah populations subject to different lion densities. Lion minimum was 72 adult lionesses, corresponding to the minimum recorded over 20 years in the study area, while lion maximum was 120 female lions, corresponding to the maximum density recorded. Average lion density was 98 lionesses.

of the coat pattern to act as a "fingerprint." The computer program compensated for differences in lighting by removing gray-level patterns. Fingerprints were stored as a matrix of numbers termed an identifier array. To determine whether two sightings are of the same or different cheetahs, the computer compared four different areas of the animal's pelage, calculated a correlation coefficient between array elements, and then generated a weighted average. All potential matches greater than 0.370 were inspected by eye to determine a true match. MJK found that the percentage of positive matches as determined by eye increased with increasing correlation coefficients (Fig. 25.4). With a coefficient of 0.500 or higher, the computer program was almost 100 percent accurate in matching different sightings of the same cheetah.

The advantage of using this technique is that it is noninvasive, accurate, can be used for any species with pelage patterns remaining constant through life, and is suitable for a laptop computer in the field. Moreover, it can be used to explore quantitatively the extent to which relatives have similar morphological phenotypes: we found an increasing proportion of mother-offspring and sibling pairs as correlation coefficients of match probability increased from 0.370 to 0.499 (see Fig. 25.4). With hindsight, we found a way to work around our mistake by developing a new methodology. There is an important lesson here. Faced with an obstacle that occurs sooner or later

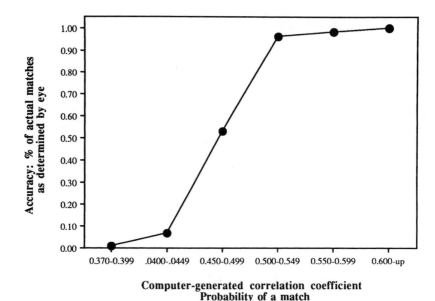

Figure 25.4. Accuracy of computer-generated correlation coefficients between two cheetah photographs in determining whether two photographs of cheetahs are the same animal. Accuracy was determined by examining a subset of a thousand potential matches with correlation coefficients ranging from 0.370 to 0.600 and above, and then inspecting these by eye to determine a true match. The proportion of true matches in each correlation coefficient category is plotted on the y-axis.

with most data sets, researchers need to use it as a springboard to generate new approaches or analyses to overcome the problem rather than allow it to demoralize them or cause them to abandon a piece of research.

A second difficulty that we encountered were inconsistencies among observers' records over the twenty-five-year history of the project (Kelly et al. 1998). While observers uniformly paid attention to taking rigorous demographic notes at each sighting (such as number and sex of cheetahs, and date and location), they differed in the amount of environmental information that they collected, such as presence of prey and predators. In addition, some observers would rigorously go through cleaning up question marks in their sightings notebooks at the end of the day but others forget to do this, making interpretation difficult for MJK years later. Finally, our differing foci of research interests, an undoubted strength of our project, resulted in researchers devoting different amounts of effort to searching different parts of the study area. This meant that we missed reproductive events for several females because females had temporarily quit the core of the study area. This greatly reduced the numbers of females for which we could calculate LRS. This parameter is difficult to determine in long-lived species and we regret the number of missed opportunities due to observer absence during six months to a year of a particular female's life. Additionally, LRS has important conservation ramifications for effective population size (Kelly, in press). In short, consistency is absolutely critical in long-term data sets.

One advantage of inheriting such a data set is that population size trends and LRS can only be examined with long-term data and such a data set could not have been obtained for long-lived species during the course of a single Ph.D. dissertation. On the other hand, incomplete or inconsistent data collection over time makes analyses difficult and, at times, frustrating.

## The Future

The most important aspect of the cheetah's mating system is that female movements, themselves dependent on migratory Thomson's gazelle prey, result in temporary aggregations of females on the Serengeti Plains. Males compete intensely over areas where these aggregations occur and increase their competitive ability by being in a coalition. A key question is whether we will observe male coalitions in other ecological circumstances where females are more evenly dispersed or live at lower densities. In short, we need a new study of cheetahs in a different part of Africa where prey densities are lower and less aggregated. If males do form coalitions under these circumstances, it will point to other factors, including perhaps phylogenetic constraints, that drive sociality in male felids. Currently we do know that males live in coalitions in other regions (Nairobi National Park, McLaughlin 1970; Kruger National Park, Bowland 1993) but we do not know whether females form temporary aggregations in these areas. Although there is currently no fieldwork being done on this topic, MJK is using the long-term data set to examine philopatry, or the extent to which related individuals overlap in space and time.

A second important issue that needs to be addressed is paternity: how reproductive success is shared among males in general, and between coalition members in particular (Packer et al. 1991; Feh 1999). While DNA paternity exclusion studies are constrained by difficulties in collecting blood and tissue samples of many cubs and putative fathers, we believe that it might be possible to collect DNA from hair in scats deposited by cheetahs (Hughes 1998). The chief difficulty is that whereas resident males defecate frequently (Caro et al. 1989), nonresidents, cubs and, to some extent, mothers do not do so, forcing prolonged follows of these individuals in the field. This is also an open field of study as no work is currently being carried out to determine paternity of cheetahs in the wild.

Our other avenues of research on cheetahs have raised a number of issues too. Laurenson's discovery that lions are the chief source of mortality of cheetah cubs suggests that in protected areas where adult survival is high, it will be necessary to monitor the relationship between cheetahs and lions, particularly juvenile cheetah survival, as this parameter exerts strong control over population growth. Outside protected areas, where lions are hunted, cheetahs may have reasonably robust population sizes, although evidence for this is controversial (Laurenson 1995; Gros 1998). In other areas, where both large predators and cheetahs are hunted, it is likely that protection of adult cheetahs would be the most effective conservation strategy. Fieldwork is being conducted outside the Serengeti Plains in woodland areas to determine cheetah population size and reproductive rates, but nevertheless this is still a protected area.

MJK has also found that certain matrilines in Serengeti are far more successful than others (Kelly, in press) with five lineages contributing 45 percent of the cheetah population. Although we do not yet understand the reason why reproductively successful mothers produce reproductively successful daughters and granddaughters (Kelly, in press), Durant (in press) has used playbacks of lion roars to demonstrate that female cheetahs that react most strongly to lion roars by moving away from the speaker are those with the highest RS. Since lion predation is so important in affecting RS in this species, her finding raises the interesting possibility that daughters inherit some aspects of their mothers' wariness or antipredator tactics. This in turn generates questions about the extent to which such behavior is environmentally, genetically, or culturally transmitted. These sorts of questions are best answered in zoos, where cross-fostering is possible. Moreover, work on cheetahs' temperaments has actually been carried out in zoos (Wielebnowski 1999). Nevertheless, Wielebnowski's work has shown quite the opposite result. In zoos, female and male cheetahs that had never bred successfully scored higher on tense-fearful components of personality than did breeders! Although the nature of stress is likely to be different in captivity, we have contradictory findings on the association between temperament and RS which need to be resolved in order to inform managers about which subjects to use in reintroduction programs and to inform behavioral ecologists of which behavior patterns are under strong selection in the wild. There is still plenty to do!

## Acknowledgments

We thank Lex Hiby and Phil Lovell for developing the cheetah matching program; Brian Bertram, Anthony Collins, Sarah Durant, Clare FitzGibbon, George Frame, and Karen Laurenson for help in collecting long-term demographic data; the Government of Tanzania for permissions; and Judy Stamps and an anonymous reviewer for comments.

# References

Arcese P, 1999. Effects of auxillary males on territory ownership in the oribi and attributes of multimale groups. Anim Behav 57:61–71.

Arcese P, Jongejan G, Sinclair ARE, 1995. Behavioural flexibility in a small African antelope: group size and competition in the oribi (*Ourebia ourebi*, Bovidae). Ethology 99:1–23.

Bercovitch FB, 1988. Coalitions, cooperation and reproductive tactics among adult male baboons. Anim Behav 36:1198–1209.

Berger J, 1986. Wild horses of the Great Basin. Chicago: University of Chicago Press.

Bowland AE, 1993. The 1990/1991 cheetah photographic survey. Unpublished report. Department of Nature Conservation, Skukuza, Kruger National Park, South Africa.

Burgman MA, Ferson S, Akcakaya HR, 1993. Risk assessment in conservation biology. New York: Chapman & Hall.

Caro TM, 1987. Cheetah mothers' vigilance: looking out for prey or for predators? Behav Ecol Sociobiol 20:351–361.

Caro TM, 1989. Determinants of asociality in felids. In: Comparative socioecology: the behavioral ecology of humans and other mammals (Standen V, Foley RA, eds). Special publication of the British Ecological Society, no 8. Oxford: Blackwell Scientific Publications; 41–74.

Caro TM, 1993. Behavioral solutions to breeding cheetahs in captivity: insights from the wild. Zoo Biol 12:19–30.

Caro TM, 1994. Cheetahs of the Serengeti Plains: group living in an asocial species. Chicago: Chicago University Press.

Caro TM, 1995. Short-term costs and correlates of play in cheetahs. Anim Behav 49:333–345.

Caro T, 1998. How do we refocus behavioral ecology to address conservation issues more directly? In: Behavioral ecology and conservation biology (Caro T, ed). Oxford: Oxford University Press; 557–565.

Caro TM, 2000. Controversy over behaviour and genetics in cheetah conservation. In: Behaviour and conservation (Gosling LM, Sutherland WJ, eds). Cambridge: Cambridge University Press; 221–237.

Caro TM, Collins DA, 1986. Male cheetah social organization and territoriality. Nat Geogr Res 2:75–86.

Caro TM, Durant SM, 1991. Use of quantitative analyses of pelage characteristics to reveal family resemblances in genetically monomorphic cheetahs. J Hered 82:8–14.

Caro TM, FitzGibbon CD, Holt ME, 1989. Physiological costs of behavioural strategies for male cheetahs. Anim Behav 38:309–317.

Caro TM, Laurenson MK, 1994. Ecological and genetic factors in conservation: a cautionary tale. Science 263:485–486.

Caro TM, Pelkey N, Borner M, Severre ELM, Campbell KLI, Huish SA, Ole Kuwai J, Farm BP, Woodworth BL, 1998. The impact of tourist hunting on large mammals in Tanzania: an initial assessment. Afr J Ecol 36:321–346.

Clutton-Brock TH, 1989. Mammalian mating systems. Proc Roy Soc Lond B 236:339–372.

Clutton-Brock TH, 1991. The evolution of parental care. Princeton: Princeton University Press.

Clutton-Brock TH, Parker GA, 1992. Potential reproductive rates and the operation of sexual selection. Q Rev Biol 67:437–456.

Conner RC, Heithaus MR, Barre LM, 1999. Superalliance of bottlenose dolphins. Nature 397:571–572.

Davies NB, 1991. Mating systems. In: Behavioural ecology: an evolutionary approach, 3rd ed. (Krebs JR, Davies NB, eds). Oxford: Blackwell Scientific Publications; 263–294.

Dunbar RIM, 1984. Reproductive decisions: an economic analysis of gelada baboon social strategies. Princeton: Princeton University Press.

Durant SM, 1998. Competition refuges and coexistence: an example from Serengeti carnivores. J Anim Ecol 67:370–386

Durant SM, 2000. Predator avoidance affects reproductive success: evidence from playback experiments and distribution patterns. Anim Behav 60:121–130.

Durant SM, Caro TM, Collins DA, Alawi RM, FitzGibbon CD, 1988. Migration patterns of Thomson's gazelles and cheetahs on the Serengeti Plains. Afr J Ecol 26:257–268.

Durant SM, Kelly MJ, Caro TM, under review. Factors affecting life and death in Serengeti cheetahs: environment, age and sociality. Behav Ecol.

Feh C, 1999. Alliances and reproductive success in Camargue horses. Anim Behav 57:705–713.

FitzGibbon CD, 1989. A cost to individuals with reduced vigilance in groups of Thomson's gazelles hunted by cheetahs. Anim Behav 37:508–510.

FitzGibbon CD, 1990. Why do cheetahs prefer male gazelles? Anim Behav 40:837–845.

FitzGibbon CD, Fanshawe JH, 1988. Stotting in Thomson's gazelles: an honest signal of condition. Behav Ecol Sociobiol 23:69–74.

Frame GW, 1984. Cheetah. In: The encyclopedia of mammals, vol. 1 (Macdonald DW, ed). London: Allen and Unwin; 40–43.

Fretwell SD, 1972. Populations in a seasonal environment. Princeton: Princeton University Press.

Gosling LM, 1986. The evolution of mating strategies in male antelope. In: Ecological aspects of social evolution: birds and mammals (Rubenstein DI, Wranghm RW, eds). Princeton: Princeton University Press; 244–281.

Gros PM, 1998. Status of the cheetah Acinonyx jubatus in Kenya: a field-interview assessment. Biol Cons 85:137–149.

Hanby JP, Bygott JD, Packer C, 1995. Ecology, demography, and behavior of lions in two contrasting habitats: Ngorongoro crater and the Serengeti Plains. In: Serengeti II: dynamics, management, and conservation of an ecosystem. (Sinclair ARE, Arcese P, eds). Chicago: Chicago University Press; 315–331.

Hiby L, Lovell P, 1990. Computer aided matching of natural markings: a prototype system for grey seals. Rep Int Whale Comm (special issue) 12:57–61.

Hoglund J, Alatalo RV, 1995. Leks. Princeton: Princeton University Press.

Hrdy SB, 1977. The langurs of Abu. Cambridge, MA: Harvard University Press.

Hughes C, 1998. Integrating molecular techniques with field methods in studies of social behavior: a revolution results. Ecology 79:383–399.

Kelly MJ, 2001. Lineage loss in Serengeti cheetahs: consequences of high reproductive variance and heritability of fitness on $N_e$. Cons Biol 15:137–147.

Kelly MJ, Laurenson MK, FitzGibbon CD, Collins DA, Durant SM, Frame GW, Bertram BCR, Caro TM, 1998. Demography of the Serengeti cheetah (*Acinonyx jubatus*) population: the first 25 years. J Zool 244:473–488.

Kelly MJ, Durant SM, 2000. Viability of the Serengeti cheetah population. Cons Biol 114:786–797.

Laurenson MK, 1993. Early maternal behavior of wild cheetahs: implications for captive husbandry. Zoo Biol 12:31–43.

Laurenson MK, 1994. High juvenile mortality in cheetahs (*Acinonyx jubatus*) and its consequences for maternal care. J Zool 234:387–408.

Laurenson MK, 1995. Implications of high offspring mortality for cheetah population dynamics. In: Serengeti II: dynamics, management, and conservation of an ecosystem. (Sinclair ARE, Arcese P, eds). Chicago: Chicago University Press; 385–399.

Laurenson MK, 1996. Cub growth and maternal care in cheetahs. Behav Ecol 6:405–409.

Laurenson MK, Caro TM, Borner M, 1992. Female cheetah reproduction. Nat Geogr Res Expl 8:64–75.

Macdonald DW, Yamaguchi N, Kerby G, in press. Group-living in the domestic cat: its sociobiology and epidemiology. In: The domestic cat: the biology of its behaviour, 2nd ed. (Turner DC, Bateson P, eds). Cambridge: Cambridge University Press.

McCarthy MA, Burgman MA, Ferson S, 1995. Sensitivity analysis for models of population viability. Biol Cons 73:93–100.

McLaughlin RT, 1970. Aspects of the biology of the cheetah (*Acinonyx jubatus*, Schreber) in Nairobi National Park. Master's thesis, University of Nairobi.

Noe R, 1990. A veto game played by baboons: a challenge to the use of Prisoner's dilemma as a paradigm for reciprocity and cooperation. Anim Behav 39:78–90.

Owen-Smith RN, 1972. Territoriality: the example of the white rhinoceros. Zool Africana 7:273–280.

Owen-Smith RN, 1975. The social ethology of the white rhinoceros *Ceratotherium simum* (Burchell 1817). Zeit Tierpsychol 38:337–384.

Packer C, 1986. The ecology of sociality in felids. In: Ecological aspects of social evolution: birds and mammals (Rubenstein DI, Wranghm RW, eds). Princeton: Princeton University Press; 429–451.

Packer C, Gilbert DA, Pusey AE, O'Brien SJ, 1991. A molecular genetic analysis of kinship and cooperation in African lions. Nature 351:562–565.

Packer C, Herbst L, Pusey AE, Bygott JD, Hanby JP, Cairns SJ, Borgerhoff Mulder M, 1988. Reproductive success in lions. In: Reproductive success: studies of individual variation in contrasting breeding systems. (Clutton-Brock TH, ed). Chicago: Chicago University Press; 363–383.

Rachlow JL, Berkeley EV, Berger J, 1998. Correlates of male mate strategies in white rhinos (*Ceratotherium simum*). J Mammal 79:1317–1324.

Shaffer ML, 1990. Population viability analysis. Cons Biol 4:39–40.

Sinclair ARE, 1995. Serengeti past and present. In: Serengeti II: dynamics, management, and conservation of an ecosystem. (Sinclair ARE, Arcese P, eds). Chicago: Chicago University Press; 3–30.

Trivers RL, 1972. Parental investment and sexual selection. In: Sexual selection and the descent of man, 1871–1971 (Campbell B, ed). Chicago: Aldine; 136–179.

de Waal FBM, 1992. Coalitions as part of reciprocal relations in the Arnhem chimpanzee colony. In: Coalitions and alliances in humans and other animals (Harcourt AH, de Waal FBM, eds). New York: Oxford University Press; 233–257.

Wielebnowski N, 1999. Behavioral differences as predictors of breeding status in captive cheetahs. Zoo Biol 18:335–349.

Wirtz P, 1982. Territory holders, satellite males, and bachelor males in a high-density population of waterbuck (*Kobus ellipsiprymnus*) and their associations with conspecifics. Zeit Tierpsychol 58:277–300.

# Closing Thoughts

Lee Alan Dugatkin

A few years back, I was invited to give a seminar at the University of California at Davis. I had always thought of Davis as a hotbed of behavioral ecology, and so I was really surprised to hear of a recent graduate seminar there on the question "Is Behavioral Ecology Dead?" I argued with folks that not only was behavioral ecology *not* dead, it was alive and vibrant. I hope that this volume vindicates my views.

It is worth noting, however, that some of the arguments that I used in my over-dinner defense of behavioral ecology did not focus directly on the subject of this book—the integration of conceptual, theoretical, and empirical approaches. In my somewhat spirited defense, I presented the case that behavioral ecology, broadly defined, was poised to have an impact on many different fields. That is, while the figure I displayed in the Introduction to this volume depicted how behavioral ecology had built on the foundations of other disciplines (ecology, evolution, ethology, community and population ecology, and population genetics), I believe we are now poised to give something back. For example, game theory models developed in behavioral ecology can now be found in textbooks and journal articles in the areas of mathematical economics and political science, as well psychology and anthropology (see Dugatkin & Reeve 1998 for a review). While evolutionary game theory owes its start to models developed in these other areas, the flow is now two-way. Behavioral ecologists should feel good about that.

Another example of how the work behavioral ecologists do can influence other disciplines comes from the newly emerging (and I'd argue extremely important) area of Darwinian medicine. This field attempts to apply evolutionary thinking to medical issues in novel ways. Much of the work in Darwinian medicine focuses on how "natural selection thinking" can reshape our view of important medical issues such as cancer, sexually transmitted diseases, morning sickness, allergies, aging, and mental illness (Williams & Nesse 1991; Nesse & Williams 1995).

Scanning the chapters in Nesse & Williams's *Why We Get Sick: The New Science of Darwinian Medicine*, it is clear that behavioral ecology has had a strong influence on this fledgling discipline. A personal example may illustrate this point. At a seminar in 1996, I learned that Randy Nesse was using some work I did on predator inspection and survival in the guppy (Dugatkin 1992). I was flattered, but surprised. The work he spoke of demonstrated that guppies that inspected predators were eaten more often than their non-inspecting peers. How on earth, you might ask yourself, does that finding have any place in the world of Darwinian medicine? It turns out that one can argue that the guppy work was the first experimental study demonstrating clear fitness differences between "anxious" (noninspecting) and "bold" (inspecting) individuals. For mental health researchers interested in personality, such findings are hard to come by.

My interactions with Randy Nesse since that seminar have had a large impact on my research. For example, after I realized that the guppy work had implications for medical conditions such as "anxiety disorders" (which itself can be broken down into many conditions), Jean-Guy Godin and I examined the flip side of the coin and learned that bold individuals are more attractive to mates. So, anxious individuals may live longer and mate less, while risk-takers live shorter, but sex-filled lives. Such work may, in the long run, explain the coexistence of two very different personality types and shed light on whether anxiety is a disorder or an evolutionarily stable strategy.

It is this sort of give and take between behavioral ecology and other areas (in this case Darwinian medicine) that I hope will characterize the next stage in the evolution of the field of behavioral ecology. The door is wide open for behavioral ecologists to share their ideas and work with evolutionary psychologists, social psychologists, mathematical biologists, molecular biologists, cultural anthropologists, economists, and sociologists. This process, which is already under way in some of the aforementioned areas, will not always be an easy one. Both sides, in an interchange between disciplines, will likely want to take what they can get from the other, and at the same time hold on to their own sacred ideas. All in all, though, in the long run, the effort to expand the reach of behavioral ecology and at the same time learn from other disciplines, will, no doubt, be worth the effort.

# References

Dugatkin LA, 1992. Tendency to inspect predators predicts mortality risk in the guppy, *Poecilia reticulata*. Behav Ecol 3:124–128.

Dugatkin LA, Reeve HK (eds), 1998. Game theory and animal behavior. New York: Oxford University Press.

Nesse R, Williams GC, 1995. Why we get sick: the new science of Darwinian medicine. New York: Vintage Books.

Williams GC, Nesse R, 1991. The dawn of Darwinian medicine. Q Rev Biol 66:1–22.

# Contributors

Marc Bekoff
University of Colorado
Environmental, Population and Organismic Biology
Boulder, CO 80309-0334

Christophe Boesch
Max-Planck Institute for Evolutionary Anthropology
Inselstrasse 22
04 103 Leipzig
Germany

Alan B. Bond
School of Biological Sciences
Nebraska Behavioral Ecology Group
University of Nebraska
Lincoln, NE 68588-0118

Jerram L. Brown
Department of Biology
State University of New York at Albany
Albany, NY 12222

T. M. Caro
Wildlife, Fish and Conservation Biology
University of California
Davis, CA 95616

Richard C. Connor
Department of Biology
University of Massachusetts at Dartmouth
North Dartmouth, MA 02747

Scott Creel
Department of Biology
Montana State University
Bozeman, MT 59717

Lee Alan Dugatkin
Department of Biology
University of Louisville
Louisville, KY 40208

H. Carl Gerhardt
Division of Biological Sciences
215 Tucker Hall
University of Missouri, Columbia
Columbia, MO 65211

A. H. Harcourt
Department of Anthropology
University of California
One Shields Avenue
Davis, CA 95616

Bert Hölldobler
Theodor Boveri-Institut (Zool. II)
Am Hubland, Biozentrum
D-97074 Wuerzburg
Germany

Alan C. Kamil
School of Biological Sciences
Nebraska Behavioral Ecology Group
University of Nebraska
Lincoln, NE 68588-0118

M. J. Kelly
Wildlife, Fish and Conservation Biology
University of California
Davis, CA 95616

Manfred Milinski
Max-Planck-Institute for Limnology
Department of Evolutionary Ecology
August-Thienemann-Strasse 2
D-24306 Ploen
Germany

Anders Pape Møller
Laboratoire d'Ecologie
CNRS URA 258
Universite Pierre et Marie Curie
Bat. A, 7eme etage
7, quai St. Bernard, Case 237
F-75252 Paris Cedex 5
France

Geoff A. Parker
Population Biology Research Group
Nicholson Building
School of Biological Sciences
University of Liverpool
Liverpool L69 3BX
United Kingdom

Naomi E. Pierce
Museum of Comparative Zoology
Harvard University
26 Oxford Street
Cambridge, MA 02138

Hudson Kern Reeve
Section of Neurobiology and Behavior
Cornell University
Ithaca, NY 14853

Flavio Roces
Theodor Boveri-Institut (Zool. II)
Am Hubland, Biozentrum
D-97074 Wuerzburg
Germany

Gil G. Rosenthal
Section of Integrative Biology C0930
School of Biological Sciences
University of Texas
Austin TX 78712

Michael J. Ryan
Section of Integrative Biology C0930
School of Biological Sciences
University of Texas
Austin, TX 78712

Thomas D. Seeley
Section of Neurobiology and Behavior
Cornell University
Ithaca, NY 14853

Paul W. Sherman
Section of Neurobiology and Behavior
Cornell University
Ithaca, NY 14853

Joan B. Silk
Department of Anthropology
University of California
Los Angeles, CA 90095

Barry Sinervo
Biology Department
University of California
Santa Cruz, CA 95064

Judy Stamps
Department of Ecology and Evolution
University of California at Davis
Davis, CA 95616

George W. Uetz
Department of Biological Sciences
University of Cincinnati
Cincinnati, OH 45221-0006

Robert R. Warner
Ecology, Evolution, & Marine Biology
University of California
Santa Barbara, CA 93106

David F. Westneat
Center for Ecology, Evolution, and Behavior
T. H. Morgan School of Biological Sciences
University of Kentucky
Lexington, KY 40506-0225

Gerald S. Wilkinson
Department of Zoology
University of Maryland
College Park, MD 20742

# Index